Springer Proceedings in Mathematics & Statistics

Volume 39

For further volumes:
http://www.springer.com/series/10533

Springer Proceedings in Mathematics & Statistics

This book series features volumes composed of select contributions from workshops and conferences in all areas of current research in mathematics and statistics, including OR and optimization. In addition to an overall evaluation of the interest, scientific quality, and timeliness of each proposal at the hands of the publisher, individual contributions are all refereed to the high quality standards of leading journals in the field. Thus, this series provides the research community with well-edited, authoritative reports on developments in the most exciting areas of mathematical and statistical research today.

Eberhard Knobloch • Hikosaburo Komatsu • Dun Liu
Editors

Seki, Founder of Modern Mathematics in Japan

A Commemoration on His Tercentenary

Springer

Editors
Eberhard Knobloch
Professor
Technische Universität Berlin
Berlin 10623, Germany

Hikosaburo Komatsu
Professor Emeritus
The University of Tokyo

Dun Liu
Professor
Institute for the History of Natural Science
Beijing 100190, P. R. China

ISSN 2194-1009 ISSN 2194-1017 (electronic)
ISBN 978-4-431-54724-2 ISBN 978-4-431-54273-5 (eBook)
DOI 10.1007/978-4-431-54273-5
Springer Tokyo Heidelberg New York Dordrecht London

Mathematics Subject Classification (2010): 01-06, 01A05, 01A27, 11C20, 11D72, 13P15, 01A25, 11D41, 01-08

Typesetting by volume editors using a Springer TeX macro package

Printed on acid-free paper

Springer is part of Springer Science+Business Media (www.springer.com)

Dedicated to the spirit of Takakazu Seki

Postal Stamp (Japan)
T. Seki (mathematician, 1642–1708)
The background shows his expansion of a determinant of order four.

To the best knowledge of Hikosaburo Komatsu, one of the editors, there are no authentic portraits of T. Seki, even though there are two existing portrayals of him. The image on the postal stamp shown on the preceding page is based on a painting of T. Seki that is now in the Ichinoseki City Museum. The image seen above is a scroll painting (*kakejiku*) of T. Seki that Ishikuro Nobuyoshi (1760–1836) had an artist draw for ceremonies in his school. (Owned by the Kōju Foundation; now in the Imizu City Shinminato Museum, Toyama, Japan)

近代科学資料館
関孝和と和算の世界
近代科学資料館

(*) Seki Takakazu
1. Miyazaki Kazumi
2. Matsushita Toshiko
3. Alexei Volkov
4. Miyajima Shigeko
5. Morimoto Mitsuo
6. Sato Ken'ichi
7. Sa Rina
8. Karine Chemla
9. Muroi Kazuo
10. Qu Anjing
11. Tsuchikura Tamotsu
12. Ozaki Fumiaki
13. Feng Lisheng
14. Wenlin Li
15. Tamura Makoto
16. Luo Jianjin
17. Guo Shuchun
18. Kawahara Hideki
19. Kota Osamu
20. Eberhard Knobloch
21. Silke Wimmer-Zagier
22. Kanauchi Tadahiro
23. Ha Huy Khoai
24. Don Zagier
25. Yamaji Katsunori
26. Xu Zelin
27. Takenouchi Osamu
28. Ohashi Yukio
29. Yano Michio
30. Komatsu Hikosaburo
31. Fujii Yasuo
32. Liu Dun
33. Guo Shirong
34. Majima Hideyuki
35. Sugimoto Toshio
36. Makino Masahiro
37. Muramatu Tosinobu
38. Nakamura Yukio
39. Habuka Takashi
40. Matumoto Takao
41. Kim Yong-Woon

The Use of Sans-serif Type

Emphasized texts in Japanese are usually set in sans-serif type (see [1], p. 226 or [2], p. 160.) in place of boldface type, which is too thick for Japanese texts.
In this book sans-serif type is used for the same purpose, but with two additional uses:
One is as an identifier in English of the title of a book written in a language other than English, or of the name of a person, place, etc. The reader may find the original (title, etc.) in the Index at the end of the book.
In the Supplements, five of Seki ' s works are published for the first time as a collated edition. Here a character of classical Chinese in sans-serif type means that the character in question is not taken from any of the documents listed in the notes on collation following the text, but is chosen at the discretion of the collators.

Editor

References

[1] L. Lamport: LaTeX: A Document Preparation System, User's Guide and Reference Manual, Addison-Wesley Publ. Co., 1994.
[2] M. Goossens, F. Mittelbach and A. Samarin: The LaTeX Companion, Addison-Wesley Publ. Co., 1994.

Foreword

These proceedings are a collection of papers written by lecturers at the International Conference on the History of Mathematics in Memory of Seki Takakazu (1642?–1708), held August 25–31, 2008. The conference was one of the main events of Seki's 300th memorial. The papers are faithfully based on the lectures, except the first paper, entitled "Seki Takakazu, His Life and Bibliography", which contains new and significant facts discovered later by studying some documents that were unknown at the time of the original lecture. We are deeply grateful to the document-holder.

Seki Takakazu was a Japanese mathematician in the early part of the Edo era and was known as the first to discover the so-called determinants in the world. He passed away on October 24 in the fifth year of the Hôei period, which corresponds to December 5, 1708, in the current solar calendar. The year 2008 thus corresponds to his 300th posthumous anniversary.

In 1907, according to Buddhist custom, the Tokyo Mathematico-Physical Society observed his 200th posthumous anniversary with a commemorative ceremony on the occasion of an annual meeting, publishing a book of seven main articles of his school, and with a conference on Japanese mathematics for a general audience and publishing the conference proceedings.

In 1958, the Mathematical Society of Japan observed his posthumous 250th anniversary as a memorial by organizing a commemorative conference. After a commemorative service in front of his tombstone at Jôrinji temple in Benten-chô, Shinjuku Ward, Tokyo, a conference with lectures on the old Japanese mathematics was held by the society jointly with the Tokyo Metropolitan Board of Education, the Shinjuku Ward Board of Education, and others. In the same year, the small area surrounding his tombstone including the monument was designated as a historic site by the Tokyo Metropolitan Government.

A year later, in 1959, an association of Japanese traditional mathematics was formed and, in three years, was reorganized as the History of Mathematics Society of Japan.

In order to commemorate his 300th posthumous anniversary, the executive committee of Seki's 300th memorial was organized on May 6, 2007, and decided to organize the following commemorative events: a commemorative Buddhist service at Jôrinji temple; exhibitions in several museums; the International Conference on the History of Mathematics on the Kagurazaka Campus of the Tokyo University of Science; lectures and events in connection with the exhibitions; and other events. The aim of these commemorative events was to honor Seki Takakazu and his mathematical works, to inform people about the mathematical achievements in the Edo era and their later developments, and thus to encourage young mathematical talent in Japan.

The following is a list of our activities in addition to the International Conference:

- On December 2, 2007, a commemorative Buddhist service at Jôrinji temple, followed by a conference on mathematical power in Japan, where a report on the preparation of commemorative activities was given by representatives of the executive committee, and lectures on education and research in mathematics were given;
- From January 19 to March 2, 2008, a special exhibition entitled "Seki Takakazu and the old Japanese mathematics" in the Shinjuku Historical Museum;
- On March 9, 2008, the 10th commendation ceremony of the Sangaku (mathematical tablet) contest and lectures on old Japanese mathematics in the Edo-Tokyo Museum;
- From March 6 to April 10, 2008 (the first period), a special exhibition entitled "Seki Takakazu and the old Japanese mathematics" in the Museum of Science, Tokyo University of Science;
- From April 15 to May 18, 2008, a special exhibition entitled "Seki Takakazu and the old Japanese mathematics" in the sea adventure pavilion, Osaka Maritime Museum;
- From April 26 to June 8, 2008, a special exhibition entitled "Seki Takakazu and the old Japanese mathematics" in the Ichinoseki City Museum;
- From August 21 to November 3, 2008 (the second period), a special exhibition entitled "Seki Takakazu and the old Japanese mathematics" in the Museum of Science, Tokyo University of Science;
- Repair of the tomb and the monument of Seki Takakazu at Jôrinji temple;
- From November 22, 2008, to January 12, 2009, the 7th exhibition on Japanese scientists and technicians entitled "Japanese Pioneers in Mathematics" in the Japan Gallery, National Museum of Nature and Science, Ueno, Tokyo;
- On December 6, 2008, a commemorative Buddhist service at Jôrinji temple.

We are proud to report that all of these activities have been successfully carried out and we are grateful to all people concerned for their heartfelt support and cooperation.

Tokyo, October 2011

Hideyuki Majima
Chairman of the Executive Committee of
Seki's 300th Memorial

Preface

Seki Takakazu (1642?–1708) was a mathematician of the Edo period (1603–1868) of Japan who made outstanding contributions to the mathematics of the world — the first time for a Japanese to do so. The Mathematical Society of Japan and the History of Mathematics Society of Japan hosted the International Conference on the History of Mathematics in Commemoration of the 300th Posthumous Anniversary of Seki Takakazu August 25–31, 2008, on the Kagurazaka Campus of the Tokyo University of Science in Tokyo. This book comprises the proceedings of the conference supplemented with collated texts of five of Seki's writings and reprints of papers on these texts.

In ancient Japan, funerals lasted a very long time. It often took more than two years before the body was finally placed in the tomb. This period was called *mogari* [殯], during which relatives and others regularly met in front of the coffin and recalled with tears [誄 *shinobigoto wo su*] the deeds and wishes of the departed. During these ceremonies the story of the deceased person was fixed and was later recorded as his history.

This custom was lost after the introduction of Buddhism and cremation in the eighth century. However, even today Buddhist families in Japan commonly continue to have anniversaries 1, 3, 7, 13, 17, 23, and 33 years after a person's death. The deceased cannot sleep quietly but are remembered regularly. For a great person similar ceremonies are held again every 50 or 100 years.

In the case of Seki Takakazu, the commissioner of finance [勘定奉行] in the Shogunate presided over the 100th posthumous anniversary. Some years earlier, Honda Toshiaki [本多利明] and other mathematicians of Seki's School erected a memorial epitaph beside the tomb. The 200th posthumous anniversary was held by the Tokyo Mathematico-Physical Society on April 6, 1907. They held a memorial conference for citizens and published its proceedings and the reprint of the Septenary [七部書] of Seki's advanced-level papers.

The purpose of our conference on this 300th posthumous anniversary is not different from that of the preceding two.

Mencius [孟子] (ca. 372–289 B.C.) is regarded by many Confucians as the most important successor of Confucius [孔子] (551–479 B.C.). He left the words: "The legacy of a great person is suddenly lost after five generations. The same is true of a minor person. I have never been able to be a student of Confucius. I secretly improve myself by learning from the learned people." [孟子曰君子之澤五世而斬小人之澤五世而斬予未得為孔子徒也予私淑諸人也].

The expression "secretly improve myself" [私淑] was understood by Japanese in the old days to mean choosing a person, not necessarily a living person, as one's teacher and learn from the teacher's writings.

The theories that Seki and his pupils established are far more advanced than was long supposed. They applied those theories to various problems in geometry and so on, and left many manuscripts in classical Chinese, which was an international language at that time, but few of them had readers for these 300 years. The theoretical parts, some of which were discovered 80 years before Europeans did, are readable but their formula calculations in problem solving are almost beyond human abilities. The recent progress in computers and their usage makes it practical to continue their investigations. We hope that the reader will improve himself by reading these proceedings.

The Conference was organized by the Organizing Committee consisting of Professor Henk Bos (Utrecht, the Netherlands), Professor Karine Chemla (CNRS, France), Professor Annick Horiuchi (Paris 7, France), Professor Hideki Kawahara (Tokyo, Japan), Professor Eberhard Knobloch (T. U. Berlin, Germany), Professor Hikosaburo Komatsu (Tokyo, Japan), Professor Liu Dun (CAS, China) and Professor Michio Yano (Kyoto Sangyo University, Japan), and by the Local Committee consisting of Mr. Ken'ichi Sato (President of the History of Mathematics Society of Japan), Professor Katsuhiko Shimizu (Tokyo Universuty of Science, Japan), Professor Keitaro Sekine (Director of the Science Museum of TUS, Japan), and Mr. Kazuhiko Masuda (Science Museum TUS).

All speakers were invited by either of these Committees.

The Conference received the financial or moral support of the following organizations: The Tokyo Club; Inoue Foundation for Science; Japan–China Science and Technology Exchange Association; Tokyo University of Science; The Ministry of Education, Culture, Sports, Science and Technology, Grant-in-aid for Scientific Research on Priority Areas 17083006; Japan Society for the Promotion of Science, Grant-in-Aid for Scientific Research (C) 20540107.

We would like to thank them for their generous assistance.

Tokyo,
July, 2010
Hikosaburo Komatsu

Contents

Program xix

Part I Contributed papers

Seki Takakazu, His Life and Bibliography 3
Hideyuki Majima

Some Reflections on Main Lines of Mathematical Development 21
Wenlin Li

Babylonian Number Theory and Trigonometric Functions: Trigonometric Table and Pythagorean Triples in the Mathematical Tablet Plimpton 322 31
Kazuo Muroi

Archimedes in China: Archimedes and His Works in Chinese Literature of the Ming and Qing Dynasties 49
Liu Dun

The Nine Chapters on the Mathematical Procedures and Liu Hui's Mathematical Theory 63
Guo Shuchun

On the Alternative Algorithm of the 7th Problem in the Sea Island Mathematical Canon 89
Hideki Kawahara

A Comparative Study on Traditional Mathematics of Korea and Japan . . 93
Kim Yong-Woon

The Axes of Mathematical Methodology in the Song and Yuan Dynasties: The Construction of Mathematical Models 105
Guo Shirong

The *Suanxue Qimeng* and Its Influence on Japanese Mathematics 119
Mitsuo Morimoto

Power Series Expansions in India Around A. D. 1400 133
Setsuro Ikeyama

An Early Japanese Work on Chinese Mathematics in Vietnam: Yoshio Mikami's Study of the Vietnamese Mathematical Treatise *Chi Minh Toan Phap* 149
Alexei Volkov

The *Jinkōki* of Yoshida Mitsuyoshi 173
Ken'ichi Sato

Résumé of Works on Mathematics of Seki Takakazu 187
Osamu Takenouchi

Seki Takakazu's Measuring Process of the Volume of Solids Derived from Spheres 193
Toshio Sugimoto

Seki Takakazu's Method on the Remainder Problem 207
Sumie Tanabe

Seki Takakazu's Method of Calculating the Volume of Solids of Revolution and His Mathematical Object 219
Fumiaki Ozaki

Leibniz's Theory of Elimination and Determinants 229
Eberhard Knobloch

Algebra, Elimination and the Complete Book of Mathematics 245
Hikosaburo Komatsu

Some Questions and Observations Around the Mathematics of Seki Takakazu 275
Silke Wimmer-Zagier and Don Zagier

Ming Antu and His Power Series Expansions 299
Luo Jianjin

Standing on the Shoulders of the Giant Influence of Seki Takakazu on Takebe Katahiro's Mathematical Achievements 311
Xu Zelin and Zhou Chang

Takebe Katahiro's Algorithms for Finding the Circular Arc Length 331
Mitsuo Morimoto

The Method of Successive Divisions by Takebe Katahiro and Nakane Genkei 343
Tamotsu Tsuchikura

Manuscripts in the Edo Period: Preliminary Study on Manuscripts Written by Seki Takakazu 353
Tomomi Nishida

Influence of European Mathematics on Pre-Meiji Japan 357
Tatsuhiko Kobayashi

On Contemporary Mathematics in Vietnam 375
Ha Huy Khoai

Part II Supplements

Notes on Complete Book of Mathematics Vol. 4: Three Essentials 387
Mitsuo Morimoto

Complete Book of Mathematics Vol. 4: Three Essentials, by Seki Takakazu, Takebe Kataakira and Takebe Katahiro, collated by Fumiaki Ozaki and Hikosaburo Komatsu 393
Seki Takakazu, Takebe Kataakira and Takebe Katahiro

Seki's Trilogy: Methods of Solving Explicit Problems, Methods of Solving Implicit Problems and Methods of Solving Concealed Problems .. 437
Hikosaburo Komatsu

Methods of Solving Explicit Problems, by Seki Takakazu, collated by Hikosaburo Komatsu 441
Seki Takakazu

Methods of Solving Implicit Problems, by Seki Takakazu, collated by Hikosaburo Komatsu 457
Seki Takakazu

Methods of Solving Concealed Problems, by Seki Takakazu, collated by Hikosaburo Komatsu 469
Seki Takakazu

Notes on Complete Book of Mathematics Vol. 10: Geometry 495
Hikosaburo Komatsu

Complete Book of Mathematics Vol. 10: Geometry, by Seki Takakazu, Takebe Kataakira and Takebe Katahiro, collated by Hikosaburo Komatsu 499
Seki Takakazu, Takebe Kataakira and Takebe Katahiro

Seki's Theory of Elimination as Compared with the Others' 553
Takefumi Goto and Hikosaburo Komatsu

Index 575

Program

August 25, 2008 (Monday)

9:00–12:00, 12:30–17:00 **Registration**

Introductory Lectures in Japanese

13:00–13:50
佐藤健一 Sato Ken'ichi
吉田光由の『塵劫記』の紹介 (On Yoshida Mitsuyoshi's "Jinkōki")
14:00–14:50
川原秀城 Kawahara Hideki
九章算術 (The Nine Chapters of Mathematical Procedures)
15:00–15:50
森本光生 Morimoto Mitsuo
算学啓蒙について (On Zhu Shijie's "Suanxue Qimeng")

Lecture in Japanese

16:00–16:50
金容雲 Kim Yong Woon
韓日傳統数学の歴史と創造性の比較 (A comparative study on traditional mathematics of Korea and Japan)

18:30
Snack

August 26

9:00–12:00, 12:30–17:00 **Registration**

9:30–10:20

Opening Ceremony

Address by Takeuchi Shin (竹内伸)
President of Tokyo University of Science

Address by Komatsu Hikosaburo (小松彦三郎)
Representative of the Organizing Committee of International Conference

Address by Majima Hideyuki (真島秀行)
Director of the Executive Committee of Seki's 300th Memorial

10:30–11:20
Majima Hideyuki 真島秀行
Seki Takakazu, his life and bibliography

11:30–12:20
Jochi Shigeru 城地茂
Seki Takakazu's biography and "Kanjo-gata Wasan Era" (1674–1780)

14:00–14:50
Eberhard Knobloch
Leibniz' theory of elimination and determinants

15:00–15:50
Komatsu Hikosaburo 小松彦三郎
Algebra, elimination theory and "Complete Books of Mathematics 大成算經"

16:00–16:50
Matsumoto Takao 松本尭生
"Sanpo-Hakki 算法発揮," the first published book on determinants, and the related works by Seki and others

August 27

9:00–12:00, 12:30–17:00 **Registration**

9:30–10:20
Karine Chemla
Remarks on the polynomial algebra of the Song-Yuan dynasties and its continuation in Edo Japan

10:30–11:20
Guo Shuchun 郭書春
A study on "The Nine Chapters of Mathematical Procedures" and Liu Hui's mathematical theory
11:30–12:20
Muroi Kazuo 室井和男
Babylonian number theory and trigonometric functions: Trigonometric table and Pythagorean triples in the mathematical tablet Plimpton 322
14:00–14:50
Guo Shirong 郭世栄
Mathematics in the Song and Yuan Dynasties, "Axis of Mathematical Methods"
15:00–15:50
Takenouchi Osamu 竹之内脩
"Katsuyō Sanpō 括要算法", résumé of works on mathematics of Seki Takakazu
16:00–16:50
Sugimoto Toshio 杉本敏夫
Seki Takakazu's measuring process of the volume of solids derived from spheres

August 28

9:00–12:00, 12:30–17:00 **Registration**

9:30–10:20
Liu Dun 劉鈍
Archimedes in China — Archimedes and his works in Chinese literature of the Ming and Qing dynasties
10:30–11:20
Kobayashi Tatsuhiko 小林龍彦
Influence of European mathematics on pre-Meiji Japan
11:30–12:20
Feng Lisheng 馮立昇
Mathematics exchanges between China and Japan in modern times (1850's–1920's)
14:00–14:25
Fumiaki Ozaki 尾崎文秋
Takakazu Seki's method of calculating the volume of solids of revolution

15:00–17:00
Excursion to Jōrinji temple 浄輪寺, where Seki is buried, and Eisei Bunko 永青文庫

18:30–20:00
Banquet at Bettei torijaya 別亭鳥茶屋

August 29

9:00–12:00, 12:30–17:00 **Registration**

9:30–10:20
Li Wenlin 李文林
Some reflections on "main lines" of mathematical development
10:30–11:20
Luo Jianjin 羅見今
Ming Antu and his power series expansions
11:30–12:20
Ikeyama Setsuro 池山説郎
Power series expansions in India around 1400
14:00–14:50
Ogawa Tsukane 小川束
Theories of circles originated by Seki and Takebe Katahiro
15:00–15:50
Morimoto Mitsuo 森本光生
Takebe Katahiro's algorithms to find the circular arc length
16:00–16:25
Yokotsuka Hiroyuki 横塚啓之
"Kohai-Setsuyaku-Shū 弧背截約集" regarded as a work of Takebe Katahiro
16:30–16:55
Don Zagier
The mathematics of and around Seki Takakazu from a modern mathematician's viewpoint

August 30

9:00–12:00, 12:30–17:00 **Registration**

9:30–10:20
Alexei K. Volkov
Vietnamese mathematics and mathematics education: adaptation or invention?
10:30–11:20
Ha Huy Khoai
On the history of mathematics in Vietnam
11:30–12:20
Xu Zelin 徐澤林
Standing on the shoulders of a giant — Influence of Seki Takakazu on Takebe Katahiro's mathematical achievement
14:00–14:25
Free Discussion

14:30–14:55
Tsuchikura Tamotsu 土倉保
"Ruiyaku-Jutsu 累約術," the method of successive divisions
15:00–15:50
Qu Anjing 曲安京
How did Chinese deal with a scientific problem? — A case study of the solar eclipse theory
16:00–16:25
Ohashi Yukio 大橋由紀夫
Mathematical astronomy of Seki Takakazu 關孝和 and Shibukawa Harumi 澁川春海 — Understanding and overcoming the Chinese traditional calendars
16:30–17:00
Closing Ceremony

August 31

9:00–12:30 **Registration**

Lectures in Japanese

9:30–9:55
西田知己 Nishida Tomomi
江戸時代の写本・稿本 — 関孝和の稿本研究の予備的考察 — (Manuscripts in the Edo period)
10:00–10:25
田辺寿美枝 Tanabe Sumie
翦管術 ("Senkan-Jutsu", Seki Takakazu's method on the remainder problems)
10:30–10:55
藤井康生 Fujii Yasuo
奇零方垜と $1/\sin x$ の展開について (The sum of the power of odd numbers and the development of $1/\sin x$)

List of Contributors

Karine Chemla
Université Paris 7 Paris Diderot
& CNRS, REHSEIS UMR 7219,
Laboratoire de Philosophie
et d'Histoire des Sciences,
Case 7064, 5 rue Thomas Mann,
75205 PARIS CEDEX 13, France
e-mail: chemla@univ-paris-diderot.fr

Feng Lisheng
Institute for History of Science
and Technology and Ancient Texts,
Tsinghua University,
113 Tsinghua University Library,
Beijing 100084, P. R. China
e-mail: fls@tsinghua.edu.cn

Yasuo Fujii
Takarazuka Higashi High School,
1-12-1, Nakayama Satsukidai,
Takarazuka 665-0871, Japan
home address: 2-6-17, Funahara,
Itami 664-0896, Japan
e-mail: yfujii@pearl.ocn.ne.jp

Guo Shirong
College of the History of Science,
Inner Mongolia Normal University,
No.81, Road Zhaowuda, Huhhot,
Inner Mongolia 010022, P. R. China
e-mail: gsr@imnu.edu.cn

Guo Shuchun
Institute for the History of Natural
Science, 1-905# Building One,
Xin Ke Xian Yuan, Ke Xue Yuan
Nan Lu, Zhong Guan Cun,
Beijing 100190, P. R. China
e-mail: shuchunguo@gmail.com

Setsuro Ikeyama
Faculty of Cultural Studies,
Kyoto Sangyo University,
Kamikamo-motoyama, Kita-ku,
Kyoto 603-8555, Japan
e-mail: QWN11071@nifty.com

Shigeru Jochi
National Kaohsiung First University of
Science and Technology,
1, University Road, Yenchao,
Kaohsiung, Taiwan, R. O. China
e-mail: jochi@cc.osaka-kyoiku.ac.jp

Hideki Kawahara
Graduate School of Humanities and
Sociology, University of Tokyo,
7-3-1, Hongo, Bunkyo-ku,
Tokyo 113-0033, Japan
e-mail: kawahara@1.u-tokyo.ac.jp

Ha Huy Khoai
Institute of Mathematics, 18 Hoang Quoc Viet., Hanoi 10307, Vietnam
e-mail: hhkhoai@math.ac.vn

Kim Yong Woon
Hanyang University,
17 Haengdang-dong, Seong dong-gu,
Seoul 133-791, Korea
e-mail: kyw06@hotmail.com

Eberhard Knobloch
Technische Universität Berlin,
Strasse des 17 Juni, 135,
Berlin 10623, Germany
e-mail: Eberhard.Knobloch@tu-Berlin.de

Tatsuhiko Kobayashi
Maebashi Institute of Technology,
460-1, Kamisadori-machi, Maebashi,
Gunma 371-0816, Japan
e-mail: koba@maebashi-it.ac.jp

Hikosaburo Komatsu
Department of Mathematics,
Faculty of Science,
Tokyo University of Science,
Kagurazaka 1-3, Shinjuku-ku,
Tokyo 162-8601, Japan
e-mail: komatsu@ms.u-tokyo.ac.jp

Wenlin Li
Academy of Mathematics and System Science,
55 Zhongguancun East Road,
Beijing 100190, P. R. China
e-mail: wli@math.ac.cn

Liu Dun
Institute for the History of Natural Science, 1-905# Building One,
Xin Ke Xiang Yuan, Ke Xue Yuan Nan Lu, Zhong Guan Cun,
Beijing 100190, P. R. China
e-mail: liudun@ustc.edu.cn

Luo Jianjin
Institute of the History of Natural Science, Inner Mongolia Normal University,
No 81, Road Zhaowuda, Huhhot,
Inner Mongolia 010022, P.R. China
e-mail: lluojj@163.com

Majima Hideyuki
Ochanomizu University,
2-1-1, Otsuka, Bunkyo-ku,
Tokyo 112-8610, Japan
e-mail: majima.hideyuki@ocha.ac.jp

Takao Matumoto
Department of Mathematics,
Graduate School of Science,
Hiroshima University,
1-3-1, Kagamiyama,
Higashi-Hiroshima 739-8526, Japan
e-mail: matumoto@math.sci.hiroshima-u.ac.jp

Morimoto Mitsuo
Sophia University, 7-1, Kioicho,
Chiyoda-ku, Tokyo 102-0094, Japan
e-mail: mormormor@jcom.home.ne.jp

Kazuo Muroi
Kawai-juku, Bunri,
2-12-12, Honcho, Aoba-ku,
Sendai 980-8405, Japan
e-mail: edubakazuo@ac.auone-net.jp

Tomomi Nishida
3-7-10 Mama, Ichikawa,
Chiba 272-0826, Japan
e-mail: sancan-cufu@goo.jp

Tsukane Ogawa
Faculty of Environmental Information Sciences, Yokkaichi University,
Yokkaichi 512-8512, Japan
e-mail: ogawa@yokkaichi-u.ac.jp

Ôhashi Yukio
3-5-26, Hiroo, Shibuya-ku,
Tokyo 150-0012, Japan
e-mail: yukio-ohashi@dk.pdx.ne.jp

Fumiaki Ozaki
Faculty of Mathematics,
Kyushu University,
6-10-1, Hakozaki, Higashi-ku,
Fukuoka 819-8581, Japan
e-mail: f-ozaki@math.kyushu-u.ac.jp

Qu Anjing
Department of Mathematics,
Northwest University,
Xian 710069, P. R. China
e-mail: qaj@nwu.edu.cn

Sato Ken'ichi
Wasan Institute,
48-23, Sakae-cho, Kita-ku,
Tokyo 114-0005, Japan
e-mail: kensatoh@tt.rim.or.jp

Sugimoto Toshio
Japan Women's University,
1-1-1, Nishi-ikuta, Tama-ku,
Kawasaki 214-8565, Japan

Takenouchi Osamu
Osaka University,
1–3, Machikaneyama, Toyonaka,
Osaka 560-8531, Japan
e-mail: osamu-ta@mxs.mesh.ne.jp

Tanabe Sumie
Sacred Heart Senior High School,
4-11-1, Shirokane, Minato-ku,
Tokyo 108-0072, Japan
e-mail: tstanabe@da.airnet.ne.jp

Tamotsu Tsuchikura
Tohoku University,
6-3 Aramaki, Aoba-ku,
Sendai 980-8578, Japan
e-mail: o-tsuchikura@nifty.com

Alexei K. Volkov
Center for General Education and
Institute of History,
National Tsing-Hua University,
101, Sect. 2, Kuang-Fu Road,
Hsinchu 30013, Taiwan, R. O. China
e-mail: alexei.volkov@gmail.com

Xu Zelin
College of Mathematics,
Tianjin Normal University,
Tianjin 300387, P. R. China
e-mail: zelinxujp@yahoo.co.jp

Hiroyuki Yokotsuka
Graduate School of Mathematics and
Science Education,
Tokyo University of Science,
1-3, Kagurazaka, Shinjuku-ku,
Tokyo 162-8601, Japan
e-mail: BYS01005@nifty.ne.jp

Don Zagier and Silke Wimmer Zagier
Max-Planck-Institut für Mathematik,
Vivatsgasse 7, Bonn 53111, Germany
e-mail: don.zagier@mpim-bonn.mpg.de

Part I
Contributed papers

Seki Takakazu, His Life and Bibliography

Hideyuki Majima*

Abstract Seki Takakazu is a Japanese mathematician in the early period of the Edo era. He is known as the first in the world to study the so-called "resultants and determinants." However, one had not known so much about his life before the author's investigation started about two years ago. On the occasion of the 300th posthumous anniversary of Seki, the author has found some significant records and facts concerning his family, especially his adoptive father and his first career. Now it is possible to write a "Curriculum Vitae of Seki Takakazu."

Moreover, his professional career in the Kōfu fief is now clarified by a document added in proof at the end of the paper.

Introduction

While Seki Takakazu is a famous mathematician in the Edo era[1] as the first person to study the so-called "resultants and determinants" in the world, [2, pp. 141–158] and [1], we had not known so much about his life. After the 250th anniversary of Seki's death, some documents on his profession were introduced to researchers of history

Hideyuki Majima
Ochanomizu University, 2-1-1, Otsuka, Bunkyo-ku, Tokyo 112-8610, Japan
e-mail: majima.hideyuki@ocha.ac.jp

* This work was partly supported by Grand-in-Aid for Scientific Research (C) 19540117.

[1] The Edo era (1603–1867) is divided into several periods; e.g., the Kan'ei [寛永] period (1624–1644), the Shōhō [正保] period (1644–1648), the Keian [慶安] period (1648-1652), the Manji [万治] period (1658–1661), the Kambun [寛文] period (1661–1673), the Empō [延宝] period (1673–1681), the Tenna [天和] period (1681–1684), the Jōkyō [貞享] period (1684–1688), the Genroku [元禄] period (1688–1704), the Hōei [宝永] period (1704–1711), the Shōtoku [正徳] period (1711–1716), the Kyōhō [享保] period (1716–1736), and the Kansei [寛政] period (1789–1801). The Japanese periods do not start from 1 January; e.g. the Kambun period starts from 25 April, 1661 and the following Empō period starts from 21 September, 1673. We also remark here that the lunar-solar calendar was used during the Edo era.

of mathematics. In this 300th posthumous anniversary of Seki, the author has found some more significant records and facts about Seki's life by careful reading of the Annual Records of Tokugawa Shōgunate [S1], the Kōfu Diary [S3], the Kōfu Palace Diary [S4] and so on. ([S1], [S3] and [S4], etc., refer to Source Books listed at the end of this paper.) They, together with known facts, for example, written in [2] and [10], admit us to write a "Curriculum Vitae of Seki Takakazu" as follows:

1 Personal particulars:

1-1 Family name: Seki [關, or 関 in the simplified character], born as Uchiyama [内山],
Popular name [通称 tsūshō]: Shinsuke [新助],
Given name [諱 imina]: Takakazu [孝和].

1-2 Date of birth: not known exactly, but surely between 1640 and 1645.

1-3 Place of birth: Most probably in Edo. In that case, the place marked as the Guard at the main tower (in the Edo Castle) [御殿主御番 Gotenshu-On-ban] or the Fire Guard [火之番 Hinoban] on the Map of Edo in the Shōhō period [正保江戸図 Shōhō Edo-zu].

1-4 Marital status: Married, had two daughters, who died prematurely, and two adopted sons, Heizō [平蔵] and Shinshichi(rō) [新七 (郎)], the latter being his nephew, a son of his brother Uchiyama Shingorō (or Shōken) Nagayuki [内山新五郎 (松軒) 永行].

1-5 Home: in front of Tenryūji Temple [天龍寺前] at Yotsuya [四谷] in a period containing September Genroku 8 (1695).

1-6 Date of death: 24 October Hōei 5 (1708), which corresponds to 5 December 1708 in the Gregorian calendar, given posthumous Buddhist name [法名 hōmyō], Hōgyōin(-den)-Sōtatsu-Nisshin-Koji [法行院 (殿) 宗達日心居士] and buried in Jōrinji Temple [浄輪寺] Cemetery.

1-7 Family crest: Butterfly [蝶 chō] in a period containing September Genroku 8 (1695) and Seki-phoenix [関鳳凰丸紋 Seki-hō-ō-maru-mon] is carved on the tombstone.

2 Professional experience:

2-1 November Kambun 5 (1665)-: succeeded his adoptive father Seki Jūrōemon [十郎右衛門] of the Kōfu fief [甲府藩 Kōfu-han],
December Kambun 5 (1665)-: Member of a Team for Defense [小十人番士 Kojūnin-Ban-Shi] of the Kōfu fief, salaried by 100 *hyo* plus an allowance of 3 *nin-fuchi*[2] from January Kambun 7 (1667).

[2] The Tokugawa shōgunate paid salary to retainers by rice or gave them rice-producing land. If the salary was paid by rice, then *hyo* [俵] was used as unit and if rice-producing lands were given, then *koku* [石] was used as unit. 1 *hyo* means 40 × 1.8 liters of rice per year. 1 *koku* means a lot which produces 100 × 1.8 liters of rice, and as the rate of tax was 40%, 1 *koku* of rice-producing land brought the retainer 1 *hyo* of rice. 1 *nin-fuchi* [人扶持] means 15 × 1.8 liters of rice per month of 30 days. It is approximately 330 liters of rice per year or about 5 *hyo*. 1 *nin-fuchi* was supposed to support one servant of the retainer.

(probably, promoted to the Chief of a Team for Defense [小十人組頭 Kojūnin-kumigashira] and got a raise in his salary in some years.)

2-2 Jōkyō 1–2 (1684–1685): signed and sealed copies of some documents of land survey in Kōshu [甲州] as one of the three responsible officers in charge.

(probably, turned to a position in another for some years.)

2-3 September Genroku 8 (1695): Chief of the Division of Provisions [賄頭 Makanai-Gashira] of the Kōfu fief, salaried by 200 *hyo* plus the executive allowance of 10 *nin-fuchi*.

2-4 September Genroku 11 (1698): signed a letter to officers of the Shinshū-Matsushiro fief [信州松代藩] to draw the boundary between two fiefs as one of the three responsible officers in charge.

2-5 Genroku 14 (1701)-: Examiner of the Division of Accounts [勘定吟味役 Kanjō-Gimmi-Yaku = (勘定頭差添筋) 勘定方御用改 (役)] of the Kōfu fief, salaried by 250 *hyo* plus an executive allowance of 10 *nin-fuchi*, in some years, changed into the salary of 300 *hyo* in all or plus the executive allowance of 10 *nin-fuchi*.

2-6 December Hōei 1 (1704): Chief of a Team of Ceremonies in the Household [納戸組頭 Nando-Kumigashira] of the West Castle[3] [西の丸] of Tokugawa Shōgunate, salaried by 250 *hyo* plus the executive allowance of 10 *nin-fuchi* and later 300 *hyo*.

2-7 November Hōei 3 (1706): retired [致仕 Chishi] on a pension [小普請 Kobushin] on account of illness.

3 Education and Supervisors: not known exactly, probably self-educated by reading books published in those days, e.g., the Book on Things Old and New [塵劫記 Jinkōki], the Unquestionable Methods in Mathematics [算法闕疑抄 Sampō-Ketsugishō], the Platter of Mathematics [算俎 Sanso], the Systematic Treatise of Arithmetic [算法統宗 Suànfǎ Tǒngzōng], the Yang Hui's Methods of Mathematics [楊輝算法 YángHuī Suàn-Fǎ] and An Introduction to Mathematics [算学啓蒙 Suàn-Xué Qǐ-Méng] etc.

4 Teaching experience: there were some students like Takebe Kataakira [建部賢明] (1661–1716) and Takebe Katahiro [建部賢弘] (1664–1739) and, from 1676, he taught them mathematics and wrote the afterwords [跋 batsu] at the end of books published by students. However, not known where and how he taught them.

5 Completed Research Books:

5-1 Mathematical Methods for Exploring Subtle Points [発微算法 Hatsubi-Sampō],

5-2 Compendium of Mathematics [括要算法 Katsuyō-Sampō] (posthumously published),

5-3 Complete Book of Mathematics [大成算経 Taisei-Sankei] (not published).

---

[3] The West Castle was a part of the Edo Castle. Because the heir to Shōgun lived there, the West Castle was synonymous with the heir himself at that time.

In the following sections, we discuss each fact and, in particular, newly discovered 1-2, 1-3, 1-5, 1-7, 2-1, 2-2, 2-3, 2-5 and 2-6 on showing new evidence for researchers of the history of mathematics.

The author has to admit here that this paper is considerably different from his talk at the Conference. This contains new facts he discovered afterwards, especially concerning Seki's adoptive father and his first career. However, they are included in this paper because of their importance. In his talk at the Conference, the author claimed only that Seki's adoptive father was not listed in the Kansei Family Trees [S7] of retainers of the Shōgunate and that his name should be found in the records of the Kōfu fief. Then, he spent five months to read the Kōfu Diary [S3] very carefully and succeeded in finding only one possible name for Seki's adoptive father.

1 Personal Particulars

1.1 Family Name, Given Name and Popular Name:

Takakazu's full name is Seki Shinsuke Takakazu [關新助孝和] with family name Seki, popular name Shinsuke and given name Takakazu. Hereafter we will call him Takakazu, to avoid confusions. Takakazu was born in the Uchiyama family and Seki is the name of his adoptive family.

Takakazu is the second son of Uchiyama Shichibei Nagaakira [内山七兵衛永明], according to the records in the Extinct Family Trees [S6], the Kansei Family Trees [S7] and the Family Tree of the Uchiyama [S9, S16] written in the Kansei period (1789–1800).[4]

1.2 Date of Birth:

It is known that Takjakazu's real father Uchiyama Shichibei Nagaakira became a guard at the main tower [天守御番 Tenshu-omban] in the Edo Castle on 7 November Kan'ei 16 (1639) and salaried by 100 *koku* plus 50 *hyo* from 15 November Kan'ei 18 (1641). So, the date of Takakazu's birthday should be after 1640, for he was born in Edo according to the Extinct Family Trees [S6][5].

Shichibei Nagaakira died on 2 May Shōhō 3 (1646) and Takakazu's eldest brother Uchiyama Shichinosuke [内山七之助] (later, Shichinosuke Nagasada [七之助永貞]) succeeded his father as the head of the Uchiyama on 28 November Shōhō 3 (1646) according to the Family Tree of the Uchiyama [S9, S16].

---

[4] He is, however, the third son according to the family tree of the Seki in the History of Master Seki [關夫子之由緒 Seki Fūshi no Yuisho] which is a chapter of the Stories of Mathematicians by Fujita Teishi (1734–1807) [藤定資算家談 Tō-Teishi-Sanka-Dan] (See [9]).

[5] This is also confirmed by another Kōfu Samurai Directory added in proof at the end of this paper.

Recently the author confirmed the date of succession in an official document: a record in the Annual Records of Tokugawa Shōgunate [S1] reads as follows:

> "The following persons to be ordered to succeed their own fathers, (lines omitted).
> The late Shichibei to be succeeded,
> with 150 *koku*, by Uchiyama Shichinosuke (lines omitted).
> The order was given at the Northern Anteroom of the Room of Azalea
> to the persons concerned and their relatives by Minister Izu-no-kami.
> The ceremonies were also attended by Ministers Bungo-no-kami and Tsushima-no-kami."[6]

Takakazu had an elder sister according to articles on the Uchiyama in the Death Register [過去帳 Kako-chō] of Jōrinji Temple [4], an elder brother Shichinosuke Nagasada born in Fujioka before Kan'ei 16 (1639) and two younger brothers Shingorō (or Shōken) Nagayuki [新五郎 (松軒) 永行] and Kojūrō (or Heisuke) Nagayoshi [小十郎 (平助) 永章] according to the Family Tree of the Uchiyama [S9, S16]. Judging from the death date of his real father, we can conclude from the facts mentioned above that Seki Takakazu was born between 1640 and 1645 (See [7]).

Remark. There is another conjecture about the date of his father's death. According to the Kansei Family Trees [S7], his fourth son Nagayoshi was 65 years old when he died in the year Kyōhō 20 (1725). If this statement is true, the date of Nagaakira's death should be later than $1725 - 65 + 1 = 1661$. But this contradicts the record in the Family Tree of the Uchiyama [S9, S16]. In [9], Matsusaburō Fujiwara left the question open. Muramoto, a descendent of the Uchiyama, conjectured in [8] that Nagaakira died on 3 May Kambun 2 (1662). Young Ken'ichi Sato [11, p.39] followed it on showing an evidence from the Tokugawa Samurai Directory [S8]:

> "Uchiyama Kojūrō (or Heisuke) Nagayoshi was 45 years old and became a retainer of the Shōgunate in Hōei 2 (1705)."

Jōchi [3] also supported this conjecture.

We suppose that there might be some mistakes of numbers in the Tokugawa Samurai Directory (For the details, see Majima [7]). For example, while Takebe Katahiro was born in Kambun 4 (1664) according to the Family Tree of the Takebe [S12] written in the Kansei period, he was 43 years old in Hōei 6 (1709) according to the Tokugawa Samurai Directory; because $1709 - 1664 + 1 = 46 > 43$, there is a discrepancy in accounts.

The succession from Nagaakira to his eldest son Shichinosuke (later Nagasada) took place on 28 November Shōhō 3 (1646) as confirmed by the official record. Thus their conjecture on the date of Nagaakira's death is refuted.

---

[6] The original text reads as follows:

一　実子跡目被　仰付之面々之所謂　(中略)
七兵衛跡
一　百五拾石　内山　七之助　(中略)
右者躑躅之間次之北之間ニ而被　伝之
右之面々或其身或類親之族招　殿中伊豆守伝
上意之趣豊後守対馬守列座也

1.3 Place of Birth:

According to the Family Tree of the Uchiyama [S9, S16] written in the Kansei period, Uchiyama Shichibei Nagaakira had lived in Fujioka and came to Edo (now Tokyo) to serve the Tokugawa Shōgunate as a guard at the main tower, given a residence around Ushigome [牛込] (now in Shinjuku-ku, Tokyo). Therefore, the place of birth of Seki Takakazu is either Fujioka or Edo (See [2]); in the latter case, it is the place marked as the Guard at the main tower and the Fire Guard on the Map of Edo in the Shōhō period, near Jōrinji Temple, because his father's title, the Guard at the main tower [御天守御番 Gotenshu-on-ban] could be put in other Chinese characters [御殿主御番] with the same meaning and the same Japanese pronunciation. The place is near Kagurazaka [神楽坂] Station of Tokyo Metro at present (See [7]).

1.4 Marital Status:

Takakazu got married and had two daughters, who unfortunately died prematurely in Jōkyō 3 (1686) and Genroku 11 (1698), respectively, again according to the Death Register [4].

According to the Kansei Family Trees [S7] Takakazu had two adopted sons, Heizō and Shinshichi(rō). The latter was his nephew, i.e. a son of his brother Uchiyama Shōken Nagayuki and succeeded Takakazu after his death. Unfortunately, the Seki Family was made to be extinct by the Shōgunate because of the misconduct of Shinshichi(rō) in Kyōhō 12 (1727) according to the Kansei Family Trees [S7] or in Kyōhō 20 (1735) according to the Extinct Family Trees [S6] (See [9], [11]).

1.5 Home:

Takakazu was living in front of Tenryūji Temple [天龍寺前] according to the Kōfu Samurai Directory of Genroku 8 (1695) [S15], in which we find the following lines:

> "Chief of division of provisions, the executive allowance of 10 *nin-fuchi*
> [Family Crest] *Gotoku*, [Address] Residence at Mita,
> [Salary] 200 *hyo*, [Name] Yamori Sukejūrō,
> [Family Crest] Butterfly, [Address] Tenryūji-mae,
> [Salary] same as Yamori, [Name] Seki Shinsuke." [7]

---

[7] The original text reads as follows:
御賄頭　御役料拾人扶持
五徳　三田御屋敷
弐百俵　矢守助十郎
蝶　天龍寺前
同　関新助

"Tenryūji-mae" is the name of place meaning "in front of Tenryūji Temple" but there were two places named after "Tenryūji-mae": one in Ushigome and the other in Yotsuya. The reason of this confusion is the great fire in Tenna 3 (1683): after the fire the Tenryūji Temple located in Ushigome was obliged to move to Yotsuya. Even though the name was used for the place in Ushigome until the end of Edo era, many records and facts suggest that Takakazu was living at the place in Yotsuya at that time. The author showed some evidence in an article written in Japanese but do not enter into the details here (See [7]).

We do not know where he lived in the other periods of his life, but there were some possibilities (See also [7]).

1.6 Date of Death:

Takakazu died on 24 October Hōei 5 (1708), i.e. 5 December 1708 in the current solar calendar, was given the posthumous Buddhist name, Hōgyōin(den)-Sōtatsu-Nisshin-Koji and was buried in Jōrinji Temple Cemetery, located in Shichiken-teramachi [七軒寺町], Ushigome, Edo.[8] The tombstone is still kept in good shape along with a monument which was originally raised by Honda Toshiaki [本多 (田) 利明] et al. in Kansei 6 (1794) (See [2]), repaired on the occasion of the 250th posthumous anniversary of Seki Takakazu in 1958.

1.7 Family Crest:

The family crest carved on the tombstone is of Seki-phenix not of crane nor of butterfly, even though his adoptive family crest was of butterfly according to the Kōfu samurai Directory of Genroku 8 (1695) [S15] quoted above (cf. [10], [11], [3]).

A distinct picture of the part was taken by a professional cameraman and was reproduced in [4]. Here are an example of Seki-phenix and a photo taken by the author.

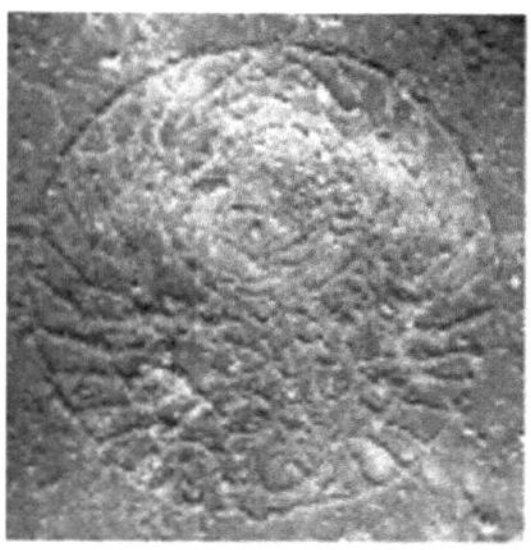

[8] The present address is Benten-chō [弁天町], Shinjuku-ku [新宿区], Tokyo.

2 Professional Experience

2.1 On the Seki Family and His First Career

Takakazu's adoptive father died on 9 August, Kambun 5 (1665), was given posthumous Buddhist name, Ungan-Sōhaku-Shinshi [雲岩宗白信士] and buried in Jōrinji Temple Cemetery according to the Death Register.

By the Kōfu Diary [S3] of the year Kambun 5 (1665), there was only one posthumous succession of a family named Seki, the popular name of the head of the family was Jūrōemon [十郎右衛門], and the popular name of adopted heir was Yaemon [弥右衛門]. The following are quoted from the articles of Kambun 5 (1665) and 7 (1667) of the Kōfu Diary [S3]:

23 November, Kambun 5 (1665)

To be ordered to succeed the late father as the head of family (lines omitted)
Adopted heir to succeed the late Seki Jūrōemon (lines omitted)

18 December, Kambun 5 (1665)

(lines omitted) Seki Yaemon expressed his gratitude to the lord (lines omitted)

21 December, Kambun 5 (1665)

Seki Yaemon be ordered to work as a member of a team for defense of the lord or the fief

18 January, Kambun 7 (1667)

(lines omitted) salaried by 100 *hyo* succeeding the late adoptive father at this moment, plus the allowance of 3 *nin-fuchi*

Seki Yaemon (lines omitted) [9]

---

[9] The original text reads as follows:
寛文五年十一月二十三日
跡目被仰付候次第
一　但馬守出雲守壱岐守淡路守番頭詰座敷ニ列座有之而被申渡之
(養子あるいは実子　父親の跡目、という記載が 6 件先にあるが略)
花房平左衛門被仰渡候
養子 関十郎右衛門跡目 (もう一人の跡目の記載があるが略)
右何茂跡目御定之通被仰付候間勘定頭両人之衆可申合旨但馬守申渡之
寛文五年十二月十八日
(何人か跡目のお願いがあるが略)
関弥右衛門
右鳥目を以中御座敷縁頬杉戸障子際ニテ御礼
申上候　殿様中之御座敷囲際ニテ御立座被遊也 (もう一人の跡目のお願いがあるが略)
寛文五年十二月二十一日
一　衝立之間壱岐守淡路守列座 関弥右衛門呼出之
小十人エ御番被為入候間勤番可仕候也
御礼御出座無之
寛文七年正月十八日
(何人かの加増があるが、前略)
養父跡目百俵被下之

These establish the following facts:

> Takakazu's adoptive father was Seki Jūrōemon. Takakazu had another popular name Yaemon and succeeded his adoptive father. He was ordered to work as a Member of a Team for Defense of the Kōfu fief with the salary of 100 *hyo* plus the allowance of 3 *nin-fuchi*.

Remark. In the literature (cf. [8], [2], [11], [3] etc.), it is traditionally told that the popular name of Seki's adoptive father was Gorōzaemon [五郎左衛門] according to the Kansei Family Trees [S7]. However, there is no popular name of Seki's adoptive father either in the Extinct Family Trees [S6] nor in the History of Master Seki by Fujita Teishi. In his lecture on 9 March 2008, Tatsuhiko Kobayashi told firstly that it was Gorōemon according to a copy of the Family Tree of the Uchiyama written in the Kansei period preserved by the National Archives of Japan [S9]. Then, the author verified that it was also Gorōemon according to a copy of the Family Tree of the Uchiyama written in the Kansei period preserved by the Tokyo Metropolitan Archives [S16], and finally found that it should be Jūrōemon [十郎右衛門] by reading the Kōfu Diary [S3] carefully.
Remark. In the year Kambun 10 (1670) Toramatsu (later, Tsunatoyo) [虎松 (後に, 綱豊)] became the heir of the lord of the Kōfu fief officially under the name of Matsudaira Tsunashige [松平綱重]), which was a celebration for the fief. The fourth Shōgun, Tokugawa Ietsuna [徳川家綱] died in Empō 8 (1680) and was succeeded by Tokugawa Tsunayoshi [徳川綱吉], who became the fifth Shōgun and gave additionally a hundred thousand *koku* to the Kōfu fief of two hundred and fifty thousand *koku*. It seems that Takakazu had a chance to be promoted to the Chief of a Team for Defense with a raise of salary on this occasion.

2.2 Land Survey

Between Jōkyō 1 and 2 (1684–1685), Seki signed and sealed copies of some documents of land survey in some villages in Kōshū ruled by the Kōfu fief as one of the three responsible officers in charge; the other two were Ogiwara Magoshirō [荻原孫四郎] (1635–1694), and Toda Kahei [戸田加 (嘉) 兵衛] (1623–1697).

Some of the copies are preserved in the Meiji University museum, the Yamanashi prefectural museum, and a private museum of abacuses [そろばん Soroban] in Yamanashi prefecture (cf. [10], [11]).

We wonder whether or not Seki went to the villages to survey lands by himself, or just signed and sealed documents. However, Ogiwara Magoshirō, the chief of deputies in the prefecture [代官触頭 Daikan-Furegashira] and Toda Kahei, the superintendent officer [目付役 Metsuke-Yaku], used to go to Kōshū in the Kambun period for their tasks according to the Kōfu Diary ([S3]) and it is suggested that Seki also used to go there to survey lands as the *youngest* responsible officer in charge; the ages of the other two are known by the Family Tree of the Ogiwara [S10] and by

---

今度三人扶持被下
関弥右衛門 (もう一人の加増があるが略、以下略)

the Family Tree of the Toda [S11], respectively. He seems capable of commanding to survey lands by himself because of his mathematical talent.

2.3 Chief of the Division of Provisions of the Kōfu Fief

In some years after the assumed promotion, it is suggested that Seki turned to a position before Genroku 8 (1695).

The Yamanashi Prefectural Museum is preserving a Kōfu Samurai Directory of Genroku 8 (1695) [S15]. In it, Seki was a Chief of the Division of Provisions of the Kōfu fief, salaried by 200 *hyo* plus the executive allowance of 10 *nin-fuchi* (cf. [10]).

In the Kōfu Palace Diary [S4], there is one item with his name, and four items with his title. One record is as follows: two confectioners asked them to offer sweets to the lord of Kōfu in front of chief senior retainers on the five seasonal festivals each. Other items are concerned with their subordinates. He seems busy with just administrative matters, not related to mathematics.

2.4 Drawing Maps

In Genroku 11 (1698), he signed a letter to officers of the Shinshū-Matsushiro fief to draw the boundary between the two fiefs as one of the three responsible officers in charge; the other two were Okumura Sakuzaemon [奥村作左衛門] (1653–1734), the superintendent officer, and Habuto Seizaemon [羽太清左衛門] (1645–1719), the Chief of Division of Accounts [勘定頭 Kanjō-Gashira] of the Kōfu fief (cf. [11]).

We wonder whether Seki went to the prefecture to draw boundaries by himself or not. However, he seemed capable of commanding to draw boundary by himself because of his mathematical talent.

2.5 Examiner of the Division of Accounts

Arai Hakuseki Kimmi [新井白石君美] (1657–1725) was employed by the lord of Kōfu as a Confucian master from December Genroku 6 (1693) with an allowance of 40 *nin-fuchi*. He got a raise of his salary on 25 December Genroku 15 (1702), namely, salaried by 200 *hyo* plus the allowance of 20 *nin-fuchi*. At that time, Seki Shinsuke (Takakazu) was one of the 7 responsible endorsers on it. Arai recorded it in his diary [新井白石日記] (cf. [10, the part of Kobayashi]). According to some Kōfu Samurai Directories [S17], [S2], [S5], [S13],[10] it was reasonable because, in

[10] There is another Kōfu samurai directory [甲府殿御分限帳 Kōfu-dono go-Bugen-chō], which is included in the library of Kai [甲斐叢書 Kai-sōsho] and in Collection of document concerning Kai [甲斐志料集成 Kai-shiryō-shūsei] published in 1935.

the year Genroku 14 (1701), Seki had been promoted to an examiner attached to chiefs of the division of accounts[11] of the Kōfu fief, salaried by 250 *hyo* plus the executive allowance of 10 *nin-fuchi*, according to the Annual Records of Tokugawa Shōgunate [S1] and the Kansei Family Trees [S7]. According to some Kōfu Samurai Directory, he, with the title, was told to be supported by the salary of 300 *hyo* (cf. [10, the part of Kobayashi]). It seems to be recorded like that because 10 *nin-fuchi* was approximately equal to 50 *hyo* and 250+50=300.

2.6 Chief of a Team of Ceremonies in the Household of the West Castle

In December Hōei 1 (1704), the lord of Kōfu Matsudaira Tsunatoyo [松平綱豊] became officially the heir of the fifth Shōgun, Tokugawa Tsunayoshi, and moved to the West Castle [西の丸] of the Edo Castle. Retainers of Tsunatoyo also moved to the West Castle of Tokugawa Shōgunate. On this occasion, Seki Takakazu was promoted to the Chief of a Team of Ceremonies in the Household [納戸組頭 Nando-kumi-Gashira] of the West Castle on 12 December Hōei 1 (1704) and salaried by 250 *hyo* plus the executive allowance of 10 *nin-fuchi* and later 300 *hyo* according to the Kansei Family Trees [S7]; His salary 200 *hyo* according to the Extinct Family Trees [S6], seems to be a mistake (cf. [10, the part of Kobayashi]).

2.7 Retirement on a Pension

On 4 November Hōei 3 (1706), according to the Annual Records of Tokugawa Shōgunate [S1], he retired on a pension on account of illness. Before his retirement, his adopted son, Shinshichi(rō) Hisayuki [久之], was recognized officially at the presence of the Shōgun Tokugawa Tsunayoshi, on 1 October Hōei 3 (1706). Hisayuki succeeded his late adoptive father Takakazu as the head of the Seki family on 29 December Hōei 5 (1708) according to the Kansei Family Trees [S7]; with the salary of 300 *hyo* according to the Tokugawa Samurai Directory [S8]. But the salary was recorded 200 *hyo* in the Extinct Family Trees [S6], which seems to be a mistake. (cf. [10, the part of Kobayashi].)

---

[11] The Examiner of the Division of Accounts [勘定吟味役 Kanjō-Gimmi-Yaku] is originally called the Examiner Attached to Chiefs of the Division of Accounts [勘定頭差添筋 Kanjō-Gashira-Sashisoe-no-suji] [S17] and an extra officer of the Division of Accounts [勘定方御用改 (役) Kanjō-kata-Goyō-Aratame (Yaku)] [S2, S5, S13] in the Kōfu fief.

3 Education and Supervisors

We know very little about his education. Probably he was self-educated by reading books published in those days, the Book on Things Old and New [塵劫記 Jinkōki], the first edition in Kan'ei 4 (1627) and the last edition with 12 open problems in Kan'ei 18 (1641) by Yoshida Mitsuyoshi [吉田光由] (1598?–1672), the Unquestionable Methods in Mathematics [算法闕疑抄 Sampō-Ketsugishō] with preface in Manji 2 (1659) published in Kambun 1 (1661), by Isomura Yoshinori [礒村吉徳] (?–1710), the Platter of Mathematics [算俎 Sanso] published in Kambun 3 (1663) by Muramatsu Shigekiyo [村松茂清] (1608–1695), the Systematic Treatise of Arithmetic [算法統宗 Suànfǎ Tǒngzōng] published in 1592 by Cheng Dawei [程大位], the Yang Hui's Methods of Mathematics [楊輝算法 Yáng Huī Suàn-Fǎ] published in 1275 in China, reprinted later in Korea and imported into Japan in Bunroku [文禄] period (1592–1596) and An Introduction to Mathematics [算学啓蒙 Suàn-Xué Qǐ-Méng] published in 1299 in China by Zhu Shijie [朱世傑], reprinted between 1419 and 1450 in Korea and in 1658 in Japan.[12]

Here, we discuss on his study of Yang Hui's Methods of Mathematics. Not only Seki Takakazu transcribed it, but also correcting mistakes of the original he left his notes as a revised edition. There are two extant copies: a copy by Ishikuro Nobuyoshi [石黒信由] is preserved in Shinminato Museum, Toyama, and the other copy had been preserved by the late Yabuuchi Kiyoshi [薮内清]. Yabuuchi's one is dated the last ten days of May, Kambun 13 (1673) [寛文癸丑仲夏下浣日] and the other one in Shinminato Museum is with the date of the last ten days of May, Kambun 1 (1661) [寛文辛丑仲夏下浣日].

The copier of the latter one seems to have amended the former one by overwriting [辛] on [癸] because there were no such days. However, the former one is of the right form, since the Kambun period lasted for approximately 13 years, followed by the Empō period, and "21 September Kambun [寛文癸丑]" became "21 September Empō [延宝癸丑]." Even though Seki had started to read Yang Hui's Methods of Mathematics before the last year of Kambun, he seems to have written down the date after studying it completely (See [12, the part of Ueno]).

Since Takakazu normally succeeded his adoptive father as the head of the Seki family on 18 December Kambun 5 (1665) and started to work for the Kōfu fief at the same time, the difference of two dates is of no importance for his career.

Concerning An Introduction to Mathematics, his student, Takebe Katahiro, reproduced the original text with appropriate commentaries as the Colloquial Commentaries of "An Introduction to Mathematics" in Genroku 3 (1690). It seems that there was, to some extent, advice from Seki to Takebe.

For calculating an approximation of π, Seki followed the method in the Platter of Mathematics published in Kambun 3 (1663) by Muramatsu Shigekiyo and developed the idea more fully. Using the so-called acceleration method, he finally

[12] By Haji Dōun [土師道雲] and Hisada Gentetsu [久田玄哲]; in 1672 as "An Introduction to Mathematics" and Comments [新編算学啓蒙註解 Shimpen Sangaku Keimō Chūkai] by Hoshino Sukeemon Sanenobu [星野助右衛門実宜] and in 1690 as Colloquial Commentaries of "An Introduction to Mathematics" [算学啓蒙諺解大成 Sangaku Keimō Gennkai Taisei] by Takebe Katahiro.

declared that he adopted "the number weakly less than 3.14159265359 as the fixed circle number [定周 teishū] instead of π. The author discussed the reason why Seki did so. See [6] and [12, the part of Ogawa].

4 Teaching Experience:

There were some students like Mitaki Shirōemon Gunchi [三滝四郎右衛門郡智], Mimata Hachizaemon Hisanaga [三俣八左衛門久長], Araki Hikoshirō Murahide [荒木彦四郎村英] and Takebe brothers (cf. the Master Araki's Comments [荒木彦四郎村英先生茶談 Araki Hikoshirō Murahide Sensei Sadan])[13].

Concerning Takebe Kataakira (1661–1716) and Takebe Katahiro (1664–1739), we knew by the Biography of the Takebe [S13] written by Kataakira, that Takakazu taught them mathematics from 1677, and wrote the afterwords [跋 batsu] at the ends of the Mathematical Methods for Clarifying Slight Signs [研幾算法 Kenki-Sampō] published in Tenna 3 (1683) and the Colloquial Commentaries of Operations in the Mathematical Methods for Exploring Subtle Points [発微算法演段諺解 Hatsubi-Sampō Endan-Genkai] published in Jōkyō 2 (1685) by Takebe Katahiro, the most brilliant student of Seki. However, it has not been known where and how he taught them.

In the afterword of the Colloquial Commentaries on Operations, Seki wrote his opinion on mathematics:

> "For what does Mathematics exist? We study it in order to know methods for solving every problem, easy and difficult. without an exception.
>
> However sophiscated a theory may be, it is unorthodox of mathematics if its solving method is roundabout one."[14]

We suppose that Seki gave his manuscripts to students, they read them making copies and asked Seki what they could not understand, then he gave answers.

Remark. The Japan Academy preserves a scroll which was told to be a license for mathematics addressed to Miyachi Shingorō [宮地新五郎] given by Seki Shinsuke Fujiwara Takakazu [関新助藤原孝和] with his seal. However, young Ken'ichi Sato doubts whether it was made by Seki and supposes that it was done by Araki (See [11]).

---

[13] There are several copies, preserved by the Japan Academy [日本学士院], Modern Science Museum of Tokyo University of Science [東京理科大学近代科学資料館] and so on.

[14] The original text reads as follows:
算学何為乎 学難題易題尽无不明之術也
雖説理高尚 解術迂闊者乃算学之異端也

5 Published Research Works

5.1 *Mathematical Methods for Exploring Subtle Points*

There is only one book published in his life: the Mathematical Methods for Exploring Subtle Points [2, pp. 103–120][15] with a preface dated 14 December Empō 2 (1674), which corresponds to 9 January 1675 in the solar calendar, and an afterword written by his students Mitaki Shirōemon Gunchi and Mimata Hachizaemon Hisanaga. This book was intended to give the answers to all the 15 open problems posed by Sawaguchi Kazuyuki [沢口一之] in his Mathematical Methods Old and New [古今算法記 Kokon-Sampōki] published in Kambun 11 (1671). In the book, however, Seki gave only rough sketches to derive algebraic equations and the degrees of the final algebraic equations to solve the problems. He did not write in detail how to derive the final algebraic equations nor gave any numerical solutions.

According to the preface of the book, the Mathematical Methods for Exploring Subtle Points was published because Seki's students insisted on reading his answers to the problems in order to learn mathematics. But, in reality, the book was too difficult for them to read and people understood Seki's method only after Takebe Katahiro published eleven years later the Colloquial Commentaries on Operations (in the Mathematical Methods for Exploring Subtle Points) in 4 volumes in Jōkyō 2 (1685). Takebe's Commentaries explained the algebraic expressions side writing method [傍書法 bōshohō] of Seki for the first time, which were necessary for the algebraic operations to derive the final equation of one unknown.

There are two versions of the Mathematical Methods for Exploring Subtle Points: (This fact was found firstly at the present day by young Ken'ichi Sato. See [11]) the difference was found in the solution to the 7th problem. Seki had corrected the solution after the first publication as a natural behavior of mathematicians and Takebe published the revised version as the first volume of the Colloquial Commentaries on Operations.

5.2 *Compendium of Mathematics*

The Compendium of Mathematics [2, pp. 267–370][16] is a posthumous publication of Seki's writings which was edited by his students Araki Hikoshirō Murahide and Ōtaka Yoshimasa [大高由昌] and was published in four volumes [元亨利亭 gen, kō, ri, tei].

---

[15] As far as we know, there are left only four copies: each of the three is preserved by the Japan Academy [日本学士院], Kansai University [関西大学] and Wasan Institute [和算研究所], respectively and one is of private possession. That in Wasan Institute is different from the others.

[16] There are relatively many books left, preserved by organizations, the Japan Academy Wasan Institute and so on.

There have been several versions (It was also pointed out at the present day by young Ken'ichi Sato. See [11]): the differences were in the description of the three persons: Seki Takakazu, Araki Murahide and Ōtaka Yoshimasa (See [11]). However, by the afterword at the end of the book written by Araki Hikoshirō Murahide, it is not doubtable that the works were done by Seki Takakazu himself. For example, the forth book explaines Seki's calculation of approximations of π.

5.3 Complete Book of Mathematics

This work was written in 1683 through 1711 by Seki and his students Takebe Kataakira and Takebe Katahiro. The project to compile a collection of mathematics started in the summer Tenna 3 (1683), with Takebe Katahiro as the head of the project. They discontinued writing in the middle of the Genroku period, and named the temporary book of 12 volumes the Complete Collection of Mathematics [算法大成 Sampō-Taisei]. In those days, Takebe Katahiro was busy with his official duties and Seki Takakazu was old, ill and not capable of deep thinking of mathematics. So, from the winter in Genroku 14 (1701), Takebe Kataakira continued to write and revised it until 1711, and finally finished it as the Complete Book of Mathematics of 20 volumes.[17]

This anecdote is based on tthe Biography of the Takebe [S13] written by Takebe Kataakira. We have no other information about it and no choice not to believe that it was true. However, we would like to point out that the following things. Firstly, the purpose of this project seems to publish the collection of Seki's works: Seki wrote many manuscripts, for the revised version of Methods of Solving Concealed Problems [解伏題之法 Kaihukudai no Hō] with the date of 9 September 1683, to give them to the Takebe brothers near 1683. See [5] by Komatsu. Secondly, we wonder whether Seki was ill or just busy with his official duties. As was seen in the previous section, Takakazu was promoted to the Examiner of the Division of Accounts of the Kōfu fief and got a raise in his salary from the winter of Genroku 14 (1701). He must have been busy with official duties, though, after all, he retired on a pension on account of illness in Hōei 3 (1706).

The author has a personal experience: he was the dean of faculty some years ago and busy with many official duties so that he had not much time to think of mathematical problems deeply.

There are many other books or manuscripts written by Seki. But, the author has not had much time to investigate them, especially books on astronomy and calendars. So, the reader is kindly referred to other articles in this volume.

17 There are about 20 sets of copies, preserved by organizations, the University of Tokyo [東京大学], the National Archives of Japan [国立公文書館] and so on [5].

Source Books:

- Books preserved by the National Diet Library [国立国会図書館]

  S1. Annual Records of Tokugawa Shōgunate [年録 Nenroku].

  S2. Kōfu Samurai Directory [甲府御分限帳 Kōfu go-Bugen-chō], i.e., directory of retainers belonging to the Kōfu fief.

- Books preserved by the National Archives of Japan [国立公文書館]

  S3. Kōfu Diary [甲府日記 Kōfu-Nikki], i.e., Daily records of the Kōfu fief.

  S4. Kōfu Palace Diary [甲府御館記 Kōfu-Oyakata-ki], i.e., Daily records in the palace of the lord of Kōfu.

  S5. Kōfu samurai Directory of Saishō [甲府宰相綱重卿之御事並御分限帳 Kōfu-Saishō Tsunashige-kyō-no-on-koto narabini go-Bugen-chō], i.e., directory of retainers belonging to the Kōfu fief.

  S6. Extinct Family Trees [断家譜 Dankafu], i.e., Collection of family trees of extinct families of retainers.

  S7. Kansei Family Trees [寛政重修諸家譜 *Kansei-chōshū-shokafu*], i.e., Collection of family trees of retainers belonging to the Tokugawa Shōgunate inquired in the Kansei period (1789–1800).

  S8. Tokugawa Samurai Directory [御家人分限帳 Gokenin Bugen-chō], i.e., directory of retainers belonging to Tokugawa Shōgunate.

  S9. Family Tree of the Uchiyama [内山家系譜 Uchiyama Kakei-fu] written in the Kansei period.

  S10. Family Tree of the Ogiwara [荻原家系譜 Ogiwara Kakei-fu] written in the Kansei period.

  S11. Family Tree of the Toda [戸田家系譜 Toda Kakei-fu] written in the Kansei period.

  S12. Family Tree of the Takebe [建部家系譜 Takebe Kakei-fu] written in the Kansei period.

- Books preserved by the Japan Academy [日本学士院]:

  S13 Biography of the Takebe [建部氏伝記 Takebe-shi denki].

- Books preserved by the National Institute of Japanese Literature [国文学研究資料館]:

  S14. Kōfu samurai Directory of Kōmon [甲府黄門次郎様臣下録 Kōfu-kōmon-jirō-sama-shinka-roku], i.e., directory of retainers belonging to the Kōfu fief.

- Books preserved by the Yamanashi Prefectural Museum [山梨県立博物館]:

  S15. Kōfu samurai Directory of Genroku 8 (1695) [甲府様御人衆中分限帳 Kōfu-sama-Goninjūchū Bugen-chō], i.e., directory of retainers of the Kōfu fief in a period containing September Genroku [元禄] 8 (1695).

- Books preserved by the Tokyo Metropolitan Archives [東京都公文書館]:

S16. Family Tree of the Uchiyama [内山家系譜 Uchiyama Kakei-fu] written in the Kansei period.

- Books preserved by the Tokyo Metropolitan Library (the Central Library) [東京都立図書館 (中央図書館)]:

S17. Kōfu Samurai Directory [甲府分限帳 Kōfu-Bugen chō], i.e., directory of retainers belonging to the Kōfu fief.

Added in proof

After finishing this article, the author was permitted to read another Kōfu Samurai Directory [甲府分限帳 Kōfu Bugen-chō] and to print the part of Seki Shinsuke (Takakazu)'s record, which reads as follows:

[Prefecture of his homeland] Hitachi;
Adoptive father [given Name] Jūrōemon; Real father [name] Uchiyama Shichibei;
[Salary] 250 *hyo*; [Prefecture of Birth] Musashi (⊃ Edo) ; [Name] Seki Shinsuke;
The executive allowance of 10 *nin-fuchi*; [Age] 57 in Genroku 14 (1701).

Kambun 5 (1665) His adoptive father Jūrōemon died of a disease. In the same year, ordered to succeed the late father as the head of family, salaried by 100 *hyo* out of the salary 130 *hyo* of the late father, ordered to work as a member of a team for defense of the Kōfu fief.

Kambun 7 (1667) Given the allowance of 3 *nin-fuchi*.

Kambun 10 (1670) Given the salary of 10 *hyo* in addition.

Empō 8 (1680) Ordered to work as the chief of a team for defense of the Kōfu fief, given the salary of 90 *hyo* in addition and canceled the allowance of 3 *nin-fuchi*.

Genroku 5 (1692) Ordered to work as a chief of the division of provisions.

Genroku 14 (1701) Ordered to work as an examiner attached to the chiefs of the division of accounts, given the salary of 50 *hyo* in addition and granted the executive allowance of 10 *nin-fuchi*.

His adoptive father was listed to belong to the Kōfu fief in Keian 4 (1651).
When he died of a disease, he was being ordered to work as a member of a team for defense of the Kōfu fief.[18]

---

[18] The original text reads as follows:
本国常陸　養父関十郎右衛門　実父内山七兵衛
弐百五拾俵　生国武蔵　　関　新助
御役料拾人扶持　　　　　辛巳五十七
寛文五乙巳年　養父十郎右衛門病死同年跡式被　仰付御切米高
百三拾俵之内百俵被下之小十人組御番被　仰付
同七丁未年　三人扶持被下之
同十庚戌年　御足米拾俵被下之
延宝八庚申年　小十人組与頭被　仰付御加増九拾俵被下之　三人扶持者上ル
元禄五壬申年　御賄頭被　仰付
同十四辛巳年　御勘定頭ニ差添可相勤旨被　仰付　御加増五拾俵
御役料拾人扶持被下之
養父十郎右衛門儀慶安四辛卯年御帳面ニ而被為附之
病死之節者小十人組御番相勤申候

In the main text, the author explained the facts that he found previously by reading documents carefully. These records guarantee the corresponding facts cited above from the Kōfu Samurai Directory, in which there are more information. For example, Seki was 57 years old in Genroku 14 (1701) according to the above record. If this is true, he was born in 1645.

According to this Kōfu Samurai Directory, Takebe Katahiro was also recorded to be 38 years old in Genroku 16 (1703) so that he was born in 1666 by calculation. However, according to the Family Tree of the Takebe, Takebe Katahiro was born in Kambun 4 (1664). The author will discuss on this issue of birth years elsewhere very soon. He will also discuss on the prefecture of Seki's homeland. Hitachi, which is the prefecture of the Hanabusa [花房] family of Hanabusa Heizaemon [花房平左衛門], the messenger between the Seki family and the chief retainers of the Kōfu fief in Kambun 5 (see 2-1 on p.10).

The author expresses his deep gratitude to the person and his wife who permitted him to publish Seki Shinsuke (Takakazu)'s record in the Kōfu Samurai Directory of their possession.

References

1. T. Goto and H. Komatsu: Determinants, resultants and discriminants in Japan in the seventeenth century and in Europe in the eighteenth and nineteenth centuries, *J. Northwest University (Natural Science Edition)*, **33** No.3, pp. 363–367 (2003); these Proceedings, pp. 553–559.
2. A. Hirayama [平山諦], K. Shimodaira [下平和夫] and H. Hirose [広瀬秀雄] (eds.): *Collected Works of Seki Takakazu edited and explained by the editors* [関孝和全集], Ōsaka Kyōiku Tosho [大阪教育図書], Osaka (1974).
3. S. Jochi [城地茂]: *History of Cultural Exchange in Mathematics of Japan* [日本数理文化交流史—関孝和と『楊輝算法』—], Jillion-Publisher [致良出版社], Taibei (2005).
4. T. Kobayashi [小林龍彦], (old) Ken'ichi Sato [佐藤健一], H. Majima [真島秀行] and K. Yamaji [山司勝紀]: *Collection of Data in Jōrinji Temple* [浄輪寺調査資料集], Jōrinji [浄輪寺] (2007).
5. H. Komatsu: An interim report on collation of the texts of Complete Book of Mathematics [「大成算経」校訂本作成の中間報告], *RIMS Kôkyûroku*, **1546**, pp. 140–156 (2007).
6. H. Majima [真島秀行]: Remarks on calculation of π by Seki Takakazu. [関孝和の円周率の計算についての注意], *RIMS Kôkyûroku*, **1625**, pp. 192–199 (2009).
7. H. Majima [真島秀行]: Newly discovered facts concerning Seki Takakazu on the occasion of the 300th Anniversary of his death [関孝和三百年祭に明らかになったこと], *Journal of the History of Mathematics, Japan* [日本数学史学会『数学史研究』], **200**, pp. 5–15 (2009).
8. K. Muramoto [村本喜代作]: Seki Takakazu and the Family Tree of the Uchiyama [関孝和と内山家譜考], Uchiyama trading company [内山商事] (1963).
9. The Japan Academy (ed.) [日本学士院編]: *History of Mathematics of Japan before Meiji Era* [明治前 日本数学史], Vol. 2 [第 2 巻], Iwanami Shoten [岩波書店] (1956).
10. (Old) Ken'ichi Sato [佐藤健一] and H. Majima [真島秀行] (eds.): *Seki Takakazu, his Life and Works* [関孝和の人と業績], Kenseisha [研成社], Tokyo (2008).
11. (Young) Ken'ichi Sato [佐藤賢一]: *History of Mathematics of Modern Japan* [日本近世数学史], University of Tokyo Press [東京大学出版会] (2005).
12. K. Ueno [上野健爾], T. Ogawa [小川束], T. Kobayashi [小林龍彦] and (young) Ken'ichi Sato [佐藤賢一]: *Introduction to Life and Works of Seki Takakazu* [関孝和論序説], Iwanami-shoten [岩波書店], Tokyo (2008).

Some Reflections on Main Lines of Mathematical Development

Wenlin Li

Abstract There are two major activities of mathematics—theorem-proving and algorithm-creating. While theorem-proving, which originated in ancient Greece, had been being the backbone of the deductive tradition in the history of mathematics, the algorithm-creating, which flourished in ancient and medieval East, formed a strong algorithmic trend in the evolution of mathematics. The main purpose of this paper is to argue the indispensable role in advancing the development of mathematics played by the algorithmic tradition as one of so-called *main lines of mathematical development.*

Description of the algorithmic characters of the Oriental mathematics constitutes the first part of this paper. Some representative Chinese and Wasan algorithms in ancient and medieval times are observed and their modern implication is discussed. The second part of the paper analyzes the algorithmic tendency in the origin of the modern Western mathematics by taking Descartes' geometry as a case study. Finally the author proposes some questions about the relation of ancient Oriental mathematics with main lines of mathematical development for further discussion.

1 Algorithms in Chinese Mathematics

The ancient Oriental mathematics has a strong algorithmic tendency, which paid more attention to algorithm-creating, especially algorithms for solving equations, and was therefore very different from the Greek type of mathematics characterized by the major activity of deductive theorem-proving.

Wenlin Li
Academy of Mathematics and Systems Science, Chinese Academy of Sciences,
55 East Road, Zhonguancun, Beijing 100190, China
e-mail: wli@math.ac.cn

1.1 Nine Chapters on the Mathematical Art

In the case of ancient Chinese mathematics, such a tendency might have originated from the Nine Chapters on the Mathematical Art [九章算術 Jiuzhang Suanshu], in which there is always a paragraph called procedure [術 shu] for each class of problems at their first appearance. That was, in fact, a method for solving problems of the class, which was described in an algorithmic form, step by step. Among the algorithms appeared in the text of Nine Chapters on the Mathematical Art as well as in the annotations by Liu Hui [劉徽], the following four seem to be fundamental and of far reaching influence.

The rectangular array method [方程術 Fangcheng Shu]

This method for solving simultaneous linear equations was of three significant ideas:

1. The formal expression of an algebraic equation,
2. Elimination,
3. Negative numbers,

in which rooted more general approach to express algebraic equations and more complicated elimination of systems of higher equations later in Song and Yuan Dynasties.

The root-extracting method [開方術 Kaifang Shu]

The method listed in Nine Chapters for extracting square and cubic roots is by nature applicable to solving any quadratic and cubic equations and had been studied enormously and led eventually to the establishment of an algorithm for numerical solution of the higher degree equations.

The excess and deficiency method [盈不足術 Ying Buzu Shu]

As an approximating solution of linear equation it was a starting point of a long series of developments of interpolation which culminated in higher interpolation [招差術 Zhaocha Shu] companied by the advance of summation of higher order arithmetic series [垛積術 Duoji Shu] in Song-Yuan dynasty.

The circle dividing method [割圓術 Geyuan Shu]

Though being different from the above three, not in the category of equation-solving, but it was a typical algorithm of treating geometrical problems and offered an infinite approximating procedure for calculating π, arc lengths and the area of circle segments.

Rectangular array → Celestial elements, Four elements
Root-extraction, celestial elements → Positive-negative root-extraction
Excess and deficiency → Higher interpolation, Summation of high powers
Circle dividing → Infinite approximating algorithms

The algorithmic tradition originated from Nine Chapters and developed around equation-solving reached its peak in Song and Yuan dynasties in China. To illustrate the algorithmic feature of such a development we observe mainly two cases here.

1.2 *Qin Jiushao's Book of Mathematics in Nine Chapters*

In his Book of Mathematics in nine Chapters [数書九章 Suanshu Jiuzhang] Qin Jiushao [秦九韶] (1202?–1261) solved the equation of higher degree

$$f(x) = a_0x^n + a_1x^{n-1} + \cdots + a_{n-1}x + a_n = 0 \tag{1}$$

with positive and negative numerical coefficients a_i. His method is known as the positive-negative root-extraction method [正負開方術 Zhengfu Kaifang Shu].

Let $a_0 > 0$, and $a_n < 0$. He first obtained the initial figure c of the root x (in Qin's term, initial quotient [初商 Shou Shang] by trial and error.

Let $x = c + h$. Then we have a new equation for h:

$$f(h) = \overline{a}_0h^n + \overline{a}_1h^{n-1} + \cdots + \overline{a}_{n-1}h + \overline{a}_n = 0, \tag{2}$$

from which Qin suggested the next figure of the root, succeeding quotient [續商 Xu Shang] and starts the operations all over again. Qin gave a mechanical algorithm for calculating the coefficients $\overline{a}_i$ of the reduced equations, which is in fact as follows:

實 shi	a_n	$r^1_{n-1}c + a_n = r^1_n$ $(= \overline{a}_n)$				
方 fang	a_{n-1}	$r^1_{n-2}c + a_{n-1} = r^1_{n-1}$	$r^2_{n-2}c + r^1_{n-1} = r^2_{n-1}$ $(= \overline{a}_{n-1})$			
上廉 shanglian	a_{n-2}	$r^1_{n-3}c + a_{n-2} = r^1_{n-2}$	$r^2_{n-3}c + r^1_{n-2} = r^2_{n-2}$	$\ddots$		
廉 lian	$\vdots$ $\vdots$ a_3 a_2	$\vdots$ $\vdots$ $r^1_2c + a_3 = r^1_3$ $r^1_1c + a_2 = r^1_2$	$\vdots$ $\vdots$ $r^2_2c + r^1_3 = r^2_3$ $r^2_1c + r^1_2 = r^2_2$	$\cdots$		
下廉 xialian	a_1	$a_0c + a_1 = r^1_1$	$a_0c + r^1_1 = r^2_1$	$\cdots$	$a_0c + r^{n-1}_1 = r^n_1$ $(= \overline{a}_1)$	
隅 yu	a_0	a_0	a_0	a_0	a_0	a_0 $(= \overline{a}_0)$

The method contained two important conceptual developments of algorithmic implication:

1. Binomial expansion for integral exponents (Yang Hui triangle);
2. Infinite approximation of irrational number in decimal fractions.

Qin Jiushao's procedure was an iterative infinite approximating process which allows to approximate a real number at any given accuracy. The idea had already appeared in the root-extraction method itself in Nine Chapters. The text for the root-extraction method pointed out that there exists unextractable case, for which Liu Hui added an annotation as follows:

> "To find the successive digits of a decimal fraction, set the digits as numerators, and take ten as the denominator at the first step, one hundred for the second step and so on. The more the steps, the finer the fractions, till the number omitted ... is negligible."

1.3 Zhu Shijie's Four Elements Method

Let's simply cite an instance from Zhu Shijie's work Jade Mirror of Four Elements [四元玉鑑 Siyuan Yujian] (1303):

Problem Suppose we have a right triangle such that the sum of the shorter leg [勾 Gou], the longer leg [股 Gu] and the hypotenuse [弦 Xian], divided by the difference Xian minus Gu is equal to the product Gou times Gu, and that the sum of the difference Gu minus Gou, and Xian, divided by the difference Xian minus Gou is equal to Gou. Find Xian.

Zhu did as follows:

1. First this geometrical problem is reduced to a system of algebraic equations (this is what the ancient Chinese mathematicians often did). Using three Chinese characters Heaven [天 Tian], Earth [地 Di] and Human being [人 Ren] for three unknowns Gou, Gu and Xian, Zhu expressed the system of equations which amounts to :
$$xyz - xy^2 - z - x - y = 0,$$
$$xz - x^2 - z - y + x = 0,$$
$$z^2 - x^2 - y^2 = 0$$
in modern notations.
2. Applying his eliminating procedure, Zhu eliminated first y, then x and obtained the single equation in one unknown z:
$$z^4 - 6z^3 + 4z^2 + 6z - 5 = 0.$$
The last equation could be solved by the standard procedure for numerical solution of equations of higher degrees which was well developed by Song mathematicians as mentioned above. Zhu finally obtained $z = 5$ as the solution required.

In summary, the algorithmic tradition from Nine Chapters time down to Song-Yuan Dynasty showed the following main characters:

1. Arithmetization or algebrization of geometrical problems. Geometrical problems were usually reduced to algebraic equations, or solved simply by a set of standard iterative calculation;
2. Paying great attention to seek mechanical algorithms for solving equations;
3. Sophisticated in infinite approximating process.

We may say that it was the activities of above characters constituted the main line of Chinese mathematics from Han to Song–Yuan, although there appeared much endeavor for demonstration during Liu (Hui)–Zu (Chongzhi) times which came to an abrupt end during the sixth century.

2 Algorithms in Wasan

What's the potential of Song–Yuan mathematics with such characters? There is no answer in classic Chinese mathematics itself, since it eventually declined after Ming Dynasty. However, where Chinese mathematics stagnated, Japanese mathematicians took up and made their breakthrough. It seems that one of original characters of Wasan mathematicians was the synthetic use of methods they learnt from Chinese mathematical texts to create more advanced methods. We observe also some examples.

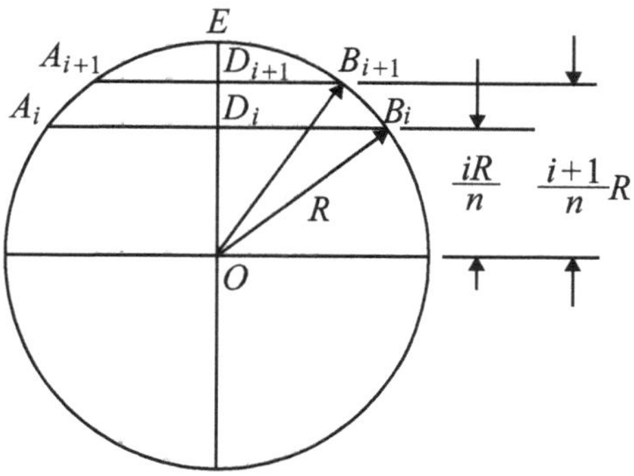

2.1 Theory of Circles

In calculating volume of a sphere, Japanese mathematicians in the 17th extended the circle dividing method to the cases of solid figures (see the figure above). Instead of cutting a circle in plane, they cut a sphere of radius R into n slices of equal height

$$D_iD_{i+1} = \frac{R}{n}$$

Considering each slice as a frustum, which was in turn considered as a cylinder with the same height and the average bottom, calculating volumes of each cylinder and summing up all together, they obtained the approximate volume of the sphere:

$$V_{\text{sphere}} \approx \sum_{i=1}^{n} \frac{\pi}{4} \cdot \frac{A_i B_i^2 + A_{i+1} B_{i+1}^2}{2} \cdot \frac{R}{n}$$

In his Essential Mathematics [括要算法 Katsuyō Sanpō] Seki Takakazu found that the differences of neighboring cylinder volumes constitute a geometrical progression, and used further his accelerated approximation method [增約術 zōyaku jutsu] to get his result of volume of the sphere more effectively.

2.2 Higher Interpolation Formulas

Seki applied the interpolation creatively to calculate the length of an arc, which required differences of higher degree and led him to his discovery of Chaotic Interpolation Method [混沌招差術 konton shōsajutu]:

$$f(x) = f(x_1) + k_1(x - x_1) + k_2(x - x_1)(x - x_2) + \cdots + k_n(x - x_1)(x - x_2)\ldots(x - x_n)$$

amounting to Newton's formulae and enlightened eventually Takebe Katahiro to introduce infinite series such as sine series.

2.3 Elimination theory for higher equations of multi-unknowns

Integration of rectangular array method and celestial element method gave birth to the elimination theory for higher equations of multi-unknowns. For the simplest case of two unknowns

$$\begin{cases} F(x_1, x_2) = A_0 + A_1 x_1 = 0, \\ G(x_1, x_2) = B_0 + B_1 x_1 = 0, \end{cases}$$

A_0, A_1, B_0, B_1 being polynomials of unknown x_2 only, Zhu Shijie used a process called Elimination by Out-In Mutual Multiplication [内外互乘相消 Zhongwai Hucheng Xiangxiao] to eliminate x_1 and obtained

$$H(x_2) = A_0 B_1 - A_1 B_0 = 0.$$

$H(x_2)$ is a polynomial of x_2 only, and can be solved by celestial element method.

More complicated cases can be solved in principle by repeatedly using the above Elimination by Mutual Multiplications [互乘相消] process though the calculations became more and more heavy as the number of unknowns increased. Zhu Shijie did the work for systems of polynomial equations of up to four unknowns [四元

術 Siyuan Shu], but the limitation of the number of unknowns expressed on the counting board prevented him from going further to formalize his method. Japanese mathematicians did not know Zhu's work, they used jointly their knowledge of rectangular array method and celestial element method to create their own theory of elimination with two breakthroughs :

1. Their method is applicable in principle to polynomial equations of any number of unknowns, of which the motivation might be attributed to stronger tendency of algebrization of geometric problems among Wasan mathematicians such as [六斜術 rokusha jutu] that made them encounter often cases of multi-unknowns.
2. they formalized the whole operation of elimination that led them to the discovery of determinants (see [3] or [1]).

割圓術 (summation of sequences)	→	Theory of circles [圓理] (naïve calculus)
方程術 (rectangular array method) 天元術 (celestial element method)	→	Elimination theory for higher equations of multi-unknowns with the introduction of determinants
招差術 (interpolation) 垜積術 (sums of powers)	→	Infinite series, Bernoulli numbers

3 René Descartes' Geometry

Seki Takakazu was contemporary of Newton and Leibniz, a period which was crucial for European mathematics and two events have been considered as milestones—the invention of Analytic Geometry and establishment of Calculus. Researches have shown that in both cases strong algorithmic tendency was embodied. Establishment of calculus was in fact an achievement of long effort of mankind in search for infinitesimal algorithms. The role algorithmic thought played in the process has been argued convincingly by K. A. Rybnikov [4]. I shall discuss here mainly Descartes' creation of the analytic geometry as an example.

It is well known that Descartes' analytical geometry was in fact an algebrization of geometry, that is already poles apart from the Euclidean way. The further question is how Descartes created his analytical geometry. By reading of Descartes' sources including his methodological works, we could find that the motivation which led Descartes to his new geometry was even more along with the algorithmic tradition.

In his unfinished work titled *Rules for the Direction of the Mind* (1619–1628) (abbreviated here after as *Rules*), Descartes criticized strongly the "superficial demonstration" of the Greek which did not seem to teach people "why those thing are so and how they discovered them." Being convinced that the Greek had made their discoveries "more frequently by chance," Descartes felt that there is the need of a method for finding out the truth. The major objective of the *Rules* was right to search for a general method for finding out the "truth," which Descartes called "universal mathematics" (in Latin, "mathesis universalis.") There is no definition of what the so-called "universal mathematics" is in Descartes' works, and it has aroused much debate among Cartesian scholars. According to Chikara Sasaki, "mathesis universalis" meant in Descartes' symbolic algebra as the program of algebraic analysis (see Chikara Sasaki [5]). The rest texts of the *Rules* after rule IV shows clearly the algebraic implication of Descartes' universal mathematics where Descartes offered in fact a universal problem-solving program which can be summarized as follows:

1. Reduce any kind of problem to a mathematical problem;
2. Reduce any kind of mathematical problem to a problem of algebra;
3. Reduce any problem of algebra to a system of algebraic equations;
4. Reduce the system of algebraic equations to a single equation.

The extant version of the *Rules* does not offer readers any detail about what happens after reducing a problem to a single algebraic equation, since the text of the book which should have originally contained thirty six rules breaks off abruptly after rule XXI. Nevertheless, a comparative reading of the *Rules* and Descartes' *Geometry* (1637) shows that Descartes realized definitely his program of "universal mathematics" in the realm of geometry in his classic work *Geometry*.

The *Geometry* started from where the *Rules* broke off, that was reduction of all geometrical problems to a system of algebraic equations and finally to one single equation with one single unknown z. Descartes wrote the equations in order of the highest powers of the unknown:

$$\begin{aligned} z &= b \\ z^2 &= -az + b \\ z^3 &= -az^2 + b^2 z - c \\ z^4 &= az^3 - c^3 z + d^4 \\ &\text{etc.} \end{aligned}$$

Descartes classified equations by their degrees. Then the whole book of the *Geometry* was devoted to the construction of roots of each class of equations by a standard procedure (see Li Wenlin [2]).

It was to treat constructions of equations of the third and higher degrees that led Descartes to investigate the nature and classification of both curves and equations and to the coordinate geometry. Coordinate geometry established firmly Descartes' position as one of the founders of modern mathematics. For Descartes himself, however, it was little more than a means for his major concern—the standard construction of equations of higher degree, by which he was able to algebrize the construction procedures, i.e. to base them on the step-by-step algebraic calculations. At the end of his *Geometry*, Descartes wrote:

"I hope that posterity will judge me kindly, not only as to the things which I have explained, but also to those which I have intentionally omitted so as to leave to others the pleasure of discovery. "

From the above argument we may conclude that the motivation of Descartes' creation of analytic geometry had little to do with the deductive and axiomatic approach, but was rather along the algorithmic and mechanical tradition. In comparing Descartes' "universal mathematics" program with the way of ancient Chinese and Japanese mathematicians for problem-solving by establishing equations as introduced above, it seems that both in character belong to the same line of mathematical development, though they used different mechanical procedure for solving final algebraic equations of higher degree (Descartes used the standard construction process while Chinese and Japanese mathematicians relied on the mechanical algorithm for obtaining the numerical solutions) and classified equations according different principle (Descartes classified equations according to their degree and seemed underestimate the difficulty of reducing a system of algebraic equations to a single equation since he said no words about how he could do that, while Chinese and Japanese mathematicians distinguished the complexity of equations by the number of unknowns).

Above argument does never mean that Descartes had known any works of ancient Chinese or Japanese mathematicians, but it evokes some essential historiographic questions.

4 Deductive vs. Algorithmic Mathematics

One of the essential questions is:

What's the main line of mathematical development?

The research on this regard has been greatly advanced by Professor Wu Wen-Tsun's work since the mid-seventies of the last century (see Wu Wen-Tsun [6]). Wu's research on the history of mathematics strongly suggests two main lines of mathematics development:

- Deductive (Greek) line—Theorem-proving;
- Algorithmic (Oriental) line—Equation solving and algorithmic creating.

Both are important levers of progress of mathematics and the parts they played in development of mathematics cannot be considered interchangeable. However, it appears that, in contrast with the Greek mathematics, the algorithmic tradition in mathematics received no enough exploration so far as the aspect of its influence and transmission. To inspire researches in this field, Wu Wen-Tsun proposed and sponsored to establish the Silk Road Program on the Mathematical and Astronomical Transmission in the History in 2001. The program encourages and supports potential young Chinese scholars to work on the mathematical and astronomical exchanges between China and other Asian countries in ancient and medieval times. In the passed years we have sent scholars to Japan, South Korea and Uzbekistan, promoted international cooperation, and published works (see the series of [8]). The program is really a long serious pursuit. We cite Professor Wu's words in his Chairman Address for the opening ceremony of the International Congress of Mathematicians 2002 (see Wu Wen-Tsun [7]) as a conclusion:

> "Modern mathematics has historical roots of diverse civilizations. ... Today we have railways, airlines and even information highway instead of the Silk Road, the spirit of Silk Road-knowledge exchanges and cultural mergence ought to be greatly carried forward."

References

1. Takefumi Goto and Hikosaburo Komatsu: Determinants, Resultants and Discriminants in Japan in the Seventeenth Century and in Europe in the Eighteenth and Nineteenth Centuries, *Journal of Northwest University (Natural Science Edition)*, vol. 33 No. 3, pp. 363–367 (2003).
2. Li Wenlin [李文林] : *On the Mechanical Character of Descartes' Geometry*, " Studies in the History of Natural Science", Science Press, vol. 12, No. 3 (1993).
3. Yoshio Mikami [三上義夫] : *The Development of Mathematics in China and Japan*, Teubner, Leipzig (1913).
4. K. A. Rybnikov: On the Role of Algorithms in the History of Mathematical Analysis, *Actes du VIII Congress International d'Historie des Sciences, Firenze & Paris,* vol. 1 (1958).
5. Chikara Sasaki [佐々木力]: *Descartes' Mathematical Thought*, Kluwer Academic publishers (2003).
6. Wu Wen-Tsun [吳文俊] : *Mechanization of Mathematics*, Science Press & Kluwer (1998).
7. Wu Wen-Tsun [吳文俊] : Chairman address for the opening ceremony of the International Congress of Mathematicians 2002, *Proceedings of the International Congress of Mathematicians*, vol. 1, Beijing, pp. 21–22 (2002).
8. Xu Zelin [徐澤林]: *Highlights of Wasan Mathematics* [和算選粹], Series of Excellent Mathematical Books with Translations along the Silk Road [絲綢之路数学名著譯叢], Science Press [科学出版社], Beijing (2008).

Babylonian Number Theory and Trigonometric Functions: Trigonometric Table and Pythagorean Triples in the Mathematical Tablet Plimpton 322

Kazuo Muroi

Abstract The mathematical cuneiform tablet Plimpton 322 is one of the most important source materials in history of mathematics. It lists fifteen Pythagorean triples together with a certain table which suggests that one of the angles of a right-angled triangle decreases from 45° to 31°, as O. Neugebauer explained in detail. However, we do not know the principle of constructing the fifteen triples and the true purpose of the table.

In this paper I shall clarify the mathematical meanings of a few technical terms which occur in the headings of the four columns of the tablet, and also the constructing principle of the listed numbers by analyzing Babylonian calculation methods. As a result, we can conclude that the Babylonian scribe of our tablet calculated the fifteen Pythagorean triples using the trigonometric table of the first column which was made by a kind of linear-interpolation.

1 Introduction

In measuring of angles and time in our daily life, we usually use the sexagesimal system of measurement: degrees for angle, minutes and seconds for both angle and time. Although it is well known that it has come down to us from Babylonia by way of Greece, it is not entirely clear when and why the sexagesimal system of units, such as degrees, minutes and seconds, as well as talents, minas, and shekels in weight, was invented in Mesopotamia. In order to answer the question with confidence we would have to study tens of thousands of administrative cuneiform tablets of the third millennium B.C., which vary in form and content from site to site where the tablets have been excavated. Regrettably or naturally, none of us has undertaken the task yet. Given the difficulty of the task, we had better leave the question aside

Kazuo Muroi
2-23-5-902, Chūō, Tagajō, Miyagi 985-0873, Japan
e-mail: edubakazuo@ac.auone-net.jp

for the moment and concentrate on the decipherment of individual tablets that may give us some information about the concept of an angle, for example.

The mathematical cuneiform tablet Plimpton 322, which was published by O. Neugebauer and A. Sachs (Mathematical Cuneiform Texts) [15] in 1945, is one of the most important source materials in history of mathematics because it clearly shows that the Babylonian scribes of the Old Babylonian period (about 1900–1600 B.C.) had a good knowledge of Pythagorean triples [15, pp. 38–41]. In fact, fifteen Pythagorean triples are listed in the tablet together with a certain table which suggests that one of the angles of a right-angled triangle decreases from 45° to 31°, as Neugebauer once more emphasized [14, pp. 36–40]. For this reason Plimpton 322 has attracted the attention of both mathematicians and historians of mathematics since its publication. For example, O. Ore says in his book [16, pp. 170–179] published in 1948 as follows:

> In a new publication of cuneiform texts by Neugebauer and Sachs (1945), there is included a description of a clay tablet from the Plimpton Library at Columbia University, which bids fair to be one of the most crucial records in the history of mathematics.

He points out a possibility that a table of the trigonometric function cosec θ could be constructed by the values given in the tablet.

Another example is the comment on the tablet by V. Katz [4, p. 31]. After having attempted to explain how to obtain the Pythagorean triples he says:

> Why were the particular Pythagorean triples on this tablet chosen?
> Again, we cannot know the answer definitively.

As opposed to these careful opinions, E. Robson [17, p. 167] has confidently formed a judgment on the contents of Plimpton 322:

> I show that the popular view of it as some sort of trigonometric table cannot be correct, given what is now known of the concept of angle in the Old Babylonian period.

Thus we are not completely successful in interpreting the Pythagorean triples or the closely related table in the tablet. There are a couple of reasons for it. First, a few obscure technical terms occur in the headings of the columns and in the fourth column itself, as we will see in the next section. Secondly, we do not know the principle of constructing the fifteen Pythagorean triples of our tablet, although we know the fact that in another tablet the Babylonians derived Pythagorean triples (3, 4, 5), (5, 12, 13), (7, 24, 25), and (19, 180, 181) from the famous formula attributed to *Pythagoras* by Proclus [11]:

$$m^2 + \{(m^2 - 1)/2\}^2 = \{(m^2 + 1)/2\}^2, \text{ where } m \text{ is odd.}$$

Thirdly, the left side of the tablet is broken away and so one or two columns must have been lost forever. In all probability we can reconstruct the lost part if we understand the constructing principle of the triples.

2 The technical terms in the headings

On the obverse of the tablet there are four columns, each of which consists of fifteen lines, which we number I to IV from left to right, while nothing is inscribed on the reverse. The size of the tablet, 12.7 by 8.8 cm, seems rare since the length between the sides is longer than the length between the top and the bottom. See Fig. 1 and the hand-copy of the tablet by Robson [17, p. 171] where two figures are erroneously copied.

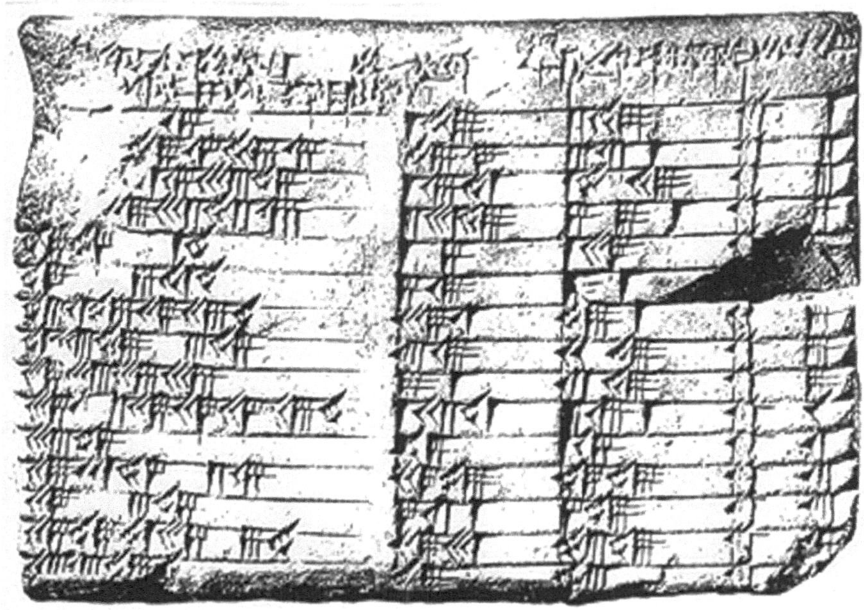

Fig. 1

In the heading of Column I an Akkadian abstract noun *takīltum* occurs, which had been obscure for about seventy years and has recently been identified by Muroi [9] as "completing the square" :

[*ta-k*]*i-il-ti ṣi-li-ip-tim*

[*ša* 1 *in*]*-na-as-sà-ḫu-ú-ma* sag *i-il-lu-ú*

"Completing the square of the hypotenuse from which 1 is subtracted and the width comes up," where the restored parts of the text are indicated by square brackets and Akkadian words are in italics and Sumerian words in roman letters.

The term *takīltum*, which never occurs in non-mathematical texts, is the derivative of the verb *kullum* "to contain, hold" and its literal meaning is "the one which contains (something)." The Babylonian scribes must have visualized completing the square or the factorization of $a^2+2ab+b^2=(a+b)^2$ as involving the fact that the square contains two smaller squares and two congruent rectangles. See Fig. 2.

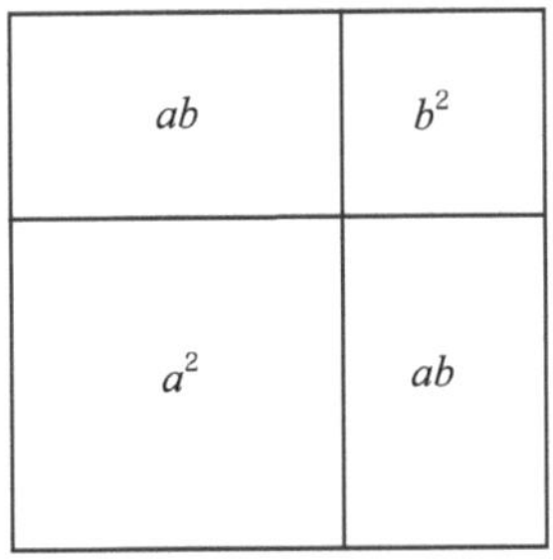

Fig. 2

As will be seen in the next section, the phrase of the heading has to be understood as:

$(x-y)/2=\sqrt{\{(x+y)/2\}^2-1}$, where $(x+y)/2$ is the hypotenuse and $(x-y)/2$ is the width.

Judging from the established meaning of the phrase it is improbable that the Pythagorean triples in Plimpton 322 were constructed by the general solution for the indeterminate equation $x^2+y^2=z^2$:

$x=2pq, y=p^2-q^2, z=p^2+q^2$, where p and q are integers of the form $2^\alpha 3^\beta 5^\gamma$ (α,β,γ: non-negative integers),

which was first proposed by Neugebauer and Sachs.

The headings of the second and the third columns are respectively the following:

íb-si₈ sag "The square root on the width." and
íb-si₈ *ṣi-li-ip-tim* "The square root on the hypotenuse."

The literal translations of these terms, which are not found in any other mathematical tablet, might be "the square root of the width" and "the square root of the hypotenuse," but it is impossible to understand them to be $\sqrt{\text{the width}}$ and $\sqrt{\text{the hypotenuse}}$ because the numbers listed in those columns definitely reject these translations. In fact, under the heading of Col. II are listed the integers of the widths of the right-angled triangles that were obtained by multiplying the *square root* of the square of the *width* by certain integers. Similarly, the integers of the hypotenuses are listed in Col. III. See §4, below.

Moreover I would like to comment on the technical term íb-si₈ which usually means "square root" or "side of a square." Neugebauer and Sachs proposed, however, "solving number" for a translation of this term in order to cover the several meanings of the term in their commentary of Plimpton 322. Since this is too vague and in some cases improper, I have proposed in [7] the translation:

x-e y íb-si$_8$ "x corresponds to y."

The original meaning of the Sumerian verbal phrase íb-si$_8$ is "it corresponds to, it equates to" as many Old Babylonian tables of square roots and one table of exponents and logarithms clearly show: where $y = \sqrt{x}$, $y = 16^x$, or $y = \log_2 x$ is implicitly assumed [7]. Therefore we can definitely state that "functions" were well known in Babylonian mathematics. Bearing in mind the basic idea of íb-si$_8$ of the Babylonians, that is, "one-to-one correspondence" in modern terminology, we had better translate it into an appropriate modern term instead of "solving number" depending upon mathematical circumstances.

The heading of Col. IV is simple and has no problem:

mu-bi-im "Its name."

This phrase is not a mathematical one and frequently occurs as a heading in lists of items [18, pp. 284–297]. Under this heading fifteen similar *items* are written down:

ki-n "the place of n"($n = 1, 2, 3, \cdots, 15$).

It has been assumed that these *items* only refer to numbering the lines from 1 to 15 common to the four columns. Neugebauer [14, pp. 36–37] says:

> The last heading is "its name" which means only "current number," as is evident from the fact that the column of numbers beneath it counts simply the number of lines from "1st" to "15th." This last column is therefore of no mathematical interest.

Similarly Ore [16, p. 176] says:

> Clearly the last column only enumerates the lines.

But I doubt that Col. IV is of no mathematical importance because of two reasons. First it was not common practice for the Babylonian scribes to number the lines of a text one by one. Secondly we have no example of ki-n ($n \geq 3$) in non-mathematical texts [1, p. 412]. In other words, there is a possibility that ki-n is a mathematical term. After having reexamined several intelligible mathematical problems in which ki-n occurs, I have reached the conclusion that it is a mathematical term and it means "the n-th term" of an arithmetic sequence or of the like. I cite two examples from Old Babylonian mathematical texts.

(1) Str 364 [12, pp. 248–256]

In the first three problems of Str 364, a triangle is substantially divided into five trapezoids and one small triangle by five transversals parallel to the base. See Fig. 3. Their areas from left to right in the figure form an arithmetic sequence:

$$18,20 \quad 15,0 \quad 11,40 \quad 8,20 \quad 5,0 \quad 1,40\,.$$

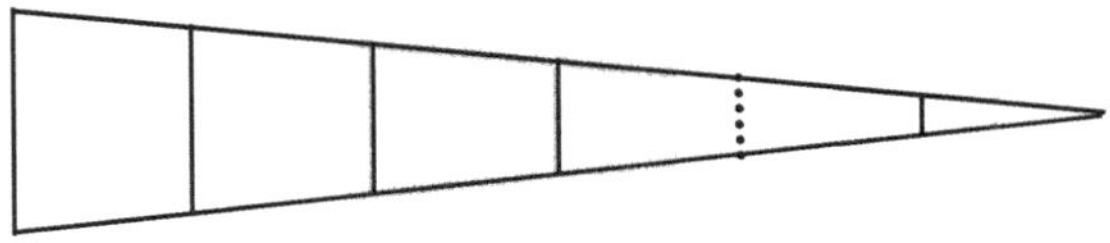

Fig. 3

The first number 18,20 is named "the upper area" as is usual in Babylonian mathematics since the beginning of a line in cuneiform writings is called "the upper" in contrast to the end "the lower." The numbers 15,0 and 11,40 are named "the area of ki-2" and "the area of ki-3" respectively. In the problems the fourth and fifth trapezoids are actually treated as one trapezoid and therefore "the area of ki-4" is not 8, 20 but 13,20 (= 8,20 + 5,0). The last number 1,40, "the area of ki-5," is the area of the small triangle.

(2) BM 13901 [13, pp. 1–14]

In the eighteenth problem of this large tablet, the sides of three squares under certain conditions are asked for. Of the three sides 10, 20, and 30, the latter two numbers in particular are modified by ki-2 and ki-3 respectively. On the other hand, in the twenty-fourth problem the three sides 30, 25, and 15 are modified by not ki-n but by Akkadian ordinals:

ištiat "the first," *šanītum* "the second," *šaluštum* "the third."

Similarly the three sides of the fifteenth and seventeenth problems do not form an arithmetic sequence and they are modified by the same Akkadian ordinals.

Thus there existed a specific technical term for the n-th term of an arithmetic sequence in Babylonian mathematics. The Babylonian scribe of Plimpton 322 must have used the term ki-n in order to describe the fact that the fifteen numbers in Col. I are linearly decreasing.

3 How to construct the numbers in Column I

In the first column of our tablet the following fifteen numbers, which are from three to nine figures in sexagesimal place value notation, are listed. Neither the sexagesimal point (;) nor the number zero is written down on the tablet, while a blank is occasionally used to indicate a vacant place.

Col. I	notes
[1;59], 0, 15	
[1;56, 56], 58, 14, 50, 6, 15	50, 6 seems to be 56.
[1;55, 7], 41, 15, 33, 45	
1;53, 10, 29, 32, 52, 16	
1;48, 54, 1, 40	A meaningless blank between 54 and 1.
1;47, 6, 41 40	A meaningless blank between 47 and 6.
1;43, 11, 56, 28, 26, 40	
1;41, 33, 45, 14, 3, 45	45, 14 is erroneously written 59. A meaningless blank between 59 and 3.
1;38, 33, 36, 36	
1;35, 10, 2, 28, 27, 24, 26, 40	A blank between 10 and 2.
[1];33, 45	
1;29, 21, 54, 2, 15	A meaningless blank between 54 and 2.
1;27, 0, 3, 45	
1;25, 48, 51, 35, 6, 40	
1;23, 13, 46, 40	

The task assigned to us is to clarify the principle by which the Babylonian scribe calculated these numbers and the practical purpose of them. First of all, we had better briefly discuss the explanation for the construction of the numbers in Col. I which was first proposed by E. M. Bruins [2, pp. 191–194] and later followed by Robson.

In order to obtain the Pythagorean triples of our tablet, the Babylonian scribe must have used an identity familiar to them, that is:

$$xy + \{(x-y)/2\}^2 = \{(x+y)/2\}^2 ,$$

and he considered xy as the length of a right-angled triangle assuming it to be 1, $(x-y)/2$ as the width, and $(x+y)/2$ as the hypotenuse. Although we know the fifteen reciprocal pairs of x and y that produce all the numbers in Cols. I, II and III, we have not understood as yet why such pairs were chosen by him. Having examined the Babylonians' calculation methods, the system of the reciprocal tables, and the meanings of the technical terms occurring in the headings, I have come to the conclusion that Col. I is a trigonometric table for:

$$1 + \tan^2\theta \ (\theta = 45°, 44°, \cdots, 31°) ,$$

whose actual values were obtained by *linear interpolation* or the like where a certain table of reciprocals must have been used.

We will analyze in more detail the method by which the Babylonians made this table. As remarked above, they called the side AB $(=xy)$ of the right-angled triangle

ABC in Fig. 4 uš "the length," which was assumed to be 1, the side CA $(= (x-y)/2)$ sag "the width," and the side BC $(= (x+y)/2)$ *ṣiliptum* "the hypotenuse." It is most likely that the two sides *AB* and *CA* that contain a right angle were chosen out of the three sides in order to construct the table in Col. I, because we can see the same practice of Babylonian mathematics in the definition of the inclination of a plane. They described the degree of the inclination of a plane, for example, as follows:

In 1 kùš (in height) it ate x kùš of fodder (1 kùš $\doteqdot$ 50 cm) [5, p. 6], which corresponds to $\tan\varphi = x$ in Fig. 5, if we use modern symbols. In other words, the Babylonians calculated CA/AB, that is, $\tan\theta$ $(\theta = \angle ABC)$ in the right-angled triangle *ABC*. Therefore it is not appropriate to explain according to Neugebauer that Col. I is a table for $(d/l)^2$, where d is the diagonal $(= BC)$ and l is the length $(= AB)$, although $(d/l)^2$ is numerically equal to $1+\tan^2\theta$.

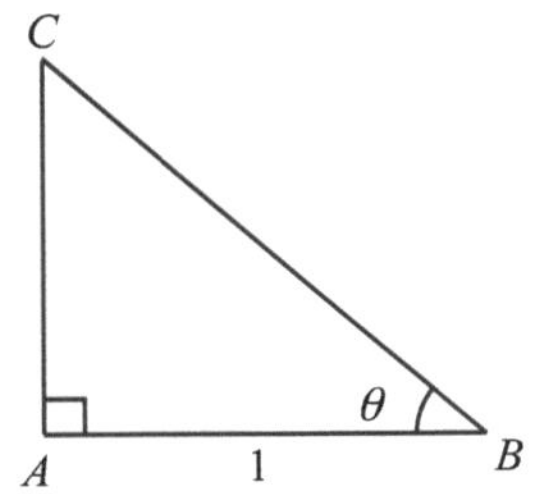

Fig. 4

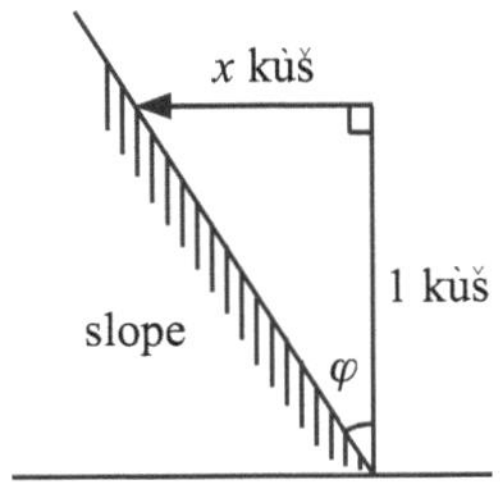

Fig. 5

The next step is to calculate $\tan\theta = CA/AB$, that is, to obtain such pairs of reciprocals as $xy (= 1)$ and $(x-y)/2 (= \tan\theta)$ for a few particular values of θ. In the calculation, the scribe must have used a table of reciprocals of many-place numbers which lists more numbers in an interval than the standard tables. A typical example of the tables of this kind is the cuneiform tablet AO 6456 [12, pp. 14–22] from the ancient city Uruk of the third century B.C., in which 157 numbers from 1 to 3 and their reciprocals are listed in pairs. This tablet probably is a traditional one since we know the fact that the several themes of the Old Babylonian mathematics had come down to the Seleucid period (last three centuries B.C.) and we have a few fragments of the tables of the same kind of the Old Babylonian period [15, pp. 13–16]. In addition the colophon [6] of AO 6456 seems to suggest that the tradition of mathematics, at least in Uruk, had been passed on from generation to generation:

per-su reš-tu-u : 1 : *a-mu-ú* : 2 : *a-mu-ú*
nu al-til / [i]m *ᵐNidintu-Anu* a *šá ᵐIna-qí-bit-Anu*
a ᵐḪun-zu-u ˡᵘ́maš-maš Anu u An-tum
Uruk^{ki}-*u* / *qàt ᵐIna-qí-bit-Anu* duru-a-ni-*šú*

"The first section. '1' is a head number. '2' is a head number. It is not completed. / The tablet of Nindintu-Anu who is a son of Inaqibit-Anu (who is) a son of Ḫunzû, the incantation priest of (the gods) Anu and Antum in Uruk. / (By) the hand of Inaqibit-Anu, his son, (it was written)."

In the following, a part of the tablet of AO 6456 necessary for our understanding of Plimpton 322 is given, in which some numerals erroneously written down by Inaqibit-Anu have been corrected, and the bold-faced pairs of numbers concern Col. I. The numbers in square brackets are the ones supplemented by Neugebauer.

x	$y(=\bar{x})$
1;40	0;36
1;41, 8, 8, 53, 20	0;35, 35, 44, 31, 52, 30
1;41, 15	0;35, 33, 20
[1;41, 43, 30, 56, 15]	[0;35, 23, 21, 59, 2, 24]
1;42, 24	0;35, 9, 22, 30
[1;42, 30, 56, 15]	[0;35, 6, 59, 45, 11, 6, 40]
1;42, 52, 50, 22, 13, 20	0;34, 59, 31, 12
[1;43, 33, 47, 1, 20]	[0;34, 45, 41, 8, 37, 40, 32, 48, 45]
1;43, 40, 48	0;34, 43, 20
1;44, 10	0;34, 33, 36
[1;44, 51, 27, 21, 36]	[0;34, 19, 56, 11, 29, 3, 45]
[1;44, 58, 33, 36]	[0;34, 17, 36, 47, 24, 26, 40]
1;45, 20, 59, 15, 33, 20	0;34, 10, 18, 45
1;45, 28, 7, 30	0;34, 8
[1;46, 17, 17, 31, 12]	[0;33, 52, 12, 37, 55, 59, 40, 14, 48, 53, 20]
1;46, 40	0;33, 45
1;48	**0;33, 20**
1;48, 30, 25	0;33, 10, 39, 21, 36
1;49, 13, 36	0;32, 57, 32, 20, 37, 30
1;49, 21	0;32, 55, 18, 31, 6, 40
[1;49, 51, 47, 48, 45]	[0;32, 46, 4, 48]
1;50, 35, 31, 12	0;32, 33, 7, 30
1;50, 43, 0, 45	0;32, 30, 55, 19, 36, 57, 17, 2, 13, 20
1;51, 6, 40	**0;32, 24**
1;51, 14, 11, 39, 36, 33, 45	0;32, 21, 48, 26, 40
[1;51, 58, 27, 50, 24]	[0;32, 9, 0, 44, 26, 40]
1;52, 30	**0;32**
[1;53, 1, 41, 2, 30]	[0;31, 51, 1, 47, 8, 9, 36]
1;53, 46, 40	0;31, 38, 26, 15
1;53, 54, 22, 30	0;31, 36, 17, 46, 40
1;55, 12	**0;31, 15**
1;55, 44, 26, 40	0;31, 6, 14, 24
[1;56, 30, 30, 24]	[0;30, 53, 56, 34, 20, 9, 22, 30]
1;56, 38, 24	0;30, 51, 51, 6, 40
1;57, 3, 19, 10, 37, 2, 13, 20	0;30, 45, 16, 52, 30

1;57, 11, 15
1;57, 57, 53, 16, 48
[1;58, 5, 52, 48]
1;58, 31, 6, 40
1;58, 39, 8, 26, 15
[1;59, 34, 27, 12, 36]

[2]
2;0, 25, 38, 14, 52, 25, 29, 46, 0, 29, 37, 46,40
[2;0, 33, 47, 46, 40]
2;1, 4, 8, 3, 0, 27

[2;1, 21, 46, 40]
2;1, 30
[2;2, 4, 13, 7, 30]
[2;2, 52, 48]
2;3, 1, 7, 30
2;3, 27, 24, 26, 40
[2;4,16,32,25,36]
2;4, 24, 57, 36
2;5
[2;5, 58, 16, 19, 12]
[2;6, 25, 11, 6, 40]
2;6, 33, 45
2;8
[2;8, 8, 40, 18, 45]
[2;9, 27, 13, 46, 40]
2;9, 36
2;10, 12, 30
[2;11, 4, 19, 12]
[2;11, 13, 12]
2;11, 41, 14, 4, 26, 40
[2;11, 50, 9, 22, 30]
[2;12, 42, 37, 26, 24]
[2;12, 51, 36, 54]
2;13, 20
2;15
[2;15, 38, 1, 15]
2;16, 32
2;16, 41, 15
[2;18, 14, 24]

0;30, 43, 12
0;30, 31, 3, 16, 52, 30
[0;30, 28, 59, 22, 8, 23, 42, 13, 20]
0;30, 22, 30
0;30, 20, 26, 40
[0;30, 6, 24, 33, 43, 6, 22, 26, 30, 7, 24, 26,40]
[0;30]
0;29, 53, 36, 48, 9

[0;29, 51, 35, 25, 26,24]
0;29, 44, 6, 28, 51, 27, 46, 36, 32, 42, 52, 17, 26, 54, 48, 53, 20
[0;29, 39, 47, 6, 33, 45]
0;29, 37, 46, 40
[0;29, 29, 28, 19, 12]
[0;29, 17, 48, 45]
0;29, 15, 49, 47, 39, 15, 33, 20
0;29, 9, 36
[0;28, 58, 4, 17, 11, 23, 47, 20, 37, 30]
0;28, 56, 6, 40
0;28, 48
[0;28, 34, 40, 39, 30, 22, 13, 20]
[0;28, 28, 35, 37, 30]
0;28, 26, 40
0;28, 7, 30
[0;28, 5, 35, 48, 8, 53, 20]
[0;27, 48, 32, 54, 54, 8, 26, 15]
0;27, 46, 40
0;27, 38, 52, 48
[0;27, 27, 56, 57, 11, 15]
[0;27, 26, 5, 25, 55, 33, 20]
0;27, 20, 15
[0;27, 18, 24]
[0;27, 7, 36, 15]
[0;27, 5, 46, 6, 20, 47, 44, 11, 51, 6, 40]
0;27
0;26, 40
[0;26, 32, 31, 29, 16, 48]
0;26, 22, 1, 52, 30
0;26, 20, 14, 48, 53, 20
[0;26, 2, 30]

[2;18, 23, 45, 56, 15]	[0;26, 0, 44, 15, 41, 33, 49, 37, 46, 40]
2;18, 53, 20	**0;25, 55, 12**
[2;19, 48, 36, 28, 48]	[0;25, 44, 57, 8, 36, 47, 48, 45]
[2;19, 58, 4, 48]	[0;25, 43, 12, 35, 33, 20]
2;20, 37, 30	**0;25, 36**
[2;21, 43, 3, 21, 36]	[0;25, 24, 9, 28, 26, 59, 45, 11, 6, 40]
[2;22, 13, 20]	**[0;25, 18, 45]**
2;22, 22, 58, 7, 30	0;25, 17, 2, 13, 20
2;24	**0;25**
[2;24, 40, 33, 20]	[0;24, 52, 59, 31, 12]
2;25, 38, 8	0;24, 43, 9, 15, 28, 7, 30
[2;25, 48]	[0;24, 41, 28, 53, 20]
[2;26, 29, 3, 45]	[0;24, 34, 33, 36]
[2;27, 27, 21, 36]	[0;24, 24, 50, 37, 30]
2;27, 37, 21	0;24, 23, 11, 29, 42, 42, 57, 46, 40
2;28, 8, 53, 20	0;24, 18
[2;29, 17, 57, 7, 12]	[0;24, 6, 45, 33, 20]
[2;29, 28, 4, 0, 45]	[0;24, 5, 7, 38, 58, 29, 5, 57, 12, 5, 55, 33, 20]
[2,30]	[0;24]

In Fig. 4 if we assume $(x-y)/2 = 1$ for $\theta = 45°$, both x and y will be irrational numbers and they do not suit the purpose of obtaining $(x+y)/2$ as a rational number. So we had better choose, from the table above, the largest number x and its reciprocal y that satisfy the inequality:

$$(x-y)/2 < 1\ .$$

In choosing x, a two figure number in the sexagesimal system would be convenient for calculation. The most suitable pair of x and y is $x = 2;24$ and $y = 0;25$, because

$$(x-y)/2 = 0;59,30 < 1\ ,$$

whereas $x = 2;25, 2;26, 2;27, 2;28$ and $2;29$ are all so-called irregular numbers, that is, the numbers whose reciprocals can not be expressed in finite sexagesimal fractions, and another pair, $x = 2;30$ and $y = 0;24$ does not satisfy the inequality:

$$(x-y)/2 = 1;3 > 1\ .$$

In this way if we choose $x = 2;24$ and $y = 0;25$ for $\theta = 45°$ we obtain the number of the first line in Col. I:

$$1+\tan^2\theta = \{(x+y)/2\}^2 = 1;24,30^2 = 1;59,0,15\ .$$

Similarly if we assume $(x-y)/2 = \sqrt{3}/3$ for $\theta = 30°$, x and y will be irrational numbers and again they are unsuitable for our purpose. Since an approximation to $\sqrt{3}$ is 1;45 in Babylonian mathematics, we may choose the smallest number x and its reciprocal y that satisfy an inequality:

$$(x-y)/2 > \sqrt{3}/3 \doteqdot 1;45/3 = 0;35\,,$$

and we obtain an approximation to $\tan 31°$. This pair of numbers is $x = 1;48$ and $y = 0;33,20$, because

$$(x-y)/2 = 0;37,20 > 0;35\,,$$

whereas $x = 1;47, 1;46, 1;45, 1;44, 1;43, 1;42$, and 1;41 are all irregular numbers, and another pair of $x = 1;40$ and $y = 0;36$ does not satisfy the inequality:

$$(x-y)/2 = 0;32 < 0;35\,.$$

The number of the last line in Col. I can be obtained like this from the pair of $x = 1;48$ and $y = 0;33,20$ for $\theta = 31°$:

$$1+\tan^2\theta = \{(x+y)/2\}^2 = 1;10,40^2 = 1;23,13,46,40\,.$$

After the upper and lower bounds of x have been decided, thirteen values of x between them were carefully chosen so that $\{(x+y)/2\}^2$ was almost linearly decreasing, and the table of $1+\tan^2\theta$ ($\theta = 45°, 44°, \cdots, 31°$) was finally made up in Col. I. Of the thirteen pairs of x and y the following four pairs surely occur in the standard table of reciprocals or the like:

$$(x,y) = (2;13,20,\ 0;27), (2;5,\ 0;28,48), (2,\ 0;30) \text{ and } (1;52,30,\ 0;32).$$

Each number of the thirteen pairs is at most four figures disregarding the sexagesimal point, and each is easy to deal with in calculation.

The Babylonians must have noticed the fact that the value of the function $1+\tan^2\theta$ is gradually increasing with the angle θ, since they used in particular the term ki-*n* ($n = 1,2,3,\ldots,15$) which is typical of arithmetic sequences in Babylonian mathematics. See Fig. 6.

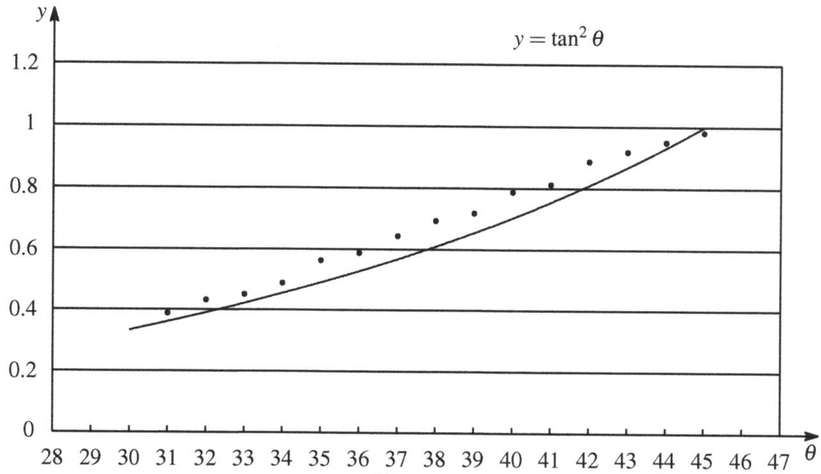

Fig. 6

4 How to construct the numbers in Columns II and III

If we calculate $(x-y)/2$ and $(x+y)/2$ for each of the fifteen pairs of x and y occurred in Col. I, and if we multiply the results by the smallest positive integer $l(n)$ $(n = 1, 2, \cdots, 15)$ that makes them integers, we can obtain the fifteen numbers which are written down in Col. II and Col. III respectively. As is evident from the heading of Col. I:

> Completing the square of the hypotenuse from which 1 is subtracted and the width comes up, or $(x-y)/2 = \sqrt{\{(x+y)/2\}^2 - 1}$,

the value of $l(n)$ can be determined by the value of $(x+y)/2$ only.

Now let us obtain the Pythagorean triple $l(n), b(n)$ and $d(n)$, for each of the fifteen pairs of x and y, where $b(n) = l(n) \times \{(x-y)/2\}$ and $d(n) = l(n) \times \{(x+y)/2\}$.

(1) $x = 2;24 \quad y = 0;25$

$(x+y)/2 = 1;24,30 = 2^{-3} \times 3^{-1} \times 5^{-1} \times 13^2$
$\therefore l(1) = 2^3 \times 3 \times 5 = 2,0$
$b(1) = 2,0 \times 0;59,30 = 1,59$
$d(1) = 2,0 \times 1;24,30 = 2,49$.

(2) $x = 2;22,13,20 \quad y = 0;25,18,45$

$(x+y)/2 = 1;23,46,2,30 = 2^{-7} \times 3^{-3} \times 5^2 \times 193$
$\therefore l(2) = 2^7 \times 3^3 = 57,36$
$b(2) = 57,36 \times 0;58,27,17,30 = 56,7$
$d(2) = 57,36 \times 1;23,46,2,30 = 1,20,25$.

(3) $x = 2;20,37,30 \quad y = 0;25,36$

$(x+y)/2 = 1;23,6,45 = 2^{-6} \times 3^{-1} \times 5^{-2} \times 61 \times 109$
$\therefore l(3) = 2^6 \times 3 \times 5^2 = 1,20,0$
$b(3) = 1,20,0 \times 0;57,30,45 = 1,16,41$
$d(3) = 1,20,0 \times 1;23,6,45 = 1,50,49$.

(4) $x = 2;18,53,20 \quad y = 0;25,55,12$

$(x+y)/2 = 1;22,24,16 = 2^{-2} \times 3^{-3} \times 5^{-3} \times 18541$
$\therefore l(4) = 2^2 \times 3^3 \times 5^3 = 3,45,0$
$b(4) = 3,45,0 \times 0;56,29,4 = 3,31,49$
$d(4) = 3,45,0 \times 1;22,24,16 = 5,9,1$.

(5) $x = 2;15 \quad y = 0;26,40$

$(x+y)/2 = 1;20,50 = 2^{-3} \times 3^{-2} \times 97$
$\therefore l(5) = 2^3 \times 3^2 = 1,12$
$b(5) = 1,12 \times 0;54,10 = 1,5$
$d(5) = 1,12 \times 1;20,50 = 1,37$.

(6) $x = 2;13,20 \quad y = 0;27$

$(x+y)/2 = 1;20,10 = 2^{-3} \times 3^{-2} \times 5^{-1} \times 13 \times 37$
$\therefore l(6) = 2^3 \times 3^2 \times 5 = 6,0$
$b(6) = 6,0 \times 0;53,10 = 5,19$
$d(6) = 6,0 \times 1;20,10 = 8,1$.

(7) $x = 2;9,36 \quad y = 0;27,46,40$

$(x+y)/2 = 1;18,41,20 = 2^{-2} \times 3^{-3} \times 5^{-2} \times 3541$
$\therefore l(7) = 2^2 \times 3^3 \times 5^2 = 45,0$
$b(7) = 45,0 \times 0;50,54,40 = 38,11$
$d(7) = 45,0 \times 1;18,41,20 = 59,1$.

(8) $x = 2;8 \quad y = 0;28,7,30$

$(x+y)/2 = 1;18,3,45 = 2^{-6} \times 3^{-1} \times 5^{-1} \times 1249$
$\therefore l(8) = 2^6 \times 3 \times 5 = 16,0$
$b(8) = 16,0 \times 0;49,56,15 = 13,19$
$d(8) = 16,0 \times 1;18,3,45 = 20,49$.

(9) $x = 2;5 \quad y = 0;28,48$

$(x+y)/2 = 1;16,54 = 2^{-3} \times 3^{-1} \times 5^{-2} \times 769$
$\therefore l(9) = 2^3 \times 3 \times 5^2 = 10,0$
$b(9) = 10,0 \times 0;48,6 = 8,1$
$d(9) = 10,0 \times 1;16,54 = 12,49$.

(10) $x = 2;1,30 \quad y = 0;29,37,46,40$

$(x+y)/2 = 1;15,33,53,20 = 2^{-4} \times 3^{-4} \times 5^{-1} \times 8161$

$\therefore l(10) = 2^4 \times 3^4 \times 5 = 1,48,0$

$b(10) = 1,48,0 \times 0;45,56,6,40 = 1,22,41$

$d(10) = 1,48,0 \times 1;15,33,53,20 = 2,16,1$.

(11) $x = 2 \quad y = 0;30$

$(x+y)/2 = 1;15 = 2^{-2} \times 5$

$\therefore l(11) = 1,0$ is only one exception to the rule.

$b(11) = 45 \quad d(11) = 1,15$.

A right-angled triangle whose sides are 45, 60, and 75 frequently occurs in Babylonian mathematics.

(12) $x = 1;55,12 \quad y = 0;31,15$

$(x+y)/2 = 1;13,13,30 = 2^{-5} \times 3^{-1} \times 5^{-2} \times 29 \times 101$

$\therefore l(12) = 2^5 \times 3 \times 5^2 = 40,0$

$b(12) = 40,0 \times 0;41,58,30 = 27,59$

$d(12) = 40,0 \times 1;13,13,30 = 48,49$.

(13) $x = 1;52,30 \quad y = 0;32$

$(x+y)/2 = 1;12,15 = 2^{-4} \times 3^{-1} \times 5^{-1} \times 17^2$

$\therefore l(13) = 2^4 \times 3 \times 5 = 4,0$

$b(13) = 4,0 \times 0;40,15 = 2,41$

$d(13) = 4,0 \times 1;12,15 = 4,49$.

(14) $x = 1;51,6,40 \quad y = 0;32,24$

$(x+y)/2 = 1;11,45,20 = 2^{-2} \times 3^{-3} \times 5^{-2} \times 3229$

$\therefore l(14) = 2^2 \times 3^3 \times 5^2 = 45,0$

$b(14) = 45,0 \times 0;39,21,20 = 29,31$

$d(14) = 45,0 \times 1;11,45,20 = 53,49$.

(15) $x = 1;48 \quad y = 0;33,20$

$(x+y)/2 = 1;10,40 = 3^{-2} \times 5^{-1} \times 53$

$\therefore l(15) = 3^2 \times 5 = 45$

$b(15) = 45 \times 0;37,20 = 28$

$d(15) = 45 \times 1;10,40 = 53$.

There are four mistakes in writing of the triples in the tablet; $d(2) = 3,12,1$ is a mistake for 1,20,25, $b(9) = 9,1$ for 8,1, $b(13) = 7,12,1\,(= 2,41^2)$ for 2,41, and $b(15) = 56(= 2 \times 28)$ for 28.

It should be emphasized that the factorization into prime factors like that performed in the above was familiar to the Babylonians. In fact there existed a few technical terms related to this technique:

makṣarum "factorization (<bundle)" [3, p. 548], [8, p. 128], a-rá-gub-ba "normal factor"

which was exclusively used for the numbers 1, 2, 3, and 5 [10, pp. 1–8].

5 Conclusion

Judging from what we have analyzed, we conclude that the cuneiform tablet Plimpton 322 is a table of the Pythagorean triples which was constructed by making use of a trigonometric table, $1+\tan^2\theta$ ($\theta = 45°, 44°, \cdots, 31°$), and therefore the concept of angle that a right angle is 90 degrees is discerned in this Old Babylonian tablet. It is most likely that the lost left part of the tablet listed the lengths of the triples only, as Neugebauer supposed. I am very pleased if I were a legitimate heir of Neugebauer and Sachs, because I think that the alleged statement by Robson [17, p. 179] has now been disproved:

> This interpretation [Plimpton 322 represents a trigonometric table of some type] ··· seems to be the bastard offspring of a passing remark made by Neugebauer and Sachs.

Finally, I would like to pay respect both to the anonymous Babylonian scribe who had written this marvelous table and to Neugebauer who grasped the true nature of this mathematical tablet. The following is a summary of the table.

$l(n)$	I	II	III	IV
2, 0	1;59, 0, 15	1, 59	2, 49	1
57, 36	1;56, 56, 58, 14, 50, 6, 15	56, 7	1, 20, 25	2
1, 20, 0	1;55, 7, 41, 15, 33, 45	1, 16, 41	1, 50, 49	3
3, 45, 0	1;53, 10, 29, 32, 52, 16	3, 31, 49	5, 9, 1	4
1, 12	1;48, 54, 1, 40	1, 5	1, 37	5
6, 0	1;47, 6, 41, 40	5, 19	8, 1	6
45, 0	1;43, 11, 56, 28, 26, 40	38, 11	59, 1	7
16, 0	1;41, 33, 45, 14, 3, 45	13, 19	20, 49	8
10, 0	1;38, 33, 36, 36	8, 1	12, 49	9
1, 48, 0	1;35, 10, 2, 28, 27, 24, 26, 40	1, 22, 41	2, 16, 1	10
1, 0	1;33, 45	45	1, 15	11
40, 0	1;29, 21, 54, 2, 15	27, 59	48, 49	12
4, 0	1;27, 0, 3, 45	2, 41	4, 49	13
45, 0	1;25, 48, 51, 35, 6, 40	29, 31	53, 49	14
45	1;23, 13, 46, 40	28	53	15

References

1. R. Borger: *Mesopotamisches Zeichenlexikon*, Ugarit (2003).
2. E. M. Bruins: On Plimpton 322. Pythagorean numbers in Babylonian mathematics, *Indagationes Mathematicae*, **11**, pp.191–194 (1941).
3. J. Friberg: *Reallexikon der Assyriologie und Vorderasiatischen Archäologie,* Band 7, De Gruyter (1990).

4. V. J. Katz: *A History of Mathematics*, second edition, Addison-Wesley (1998).
5. K. Muroi, *Studies in Babylonian Mathematics* No. 2, published by the author (1992).
6. K. Muroi: The colophon in the table of reciprocals of the Seleucid period [セレウコス朝時代のある逆数表の奥付けについて], *Kagakusi Kenkyu* [科学史研究], 32, pp. 29–31 (1993).
7. K. Muroi: An enigmatic sentence in the Old Babylonian table of exponents and logarithms, *Historia Scientiarum*, 6, pp. 229–230 (1997).
8. K. Muroi: Extraction of square roots in Babylonian mathematics, *Historia Scientiarum*, 9, pp. 127–133 (1999).
9. K. Muroi: Mathematical term *takīltum* and completing the square in Babylonian mathematics, *Historia Scientiarum*, 12, pp. 254–263 (2003).
10. K. Muroi: *Studies in Babylonian Mathematics*, No. 3, published by the author (2007).
11. K. Muroi: Prime numbers and Pythagorean triples in Babylonian mathematics, *read at the Meeting of Mathematical Society of Japan held at Tohoku University on 23 September, 2007.*
12. O. Neugebauer: *Mathematische Keilschrift–Texte* [=MKT] I, Springer (1935).
13. O. Neugebauer: *Mathematische Keilschrift–Texte* III, Springer (1937).
14. O. Neugebauer: *The Exact Sciences in Antiquity*, second edition, Dover (1957).
15. O. Neugebauer and A. Sachs: *Mathematical Cuneiform Texts* [=MCT], American Oriental Society (1945).
16. O. Ore: *Number Theory and Its History*, McGraw-Hill Book Co. (1948).
17. E. Robson: Neither Sherlock Homes nor Babylon: A reassessment of Plimpton 322, *Historia Mathematica*, 28, pp. 167–206 (2001).
18. E. Reiner and others: *The Assyrian Dictionary*, vol.17, ŠIII (1992).

Archimedes in China: Archimedes and His Works in Chinese Literature of the Ming and Qing Dynasties

Liu Dun

Abstract Some parallels can be drawn between Archimedes and Liu Hui, Seki Takakazu, among other oriental mathematicians in ancient and medieval times, especially in terms of the concepts and methods concerning infinitesimal. It would be fantastic to make a comparative study along this direction, but this is not my task in this short talk. Instead, a focus will be put on Archimedes and his works as portrayed in Chinese literature of the Ming and Qing Dynasties.

The talk is based on a holistic investigation of the literature of the Ming and Qing Dynasties, with a timeline focused on two phases: first, the late Ming and early Qing period, i.e. the first half of the 17th century, when interest in calendar reform afforded an opportunity for the introduction into China of classical Western astronomical and mathematical knowledge; and second, following the conclusion of the Second Opium War in the late Qing Dynasty, that is, from 1861 to the onset of the 20th century, contemporary Western science was widely introduced into China. This period was also for Chinese mathematics a transition from the traditional to the modern phase.

Therefore the exact content of the talk should be "Archimedes and his Works in Chinese Literature of the Ming and Qing Dynasties."

I will begin with mathematics, and then discuss mechanics and legends, before making a brief conclusion.

One point should be kept in mind : in the Chinese of the Ming-Qing period, the name of Archimedes appears in many different ways, such as [亞奇黙德, 亞幾默德, 黙德, 阿桔弥提斯, 亞基米得, 亞基米德, 亞希黙得, 亞氏, 亞基美特斯, 亞爾日白腊, 亞及密底 ……].

Liu Dun
Institute for the History of Natural Science, Chinese Academy of Sciences, Xin Ke Xiang Yuan, Zhong Guan Cun, Beijing 100190, China
e-mail: liudun@ustc.edu.cn

1 Mathematics

1.1 The Circle

Volume 5 of Complete Explanation of Measurements [測量全義] (1631) [5] by Xu Guangqi [徐光啓] etc. gives an elaborate account of Archimedes' *Measurement of a Circle*, introducing the method of exhaustion devised by "an ancient sage" and giving π an approximate value between $3\frac{10}{71}$ and $3\frac{10}{70}$.

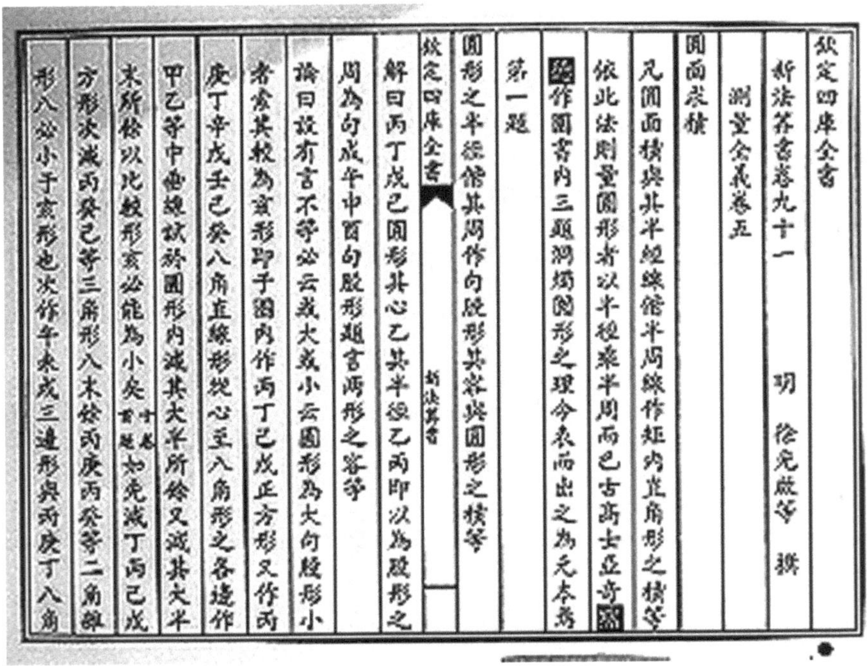
欽定四庫全書
新法算書卷九十一　明　徐光啟等　撰
測量全義卷五
圜面求積
凡圜面積與其半徑線偕半周線作矩內直角形之積等
依此法則量圜形者以半徑乘半周而已古高士亞奇黙
德作圜書內三題洞燭圜形之理今表而出之為元本焉
第一題
圜形之半徑線偕其周作句股形其容與圜形之積等
欽定四庫全書　新法算書
解曰丙丁戊己圜形其心乙其半徑乙丙即以為股形之
周為句成午甲丙句股形題言兩形之容等
論曰設有言不等必云或大或小云圜形為大句股形小
者當其較為寅形即于圜內作丙丁己戊正方形又作丙
庚丁辛戊壬己癸八角直線形從心至八角形之各邊作
甲乙等中垂線試於圜形內減其大半所餘又減其大半
末所餘以比較形寅必能為小矣十卷首題如先減丁丙己戊
方形次減丙癸己等三角形八末餘丙庚丙癸等二角雜
形八必小于寅形也次作午未戊三邊形與丙庚丁八角

Fig. 1 Measurement of a circle

In Volume 2 of The Key of Mathematics [数学鑰] (1681) [6], the early Qing [清] mathematician Du Zhigeng [杜知耕] used $3\frac{1}{7}$ to calculate the circumference of a circle given its diameter and acknowledged this as the "Westerner Archimedean method."

Volume 43 of The Biographies of Mathematicians and Astronomers [疇人傳] (1799) [7], edited by Ruan Yuan [阮元] and Li Rui [李鋭], is called "Westerners," where original texts from [5], among other books, are quoted in explaining Archimedes' work on the circle and sphere; but it also calls $3\frac{10}{70}$ the accurate ratio [密率] of Zong Chongzhi [祖冲之], i.e. the Chongzhi Legacy Method [冲之遺法].

1.2 The Sphere

Volume 6 of Complete Explanation of Measurements [5] gives an introduction to Archimedes' *On the Sphere and Cylinder*, including formulas for the volume and surface of a sphere. It also mentions relevant references in volume 14 of Elements [幾何原本] (1607) [1], even though the Chinese version of [1] only covers the first six books at that time.

Volume 43 of [7] states that the surface of the sphere equals four times (the area of) its great circle, while the volume of a sphere is two-thirds the volume of a cylinder that circumscribes the sphere. The Fundamental Principles of Calculus [微積溯源] (1874) [10], translated by John Fryer and Hua Hengfang [華衡芳], says that the surface of a sphere is four times the area of its great circle, and that is what Archimedes tried to verify.

1.3 Conics

[5], [6], Elements of Analytical Geometry and of the Differential and Integral Calculus [代微積拾級] (1859) [8], [10] and The Conics [円錐曲線] (1893) [17], all touch on the topic of conics. Among them [17] states that Archimedes was a contemporary of the King She [赦王, 314–256 BC] of the Zhou Dynasty [周代]; that ancient scholars knew only some of the principles of the geometrical subjects, that it was not until recently that the significance of conics in astronomy and physics was recognized, and that conics was the most important topic in geometry.

序
嘗攷形學一書，於中朝周赧王時，泰西已有傳焉，然當時所
用之線，惟直線與平圜周之曲線而已，故有數種習題，不能
憑此直線與平圜周一一推算出之者，是不得不另求他曲
線以配之，即所謂圜錐曲線之三種也，造赧王時西方有阿
基彌提斯者，始行而效也，証拋物線所割面積，等於外切矩
形三分之二，並証撱圜面積與外切平圜面積有何比例，但
古時惟究圜錐曲線之理，未明其用，邇來知圜錐學之用為
形學中極廣大而至要者也，比如人擲物，其受拋力與地之
吸力時同，所行之路，必為圜錐學中之拋物線，行星繞行太
圜錐曲線 序

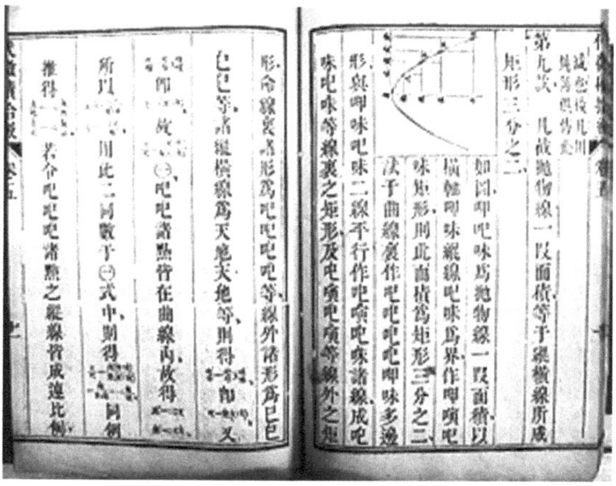

Fig. 2 Conics [円錐曲綫]

1.4 Spirals

The subject of spirals appears in volumes 9 and 18 of [8], and volume 9 of The Origin of Universe [万象一原] (1862) [9] by Xia Luanxiang [夏鸞翔]. In addition, Illustrative Hydraulics [水学圖説] (1890) [16] translated by John Fryer [傅蘭雅], says that "the machine designed on a spiral frame was created by a great master of science in Greece 1200 years earlier and was used to lift water in Egypt. The machine refers to the hydraulic screw or helix (Gao-Li-Yin [高里因] in Chinese transliteration)"; The number 1200 is no doubt mistaken for 2200.

圖之向、其水從丁管入庚管流下、至水箭甲乙之下半、將
[illegible]丙抵使向上、又前次入箭上半之水、行過己管、自辛
孔放出、因此時辛孔與己管相通、如此則槓桿子卯上升、
而凸釘寅令塞門仍歸原向、此後各事俱照前作用、而機
器可行動不息、

亞基米得螺絲龍 又名高里因或名攪水龍 泰西水法名曰龍尾車、

此機器為一千二百年前希臘格致名家所作、用於埃及
起水之工、作法如第十六圖、用圓柱丙丁、立軸戊己、軸上
端入於領戊、下端入於凹己、圓柱外有丁子丑管、繞成螺
絲形、下端入於水池內、在戊處有搖桿、使之轉動、或如本

Fig. 3 Archimedean spirals

1.5 Spheroids and Conoids

Volume 6 of [5] introduces a method to calculate the volume of a spheroid, which is purely derived from Euclidian geometry. [8] and [10], on the other hand, use analytical geometry and calculus to introduce ways to calculate the volumes of other conoids including the paraboloid of revolution and hyperboloid of revolution.

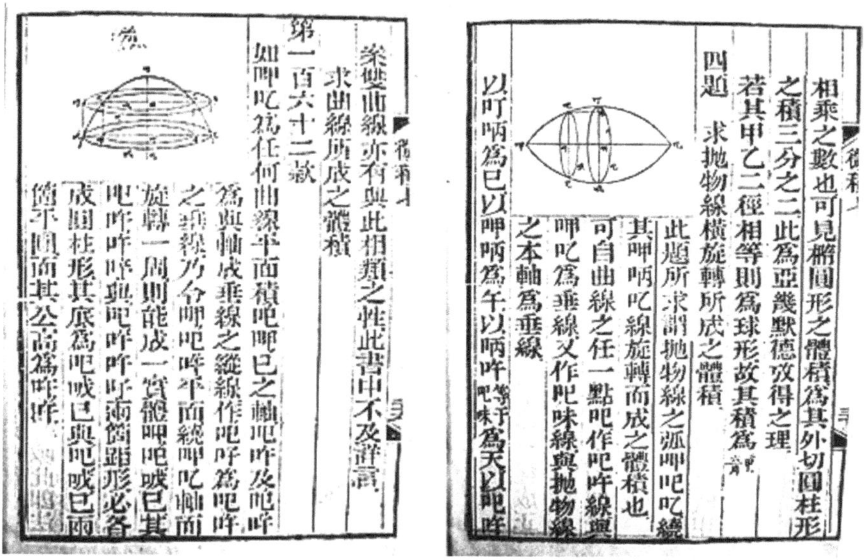

相乘之數也可見楕圓形之體積為其外切圓柱形
之積三分之二此為亞幾默德攷得之理
若其甲乙二徑相等則為球形故其積為
四題　求拋物線橫旋轉所成之體積
此題所求謂拋物線之弧呷吧吃繞
其呷呥吃線旋轉而成之體積也
可自曲線之任一點吧作吧咋線與
呷吃為垂線又作吧味線與拋物線
之本軸為垂線
以吁呥為巳以呷呥為午以呥咋吧味等于為天以吧咋

案雙曲線亦有與此相類之性此書中不及詳言
求曲線所成之體積
第一百六十二款
如呷吃為任何曲線平面積吧呷巳之軸吧咋及吧咋
為與軸成垂線之縱線作吧呼為吧咋
之垂線乃令呷吧咋平面繞呷吃軸而
旋轉一周則能成一實體呷吧成巳其
吧咋咋呼與吧咋咋呼兩個距形必各
成圓柱形其底為吧成巳與吧成巳而
簡平圓而其公高為咋咋

Fig. 4 Paraboloid and hyperboloid of evolution

2 Mechanics

2.1 The Lever Principle

Wang Tao's [王韬] preface to Six Books of Selected Western Knowledge [西学輯存六種] (1889) [13] points out that mechanics was launched with Archimedes' study of the lever, while the balance point must first be determined; and all further mechanical theories are derived from that.

The Chapter "Origins" in Elemental Physics [格物質学] (1902) [20] states that Archimedes defined the principle of the lever 300 years before the birth of Christ, but at the time people ignored its significance, while later generations realized its importance. In the Chapter on the "Principle of Machines," the principle of the lever is accurately described.

西學輯存六種自序
嗚呼學問之道無窮上而天文下而地理中而人事紛賾變化
莫可端倪天算之學行於中國久矣顧執是業者寥寥無幾西
國自明天算以來代有沿革其源流固可得而言焉古時閉他
歌刻獨得眞諦言地球環日而行後多疑某反其說以地為中
心謬宗之者幾千數百年明時歌白尼測得新法興閉說吻合
惜其法未行於生前有第谷者非信之言地球在中五星環日
日率之環地球然未得確據其徒刻白爾始知行星軌道是橢
圜非平圜其行有遲速留逆都歸一理且知地亦環日行星類
也至今墨守其說無有異議後奈端出於數尤深而其理愈明
於以見精益求精西學之未有止境也數學之中以重學為最
繁亦以重學為最適於用重學之舉端實創於亞奇默德先以
總處停勻之杆置當其中倚點處兩復權取重無少差之物分
西學輯存六種 序

己丑秋淞隱
廬遯叟校印

Fig. 5 Six books of selected Western knowledge

2.2 *Specific Gravity*

Volume 2 of Initiatory Textbook of Natural Sciences [格致啓蒙] (1875) [11] states that when weighed in water, the gold that weighs 19 grams in air weighs 18 grams, which means that gold weighs 19 times the same volume of water. Therefore, physicists assign the specific gravity of water as 1, while that of gold as 19.

第二十六 論較量輕重 某物較重於水某物較輕於水均以物與水體積大小相同而論。試驗第二十三 今有一小塊黃金在空氣內秤之重十九格零若置於水秤之重十八格零不既失一格零之重乎蓋水讓黃金體積之方寸水止一格零重故物之重數抵去一格零也以此計之則一小塊黃金重數勝於同體積之水十九格零則是水須照此黃金體積加十九倍乃得與此小塊黃金分兩相均故云黃金較重於水几十九倍也格致家稱黃金重數則云十九無論大小長短厚薄總以一分金對十九分水爲率倘有人持物至靠非淨是黃金則置於水秤之必無十九倍重也二千年前希拉古斯王名海伊羅命工人製黃金冠冕王慮工人雜有他金類其由試驗因屬格致師阿桔

格致啟蒙二 格物學

Fig. 6 Specific gravity in water

2.3 Barycenter

Volume 8 of Textbook of Mechanics [力学課編] (1906) [21] specifically discusses this issue.

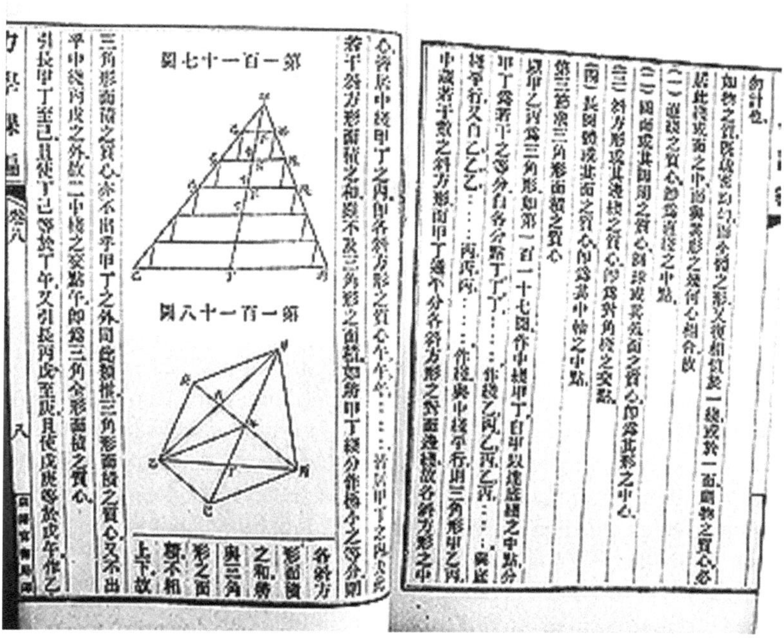

Fig. 7 Centers of gravity

2.4 Machine Designed

Volume I of Collected Diagrams and Explanations of Wonderful Machines from the Far West [遠西奇器圖説] (1627) [4], co-translated by Johann Terrenz [鄧玉函] and Wang Zheng [王徵], says that a great man named Archimedes newly made a dragon tail screw and spirals etc, and was able to explain the working principle of all machinery. [20] and Outline of Western Learning [西学考略] (1883) [12] are alike, with the latter stating that Archimedes lived abroad in Italy.

分爲二在體大力大之物者謂之格物在
微渺之質隱感而變化者謂之化學格物之學亦從
埃及而始古時有川機器灌溉者足證稍達於水學
有希臘人僑寓埃及名希羅者以蒸汽作旋機足證
於火氣二學已得端倪古之造機者莫如亞氏
既進於算學尤力求物理以螺絲管汲水以鏇螺
火均由亞氏而作引而伸之水學能辨金之眞僞火
學能製巨鏡焚敵船物之浮沈
及水中權轄重之理亦由亞氏創論晉書之際有希
西學考略 卷下

Fig. 8 Machines

2.5 Flotage

Jointly translated by Alexander Wylie [偉烈亞力] and Wang Tao, Preliminary Introduction to Mechanics [重学淺説] (1890) [14] states that mechanics includes dynamics and statics, and there is a branch of mechanics that specifically studies gas and liquids, and that Archimedes once studied the latter; he suggested that for stable equilibrium to be established in a container filled with liquid, all forces at any particular point, from all directions, must be the same. In Textbook of Modern Physics [近世物理学教科書] (1906) [22], compiled by the Ministry of Education, the principle of flotage is formally called Archimedes' Principle.

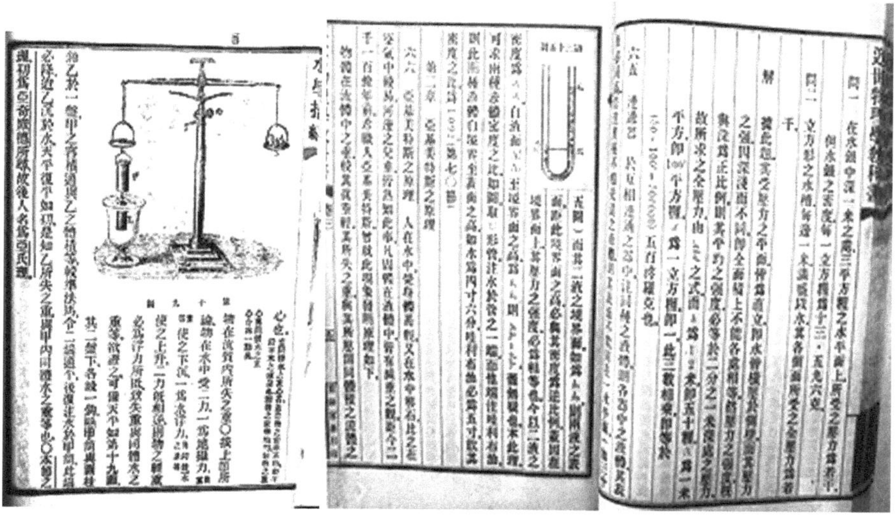

Fig. 9 On floating bodies

3 Legends

3.1 The Gold Crown of King Hiero

Different legends about Archimedes were wide-spread in China in the Ming [明] and Qing Dynasties, the most famous of which was concerned with the determination of the purity of the Cold Crown of the King of Syracuse using Archimedes' Law of Buoyancy. Originally described by the Roman architect Vitruvius, the story is widely known in the West as it is both vivid and highly educational, and thanks to the fact that Archimedes indeed was credited with the work *On Floating Bodies*. The first Chinese literature that relates this story is Combined Learning Mathematical Indicator-General Part [同文算指通編] (1614) [2], compiled by Li Zhizao [李之藻] and Matteo Ricci. Volume 4 has the following description: 100 units of gold have been used in manufacturing a gold incense burner. Upon completion, it was suspected that the goldsmith might have stolen some of the gold through the fraudulent replacement of silver, and yet the burner could not be destroyed to verify its ingredients. What then? The solution lies in the use of buoyancy of water. Even though Archimedes' name was not mentioned and the gold crown has become a gold incense burner, the story has all the necessary Archimedean components.

Collected Diagrams and Explanations of Wonderful Machines from the Far West [遠西奇器圖説] (1627) [4] and The Western Mirror of European Learning [欧羅巴西鏡録] (c. 1620) [3], another book of approximately the same time that deals with

Western Science, both mention that once while having a bath, Archimedes suddenly perceived the Law of Buoyancy.

In Qing literature, there are more accounts of this story. These include Initiatory Textbook of Natural Sciences [格致啓蒙] (1875) [11], Elemental Physics [格物質学] (1902) [20], and Illustrative Hydraulics [水学圖説] (1890) [16]. In [11] and [20], Archimedes is described as running naked through the street, excited and shouting “Eureka,” a Greek word that means “I have found it.”

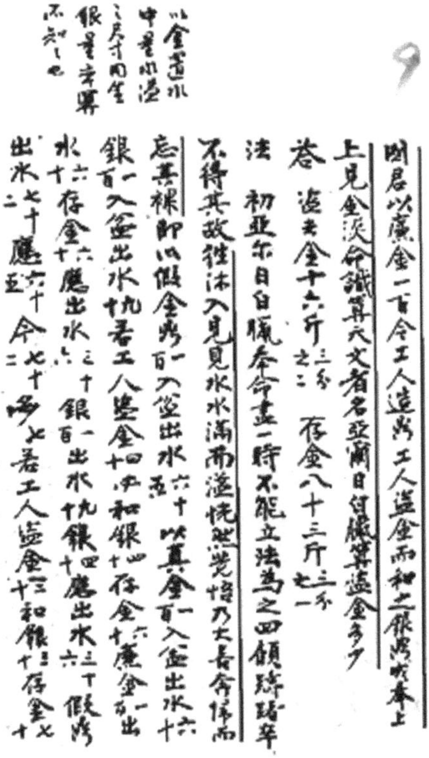

Fig. 10 King Hieron’s crown

3.2 Defending Syracuse

In his preface to Elements [1], Matteo Ricci [利瑪竇] says: 1600 years before Christianity had not been popular in the West, wars frequently occurred between the countries. There emerged a skillful man who was able to defeat the enemies with fewer soldiers, to defend a solitary city and to withstand enemies from both land and sea, just like Mo-tse [墨子] in ancient China. What skills did he possess? In fact, he was only skilled at geometry. These lines obviously refer to the legend of Archimedes and the defense of Syracuse.

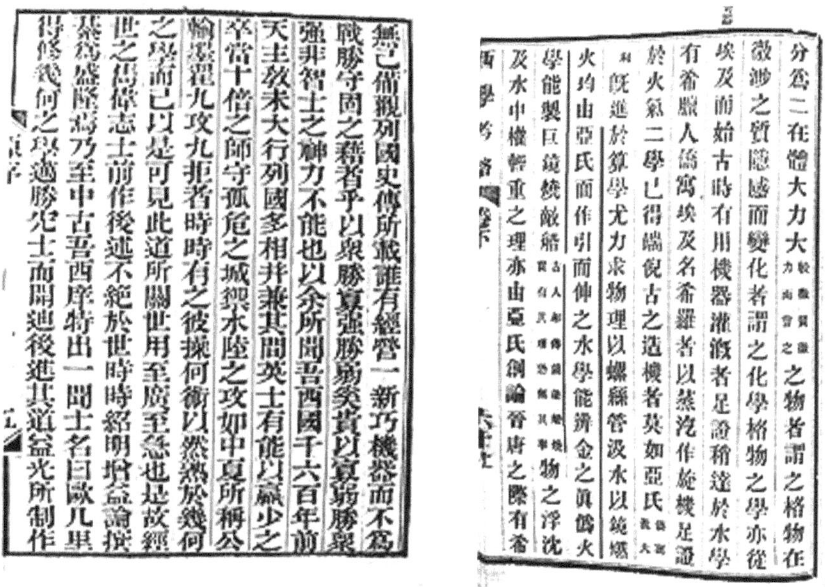

Fig.11 Defense of Syracuse

At the end of the Qing Dynasty, China had to grapple with internal instability and foreign aggression. The story of Archimedes using his invented machines in defense of the city and in resisting aggressive enemies became widespread. In Preliminary Introduction to Mechanics [重学淺説] (1890) [14] and The Origin of Western Learning [西学原始考] (1890) [15], Wang Tao [王韜] mentions the making of iron hooks at the command of Archimedes, and the capsizing of enemy ships using such mechanical devises as levers and pulleys. Archimedes was described in Elemental Physics [20] as using concave mirrors to concentrate the sun's rays to ignite the enemy ships. Outline of Western Learning [西学考略] (1883) [12] also states that Archimedes was able to produce huge mirrors to destroy enemy ships, with a note saying that although founded on a logical basis, the burning of ships with mirrors in antiquity could not possibly have happened in reality.

In the preface to Six Books of Selected Western Knowledge [西学輯存六種] (1889) [13], Wang Tao talks about the death of Archimedes: he was eventually killed by the enemy, bringing his invention of magical devices to an end, which is a great shame; in contrast, Archimedes is well-known throughout the whole West for his work on mechanics.

What is most fascinating is the story of the mathematician Xu Youren [徐有壬], who was also governor of Suzhou [蘇州] during the Taiping Rebellion [太平天国造反]. In 1860 he was responsible for the colossal task of defending the city. When the battle was perilous, he called to mind his friend Li Shanlan [李善蘭], the most famous mathematician of that time. He invited Li to discuss the strategy in defeating the enemy. According to the writings of Li's nephew, Governor Xu thought that anyone skilled in mathematics must be good at the art of war, and that if Li came to him, they would be able to devise tactics to defeat the enemy. So at the brink of despair, Xu Youren imagined Li Shanlan to be Archimedes, who was described a hero for successfully defending Syracuse.

The death of Xu was also dramatic. According to The Biographies of Mathematicians and Astronomers, Series iii [疇人傳三編] (1896) [18], when the rebel forces broke into the city of Suzhou, Xu tidied up his attire, put on his cap, and went out to oversee the fighting. He was stabbed in the forehead by the enemy and before he died, he did not forget to rearrange his official cap. The scene of Archimedes just before he was killed was again staged here.

3.3 Lifting the Earth

Volume 6 of Textbook of Mechanics [力学課編] (1906) [21] relates the famous story of lifting the earth although in this case the protagonist is Isaac Newton instead of Archimedes, with Newton claiming that "If God gives me an infinitely long lever and a firm enough fulcrum, I can raise the entire Earth."

與重向之直垂距。與二力之數。有反比之例。則無論二力相差何等。皆能使之相敵。

奈端有言曰。使天能假吾一無極長之桿。并一無極堅之倚點。吾一指之力。可以運動全球。此雖事所必無。然足以發明槓桿之用。學者盍一思之。

第七節設喻

或以爲槓桿爲機器之屬。而不識機器者。將不能意想槓桿之式。是大謬也。夫天下事無一非道。特習而見之。漫不加察。則一闢專名。往往目爲新奇。不能舉一而三隅反。試觀東洋之車。客位於後。人御於前。輪軸居其間。非即第一式之桿乎。剪刀剪物。樞扭居間。手力在其後。物力在其前。亦第一式之桿也。他如手車爲第二式之桿。市秤爲第三式之桿。皆爲通行常見之物。尤爲最精最確之喻也。

第八節論桿重

凡前所論槓桿。皆未計及桿之體重。如衆計其重。則除(力)與(重)之外。又多一力矣。

五

Fig. 12 Mechanics (1906)

3.4 Tombstone

The Roman politician Cicero paid homage to Archimedes' grave when he was appointed as the consul of Sicily in 73 BC. It was said that on Archimedes' tombstone were engraved a cylinder enclosing a sphere.

In the section "Origins of Western Science" of [12], it is stated: "Archimedes discovered that the ratio of the volume of a sphere to the volume of the cylinder that contains it is 2 : 3, and that this is an important law in geometry. He ordered to have the diagram that shows this relationship curved on his tombstone." There is also a note: several hundred years later, someone passed the tomb, and upon seeing the diagram, knew Archimedes was buried there.

[18] makes mention of a Dutch mathematician Ludolff van Ceulen, who worked out a value for π to 36 digits, and had the result carved onto his own tombstone in memory of his calculation. It goes on to say that this was done "in the same way as Archimedes had a sphere and a cylinder engraved on his tombstone."

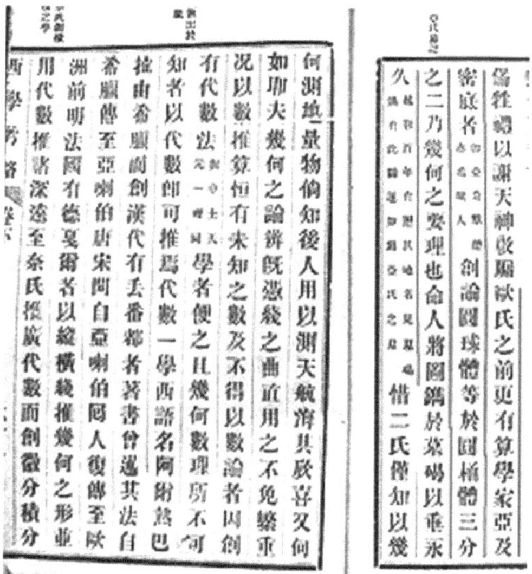
何測地量物倘知後人用以測天航海其欣喜又何
如耶夫幾何之論辯既憑幾之曲直用之不免繁重
況以數推算恒有未知之數及不得以數論者因創
有代數法學者便之且幾何數理有不可
知者以代數即可推焉代數一學西語名阿爾熱巴
攄由希臘國創漢代有丟番都者著書曾遺其法自
希臘傳至亞喇伯唐宋間自亞喇伯國人復傳至歐
洲前明法國有德戛爾者以縱橫幾推幾何之形並
用代數推諸深遠至奈氏擴廣代數而創微分積分
西學略述

其性竭以謝天神敬賜歐氏之前更有算學家亞及
密底者創論圓球體等於圓桶體三分
之二乃幾何之要理也命人將圖鐫於墓碣以垂永
久惜二氏僅知以幾

Fig. 13 Outline of Western Learning (1883)

Conclusion

With the diffusion of Western science and technology in China, the deeds and works of the greatest scientist of antiquity were made known to the intelligentsia of China. Receiving most attention were his discoveries on conical curves, which are directly related to the laws of planetary motion and to the trajectories of projectiles, and mechanical knowledge which is related to the construction of machines. The magnificent backdrop in China was the Calendar Reform during the Emperor Chongzhen reign [崇禎改曆], and the self-strengthening movement [自強運動], corresponding to the Ming-Qing transition period and the late Qing period, respectively. Concurrently, Archimedes became a heroic symbol of knowledge, and an example held in high esteem by Chinese intellectuals.

As mentioned earlier, this is only a sketch. As for why those subjects I have just mentioned were especially appreciated by Chinese, instead of others (For instance, *the Method*, *the Sand-Reckoner*, as well as the Archimedean Axiom) deserves further analysis. Likewise, comparative studies of ancient Chinese mathematics, Chinese culture and the works of Archimedes (such as the Measurements of the Circle and Sphere, the concept of infinitesimal, as well as the problem of combinations of tan-gram figures [七巧板]) should also be pursued. And the diffusion of Archimedes' theories in China after traditional Chinese science had been integrated into modern science as well as more detailed research by Chinese scholars on Archimedes' works, along with these other interesting topics, should all be further explored.

References

1. Euclid, trans. by Matteo Ricci [利瑪竇] and Xu Guangqi [徐光啓]: *Elements* [幾何原本] (1607).
2. Christopher Clavius, trans. by Matteo Ricci [利瑪竇] and Li Zhizao [李之藻]: *Combined Learning Mathematical Indicator-General Part* [同文算指通編] (1614).
3. Anonymous: *The Western Mirror of European Learning* [欧羅巴西鏡録] (c. 1620).
4. Joannes Terrenz [鄧玉函] and Wang Zhang [王徵] translate: *Collected Diagrams and Explanations of Wonderful Machines from the Far West* [遠西奇器圖說] (1627).
5. Jacques Rho [羅雅谷]: *Complete Explanation of Measurements* [測量全義] (1631).
6. Du Zhigeng [杜知耕]: *The Key of Mathematics* [数学鑰] (1681).
7. Ruan Yuan [阮元] and Li Rui [李鋭] *ed.*: *The Biographies of Mathematicians and Astronomers* [疇人傳] (1799).
8. Elias Loomis, trans. by Li Shanlan [李善蘭] and Alexander Wylie [偉烈亞力]: *Elements of Analytical Geometry and of the Differential and Integral Calculus* [代微積拾級] (1859).
9. Xia Luanxiang [夏鸞翔]: *The Origin of Universe* [万象一原] (1862).
10. John Wallis, trans. by Fryer [傅蘭雅] and Hua Hengfang [華衡芳]: *The Fundamental Principles of Calculus* [微積溯源] (1874).
11. Young John Allen [林楽知] and Zheng Changyan [鄭昌棪]: *Initiatory Textbook of Natural Sciences* [格致啓蒙] (1875).
12. Williams Martin [丁韙良]: *Outline of Western Learning* [西学考略] (1883).
13. Alexander Wylie [偉烈亞力] and Wang Tao [王韜] eds.: *Six Books of Selected Western Knowledge* [西学輯存六種] (1889).
14. Alexander Wylie [偉烈亞力] and Wang Tao [王韜] translate: *Preliminary Introduction to Mechanics* [重学淺説] (1890).
15. Wang Tao [王韜] ed.: *The Origin of Western Learning* [西学原始考] (1890).
16. John Fryer [傅蘭雅] trans.: *Illustrative Hydraulics* [水学圖説] (1890).
17. Judson[求德生] and Liu Weishi [劉維師] trans.: *The Conics* [円錐曲線] (1893).
18. Zho Kebao [諸可宝]: *The Biographies of Mathematicians and Astronomers, Series iii* [疇人傳三編] (1896).
19. Liu Guangzhao [劉光照] trans.: *Fundamental Hydraulics* [水学拾級] (1902).
20. Alvin Parker [潘慎文] and Xie Honglai [謝洪賚]: *Elemental Physics* [格物質学] (1902).
21. Yan Wenbing [嚴文炳] trans.: *Textbook of Mechanics* [力学課編] (1906).
22. Minister of Education [学部編譯局]: *Textbook of Modern Physics* [近世物理学教科書] (1906).

The Nine Chapters on the Mathematical Procedures and Liu Hui's Mathematical Theory

Guo Shuchun

Abstract When discussing ancient mathematical theories, scholars often limit themselves to Greek mathematics and, especially to its axiomatic system, which they use as the standard to evaluate traditional mathematics in other cultures: whichever failed to form an axiomatic system is considered to be without theory. Therefore, even those scholars who highly praise the achievements in ancient Chinese mathematics consider that "the greatest deficiency in old Chinese mathematical thought was the absence of the idea of rigorous proofs" and that there is no formal logic in ancient Chinese mathematics; in particular it did not have deductive logic. They further contend that, "in the flight from practice into the realm of pure intellect, Chinese mathematics did not participate," [5, p. 151][1] and conclude that Chinese mathematics has no theory.

I think that Liu Hui's commentary (263 A.D.) to the Nine Chapters on the Mathematical Procedures, hereafter Nine Chapters, completely proved the formulas and solutions in Nine Chapters. It, mainly based on deductive logics, elucidated deep mathematical theories. Even though Nine Chapters itself does not contain mathematical reasoning and proofs, which is a major flaw in the pursuit of mathematical theory in the history of Chinese mathematics, there are certain correct abstract procedures that possess a general applicability which should be considered as mathematical theories in Chinese mathematics. Sir Geoffrey Lloyd, after explaining the difference between Liu Hui's and Euclid's mathematics, said "Mais cela ne signifie pas une absence d'intérêt pour la validation des résultats ou pour la recherche d'une systématisation" [3, préface, p. xi]. Based on the Nine Chapters, this article will discuss Liu Hui's contribution to the mathematical theory in order to stimulate more fruitful discussions.

Guo Shuchun
Institute for the History of Natural Science, Chinese Academy of Sciences, 55 Zhong Guan Cun Dong Lu, Zhong Guan Cun, Beijing 100190, P. R. China
e-mail: shuchunguo@gmail.com

[1] Needham cited the claim that Chinese mathematics lacks "rigorous proofs," from his correspondence with the Japanese historian of Chinese mathematics, Yoshio Mikami [三上義夫].

1 Procedures in Nine Chapters and its Style in which Questions Associated with Procedures as Examples

Many describe Nine Chapters on the Mathematical Procedures [九章算術 *Jiuzhang suanshu*], as a collection of application problems. Generally speaking, such a simplified view is not too off target; however, it needs clarification and merits more discussions if such a view leads to the misunderstanding that the ancient Chinese mathematics had no theories. The simple truth is that many who have not studied Nine Chapters or who had but did not seek to understand fully presume that, based on this simplified view, Nine Chapters consists of collections of one question, one answer, and one computational procedure [術 *shu*]. They further assume that these procedures are in effect the concrete solutions to the application problems. Such assumption is completely off base. It does not provide proper and accurate descriptions of Nine Chapters at all.

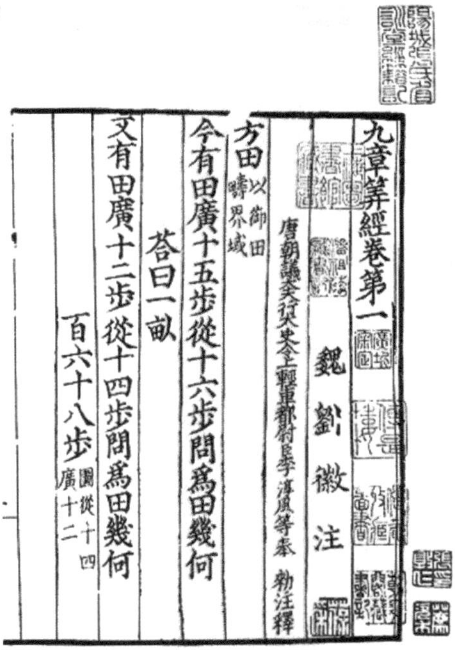
九章筭經卷第一
魏劉徽注
唐朝議大夫行太史令上輕車都尉臣李淳風等奉勅注釋
方田以御田疇界域
今有田廣十五步從十六步問爲田幾何
荅曰一畝
又有田廣十二步從十四步問爲田幾何
百六十八步圖從十四廣十二

Fig. 1 The first page of Chapter One in the Nine Chapters, the edition from Southern Song Dynasty (1127–1279)

In fact, the procedures in Nine Chapters are of different levels of abstractness and applicability. The relations among problems, answers, and procedures, or, to put it differently, the styles in the treatise are rather complicated. We first consider the relations among the problems, answers, and procedures. Roughly speaking, there are two different genres:

1. Questions associated with procedures as examples:

For this genre, there are usually multiple procedures with multiple questions, or one procedure with multiple questions, or one procedure with one question. But they can be further divided into three types of different scenarios:

(1) The text first listed one or multiple questions and then provided one or several general abstract procedures; moreover, the questions are only listed the statement and the answers without calculating procedures containing specific numeric values from the questions.[2]

Take the procedure of finding the area of the circular field in the Rectangular Field [方田 *Fangtian*] Chapter:[3]

> Now there is a circular field, the circumference of which is 30 *bu*[4] and the diameter of which is 10 *bu*. Question: what is [the area of] the field?
> Answer: 75 [square] *bu*.
> And there is another circular field, the circumference of which is 181 *bu* and the diameter of which is 60 1/3 *bu*. Question: what is [the area of] the field?
> Answer: 11 *mu*[5] 90 1/12 [square] *bu*.
> Procedure: The half-circumference multiplied by the half-diameter will yield the area in [square] *bu*.

The procedure is equivalent to the formula for the area of a circle:

$$S = \frac{1}{2}Lr, \tag{1}$$

where S, L, and r are the area, the circumference, and the radius of the circle. Here these two questions only have the statement of the questions and answers without

[2] By an "abstract" procedure, I mean a procedure containing the general description of measurements needed without any numeric values. In particular, it does not utilize the specifically numerical values appearing in the statement of a question. Instead, the procedure prescribes one or a series of operations to be performed on measurements without numerical values in words. The general description of measurements for example can be the diameter or the circumference of a circle or the number of days for a wild goose to fly from the south sea to the north sea. In this view, procedures of this kind in Nine Chapters are similar to mathematical formulas described with letters a, b, and c in the modern form.

[3] I use two references [6] and [3] for Nine Chapters and Liu Hui's [劉徽] commentary. For the circular field method [圓田術 *yuantianshu*], see [6, pp. 18–19] and [3, pp. 176–179]. Below pages from both references will be given, for all the text and examples from the Nine Chapters.

[4] *bu* [步] is a unit of length, 1 *bu* is equal to 6 *chi*. The areas and the volumes in the Chinese texts were, however, also expressed in terms of *bu*. The context made it clear whether the *bu* represents the linear *bu*, the area bu^2, or the volume bu^3. The units discussed below, *zhang*, *chi*, and *cun* all share this characteristic of representing the linear length, the area, and the volume.

[5] *mu* [畝] is a unit of area, equal to 240 *bu*.

individual procedures; the circular field method is the procedure for both questions. Some studies describe this procedure as the procedure for the 32nd question. This description is obviously not accurate. In Nine Chapters, the entire Rectangular Field Chapter, the jinglü [経率 *jinglü*] procedure, qilü [其率 *qilü*] procedure, and inverse qilü [反其率 *fanqilü*] procedure in the Millet and Rice [粟米 *Sumi*] Chapter, four root-extracting [開方 *kaifang*] procedures in the Small Width [少広 *Shaoguang*] Chapter, many procedures in the Work Discussing [商功 *Shanggong*] Chapter, the four fair labor [均輸 *junshu*] procedures in the Fair Labor Chapter, five excess-deficiency [盈不足 *yingbuzu*] procedures in the Excess-Deficiency Chapter, and five procedure in the Right Triangle [勾股 *Gougu*] Chapter are of the type. In the Nine Chapters, there are a total of seventy three procedures and one hundred and six questions of this type.

(2) The text first provided abstract procedures and then listed a few examples, which have the statements of questions and answers without procedures: Take flat-headed stack [芻童 *Chutong*] procedure for finding the volume of a frustum in the Work Discussing Chapter as an example.[6]

> Procedure: Double the upper length and add it to the lower length; also double the lower length and add it to the upper length; multiply them with their corresponding width. Add the products together. Use the height or the depth to multiply the sum and divide by 6.
>
> Now there is a frustum with the lower length equal to 2 *zhang*[7], upper length 3 *zhang*, the lower width is 3 *zhang*, the upper width is 4 *zhang*, and the height is 10 *zhang*. Question: What is the volume?
> Answer: 26500 [cubic] *chi*.[8]

The procedures and questions of finding volumes of other solids are exactly like this. The questions following the procedure also only have statement of questions and answers without individual procedures. The procedure preceding the questions is their common procedure. In this type, there are two general procedures with 10 questions.

(3) The text first provided general procedures and then listed a few questions, each of which contained the statement of question, the answer, and the calculating procedure with specific numeric values from the question. The procedure for each question is basically the application of the general procedure on that particular question. To demonstrate this type, we use the 'suppose' [今有 *Jinyou*] procedure in the Millet and Rice Chapter and some of the 31 questions of conversion involving grains and rice ([6, pp. 70–78] and [3, pp. 222–227].):

> 'Suppose' procedure: use the product of the given quantity [of grain in possession] and the ratio [of the] desired [grain] as the dividend, and the ratio [of the possessed grain] as the divisor.

[6] Here a flat-headed stack is a solid obtained by cutting a rectangular pyramid with a plane parallel to the base of the pyramid and removing the top part. The flat-headed stack procedure and the example can be found in [6, pp. 185–186] and [3, pp. 434–439].

[7] *zhang* [丈] is a unit of length, equal to 10 *chi*.

[8] *chi* [尺] is a unit of length, which is about 23–24 cm.

> Divide the dividend with the divisor. Now there is one *dou*[9] of millet and [we] want to exchange it for coarse rice [糲米 *limi*]. Ask: how much [of coarse rice] can be gotten?
> Answer: [We] get six *sheng*[10] of coarse rice.
> Procedure: Use the millet to exchange for the coarse rice. Multiply [the quantity of millet] by 3 and then divide [the product] by 5.
> Now there are 2 *dou* and one *sheng* of millet and [we] want to exchange it for polished rice [粺米 *baimi*]. Ask: how much [of polished rice] can be gotten?
> Answer: One *dou*, one and 17/50 *sheng* of polished rice.
> Procedure: Use millet to exchange for polished rice. Multiply [the quantity of millet] by 27 and then divide [the product] by 50.

Following these questions and procedures there are 29 questions of the same type. Each question is followed by its answer and a procedure; and each procedure is an application of the 'suppose' procedure, and therefore we do not repeat them here. For this type, each problem has the statement of the question, the answer, and its own procedure, which is an application of the 'suppose' procedure. Also belonged to this type are the proportional distribution [衰分 *cuifen*] procedure and inverse proportional [返衰分 *fangcuifen*] procedure with their nine examples in the Proportional Distribution Chapter, the small width procedure with its eleven examples in the Work Discussing Chapter, the eleven examples solved by excess-deficiency procedure in the Excess-Deficiency Chapter, and the rectangular array [方程 *fangcheng*] procedure, sign [正負 *zhengfu*] procedure, and loss and gain [損益 *sunyi*] procedure with their eighteen examples in the Rectangular Array Chapter. In total, there are 7 general procedures, 80 questions, and 78 sub-procedures with numerical values from the individual questions.

The above three types has a total of 82 general procedures, 196 questions, sub-procedures with numerical values, constituting 80% of Nine Chapters.

2. The collections of application problems:

In this genre, the text usually consists of one question, one procedure, and one answer. The degree of generality of the procedures varies:

(1) General procedures applicable to one kind of questions. Take the question of wild duck and wild goose in the Fair Labor Chapter as an example [6, p. 254] and [3, pp. 532–533]:

> Now a wild duck coming from the South Sea takes 7 days to reach the North Sea; a wild goose leaving from the North Sea takes 9 days to reach the South Sea. Now both birds leave [from their respective place at the same time]. Ask: when will they meet?
> Answer: 3 and 15/16 days.
> Procedure: Take the sum of the numbers of days as the divisor and the product of the days as the dividend. Divide the dividend by the divisor to get the answer.

This procedure, although did not depart from the subject of days in the question, did not include the actual numerical values from the question. It is applicable to

---

[9] *dou* [斗] is a unit of capacity, equal to 10 *sheng*.

[10] *sheng* [升] is a unit of capacity, which is about 198–210 ml.

many questions of the same nature. Many problems in the Fair Labor Chapter and the question of taking a pole and walking out of the door in the Right Triangle Chapter all belong to this type [6, p. 422] and [3, pp. 742–743].

(2) The actual computation for the concrete questions. Take the question of a door ajar with certain distance from the door threshold in the Right Triangle Chapter as an example [6, p. 413] and [3, pp. 714–717].:

> Now there is a door ajar and [the ends of the two opened panels] are 1 *chi* from the door threshold and 2 *cun*[11] apart. Ask: what is the width of the door?
> Procedure: Multiply the one *chi* by itself. Divide the result by one half of the distance between the two ends, two *cun*. Then add to the result by one-half of the distance between the two ends to get the width of the door.

The procedure incorporates the concrete numbers from the question; therefore it cannot exist independently from the question. The problems of inverse proportional distribution [非衰分 *feicuifeng*] in the Proportional Distribution Chapter, some examples in the Fair Labor Chapter, the Jade and Stone Hide Each Other [玉石互隱 *yushi huyin*] question in the Excess-Deficiency Chapter, and the solutions to right triangle questions and the measuring-the-height-of mountain-from-a-tree question in the Right Triangle Chapter are of the same type. These two types can be described as "a collection of application problems." There are 50 questions of this genre in Nine Chapters.

Obvious from the analysis above is that, it is not appropriate to summarize Nine Chapters as a collection of application problems. It is far from the truth that the text follows the "one-question-one-answer-and-one-procedure" format. In my opinion, there are at least three different formats for mathematical treatises in the history of mathematics. The first is represented by Euclid's Elements, forming an axiomatic system. The second format is collections of application problems, e.g. Diophantine' Arithmetics; to paraphrase the words of the German historian, Henkel, "one is still puzzled by the 101st question after carefully studying the first 100 questions." Obviously for the most part in Nine Chapters, the text does not conform to either of the formats in Euclid's Elements or in Diophantine's Arithmetics; therefore its format should be considered in its own right as the third type, i.e. the format centered around procedures with the questions associated with procedures as examples.

Meanwhile, it is not hard to see from the above analysis that the procedures in Nine Chapters are not of a single nature in terms of their abstractness and general applicability. At the very least, there are three varieties. The procedures of the first kind, in spite of the variations of expressions, share several common characteristics: the statement of the procedure is the core. Associated with the procedures are questions as demonstrations of the procedures. The procedures as the core are very abstract, rigorous, and with wide general applicability. When converted to the modern notion, the procedures become mathematical formulas or the operation procedures. These procedures are constructive and mechanical. Therefore, we describe the format of these procedures as "questions associated with procedures as examples." The

[11] *cun* [寸] is a unit of length, equal to one-tenth of a *chi*.

second kind of procedures can be described as abstract procedures connected to one type of questions with relatively wide applicability. The third kind consists of the actual computations of questions with actual numerical values.

The former two varieties constitute 90% of the text in Nine Chapters. These abstract procedures with general and relatively wide applicability should certainly be recognized as an expression of the mathematical theory.

2 Liu Hui's Mathematical Definitions and Deductive Reasoning

In order to describe Liu Hui's contributions to the mathematical theory, we need to examine Liu Hui's mathematical definitions and reasoning, especially his deductive reasoning.

2.1 Liu Hui's Definitions

Following the tradition of providing definitions to concepts in the Classic of Mo [墨経 *Mojing*], Liu Hui provided many rigorous definitions to many mathematical concepts. For example, the definition of power [冪 *mi*] as area:

> Whenever the width is multiplied with the length, [the result is what we] call *mi*

Another example is about the definition of *fangcheng* [方程] as the system of linear equations [6, p. 353] and [3, pp. 616–617]. *Fang* [方] means juxtaposing. For *cheng* [程] Liu Hui's annotation says:

> '*Cheng*' means to find the standard [of objects]. A group of objects is mixed together. Each row has [unknown] numbers of objects. The total sum of the products of the numbers and the objects is expressed. Set the *lü* [率] for each column.[12] If there are two unknown quantities, make the second column; If there are three unknown quantities, make three columns. Make the number of columns according to the number of unknowns. List one column after another; that is what we call *Fangcheng*. The rows do not depend on those besides them and each of them is based on the information given. [程, 課程也. 群物総雜. 各列有数. 総言其實. 令每行為率. 二物者再程. 三物者三程. 皆如物数程之. 竝列為行. 故謂之方程. 行之左右無所同存. 且為有所據而言耳.]

These are definitions of operations; that is, the definition can be carried out to obtain what is being defined.

It is worth pointing out that once the definition of a term was given, generally speaking, the term maintained the same connotation throughout the entire Nine Chapters.

---

[12] Each object might have two numbers associated with it, one known and the other unknown. For example, the number of chickens is known but the price of them isn't. But the sum of the unknown quantities is given as *shi*. The concept of *lü* in the Nine Chapters is rather complicated. I refer the readers to the discussion in [3, pp. 956–959].

2.2 Liu Hui's Deductive Reasoning

Many scholars believe that the ancient Chinese mathematics never used formal logic. This is fundamentally false. Not only did Liu Hui employ "Learning by analogy [举一反三 *juyi fansan*] Learning by consequence [告往知来 *gaowang zhilai*], and draw an analogy [触類而長 *chulei erzhang*]"[13] analogous method to expand mathematical knowledge, he also utilized formal logic in his general reasoning. He not only used inductive but also deductive reasoning.
Examples:

(1) Syllogism

There are abundant examples in which Liu Hui employed syllogism. For example, in the excess-deficiency procedure, Liu Hui described the situation when both guessed answers were assumed to be fractions ([6, p. 308] and [3, pp. 560–561]):

> Commentary: if both guessed answers are fractions, find the common denominator and arrange the numerators accordingly. For this question, both guessed answers are fractions, and therefore find the common denominator and arrange the numerators accordingly.

Take M to be "both guessed answers are fractions;" P, "one should find the common denominator and arrange the numerators accordingly;" and S, as this particular problem to be solved. This reasoning has and only has three concepts: (1) the two assumptions are fractions (the middle item, M), (2) find the common denominator and arrange the numerators accordingly (former presupposition P), and (3) this problem (latter presupposition S). Therefore Liu Hui's reasoning fits perfectly the AAA rule of syllogism:

The former inference:	$M \to P$	(A)
The latter inference:	$S \to M$	(A)
The conclusion:	$S \to P$	(A)

(2) Relational Reasoning

As a commentator of a mathematical treatise, Liu Hui used many relational reasoning, which are special cases of syllogism. Among the relation judgments made by Liu Hui, the majority of them are of the equality relation. For example, right after Liu Hui proved one formula for the area of a circle, Liu Hui stated the following in order to prove another formula $S = \frac{1}{4}Ld$, [6, p. 23] and [3, pp. 560–561]:

> Multiply one-half of the circumference and one-half of the diameter [to get the area of a circle]. But now both the circumference and the diameter are full, therefore the two [numbers in the] denominator should be multiplied [to get] four. [That is,] use 4 to divide the quantity.

That is to say,

---

[13] These terms are used to describe the teaching method in which students when shown one example are expected to understand the analogous problems. These term were Liu Hui's words in the preface of Nine Chapters [6, p. 2] and [3, pp. 128–129].

the known formula: $S = \frac{1}{2}Lr$ (equality relation) and

$r = \frac{1}{2}d$ (equality relation), therefore

$S = \frac{1}{2}L \times \frac{1}{2}d = \frac{1}{4}Ld.$ (equality relation)

The equality relation is reflective and transitive.

Moreover, Liu Hui also used inequality relation in his reasoning. The operation of sphere volume [開立圓 *kai liyuan*] procedure in Nine Chapters utilized an incorrect formula for the volume of a sphere:

$$V = \frac{9}{16}d^3,$$

where V is the volume of a sphere and d is the diameter.[14] Liu Hui recorded how this incorrect formula was derived: Take a cube with the length of its side equal to the diameter of the sphere d. The ratio of volumes of the cube and the transcribed circular cylinder inside it is 4:3; the ratio of the volumes of a circular cylinder and the transcribed sphere inside it is also 4:3, taking 3 as the value of the circumference-diameter ratio. Therefore the ratio of the volumes of the cube and its transcribed sphere should be 16:9. That was how the above formula was derived according to Liu. Liu Hui took the intersection of two perpendicular circular cylinders of the same diameter and called it matching square cover [牟合方蓋 *mohe fanggai*] in Figure 2. Then he described [6, p. 142] and [3, pp. 378–381]:

> The square cover has a ratio related to squares; a transcribed sphere in it, a ratio related to circles. Deducing from this fact, [we ask,] how can those who described the circular cylinder as related to the ratio related to squares be not wrong?

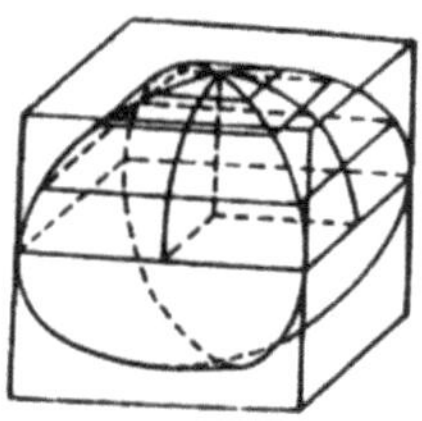

Fig. 2 matching square cover [牟合方蓋]

His lines of reasoning go as follows:

[14] This procedure describes how to find the diameter of a sphere when its volume is given. The procedure and Liu's commentary can be found in [6, pp. 141–143] and [3, pp. 378–385].

matching square cover : a sphere $= 4 : \pi$

A circular cylinder : a sphere $\neq$ matching square cover : a sphere

Hence a circular cylinder : a sphere $\neq 4 : \pi$

This fundamental overturns the formula in Nine Chapters.

(3) Hypothetical reasoning

The hypothetical reasoning is a commonly used reasoning in mathematical reasoning. First, let us examine how Liu Hui utilized the hypothetical reasoning with a sufficient condition. For example, Liu Hui's commentary to the tomb tunnel [羡除 *yanchu*] procedure in the Work Discussing Chapter [6, p. 184] and [3, pp. 432–437]:

> No [perpendicular] cross-sections of [two solids of the same height] are not equal squares, [yet] a pyramid with a square base and a yang-ma [陽馬 *yangma*] [of the same height and base] are equal in volume.

See Figure 3.

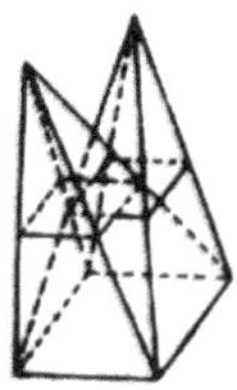

Fig. 3

The statement was written too simplistically to comprehend literally. It can be expanded to be:

If, for each level, the cross-sections of the two 3-dimensional solids of the same height are equal squares (P), then the volumes of the two solids should be equal as well (Q).

For the pyramid with a square base and yang-ma in question, the cross-sections of them for each level are the same squares (P); therefore the volumes of the pyramid with a square base and yang-ma in question should have the same volume (Q).

The reasoning format is, if P then Q; now P, therefore Q. In a true hypothetical reasoning, if the condition P is true, so is the condition Q; if the P is false, then the truth value of Q is uncertain. Liu Hui fully understood this. When we divide a rectangular parallelepiped from one edge towards its opposite, we get two moat-ends [塹堵 *qiandu*] (right triangular prisms).

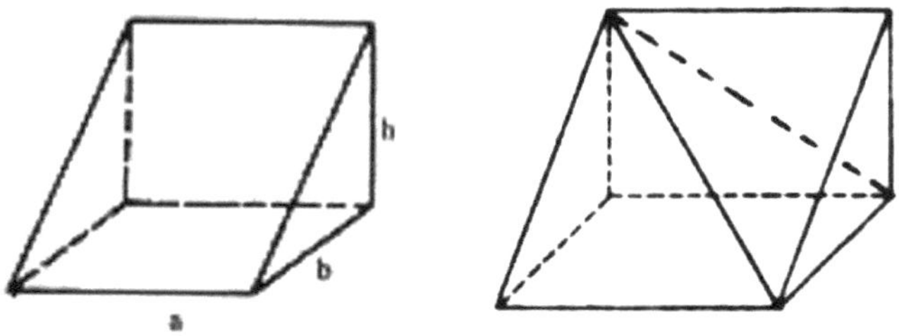

Fig. 4 moat-end [*qiandu*] (left); yang-ma [*yang-ma*] and turtle-forelimb [*bienao*] (right)

When we further divide a moat-end from one vertex towards one of its opposite edges, we get a yang-ma [陽馬] (a pyramid with a square base and four right triangular sides) and a turtle-forelimb [鼈臑 *bienao*] (a tetrahedron with all four sides being right triangles). See the right of Figure 4. Nine Chapters provided the procedure of finding the volume of a moat-end:

> Procedure says: The width and the length [should be] multiplied together; use the height to multiply [the product], then divide [the result] by 2. That is,

$$V_q = \frac{1}{2}abh. \tag{2}$$

And it also provided the volume of a yang-ma:

> Procedure says: The width and the length [should be] multiplied together; use the height to multiply [the product], then divide [the result] by 3. That is,

$$V_y = \frac{1}{3}abh. \tag{3}$$

It provided that of a turtle-forelimb as well:

> Procedure says: The width and the length [should be] multiplied together; use the height to multiply [the product], then divide [the result] by 6. That is,

$$V_b = \frac{1}{6}abh. \tag{4}$$

V_q, V_y, V_b, a, b and h are the respective volume of moat-end, yang-ma, turtle-forelimb, their width, length, and height.[15] Because a cube can be divided into 3 congruent yang-mas or 6 congruent turtle-forelimbs which form three pairs of symmetric ones (See Figure 5), Liu Hui described it as, "Observe how the cube is divided and the shape [of resulting solids] are congruent, then it is easy to verify [these formulas]."

[15] For the procedures of finding the volumes of a moat-end, yang-ma, and turtle-forelimb, see [6, pp. 182–183] and [3, pp. 428–433].

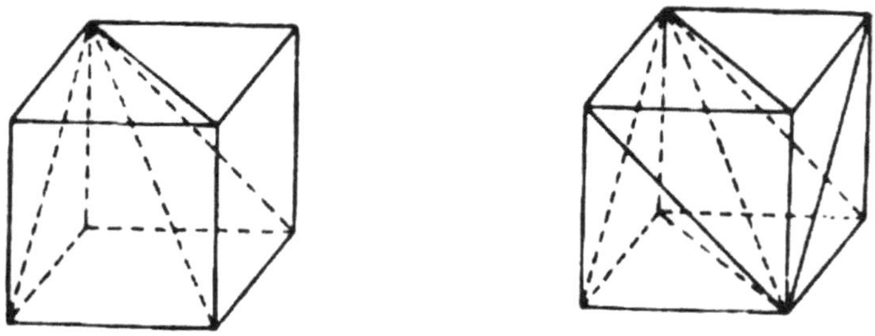

Fig. 5 A cube is divided into three yang-mas, as well as six turtle-forelimbs

That is to say, when the length, width, and height are equal, it is easy to use the geometric models to test and verify that formulas (3) and (4) are correct, as depicted in Figure 5. However, when the length, width, and height are not the same, a rectangular parallelepiped can not be divided into three congruent yang-mas or six turtle-forelimbs. It is impossible to use the testing method to verify formulae (3) and (4), just as Liu Hui commented [6, pp. 182–183] and [3, pp. 430–431]:

> The turtle-forelimbs are of the different shapes, so are the yang-mas. When yang-mas are of the different shapes, then they will not match perfectly. If they do not match perfectly, then it is difficult.

His reasoning style goes as follows:

If the polyhedrons are congruent (P), then their volumes are equal (Q). Now that the polyhedrons are not congruent (P), then it is difficult to tell. (the truth value of Q is uncertain).

That is, to prove formulas (3) and (4) for general cases, one has to find some other ways.

(4) Disjunctive reasoning

Liu Hui used the disjunctive reasoning in many places. In the basic arithmetical calculations, the order of multiplication and division can be switched, "different orders of carrying out multiplications and divisions have their meaning, but give the same result" [6, p. 187] and [3, pp. 442–443]. In the commentary to the 'suppose' procedure in the Millet and Rice Chapter, Liu Hui supported the order of operation to be multiplication first then division because "if [we] divide first and then multiply, there might create fractions [in the process]; that is why the procedure uses the other order." [6, p. 70] and [3, pp. 224–225] This is disjunctive reasoning:

> Either we carry out multiplication first and then division, or division first and then multiplication. Now it is not division first and then multiplication, therefore the order of the operation should be multiplication first and then division.

(5) Dilemma Reasoning

Dilemma Reasoning is a combination of hypothetical and disjunctive reasoning. The major presupposition consists of two hypothetical statements and the minor presupposition is a disjunctive judgment. For example, the style Liu Hui employed in disapproving the incorrect formula for the area of a circle, $S = \frac{1}{12}L^2$ is no other than reasoning by contradiction.

Liu stated [6, p. 23] and [3, pp. 186–187]:

> The ratio of the perimeter of a [regular inscribed] hexagon [inside a circle] to the diameter of the circle is three to one. Therefore, multiplying the perimeter of the hexagon by itself is like nine times the square of the diameter. Nine squares make 12 [inscribed regular] dodecagons. So divide the area of the nine squares to get the area of a dodecagon. Now if [we] make the circumference [of the circle] multiply itself, this is not like nine times the square of the diameter. Then dividing the result by 12 does not yield the area of a dodecagon. If we want to use it as the area of the circle, the discrepancy is too much.

This has two hypothetical presuppositions. One, a one-twelfth of the square of the perimeter of an inscribed regular hexagon (as the circumference of the circle) is the area of an inscribed regular dodecagon inside the circle, which is less than the area of the circle; the other, a one-twelfth of the square of the circumference is greater than the area of the circle. Moreover, there is a disjunctive presupposition: a one-twelfth of the square of the perimeter of an inscribed regular hexagon or a one-twelfth of the square of the circumference. The conclusion is, one quantity is less than and the other is greater than the area of a circle. Both proved that the above formula was incorrect.

Further more, Liu Hui utilized infinite inferences many times, which can be construed as the prototype of a mathematical induction principle.

These analyses present a very small number of many instances of deductive reasoning by Liu Hui; yet it is sufficient to demonstrate that Liu Hui in effect employed several major reasoning formats described in the textbooks of modern formal logic.

3 Liu Hui's Mathematical Proofs

The reasoning statements discussed above, due to the truthfulness of their presuppositions, form de facto proofs or part of their argument. The most beautiful proofs provided by Liu Hui are (1) a proof for the procedure of finding the area of a circle in Nine Chapters and (2) his proof for the Liu Hui Principle proposed by Liu himself.

These two proofs represent two major styles among Liu's proofs: synthesis [綜合 *zonghe*] method and a combination of synthesis and analysis [分析 *fenxi*] methods.[16]

[16] The synthesis [*zonghe*] method refers to the reasoning in which one starts with the given conditions and derive the conclusion while in the analysis [*fenxi*] method, one starts with the conclusion and figure out the condition, say A, needed to obtain the conclusion, and then continues with condi-

1. A proof for procedure of the area of a circle.

Liu believed that the traditional reasoning behind the area of a circle was based on the assumption that the ratio of the circumference of a circle to its diameter is 3 to one, which did not prove the formula properly. Therefore, he proposed a proof based on a limiting process and infinitesimally small sectors produced in the process of circle division. He stated [6, pp. 18–19] and [3, pp. 178–181]:

> Another view, draw the figure: Use one side of an [inscribed regular] hexagon to multiply the radius of the circle. Three times [the product] is the area of an [inscribed regular] dodecagon. If [we] divide the circle further [following the same principle] and use one side of the dodecagon to multiply the radius of the circle. Six times [the product] is the area of an [inscribed regular] 24-gon. The finer we divide the circle to be, the smaller the discrepancy [between the area of the inscribed regular polygon and that of the circle]. Divide the circle until it cannot be divided any more; then [the polygon obtained] will match [perfectly] with the circle and there is no discrepancy. [The line segment from the] outside of [the middle of] one side of the inscribed polygons [to the circumference of the circle] is called excess radius [余径 *yüjing*]. Use perimeter of a regular polygon to multiply its excess radius. Then [the sum of] this area [and that of the regular polygon] exceed [that of] the circle. If the division [of the circle] is so that [the polygon] match perfectly with the circle, then there is no excess radius. If there is no excess radius outside the polygon, then [the sum] of the areas will not surpass that of the circle. Use one side of this [infinite-sided] polygon to multiply the radius, [the area of which is] twice of the sector subtended by one side of the polygon. Hence use one half of the circumference of the circle to multiply the radius to produce the area of the circle.

Liu Hui first took several steps in the limiting process, starting with an inscribed regular hexagon inside a circle. Let us use S_n to denote the area of the inscribed regular 6×2^n-sided polygon obtained after the n-th step of the circle division process and S the area of the circle. Liu Hui demonstrated that

$$S_{n+1} < S < S_n + 2(S_{n+1} - S_n)$$

and

$$\lim_{n \to \infty} S_n = S,$$

and

$$\lim_{n \to \infty} [S_n + 2(S_{n+1} - S_n)] = S.$$

---

tion A and figures out the condition B needed to obtain the condition A, and etc. This reserve-styled argument aims to find the conditions in the reversed order so as to find the condition to match the ones given in the statement of the question.

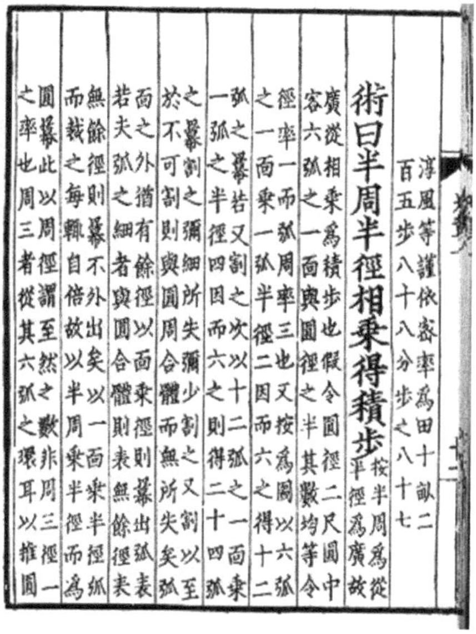
淳風等謹依密率為田十畝二百五步八十八分步之八十七

術曰半周半徑相乘得積步按半周為從半徑為廣故

廣從相乘為積步也假令圓徑二尺圓中容六弧之一面與圓徑之半其數均等令

徑率一而弧周率三也又按為圖以六弧之一面乘一弧半徑二因而六之得十二

弧之冪若又割之次以十二弧之一面乘一弧之半徑四因而六之則得二十四弧

之冪割之彌細所失彌少割之又割以至於不可割則與圓周合體而無所失矣弧

面之外猶有餘徑以面乘徑則冪出弧表若夫弧之細者與圓合體則表無餘徑表

無餘徑則冪不外出矣以一面乘半徑弧而裁之每輒自倍故以半周乘半徑而為

圓冪此以周徑謂至然之數非周三徑一之率也周三者從其六弧之環耳以推圓

(1)

(2)

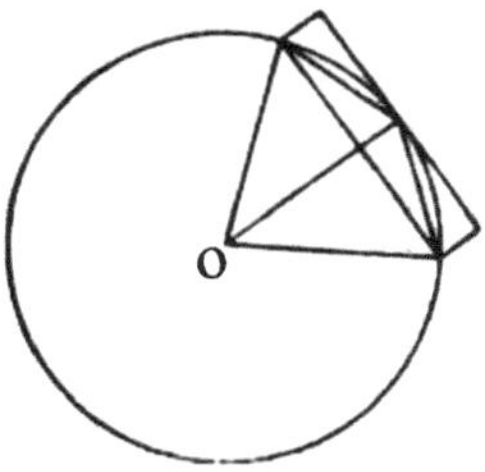

(3)

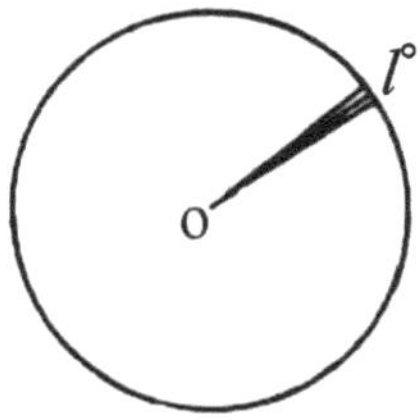

(4)

Fig. 6

And Liu Hui considered the inscribed regular infinitely many-sided polygon that matched the circle perfectly and divided it as infinitely many congruent isosceles triangles, each of which has the center of the circle as one vertex, the infinitesimally small "side" as its base, and the radius of the circle as its height. Assume the length of the base is l_i and area of the isosceles is A_i. Then, it is obvious that

$$l_i r = 2A_i.$$

The sum of the length for the base of the infinitely many isosceles triangles is equal to the circumference of the circle; the sum of their areas, the area of the circle:

$\sum_{i=1}^{\infty} A_i = S$. Therefore,

$$\sum_{i=1}^{\infty} l_i r = Lr = 2\sum_{i=1}^{\infty} A_i = 2S.$$

Then we can get S, which is Formula (1) in Section 1 (see Guo [7], [9] and K. Chemla et Guo Shuchun [3]).

This is a complete proof for the procedure of finding the area of a circle in Nine Chapters; moreover, this is a typical synthesis method: From several known conditions, Liu Hui arrived at the conclusion through reasoning.[17] This style was utilized most frequently among the proofs in Liu Hui's commentary.

2. The Proof of the Liu Hui Principle

Even more astonishing than the proof discussed above is how Liu Hui employed the concept of taking the limit and the method of dividing polyhedrons into infinitesimally small pieces to prove the Liu Hui Principle. In order to prove Formulas (3) and (4) of the volumes of a *yangma* [yang-ma] and a *bienao* [turtle-forelimb] rigorously, Liu proposed an important principle, i.e., the Liu Hui Principle:

> Divide a *qiandu* [moat-end] in a slanted manner to obtain a *yangma* and a *bienao*. The *yangma* occupies two [thirds of the volume of the *qiandu*] and the *bienao* one [-third]. This ratio will not change.

That is, in a *qiandu*, the relation

$$V_y : V_b = 2 : 1 \tag{5}$$

remains constant. Obviously, as long as the Liu Hui Principle is proved, formulas (3) and (4) can be obtained by formula (2) for the volume of a *qiandu*. To prove formula (5), Liu Hui stated [6, pp. 182–183] and [3, pp. 430–431]:

> Let *yangma* be the inside of the divided solid and *bienao* the outside. Although a geometric model (rectangular parallelepiped) can have a different length and width, this ratio [of the volumes to be 2:1] remains constant. Even if the body of a parallelepiped is different, the ratio stays the same. That is always the way. Suppose a *bienao* has its width, length, and height all equal to 2 *chi*, divide it into two *qiandu* and *bienao* and use red models [to represent these solids]. Moreover, take a *yangma* so that its width, length, and height also equal to 2 *chi*. Divide it into one cube, two *qiandu*, and two *yangma*. Use black models

[17] Liu Hui's commentary to the circular field method constitutes two distinct discussions on the circle division. One was to prove the formula for the area of a circle in Nine Chapters; the other, to find the circumference-diameter ratio. Before 1970, almost all studies on Liu Hui's commentary to the circular field method never discussed the phrases in Liu's commentary, "Use one side of this [infinite-sided] polygon to multiply the radius, [the area of which is] twice the sector subtended by one side of the polygon. Hence use one-half of the circumference of the circle to multiple the radius to produce the area of the circle. [觚而裁之, 每輒自倍, 故以半周乘半径而為圓冪]" This sentence inadvertently links together the process of taking the limit and that of finding the circumference-diameter ratio. Many scholars described that Liu's process of taking the limit was to find the circumference-diameter ratio. In fact, to find the circumference-diameter ratio did not require the process of taking limits because Liu's circumference-diameter ratio was only an approximate value. That is, his process stops after a finite number of steps.

> [to represent these solids]. Combine the red and black models to make a *qiandu*, the width, length, and height of which are all equal to 2 *chi*. Bisect its width and length, and then bisect its height. Combine one red and one black *qiandu* models to make a cube with its height equal to one and the base a square of one. Every two *bienao* models together make one *yangma* model. Take the models on the two sides together to make a cube. The portion that can be made into different kinds of parallelepiped occupied 3 while the portion that can be made into a similar parallelepiped occupied 1. Although the body of a parallelepiped might change according to the models, the ratio remains the same. If the portion that is unknown follows the ratio of 2:1 (for the volumes of divided *yangma* and *bienao*), then the ratio for the whole solid is determined. How can this as a principle be void? Take this argument of numbers to the extreme. Take the remaining portion, the width, length, and the height of which is halved. In this remaining solid, three-fourth of it can be found [by repeating the above argument]. The more halves, the tinier the remaining part becomes. When the process is pushed to the step when the remaining part becomes so fine that is called trifle [微 *wei*], which has no shape, why should we consider the remaining part?

Limited by the geometric models at hand, Liu Hui used a model with $a = b = h = 1$. However, Liu Hui explicitly stated that "Even when the measures of the width, length, and height of the parallelepiped might change, the process of division can be carried out according to the procedure". Therefore, his argument fits the general scenario in which a, b, and h are not equal. We describe the general scenario here. See Figure 4, a *qiandu* can be divided into a *yangma* and a *bienao*. Then the *yangma* can be divided further into one rectangular parallelepiped I, two smaller *qiandu* II, and III, two smaller *yangma* IV and V as in Figure 7 (1); the *bienao* in Figure 4 can be divided further into two smaller *qiandu*, II′ and III′, and two smaller *bienao*, IV′ and V′, as in Figure 7 (2). Obviously the small *qiandu* II and II′, III and III′ can be combined as rectangular parallelepipeds congruent to I, as in Figure 8 (2) and (3). The small *yangma* IV and the small *bienao* IV′ as well as V and V′ can be combined to form two *qiandu* congruent to the small *qiandu* II, III, II′ and III′, which in turn can be combined to form a 4th rectangular parallelepiped congruent to I, as in Figure 8 (4). Obviously, in the first three parallelepipeds, I, II-II′ and III-III′, the ratios of volumes for *yangma* and *bienao* are all equal to $2:1$.

In fact, as early as 1930's, Japanese scholar Mikami had solved this problem. As for the ways of combining the small *qiendu*'s, he proposed two possibilities, in addition to the aforementioned combining II with II′ and III with III′, it could also be done with II with III and II′ with III′. Mikami was leaning toward the latter [10]. His work did not catch the attentions of Li Yan or Qian Baocong. Therefore, this problem was left by Qian as an open problem to be solved in [11]. In 1980, Hideki Kawahara in his Japanese edition of *The Nine Chapters on the Mathematical Procedures commented by Liu Hui* provided much more detailed discussions on Mikami's conclusions of this problem [12]. Moreover, this question was also discussed independently by Danish scholar D. B. Wagner in 1979 [13] and Guo Shuchun in 1980 [8]. Wagner took the approach of combining II with III and II′ with III′ while Chemla and Guo in [3] II with II′ and III with III′.

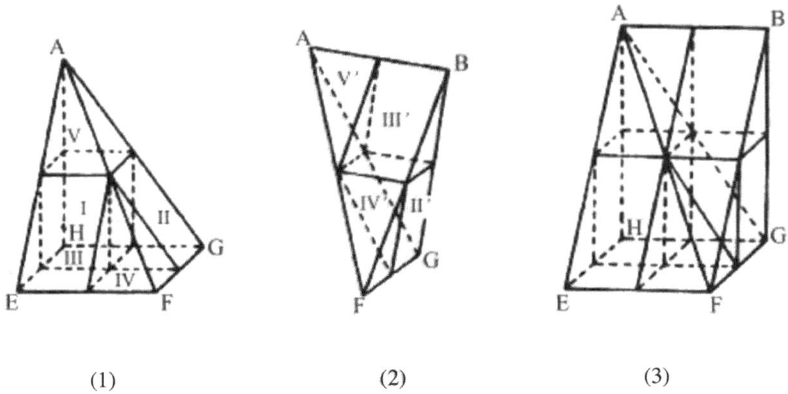

Fig. 7 Decompositions of a *yangma* and a *bienao* which compose a *qiandu*

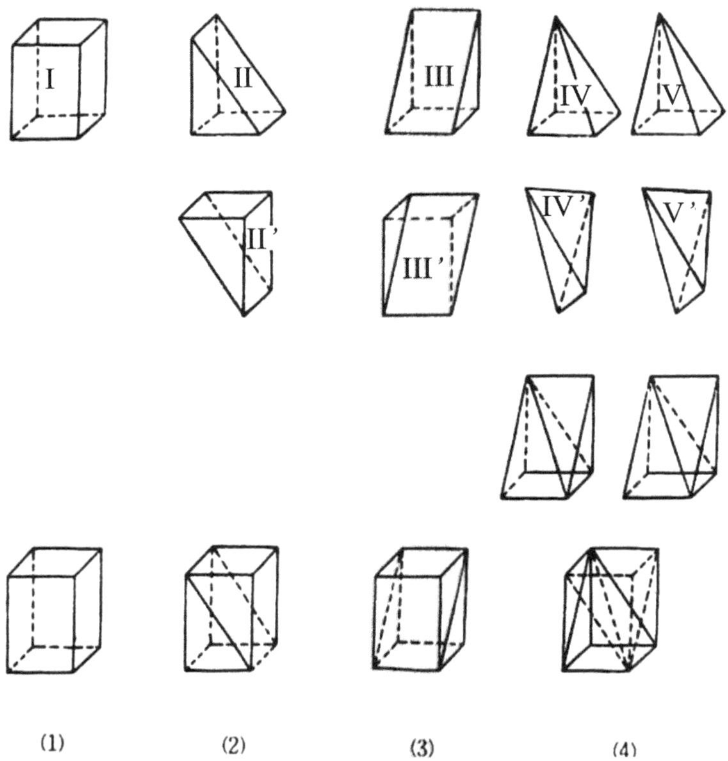

Fig. 8 The proof of the Liu Hui Principle

That is, relation (5) holds in three fourths of the original *qiandu*. Liu Hui contended that if relation (5) can be proved in the 4th rectangular parallelepiped, then (5) is proved for the entire *qiandu*. On the other hand, the two *qiandu* in the 4th rectangular parallelepiped are mathematically similar to the original *qiandu*. Consequently, the process of dividing a *qiandu* discussed above can be applied step-by-step to these two small *qiandu* in the 4th rectangular parallelepiped. Then in $\frac{3}{4}$ of these two *qiandu*, relation (5) holds while it remains to be proved in $\frac{1}{4}$ of them, i.e., $\frac{1}{4} \times \frac{1}{4}$ of the original *qiandu*. This process can be repeated indefinitely. After n divisions, Formula (5) remains to be proved in $\frac{1}{4^n}$ of the original *qiandu*. As well-known, $\lim_{n\to\infty} \frac{1}{4^n} = 0$. As a result, the Liu Hui Principle is proved for the entire *qiandu* [8, pp. 47–62], [9] and [3]. This process can be summarized as

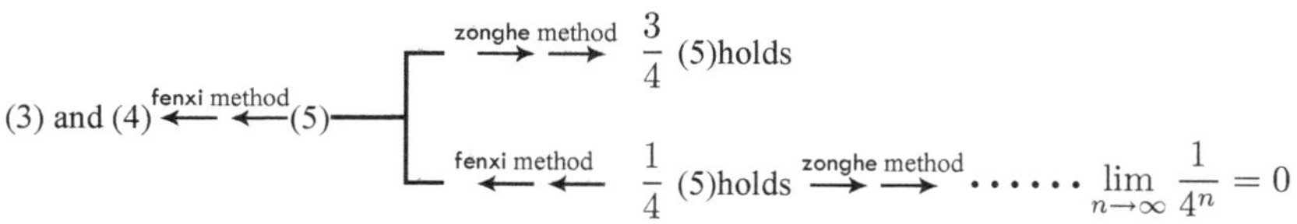

From this diagram, we can see this proof was mainly analysis [*fenxi*], interspersed with synthesis [*zonghe*].

3. Liu Hui's system of mathematical theory

Liu Hui's discussions on fractions, quantities, areas, volumes, and right triangles all constitute a part of his own system of theory. And this system is different from that of Nine Chapters. Take the discussion of the volumes as an example. When discussing his own approach to finding the volume of polyhedrons, Liu Hui stated [6, p. 179] and [3, pp. 422–423]:

> This Chapter deals with *qiandu* and *yangma*, several of which can combine to form a cube. That is why the mathematician sets up three types of geometric models in order to measure the volumes with the height or depth.

At the time of Nine Chapters, the method employed to find the volumes of polyhedrons was the testing method by geometric models. Therefore, the cube, *qiandu* and *yangma*, (see Figure 9) the width, length, and the height of which are all 1 *chi*, held a central position in this approach. The approach to find the volumes of cylinder or circular cone, polyhedrons related to circles, depended on the area of its base. The system of deriving volumes in Nine Chapters can be described in Figure 10.

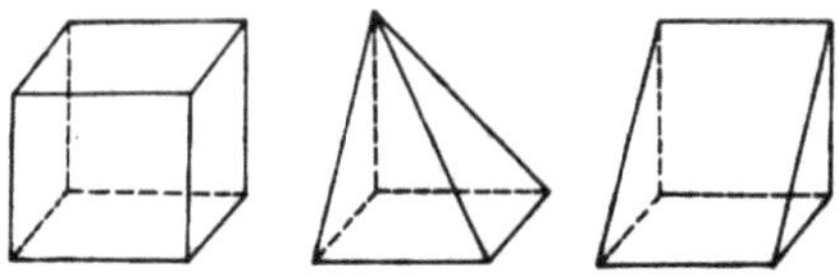

Fig. 9

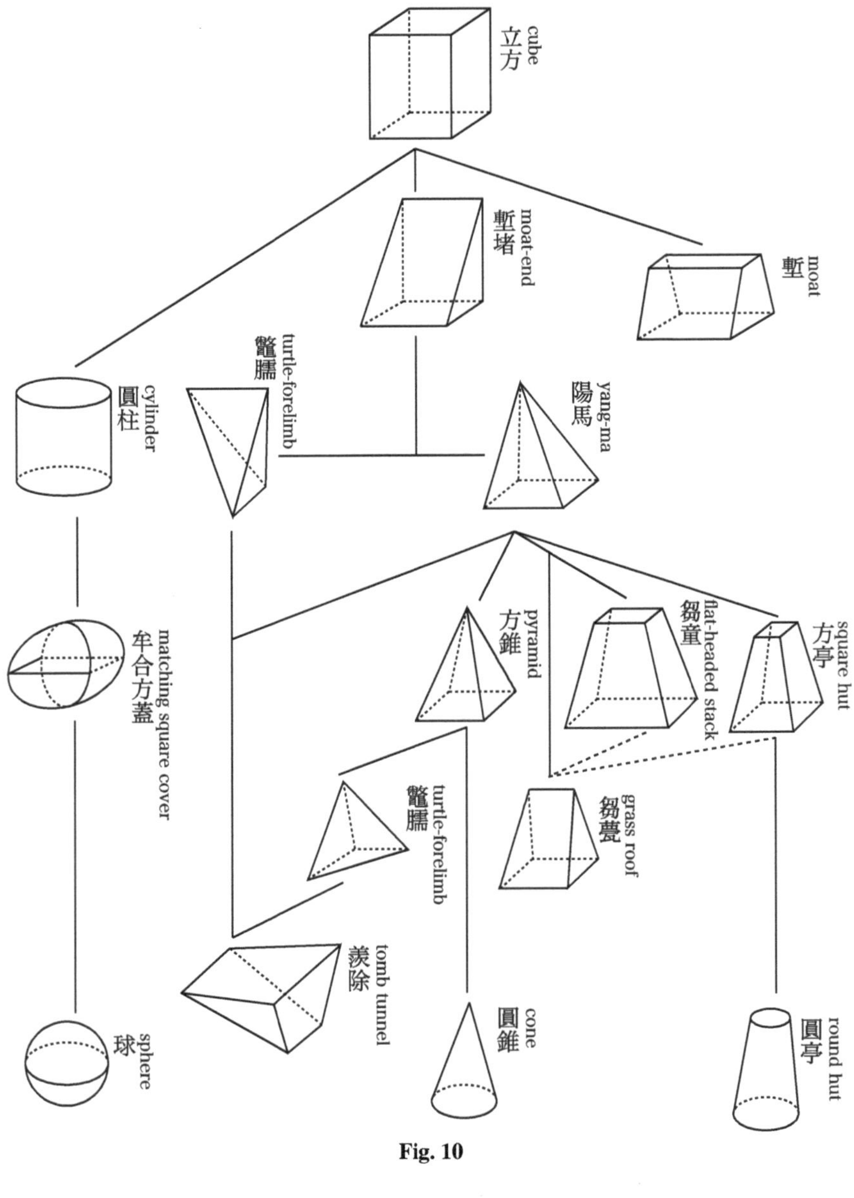
立方
cube
塹堵
moat-end
塹
moat
圓柱
cylinder
鼈臑
turtle-forelimb
陽馬
yang-ma
牟合方蓋
matching square cover
方錐
pyramid
芻童
flat-headed stack
方亭
square hut
鼈臑
turtle-forelimb
芻甍
grass roof
羨除
tomb tunnel
球
sphere
圓錐
cone
圓亭
round hut

Fig. 10

On the other hand, the basis of Liu Hui's theory on the volumes of polyhedrons is the Liu Hui Principle. After completing the proof of this principle, Liu Hui stated [6, p. 183] and [3, pp. 432–433]:

> Without *bienao*, [we] cannot verify the volume of *yangma*; without *yangma*, [we] cannot verify the solids similar to cones and frustums. Therefore, *bienao* is the main reason for finding the volumes of many polyhedrons.

Liu Hui contended that *bienao* played a key role in solving problems of volumes of polyhedrons. For many polyhedrons, their volumes can be obtained, through finitely many steps, by dividing them into a collection of rectangular parallelepipeds, *qiandu*, *yangma*, and *bienao*; since the formulas for whose volumes were established and proved, by adding these volumes together, one can find the volume of the solid in question.

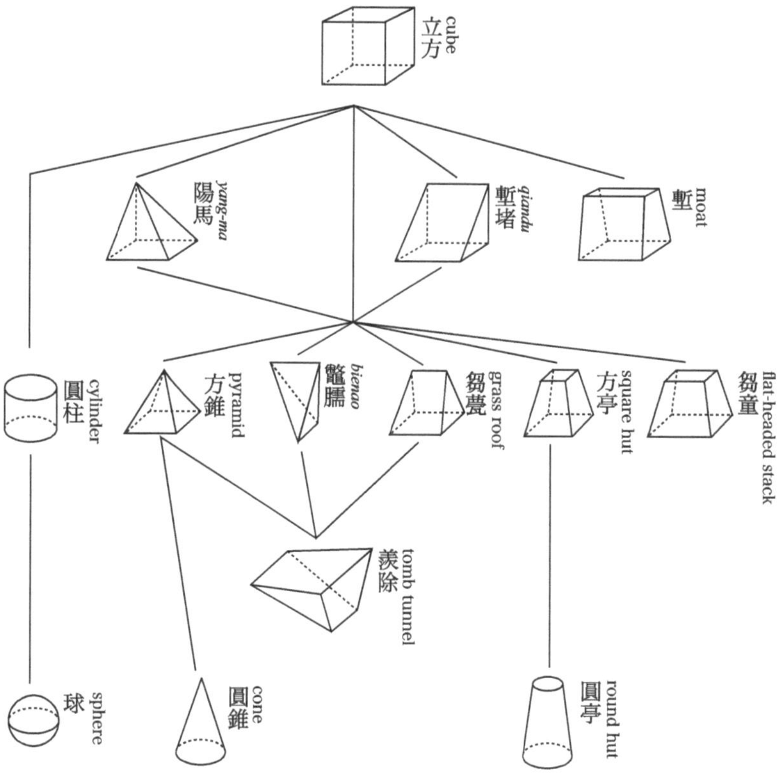

Fig. 11 Liu Hui's relation among geometric solids[18]

As for the volumes of solids related to circles (i.e. circular cylinder or sphere), their volumes can be obtained by comparing the area of cross-sections. The system of Liu Hui's theory for volumes can be described in Figure 11. Liu Hui viewed

[18] The downward lines in Fig.11 reflect the fact that the solids below were derived from the ones above.

bienao as the smallest unit in dividing polyhedrons. Both this concept of dividing volumes into smaller pieces and the procedure of finding the volume of a *bienao* have to depend on the realization of the infinitesimal division. That is, Liu's theory of the volumes for the polyhedrons in effect established on the concept of infinitesimal division, which is surprisingly in line with the modern theory of volumes. The great mathematician Gauss (1777–1855) conjectured that the solutions to finding the volumes of polyhedrons cannot be achieved without carrying out divisions of the solids into infinitesimal small pieces. Based on this conjecture, Hilbert (1862–1943) proposed the third problem in his address of 1900 to the international congress of mathematicians [2, pp. 60–84]. His follower Dehn (1878–1952) soon afterwards provided a positive answer. Liu Hui in the third century started to consider problems considered by mathematicians in the 19th and 20th centuries.

Since some years ago, the different branches in mathematics have been depicted as a tree. Located at the roots are algebra, plane geometry, trigonometry, analytic geometry, and irrational numbers. From these roots grow the strong trunk, differential and integral calculus. On the top of the trunk sprout many branches including the branches of higher mathematics [1, p. 491]. In fact, 1700 years ago, Liu Hui had the concept of the mathematical tree. He stated:

> Things of various kinds can be used to find each other, each of which can be found its location in the complex relations. The reason that the branches are apart but share the same trunk is that they came from the same root. Analyze the principles by virtue of verbal formulation; explain the substance of things using figures in the hope of achieving simplicity while remaining complete and general but not obscure, so that the reader [of the commentary] will be able to grasp more than half [6, p. 1] and [3, pp. 126–127].[19]

Liu Hui's mathematical tree starts from a point. What is this starting point? Liu Hui stated [6, pp. 1–2] and [3, pp. 126–129]:

> Although they were described as Nine Branches in Mathematics [九数 *Jiushu*], they can exhaust the fine details, get into extremely small matters, and explore and measure without any bound. As for disseminating the methods, it is just like try square and compass [規矩 *guiju*] and measurements [度量 *duliang*] can be used to find the commonality; therefore, they are not extremely difficult.

The term try square and compass represents the figures and diagrams in the space while the term measurements the relations among measured quantities. That is to say, mathematical methodology from generation to generation is the unification of geometric problems and relations among the measured quantities in the objective world. The *guiju* and *duliang* can be seen as the root of Liu Hui's tree of mathematics. Methods in mathematics are born out of the *guiju* and *duliang*. This also reflects the characteristic of ancient Chinese mathematics—the union of shapes and numbers as well as that of geometric problems and arithmetical algebraic methods.

[19] The translation of last sentence was taken from Martzloff [4, pp. 69–70].

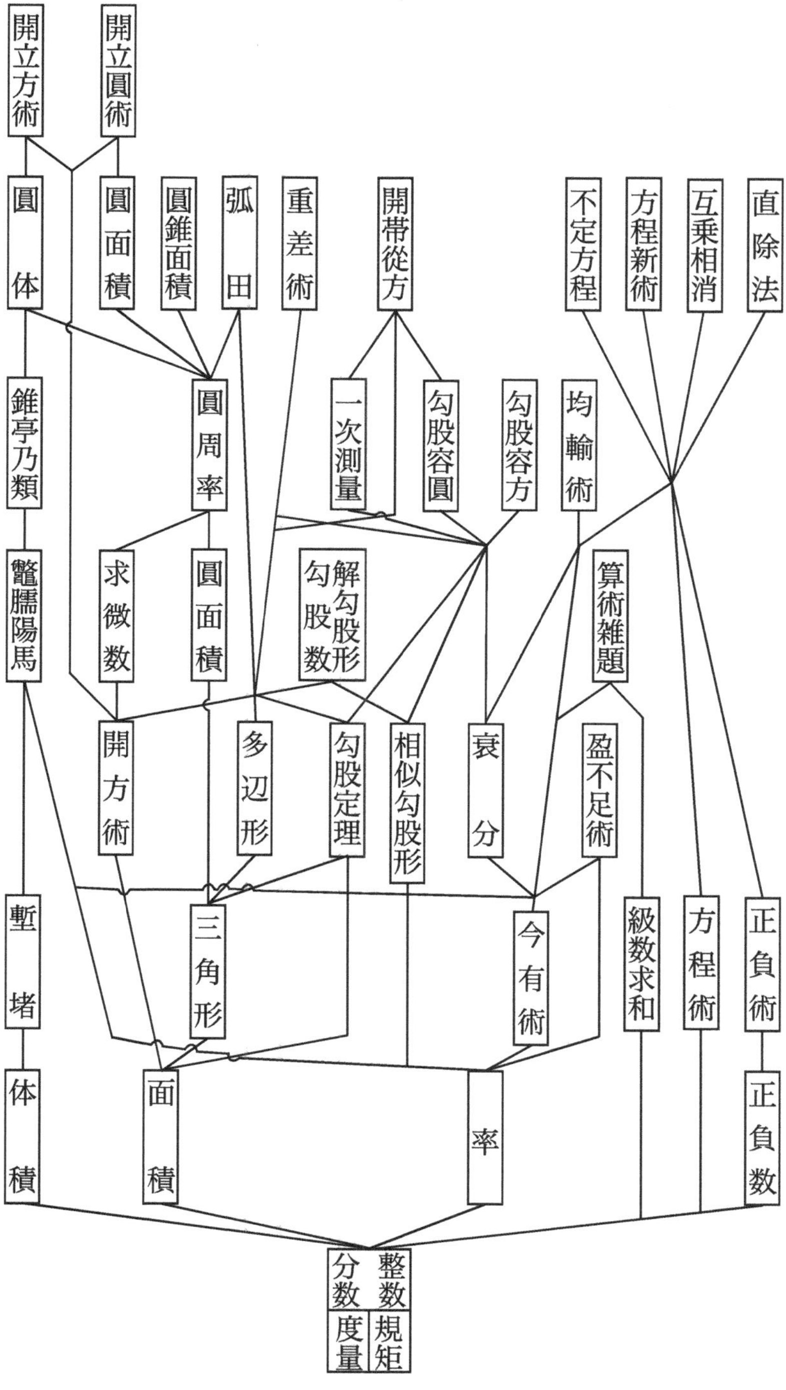

Fig. 12 Liu Hui's mathematical tree

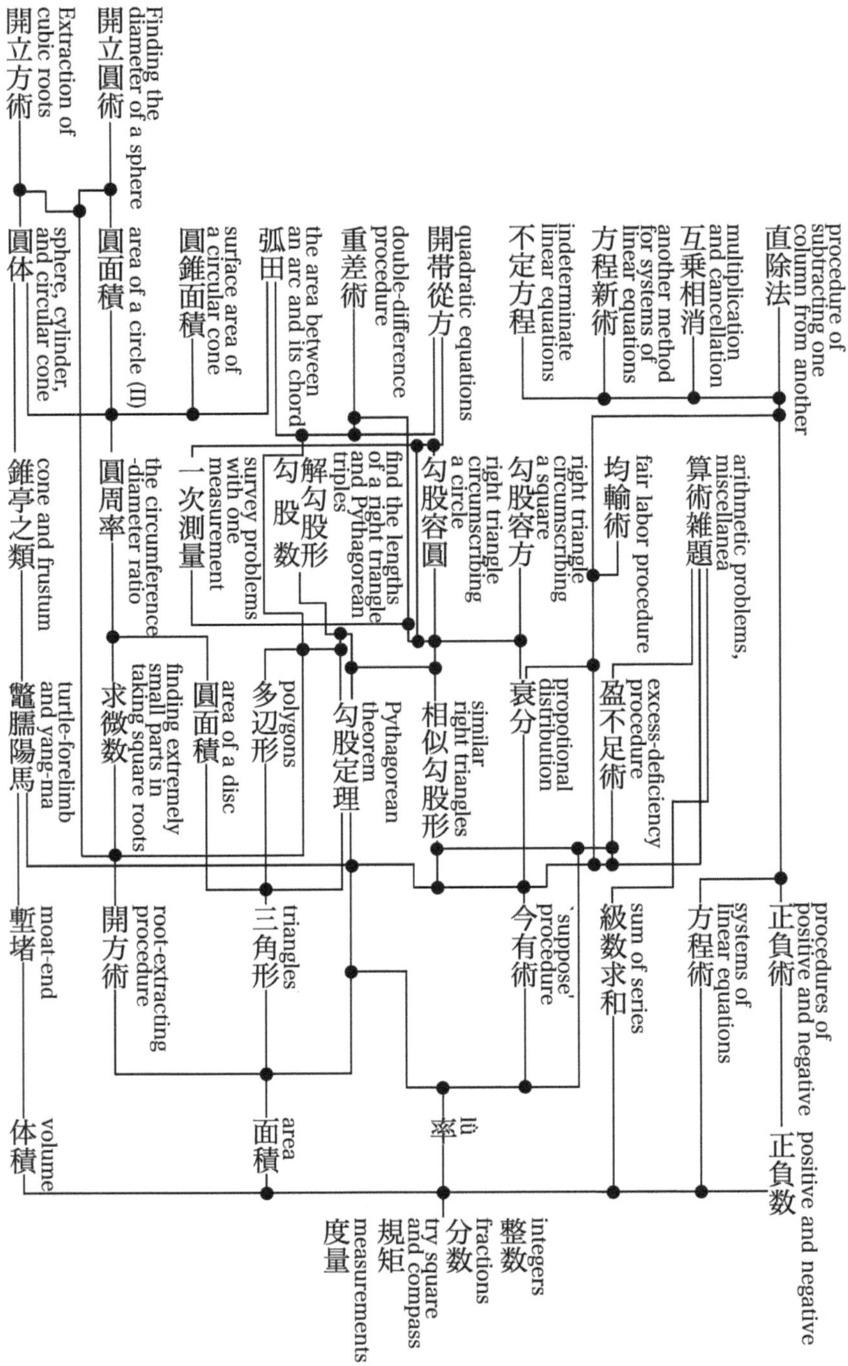
procedure of subtracting one column from another
直除法
multiplication and cancellation
互乘相消
another method for systems of linear equations
方程新術
indeterminate linear equations
不定方程
quadratic equations
開帯從方
double-difference procedure
重差術
the area between an arc and its chord
弧田
surface area of a circular cone
圓錐面積
area of a circle (II)
圓面積
sphere, cylinder, and circular cone
圓体
Finding the diameter of a sphere
開立圓術
Extraction of cubic roots
開立方術
arithmetic problems, miscellanea
算術雜題
fair labor procedure
均輸術
right triangle circumscribing a square
勾股容方
right triangle circumscribing a circle
勾股容圓
find the lengths of a right triangle and Pythagorean triples
解勾股形
勾股数
survey problems with one measurement
一次測量
the circumference -diameter ratio
圓周率
cone and frustum
錐亭之類
excess-deficiency procedure
盈不足術
propotional distribution
衰分
similar right triangles
相似勾股形
Pythagorean theorem
勾股定理
polygons
多辺形
area of a disc
圓面積
finding extremely small parts in taking square roots
求微数
turtle-forelimb and yang-ma
鼈臑陽馬
procedures of positive and negative
正負術
systems of linear equations
方程術
sum of series
級数求和
'suppose' procedure
今有術
triangles
三角形
root-extracting procedure
開方術
moat-end
塹堵
lü
率
area
面積
volume
体積
positive and negative
正負数
integers
整数
fractions
分数
try square and compass
規矩
measurements
度量

Liu Hui's tree came out of the two roots, measuring tools and measurements. They are unified under numbers, upon which grows the trunk of computations of quantities. Based on the unproved yet agreed-upon formulas of the rectangular area and volume of a rectangular parallelepiped and the definition of ratio [率 *lü*], there grows the branches of the four arithmetic operations for integers and fractions, the 'suppose' procedure, proportional distribution procedure, fair labor procedure, excess-deficiency procedure, root-extracting procedure, rectangular array procedure, solutions to area and volumes, and at least the solution to right triangles and survey, from which grows more refined branches of all kinds of mathematical methods. Eventually, all these contribute to form a leafy tree full of fruits, as depicted in Figure 12.

Liu Hui's system of mathematics is "achieving simplicity while remaining complete and general but not obscure [約而能周, 通而不黷]." That is, it is simple but complete, far-reaching without obstacles. Even though in the form of a commentary, Liu Hui cannot but separate his mathematical knowledge into the algorithms and questions in Nine Chapters. It is worth mentioning that his commentary did not contain any self-conflict paradox logically. This shows the level of his logical reasoning. Liu Hui's mathematical system was developed upon the frame of the mathematics in Nine Chapters. It inherited the correct content in Nine Chapters; moreover, it molded and complemented it. In short, Liu Hui's commentary, compared with the texts in Nine Chapters, transformed the quality of the mathematical content in it.

Acknowledgement

This article was translated into English by Dr. Jiang-Ping Jeff Chen, Department of Mathematics, St. Cloud State University, Minnesota, USA and revised with the help from Professor Komatsu. I would like to thank them for their assistance.

References

1. Eves Howard: *An Introduction to the History of Mathematics, 5th ed.*, CBS College Publishing (1983).
2. David Hilbert: *Mathematics Problems—Hilbert's address of 1900 to the International Congress of Mathematicians in Paris*, Collection of Translated Papers in the History of Mathematics [数学史訳文集 *Shuxueshi Yiwen Ji*], translated by Li Wenlin and Yuan Xiangdong, Science and Technology Press, Shanghai (1981).
3. K. Chemla et Guo Shuchun: *Les Neuf Chapitres, Le Classique mathématique de la Chine ancienne et ses commentaries*, Dunod, Paris (2004).
4. Jean-Claude Martzloff: *A History of Chinese Mathematics*, translated by Stephen Wilson, Springer-Verlag, Berlin (1997).
5. Joseph Needham: *Science and Civilization in China*, vol.III, Cambridge University Press (1959).

6. Guo Shuchun [郭書春]: Comprehensively Collated Nine Chapters of Mathematical Procedures (an enlarged edition) [匯校九章算術增補版 *Hui Jiao Jiuzhang Suanshu Zengbuban*], Education Press, Liaoning, and Jiuzhang Press, Taipei (2004).
7. Guo Shuchun: Liu Hui's Theory of Taking Limits, The Collective Papers in the History of Sciences [科学史集刊 *Kexueshi Jikan*], Dizhi Press, Beijing, vol. 11, pp. 37–46 (1984).
8. Guo Shuchun: Liu Hui's Theory of the Infinite, The Collective Papers in the History of Sciences [科学史集刊 *Kexueshi Jikan*], Dizhi Press, Beijing, vol. 11, pp. 47–62 (1984).
9. Guo Shuchun: *An Outstanding Mathematician of the Ancient World—Liu Hui*, Mingwen Book Store, Taipei (1995).
10. Yoshio Mikami [三上義夫]: *Relations and comparisons of Seki Takakazu's works with methods of mathematics by mathematicians in Kyoto and Osaka and those in China* [関孝和の業績と京阪の算家並びに支那の算法との関係及び比較], Toyo Gakuho [東洋学報], vol. 20–22 (1932–1935).
11. Qian Baocong [銭宝琮] (ed.): *The Nine Chapters on the Mathematical Procedures*, Zhonghua Shuju, Beijing (1963).
12. Hideki Kawahara [川原秀城] (trans.): *The Nine Chapters on the Mathematical Procedures commented by Liu Hui* [劉徽注九章算術], *Collections of Chinese Astronomy and Mathematics* [中国天文学数学集], Asahi Publ. [朝日出版社], Tokyo (1980).
13. D. B. Wagner: An Early Chinese Derivation of the Volume of a Pyramid: Liu Hui, Third Century A.D., *Historia Mathematica*, vol. 6 no. 2, pp. 164–188 (1979).

On the Alternative Algorithm of the 7th Problem in the Sea Island Mathematical Canon

Hideki Kawahara

Abstract The alternative algorithm of the 7th problem in Liu Hui's Sea Island Mathematical Canon is mathematically incorrect. In this short note we try to restore it, based on a characteristic of mathematical formulas by Liu Hui.

1 Introduction

Liu Hui [劉徽], who lived from the end of Wei [魏] till the beginning of Jin [晋], not only annotated the Nine Chapters on the Mathematical Arts [九章算術 *Jiuzhang Suanshu*] in 9 volumes, but also evolved the similarity of right triangles, which appears in the 9th Chapter Right-angled Triangles [句股 *Gougu*], so as to attach examples and annotations about Double Differences [重差 *Chongcha*] at the end of the 9th chapter. You see it as the *Jiuzhang Suanshu in 10 volumes selected by Liu Hui* in History of the Sui Dynasty, Record of Books [隋書経籍志 *Sui Shu, Jingji Zhi*]. Although annotations and diagrams had disappeared, examples of the 10th Chapter Double Differences survived as the Sea Island Mathematical Canon [海島算経 *Haidao Suanjing*].

The present Sea Island Mathematical Canon consists of 9 problems, a collection of examples about survey. The form as a book is the same as the Nine Chapters on the Mathematical Arts, so that every example consists of a question [問 *wen*], an answer [答 *da*] and an algorithm [術 *shu*].

There are excellent studies on the Sea Island Mathematical Canon, such as new annotations by Li Yan [1] and castigations by Qian Baocong [2], which have elucidated most of mathematical matters. Therefore, we basically depend on the cas-

Hideki Kawahara
Graduate School of Humanities and Sociology, The University of Tokyo,
Hongo, Bunkyo-ku, Tokyo 113-0033, Japan
e-mail: kawahara@l.u-tokyo.ac.jp

tigations by Qian Baocong and the diagram and mathematical signs by Li Yan, to explain the algorithm of Liu Hui.

We cite here our publications [3] and [4], and two recent works Feng [5] and Guo-Liu [6] by Chinese scholars on the Nine Chapters on the Mathematical Art and related topics. The reader may find useful two commentaries Shen-Crossley-Lun [7] and Chemla-Guo [8] in European languages.

Yet scholars who specialize in the history of Chinese mathematics have never tried to evolve detailed analysis about alternative algorithm [又術 *you shu*] of the 7th problem. The reason is that the algorithm is mathematically wrong. We think that the wrong points of the alternative algorithm of the 7th problem are caused by manipulation of later generations, and in this short essay we try to restore the alternative algorithm based on characteristics of mathematical formulas by Liu Hui.

2 Original Text

The 7th problem of the Sea Island Mathematical Canon surveys the depth z_1 of the clear water from two survey points p and n' which are four $= d$ feet away up and down, by taking water shore A' and white stone A under the depth as targets since we can't make access to the clear water.

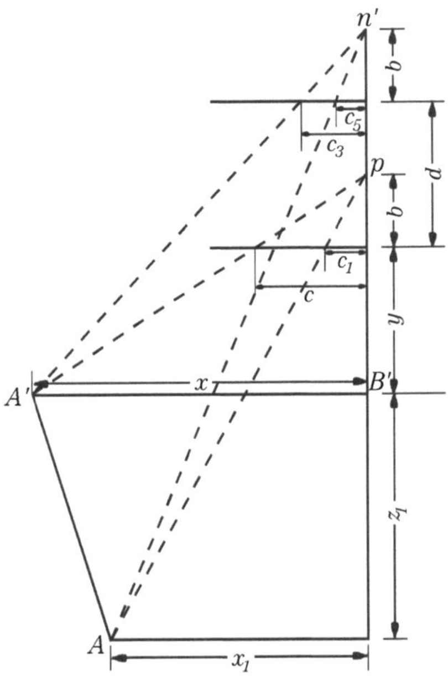

The original text reads as follows:

今有望清淵, 淵下有白石 (A). 偃矩岸上, 令句高三尺 (b). 斜望水岸 (A'), 入下股四尺五寸 (c). 望白石 (A), 入下股二尺四寸 (c_1). 又設重矩於上, 其間相去

四尺 (d). 更從句端斜望水岸 (A'), 入上股四尺 (c_3). 以望白石 (A), 入上股二尺二寸 (c_5). 問水深幾何 (z_1).

Please make reference to Diagram by Li Yan [1] about *shanggu* (上股 c_3, c_5) and *xiagu* (下股 c, c_1) and so on.

Liu Hui writes the first algorithm to the 7th problem as follows:

術曰, 置望水上下股, 相減 ($c - c_3$). 餘以乘望石上股 (c_5), 為上率. 又以望石上下股相減 ($c_1 - c_5$), 餘以乘望水上股 (c_3) 為下率. 兩率相減, 餘以乘矩間 (d) 為實. 以二差相乘為法. 實如法而一, 得水深 (z_1).

If we would write this in present mathematical formula,

$$z_1 = \frac{d[c_5(c - c_3) - c_3(c_1 - c_5)]}{(c - c_3)(c_1 - c_5)} \,. \tag{1}$$

But the alternative algorithm reads as follows:

又術, 列望水上下股及望石上下股, 相減. 餘并為法. 以望石下股減望水下股, 餘以乘矩間, 為實. 實如法而一, 得水深.

That is,

$$z_1 = \frac{d(c - c_1)}{(c - c_3) + (c_1 - c_5)},$$

which doesn't make sense.

3 Restoration of the Alternative Algorithm

The presentation of algorithm by Liu Hui has a characteristic, that is to say, he often proposes more than one algorithms to the same problem. For example, he writes two ways of arithmetical progression' formula at the 19th problem in the 7th Chapter Excess and Deficit [盈不足 *Ying Buzu*] of the Nine Chapters on the Mathematical Art.

$$S = \left(a + \frac{n-1}{2}d\right)n \quad \text{and} \quad S = an + \frac{n(n-1)}{2}d.$$

Two formulas have the same meaning as mathematical formulas, but not identical as an algorithm.

You may already know what we think from the above. We would regard expression

$$z_1 = \frac{dc_5}{c_1 - c_5} - \frac{dc_3}{c - c_3} \,, \tag{2}$$

derived from (1) as the mathematical expression of the alternative algorithm. Thus the algorithm should be corrected as follows: (12 characters are corrected.)

又術, 列望水上下股及望石上下股, 相減 $(c-c_3, c_1-c_5)$, 餘 [各] 為法. 以望 [水上] 股 (c_3) [及] 望 [石上] 股 (c_5) [乘] 矩間, [各] 為實 (dc_3, dc_5). 實如法而一, [所得相減], 得水深 (z_1).

Though we should be careful about castigation of old books, sometimes logical castigation could be allowed especially when it comes to analyze mathematical thesis, since it has universal characteristics.

References

1. Li Yan [李儼]: *New Commentary on the Sea Island Mathematical Canon* [海島算経新注], 『中国古代数学史料』, Shanghai Scientific and Technical Publishers [上海科学技術出版社] (1963).
2. Qian Baocong [銭宝琮]: *The Ten Mathematical Canons (punctuated)* in 2 vols. [校点『算経十書』(全 2 冊)], Chung Hwa Book Co. [中華書局] (1963).
3. Hideki Kawahara [川原秀城] translated: Liu Hui's Annotation on the Nine Chapters on Mathematical Art [劉徽註九章算術], Kiyoshi Yabuuchi [藪内清] ed., *Collection of Chinese Astronomy and Mathematics* [中国天文学・数学集], Asahi Publ. [朝日出版社] (1980).
4. Hideki Kawahara [川原秀城]: Section 4 Chinese Mathematics [第四節 中国の数学], Shuntarō Ito [伊藤俊太郎] ed., *Mathematics in the Medieval Age* [中世の数学], Kyōritsu Shuppan [共立出版株式会社] (1987).
5. Feng Lisheng [馮立升]: *History of Studies of Surveys in Ancient China* [中国古代測量学史], Inner Mongolia University Press [内蒙古大学出版社] (1995).
6. Guo Shuchun [郭書春] and Liu Dun [劉鈍]: *The Ten Mathematical Canons (punctuated)* 校点『算経十書』(全 2 冊), Liaoning Education Press [遼寧教育出版社] (1998).
7. Shen Kangshen, John N. Crossley and Anthony W.-C. Lun: *The Nine Chapters on the Mathematical Art, Companion & Commentary*, Oxford University Press, Oxford (1999).
8. Karine Chemla et Guo Shuchun: *Les Neuf Chapitres, Le Classique Mathématique de la Chine Ancienne et ses Commentaires*, Dunod, Paris (2004).

A Comparative Study on Traditional Mathematics of Korea and Japan

Kim Yong-Woon

Abstract Mathematics and astronomical system of ancient Japan had been influenced by those of ancient Korea since the early 5th century, Ōjin period of Japan. It was also clear that the fundamental languages, especially the numeral and arithmetic terms of both Korea and Japan were based on the common linguistic ancestors. These strongly indicate that both nations possessed the very close way of thinking and the value in mathematics. However, after the 17th century, there was a major difference in their way of thinking of mathematics.

1 Ancient Mathematics of Korea and Japan

Budo [夫道], who was very good at mechanics, writings and mathematics, was appointed as public officer of Silla [新羅] dynasty according to the record [2] below in 251 AD. It was institutionalized in 260 AD to appoint a mathematical officer in Baekjea [百済]. Moreover, mathematical officers were regularly appointed in Baekjea and Koguryo [高句麗] before Silla.

There was a report that people who were good at writings and mathematics took charge of accounting and financial affairs in Japan, the second year of Kyōtoku [孝徳], 646 AD.

The 1st period of Japanese mathematics started from Baekjea according to the official History of Nihon [日本書紀 Nihon Shoki].

1. Agicki [阿直岐] from Baekjea became a tutor of the crown prince in 400 AD.
2. Wang-In [王仁] from Baekjea instructed the Analects [論語] of Confucius and the Thousand-Character Text [千字文], then he became a tutor of the crown prince in 402 AD.

Kim Yong-Woon
Dankook University, 126 suji, yongin, Kyongi-do, 448-701, Korea
e-mail: kyw06@hotmail.com

3. Baekjea accredited Dan Yangi [段楊爾] a doctor of five Chinese classics [五経博士] of Confucianism to Japan in 513 AD, and replaced in 516 AD.
4. Japan required the regular replacement of a medicine doctor, a divination doctor and an almanac doctor, and also required astrological books, almanacs and many medical aids in 553 from Baekjea. Therefore, Si-Duk [施徳] and Wang-Doryang [王道良], who were divination doctors, and Go-Duk [固徳] and Wang-Boson [王保孫], the almanac doctors, were replaced.
5. Gwon-Rok [勸勒], the monk from Baekjea, taught almanac, astronomy, and divination to Japanese students in 602 AD.
6. Metrological system had been arranged in the period of King Jomei [舒明天皇] (629–641). It was obvious that almanacs, astrology and metrology above were related to mathematics.

2 Mathematical System in the Legal System

After the unification (669) of three kingdoms in Korea, Silla absorbed the mathematical systems of Baekjea and Koguryo, and rearranged them referring to those of Dang [唐] dynasty.

History of three Kingdoms [三国史記] showed the educational period, qualification, the course of study, etc. of mathematical doctors and their assistants. These mathematical doctors with an assistant taught Cholsul [綴術], Samgae [三開], Kujang [九章], Yukjang [六章].

Curriculum of mathematics of Silla has been revised to some degree in Koryo Dynasty [高麗王朝], such as Cholsul, Samgae, Kujang and Saga [謝家] instead of Yukjang. Cholsul which was originally from the mathematical system of Dang, had been repealed because it's all about the calculation of the circular constant π. No one wants to learn about this since it's too hard to understand, according to Suso [隋書], the history of Su [隋] dynasty.

3 Japan

Japanese mathematical system, which was included in Yōrō Ritsurei [養老律令] (718), consisted of Confucianism, writings, mathematics and phonology. It had 2 mathematical doctors and 30 students. Children of the officials higher than 5th degree, and children of the 7,8th degree officials were qualified to apply for this system.

Chinese mathematical subjects were adopted in the legal system. Japanese mathematical subjects consisted of Sonsi [孫子], Kujang [九章 Kushō in Japanese], Kaitō [海島], Cholsul [綴術 Tetsujitsu in Japanese], Samgae [三開 Sankai in Japanese], Shuhi [周髀], Kusi [九司 Kyūshi in Japanese], Goso [五曹]. Among these, Yukjang [六章 Rokushō in Japanese], Sankai, Kusi were not found in the mathematical system of Dang dynasty. Meanwhile, Rokushō, Sankai were included in the system of Silla as well. However, these subjects must have been influenced by not Silla but Baekjea because the relationship between Silla and Japan were established abruptly at the time.

4 Mathematics of Chosŏn Dynasty

In the 10th century, the reign of the central government was collapsed in Japan. Instead, it shifted to the shogunate government, which resulted in uselessness of mathematical system that was made under the central government. Meanwhile, Koryo dynasty of Korea enforced Kwagŏ [科挙], the civil examination system, which helped their mathematical system to be firm. At that time, the study of mathematics was not organized systematically because the shoguns in Japan divided the country into the local autonomy.

After the mathematical system of Silla was succeeded to Koryo, it made an epochal improvement in Chosŏn [朝鮮] dynasty, especially in the period of King Seajong [世宗].

King Seajong's policy was based on building an independent Confucian ideal country. Astronomical observations and making calendars had been originated from China. However, he revised Chiljongsan [七政算], the standard astronomical book, to make Chosŏn's geography the center of the world by means of mathematics. He also needed a great amount of mathematical knowledge to consolidate tax system and estate system as well. According to Seajong Chronicles [世宗実録], King Seajong organized a mathematical bureau [算学校正所] and a math school [習算局] to reinforce his policy of the advancement of mathematical knowledge in 1423, 1430, 1431 and 1433 and he put his effort on studying mathematics by himself. In 1438, the mathematical curriculum comprised 5 subjects such as Sangmyon-Sanbop [詳明算法], Yanghee-Sanbop [楊輝算法], Kyemong-Sanbop [算学啓蒙], Ojosan [五曹算], and Jisan [地算], which were completely different from those of Koryo influenced by Silla. Those were not only the subject also the name of the textbooks. At that time, they also revised the astronomy and astronomical system as well, which was deeply related to King Seajong's policy of the advancement of mathematics.

They specified the official status of mathematicians such as the professors of mathematics, math officers, mathematicians, accountants, and math teachers. These institutionalized mathematics affected Chosŏn's fundamental mathematical system thereafter.

Though mathematics had been succeeded as one of the fields of government controlled science since Silla dynasty, Chosŏn dynasty renovated its own mathematics in an independent way. It formed a striking contrast to Chinese or Japanese mathematical system that was cut off from time to time in their history. Chosŏn was the only country to adhere their mathematical system in the world.

5 Paradigm of Chosŏn Dynasty

The mathematicians of Chosŏn dynasty consisted of noble scholars, enlightenment scholars (practical scholars), and the intermediary mathematicians. Each has their own point of view in mathematics as below.

5.1 Mathematics of Nobles

They were based on traditional Chinese six subjects such as manners, music, archery, horse-riding, writings, and mathematics. King Seajong [世宗] (1419–1450) described the reason he studied mathematics as below;

"The mathematical knowledge does not necessarily have to be studied by Kings. However, mathematics is the study to approach to what a saint taught us."

'What a saint taught us' implied the traditional concept of Chinese philosophy. King Seajong had focused on the traditional mathematics. Nobles, based on the belief of Sung Confucianism [宋儒], concentrated on this traditional mathematics as well, which results in mixing of practical mathematics and metaphysical idea from traditional mathematical principles.

Soon after Seajong's era, Korea suffered from severe damage by two foreign invasions. The first invasion was by Japanese (1592–1598) and the second one was by Manchurians (1636–1637). Korea spent more than 60 years to restore mathematical system after these wars. Then, in the end of 17th century, the examinations for mathematical officials took place every 1 or 2 years and 1672 mathematicians were listed in the roll of successful candidates. The social status of mathematical officials was strictly restricted, and hereditary family groups in mathematics were even formed.

Choi Sukjung [崔錫鼎] (1645–1715), one of the noble scholars, had worked as the prime minister in Chosŏn dynasty. He wrote Kusuryak [九数略] on mathematics, and that book came next to Boethius (480–524), one of Roman mathematicians. Boethius was famous for monastery mathematics focused on mythical and metaphysical numeral theory regardless of the realistic problems. He also dealt with the numeral classification based on the Trinity that was very much mythical. Choi Sukjung wasn't influenced by Pythagoras or Christian numeral ideas, of course, but it was very interesting that he stated the numeral theory just like Boethius, based on the Book of Yi [易] (Changes) and the classical philosophy. Both were very aristocratic and tended to be metaphysical. In the preface of Kusuryak, it shows 'numbers are the best ...' as below:

"Numbers are originated from the morality. ... Tae-il [太一], the first being, is the beginning of numbers, Tae-geuk [太極], the Great Absolute, is the end of the morality... ."

He discussed about the origin of numbers, the ontological basis of numbers and metaphysical dogmatism by explaining names, figures, techniques and rules of numbers in this book. Besides, he classified each section of the book through connecting with the idea of the sun and the moon (Yin [陰] and Yang [陽]) into the exaggerated name like 'nine chapters in Yin-Yang theory'.

The idea of the Book of Yi and Chinese mathematics were both considered to be successful in studying of a magic square. A magic square of Chinese mathematics came from the idea of the Book of Yi such as Hado [河圖] and Rakusho [洛書]. Therefore, the study of a magic square was very natural in a traditional mathematical point of view. For instance, Mikami [7] refers to water according to the traditional 5 principal theory. It results in a very creative magic square that was able to express more complicated formed hexagon, which seemed to be affected by Yanghee-Sanbop [楊輝算法].

Choi Sukjung's academic character had been succeeded among the nobles and Choi Hanki [崔漢綺] (1803–1879) was one of them. Also, he worked as a main politician just like Choi Sukjung. He wrote a number of books on mathematics and was one of the most remarkable empiricist of Korean history. As a mathematician, he wrote Supsan Jinbol [習算津筏], but it was all the more traditional.

King Seajong, Choi Sukjung and Choi Hanki lived in a different period respectively, but they all had the common metaphysical theory of mathematics, and that was the limit of nobles who had been absorbed in Confucian culture.

5.2 *Mathematics of Practical Scholars*

Practical scholars from the nobles were those who developed the enlightenment movement, and acted academically and encyclopedically. They started to recognize the importance of mathematics, which had much in common with the French scholars of enlightenment movement.

Hwang Yoonsuk [黄胤錫] (1719–1791) was one of the representative practical scholars. He had written an encyclopedic book Isu-Sinpyon [理数新編], which included from Sung Confucianism, the doctrine of the five natural elements of Yin-Yang theory, astronomy, music, language, moral training and even how to get on in life. Among these books, Sanhak Ipumun [算学入門] and Sanhak Bonwon [算学本原] were related to mathematics. Sanhak Ipumun contained the contents of Chinese mathematical books such as Sanhak-Kemon [算学啓蒙], Sangmyon-Sanbop and other books from Chung [清] dynasty. In Sanhak Bonwon, he stated about the relationship between Chosŏn and European equation theory in the name of Chonon-sul [天元術], and Chakun-sul [開方術]. His mathematical books, however, considered too extensive topics.

Hong Daeyong [洪大容] (1731–1783) was also one of the practical scholars. He was strongly interested in science, which made him build a private astronomical observatory. He had also written the encyclopedic books including Juhae-suyong [籌解需用] on mathematics. In this book, he not only introduced traditional Chinese mathematics, but also discussed about the role of mathematics in the relationship with music, astronomy and almanacs. This showed that both French enlightenment movement and Chosŏn's practical scholars had in common with the awareness of the rationality, tendency of practicalism and the importance of mathematics. However, the former started to move toward the modern industrialization and was supported by at least their development of military science. On the other hand, there was nothing industrial or technical development in Chosŏn at that time. Even though 'the truth from the fact that practical things are the best [實事求是]' was the motto of practical scholars, they only dealt with the field of astronomy and concerned about practicality of mathematics.

5.3 Intermediary Class

Government controlled mathematics of Silla was succeeded to Koryo and Chosŏn as it was because the administrative system of each dynasty had not been changed. Kwagŏ [科挙], the civil examination system for mathematician as an technical official had been enforced since Koryo dynasty, and it was improved in Chosŏn period. Intermediary mathematicians mainly studied the techniques of administration and survey and their social position belonged to the class between the nobles and the common people. This only existed in Chosŏn dynasty of all Oriental countries.

Kyungguk-Daejon [経国大典], the constitution of Chosŏn, mentioned that only the students of observatory who dealt with astronomy and almanac could apply to astrology and astronomy.

Therefore, the intermediary class of technical officials had formed an exclusive group in the society. In other words, intermediary class was made up of noble illegitimates and there were many abuses by fixing this class into the society. Mathematics, occupied by this specific class, faded away without the self-criticism or improvement. Moreover, it had not been influenced by the change of regime, so their techniques were handed down through heredity. The existing Juhak-Iphyukan [籌学入格案], the roll of successful candidates for Chwijae [中人], the qualifying examination of mathematical officials, and their personnel record, informed that the consanguinity and the hereditary family group of 1627 candidates who passed the exam during around 400 years from the end of 15th century to the end of 19th century. According to this record, only 205 out of 1627 did not come from the family related to mathematics. The system of mathematics, especially, became more extensive as time went by and so did the mathematician's field.

The firm existence of intermediary class played an important role in maintaining of dynasty in spite of the foreign invasion such as Japanese and Manchurian invasions.

5.4 Mathematicians from Intermediary Class (Technical Officials)

Hong Jungha [洪正夏] (1684–?), whose father, grandfathers and father-in-law were all mathematicians, was a typical mathematician from intermediary class. He had written Kuilchip [九一集], which has the transformational questionaries from Kujang-Sansul, Sangmyon-Sanbop based on Sanhak-Kemon. It also has the record of the mathematical competition with Ha Gukju [何国柱], mathematician of Chung [清] dynasty, which shows the atmosphere between Korea and China at those times while most of books on mathematics in Chosŏn period were all written in Chinese characters, and they were all textbookish.

Hong Jungha met Ha Gukju, who had visited Korea as a member of a diplomatic mission, with his friend Yu Soosuk [劉寿錫] on 29th of May, 1713. It was customary to accompany the best scholars on a diplomatic mission for the purpose of showing off their cultural pride.

We could know the following from Hong's records.

1. Mathematicians from Chosŏn's intermediary class had bare minimal information of Chinese mathematics, while Chosŏn's high ranking officials, especially diplomats gathered the new information. Did the society of mathematicians in Chosŏn tend to be exclusive?
2. Chosŏn's intermediary mathematician succeeded Chonon-sul [天元術], which was almost forgotten in China. In addition, a calculation by counting rods was not used anymore in China, either. Kinnosuke Ogura [小倉金之助] indicated that Chonon-sul "had not been remembered anymore because people did not use the method by counting rods [6] while abacuses had come into use to people."
3. Ha Gukju probably went back to China with the recognition that Chosŏn still had Sanhak-Kemon [算学啓蒙], which was already gone in China, and that might have been an opportunity to be able to restore Chonon-sul later. To the casual view, Chosŏn's mathematicians preserved Cholsul, which had been already faded in China, until Koryo dynasty, Chonon-sul and Sanhak-Kemon as well. That seemed to cling to legitimacy.

5.5 Collaboration with Intermediary Class and Noble Mathematicians

As described above, mathematicians consisted of nobles, practical scholars and intermediary class, and they had their own academic characters respectively. However, both mathematicians from nobles and intermediary class shared the common point of view, because the practicality between the enlightenment movement of nobles and the intermediary class became delicately overlapped at the end of Chosŏn era.

There was a movement to drive mathematics out of its traditional character. This made both classes collaborate because it focused only on mathematics regardless of the class. Nam Byungchul [南秉哲] and Nam Byunggil [南秉吉], who were brothers and both noble high officials and scientists as well, were interested in infinite series including the calculation of the circular constant π and the trigonometric series. Nam Byunggil (1820–1869) and Lee Sanghyuk [李尚爀] (1810–?) studied together. Lee Sanghyuk, who started as a lower official of astronomy, had written the books on almanac and astronomy. In his other books such as Sansul Gwangyun [算学正義], Iksan [翼算] and Muihae [無異解], he developed mathematical theory that had not been found in the former practicality-oriented mathematical books. From this point, new mathematics started to study European geometry and trigonometry, which showed the potential of Chosŏn mathematics getting out of the traditional mathematics.

6 The Comparison with Japanese Mathematics

6.1 Abacus

It was told that Toyotomi Hideyoshi [豊臣秀吉], the shogun of Japan, sent Mōri Shigeyoshi [毛利重能] to China or Chosŏn to study mathematics. There was the oldest abacus in Maeda Han(clan) [前田藩] in Nagoya Castle [名護屋城], where was the outpost when Toyotomi Hideyoshi invaded Chosŏn. The abacus must have been made by Chosŏn in regard of the technique of its manufacture.

It was also told that Mōri Shigeyoshi came back to Japan with Sanbup-Tongjong [算法統宗], which had the directions of abacus and explanation of dividing in detail. He wrote the Book of Division [割算書 Warizansho], considering himself as 'the world best abacus operator.' Yoshida Mitsuyoshi [吉田光由], one of his disciples, wrote Jinkōki [塵劫記] based on Sanbup-Tongjong. Sanbup-Tongjong was not included in Chosŏn's mathematical system while it was the most widely read in Chosŏn period.

Japanese abacus, originally from China, were different from Korea's. There were 2 beads in the upper deck, according to the picture in Sanbup-Tongjong, which was the same as Korean abacus used by the end of Chosŏn period. However, abacus in Korea was not widely used because the nobles in Chosŏn including scholars and high officials considered it humble. Choi Sukjung [崔錫鼎], one of the noble mathematicians, even excluded abacus.

On the other hand, abacus was so well-developed in Japan that Japanese abacus had one bead on each rod in upper deck, 4 beads each in the bottom deck. That was caused by the fact that 'reading, writing and abacus' became 3 major educational subjects for common people as merchandising had been progressed in Dokukawa [徳川] period. It was an immense difference from confucian-oriented education of Chosŏn.

6.2 Chonon-Sul

Toyotomi invaders had taken most of Chosŏn's mathematical books to Japan. Above all, Sanhak-Kemon written by Ju Segul from Won [元] dynasty of China, influenced

Japanese mathematics. This book had been published many times because it was used as a regular system of mathematics in Chosŏn, while it was not seen anymore in China. It dealt with Chonon-sul, called mechanical algebra, solving high degree integer coefficient equation by the counting rods. Chonon-sul was structurally the same as Horner's approximation theory although it was invented in China at least 500 years earlier than the West. Counting rods used in calculation to solve a high degree equation, but it was also complicated to operate. Some mathematicians in the early Chosŏn were proud of being able to use Chonon-sul, which former mathematicians could not approach.

In China, no one was interested in Chonon-sul any more, then it had disappeared. As stated above, it was obviously recorded in the mathematical debate between Chosŏn's mathematician and Ha gukju, who had visited Korea as a member of a diplomatic mission from Chung dynasty.

Seki Takakazu [關孝和], the founder of Wasan [和算], started his study by copying the mathematical text books printed in Korean. Counting rods from Chonon-sul had been changed to calculation with writing down on paper, then this helped to invent Bōshojutsu [傍書術], a kind of side writing method. | 甲 [kō], | 乙 [otsu] and | 丙 [hei] ...started to be used instead of the numeral coefficient as symbolic algebra. That is to say that Japanese mathematics had improved in the middle of symbolization of letters far from the counting rods.

Mathematicians could solve very complicated calculation such as symbolic algebra repeatedly with the help of Bōshojutsu instead of the counting rods. Finally, they were able to calculate the infinite series with this method. It was also true that differential and integral calculus was just around the corner of Japanese mathematicians.

7 Paradigm of Japanese Mathematics

"The best abacus operator in the world" was the motto of Mōri Shigeyoshi. The policy of Oda Nobunaga [織田信長] and the idea of work division in the Age of Wars encouraged 'the best in the world' to be popular in every field of Japan. 'The world best kiln maker,' 'the best tea-making artist in the world,' 'the best mannered man in the world,' 'the world best seal maker' etc. showed how deep Japan had been affected by this atmosphere. Moreover, Japan had 3 kinds of phonetics such as Katakana [片仮名], Hiragana [平仮名] and Hentai Kana [変体仮名].

In addition, many schools such as Seki [關], Mogami [最上], Takuma [宅間], and Miike [三池] were derived from Japanese mathematics, which foreign mathematicians can not distinguish from each other. These schools competed with one another, which reinforced Japanese mathematics by questioning with Open Problems [遺題 Idai in Japanese], and followed by other questions. Then, with these competitions, they became to ignore the practicality. For example, they calculated the circular constant π down to 50 decimals and hundreds of answers resulted from the high degree equation, or even studied infinite circular row inscribed in triangles etc. These resulted in the factional competition, which originated the philosophy of the usefulness of uselessness [無用之用]. Aida Yasuaki [会田安明] emphasized on focusing on uselessness. He implies that traditional mathematics started from practicality, but real mathematics must be free from practicality. While the practical scholars of Chosŏn paid attention to the truth from the fact [實事求是], Japanese mathematicians ignored the practicality and started to pursue the beauty of techniques instead.

Aida Yasuaki himself was proud of being interested in ignoring the practicality. Seki Takakazu, the best Japanese mathematician, regarded 'the usefulness of uselessness' as the most important thing in mathematics. Aida Yasuaki pursued the beauty of mathematics and it was all about unrestricted idea in Japanese art and techniques including tea, drawings and judo, and so on. Therefore, Japanese mathematics was more like hobbies as if poets and Japanese mathematicians had a spiritual similarity.

Yoshio Mikami [三上義夫], one of Japanese mathematical historians, writes "Most of mathematicians were interested in Waka [和歌] or Haiku [俳句], 'Japanese poem' [7]."

Kazuo Shimodaira [下平和夫] [5] mentioned "The whole Japanese seemed to be crazy about mathematics. Some people not only teach but also take care of other poor mathematicians. What was worse, their lives lie in ruins with those hobbies."

8 The Idea of Simplification and Generalization

While mathematicians of Chosŏn were faithful to tradition and Confucian belief, Japanese mathematicians showed the opposite attitude. They never cared about the original principle. They replaced the counting rods by drawings first and then writing down on paper without hesitating just as they improved their abacus straight regardless of the existing style. It was also just the opposite to Chosŏn's mathematicians who continued using the counting rods. It could be called the symbolic algebra that Japanese replaced the counting rods by drawings and then writing down on paper as if western algebra replaced numbers by letters.

Yoshio Mikami considered Japanese mathematics as 'the spirit that respects simplification.' Also, Hajime Nakamura [中村元], one of the well-known Japanese philosophers, regarded Japanese Buddhism as simplification [7]. Those had something to do with the creativity of Kana, the improvement of abacus, and the current improvement in technology.

Meanwhile, mathematician of Chosŏn persisted in traditional confucianism and Yin-Yang theory, Chinese dualism. One of the reasons they continued using Chononsul was Confucianism that all of Chosŏn's mathematicians including intermediary class had concentrated in detailed. Especially, the philosophy of Yi began with the Great Absolute [太極] then reached to the sun, the moon and 4 items [四象]. From here, one of the origin of universe Chonwon [天元] was called the Great Absolute [太極 Taeguk]. Chosŏn's mathematicians probably seemed to feel that they put the philosophy of Yi into practice while solving Chonon-sul. However, they also had a creativity to invent Hangul, Korean phonetics and the double entry book-keeping system, and so on. Hangul, Korean phonetics, which seemed to be somewhat Cartesian scientific thinking, analyzed the phones into vowels (the sun) and consonants (the moon) first, and then integrated them again. Korean double entry system, which was well known as Sagae songdo chibubub, also divided incoming and outgoing of money by the sun and the moon, and those of goods by 4 items. (Yun Kun ho [尹根鎬], A study of Sagae songdo chibubub [四介松都治簿法研究]) .

Songdo was the capital of Koryo dynasty, and now it indicates Gaesung in Korea. Merchants of Gaesung were the most prosperous group at that time just as Oomi's [近江], the merchant of Japan, and they had traded Ginseng [朝鮮人参] from Koryo period. Sagae songdo chibubub, which had been used from the end of Koryo or the early Chosŏn, was exactly the same as the present double entry system, and the one used at the end of Chosŏn still exists. Why did not Japanese invent a book-keeping system while they improved an abacus? Why, on the other hand, did not Korean improve an abacus though they invented the double entry book-keeping system? The answers lied in the differences of their paradigm, such as Oomi's abacus against Gaesung's book-keeping [開城簿記] system and 'Kana against Hangul.'

Korean's creativity, unlikely to Japanese simplification, was more generalized and had made something useful to real life. Moreover, they dropped straight when it was against the traditional legitimacy. Japanese, however, simplified the foreign culture with their creativity regardless of tradition or ideology.

9 Conclusion

The comparison of traditional mathematics between Korea and Japan has been described as follows.

1. Korean mathematics had been maintained under the government control from Silla, Koryo through Chosŏn dynasty.
2. While most of ancient Japanese mathematics was borrowed from Baekjea, their modern mathematics improved independently by substituting calculation with writing down on paper for the counting rods.
3. Although both Korea and Japan were mostly influenced by Chinese mathematical books such as Sanhak-Kemon, Yanghee-Sanbop, Sanbup-Tongjong, and so on, their paradigm formed the contrast as below.
4. The extinction of mathematics of both Korea and Japan followed after the introduction of western mathematics. Ancient mathematical system and their thought of both countries had been very similar. However, the differences above were caused by their own paradigm derived from different social systems such as Japanese shogunate feudalism and Chosŏn dynasty's centralism.
5. Mathematics should not be restricted by philosophy. Japanese mathematics, originally derived from Korean mathematics, had been able to be improved with

unrestricted idea, the usefulness of uselessness and being free from western mathematics.

	Japan	Korea	
Social position	no relation	nobles	intermediary class
Object of mathematics	hobby, the use of uselessness	metaphysical	practicality
Successor	master and disciple, schools of study		heredity, examination
How to improve	competition	traditional	traditional
	the output, Important Problems (Idai)		
Ideal	simplification	realization of traditional philosophy	technical training
Government	shogunate	dynasty	

References

1. Kim Yong Woon [金容雲] and Kim Yong Guk [金容局]: *History of Korean Mathematics* [韓国数学史], Maki Pub. Co. [槇書店], Tokyo (1978).
2. Kim Yong Woon and Kim Yong Guk: *History of Korean Mathematics*, Salim Math. Pub. Co., Seoul (2009).
3. Kim Yong Woon and Kim Yong Guk: *History of Chinese Mathematics*, Minumsa (1996).
4. Heizaemon Kato [加藤平左ェ門]: *Japanese Mathematics History (I, II)* [日本数学史 (上, 下)], Maki Pub. Co., Tokyo (1968).
5. Kazuo Shimodaira [下平和夫]: *History of Wasan (I, II)* [数学書を中心とした和算の歴史 (上, 下)], Fuji Junior College [富士短期大学] (1970).
6. Kinnosuke Ogura [小倉金之助]: *A Study of Mathematics History (I, II)* [数学史研究 (第一輯, 第二輯)], Iwanami Pub. Co. [岩波書店] (1971).
7. Yoshio Mikami [三上義夫], Hirayama, Ōya and Shimodaira ed. : *Japanese Mathematics in the Cultural History* [文化史上より見たる日本の数学], Koseisha Pub. Co. [恒星社厚生閣] (1974).

The Axes of Mathematical Methodology in the Song and Yuan Dynasties: The Construction of Mathematical Models

Guo Shirong*

Abstract Comparative researches between eastern and western mathematics of ancient and mediaeval times revealed that there are two major activities of mathematics, that is, theorem-proving and equation-solving. Theorem-proving mainly originated from Greek mathematics, meanwhile, equation-solving was an important content of Chinese mathematics. The fact seems to lead to the conclusion that Chinese mathematics was basically based on the practical problems, and in the mean time European one on geometrical problems. Based on analyses of Chinese mathematical works, the author of the present paper argues that Chinese mathematicians paid more attentions to the design and the construction of their geometrical models different from those introduced in living practices, in particular, in the 13–14th centuries. Using their geometrical models, they constructed their mathematical contexts and problems to meet their needs in displaying their mathematical ideas and in breaking through the limitation of practical problems. The research style of Chinese mathematics, therefore, changed in some way and strengthened its theoretical aspect.

1 Introduction

When traced back the original roots of the mathematics mechanism, Wu Wenjun (or Wu Wen-Tsün) [吳文俊], a top Chinese mathematician and historian of mathematics in China, held the point of view that:

There are two major activities of mathematics, that is, theorem-proving and equation-solving. Theorem-proving mainly originated from Greek mathematics, especially so-called Euclidean geometry. Meanwhile, equation-solving was an impor-

Guo Shirong
Institute for the History of Science and Technology, Inner Mongolian Normal University, Huhhot, 010022, P.R.CHINA
e-mail: gsr@imnu.edu.cn

* This paper is supported by Natural Science Foundation of China (10561006).

tant content of Chinese mathematics. Not like ancient Greek mathematics that devoted to theorem-proving, ancient Chinese mathematics paid more attention to solving equations, and equation-solving was the main thread of mathematics in ancient China (Wu [10, p. 3]).

This generalization of the characteristics of the traditional Chinese mathematics has been generally accepted by historians of Chinese mathematics.

Talking about equation-solving, it is natural to recall René Descartes' (1596–1650) Algebraization Rules for solving any problem, which were formulated in his famous treatise *La Géométrie* (1637) and *Rules for the Direction of Mind*. As is well known, Descartes' rules can briefly be summarized as follows (see Polya [7] or Li [5]):

> Firstly, translate any practical problem into a mathematical one;
>
> Secondly, reduce the mathematical problem to an algebraic one;
>
> Lastly, reduce the algebraic problem to system of algebraic equations, then to one equation with only one variable, and then get the solution by extracting its roots.

Wu Wenjun compared Descartes' program with traditional Chinese mathematics, and concluded: According to his *Geometrie*, Descartes obviously emphasized the method using equation-solving in solving geometric problems instead of the method using theorem-proving. This is identical with the spirit of Chinese mathematics. In other words, we can say that the ancient Chinese mathematics as a whole was developed along the way of Descartes' Program; in reverse, Descartes' Program can be regarded as a summary of the way of the development of ancient Chinese mathematics.

Hikosaburo Komatsu [小松彦三郎], a Japanese mathematician and historian of mathematics, hold the same point of views in his report at the XXII International Congress of History of Science in 2005:

> It is generally believed that the Modern Mathematics was started with Descartes' Algebraization. He reduces geometric problems to systems of algebraic equations for dimensionless numbers and then solves them by reducing the systems to algebraic equations of only one variable. This is, however, exactly the same method as Chinese mathematicians had adopted since long time before [2].

Basically, I myself agree with the above analyses and points of view, especially the point that Chinese mathematical thoughts were consistent with the spirit of Descartes. At the same time, we should notice that there appear some new characteristics in the algebraization method of the Song [宋] (960–1279) and Yuan [元] (1279–1368) dynasties. One of them is that the equations of higher degree are derived. A glance at the history of traditional mathematics in China, it can be found that the degree of equation that appeared in Chinese mathematics before the Song and Yuan dynasties is not higher than cubic and all equations are derived from practical problems, as we see in the Ten Classics of Mathematics [算経十書 *Suan-jing Shi-shu*]. The situation had changed since about 11th century. Higher degree equations were engaged very frequently in the works by Li Ye [李冶] (1192–1280), Qin

Jiushao [秦九韶] (1208–?) and Zhu Shijie [朱世傑], respectively. Even a 14th degree equation was derived[1] in Zhu's Jade Mirror of Four Elements [四元玉鑑 *Si-yuan Yu-jian*] published in 1303.

The fact suggests some historical questions. Were those equations indeed based on practical problems? What were their 'practical' foundations? Were practical problems the only source of mathematics in the period of Song and Yuan dynasties? Or, in what sense do we say Chinese mathematics is based on the practical lives?

To answer these questions, it is necessary to study the new characters of mathematical methodology in the period of the Song and Yuan dynasties, for new mathematical results are, generally speaking, always connected with new methods closely. Historians have noticed that there are some changes in methods and construction of mathematics in that period (Li Di [3, pp. 219–233]). In other words, the mathematical methodology had some new development based on both its former tradition and new creation.

In this paper, we will illustrate following point of view: During the Song and Yuan dynasties, especially after appearance of the algebraic heavenly element method algebra [天元術 *tian-yuan shu*], the traditional fields of practical problems, such as computations in area and volume, metrology, commerce, transportation, architecture, and so on, had not met the demands of mathematics development. Mathematicians urgently wanted new sources of problems to develop their new mathematical thought. They created some new mathematical models to generate new resources of mathematical problems. Based on an analysis on the mathematical treatises Sea-Mirror of Circle Measurement [測圓海鏡 *Ce-Yuan Hai-Jing*] written by Li Ye in 1248 and Jade Mirror of Four Elements by Zhu Shi Jie in 1303, it can be found that the core problems in both mathematical works were not related to practical applications, but to geometrical and mathematical models. Those models played important roles both in their mathematical researches and in constructions of mathematical problems. The Illustration of A Circle Town [圓城圖式 *Yuan-Cheng Tu-Shi*] in Li Ye's treatise and the Five Sums and Five Differences [五和五較 *Wu-He Wu-Jiao*] in Zhu's treatise are examples of such models. The models can be regarded as originating from practical problems, but essentially had lost their original meaning of practices.

2 Li Ye's Algebra and his Illustration of A Circle Town

2.1 The Academic Demand for Construction of Mathematical Models

Li Ye was a mathematician who was living in the time between the Jin [金] (1115–1234) and the Mongol-Yuan [蒙古-元] (1206–1368) dynasties. He wrote many

[1] See §3.2 of this paper.

books. Among his other treatises are Sea-Mirror of Circle Measurements (1248) and Geometrical and Algebraic Interpretation to the Treatise Yi-Gu [益古演段 *Yi-Gu Yan-Duan*] (1259). The former was an academic monograph, meanwhile, and the latter was designed for students. The common topic of both treatises is the heavenly element method algebra that was created in the northern China in the 12th century or the early 13th century. His predecessor mathematicians had developed the heavenly element method deeply. They could express equations with degree between 9th and −9th. Li Ye contributed to the development of the method a lot by summarizing and formularizing the method.

The so-called heavenly element method algebra is an algebraic program that consists of several steps. Firstly, setup an unknown called a heavenly element [天元 tian yuan]. Secondly, combine the unknown with the known elements to establish an equation of numerical coefficients. Thirdly, extract a root of the established equation and obtain the solution of the problem. The third step had in general been familiarized in his time so that he did not need to discuss it.

All ancient equations in China before Li Ye were related to practical problems. For instance, the equations that appeared in the Nine Chapters of the Mathematical Arts [九章算術 *Jiu-zhang Suan-shu*] or Collected Ancient Arithmetic Classics [緝古算経 *Ji Gu Suan-Jin*] (7th century) were all derived from problems of calculations of areas or volumes.

Till the 11th century, as the literally so-called unlock by establishing table [立成釋鎖 *li-cheng shi-suo*], a method of extracting roots of equation with higher degree, was developed from the traditional method of finding roots of quadratic and cubic equations, and a new method of extraction of roots by addition and multiplication [增乘開方法 *zeng-ceng kai-fang fa*] was formalized, equations with higher degrees became wanted academically, for equations of quadratic and cubic degrees did not meet the expressions and applications of the new methods. As we know, equations derived from most practical problems are not beyond cubic degree, or we can say from traditional applied problems it is difficult to derive an equation of higher degree. Jia Xian [賈憲] and Liu Yi [劉益], both mathematicians in the 11th century, developed new methods of extraction of roots, but they did not derive equations of higher than the third degree from practical problem except one quartic equation in Liu's treatise. To construct an equation of higher degree, Qin Jiushao intended to derive a 10th degree equation from a surveying problem in 1247. Because the equation could be expressed by a cubic equation, Qin was criticized by historians of mathematics [8, p. 103] as flubdub and too ambitious to reach for a target beyond his grasp.

Li Ye faced the same situation as Qin did. He eagerly wanted to construct higher degree equations to explain his algebraic method. It was of great importance for him to design new mathematical models which could help him get more equations of higher degree. The Illustration of a Circle Town, therefore, was designed.

2.2 Illustration of a Circle Town—a Useful Mathematical Model

Scholars in the field of history of Chinese mathematics are familiar with the Illustration of a Circle Town [圓城圖式 *Yuan-cheng Tu-shi*] and study it a lot. we will explain its meaning in a new sense. Its structure is as follows (see Fig.1):

1. A right-angled triangle $\triangle ABC$ with its inscribed circle with center O.
2. DE and FG are two perpendicular diameters parallel to the two legs of the $\triangle ABC$. The circle is also inscribed in the square JKLB.
3. From intersection points I, N, M and H draws lines IP, NS, MQ, HR parallel to the legs of the triangle $\triangle ABC$ respectively.

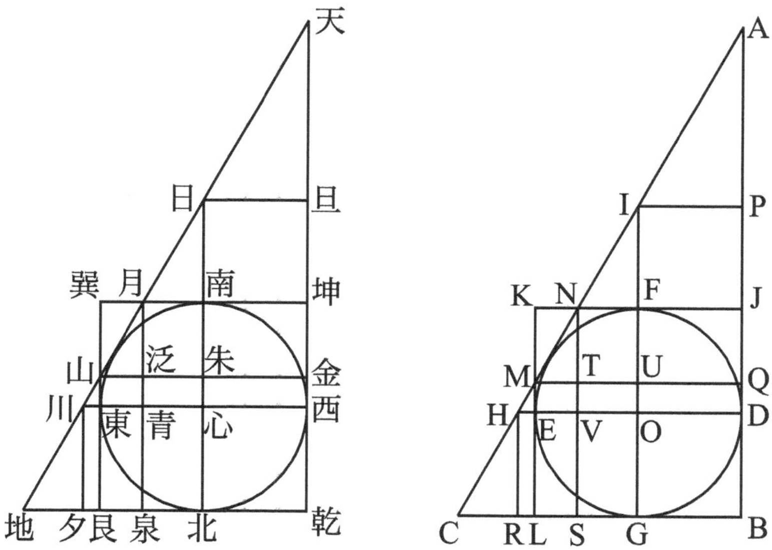

Fig. 1 Illustration of a Circle Town (*Yuan-cheng Tu-shi*)

In this way, Li Ye obtained the Illustration of a Circle Town which consists of a circle and 15 right-angled triangles. Although Li Ye said he was suggested by the nine formulae which he learned from Master Dong Yuan [洞淵], the geometric figure is actually derived from some figures that appeared in the Nine Chapters of Mathematical Arts, such as figures about the problems of the so-called a right-angled triangle containing a circle [勾股容圓 *gou-gu rong-yuan*], a right-angled triangle containing a square [勾股容方 *gou-gu rong-fang*], and a square town [邑方 *yi-fang*]. The circle in the diagram is regarded as a town with gates D, E, F and G. Hence its name comes.

2.3 The Meaning of the Illustration of a Circle Town

Li Ye made use of the figure with two purposes in his mind. One is geometrical and the other algebraic.

Geometrically, 15 similar triangles can be named in the diagram, among which two pairs are congruent. Thus, 13 triangles are the remainder. Each of them has three lines, or two legs [勾 *gou*] and [股 *gu*] and a hypotenuse [弦 *xian*]. Starting from the relations among the lines and the diameter of the circle, Li Ye got 692 identities about the lines. He engaged many kinds of reasoning methods such as geometrical proofs, analogy, parallel, number theory, and so on to derive the identities [1, pp. 122–140]. All of his results belong to geometry. The diagram, therefore, can be regarded as a geometrical model.

Algebraically, Li Ye made use of the diagram as a model to construct all kinds of problems and then to establish his equations with higher degree. Here is an example:

Ask: People C goes out of the southern gate of the town and goes forward by 135 steps, and stops. People A goes out of the eastern gate. After going forward by 16 steps, he can see C. What is the diameter of the town? [或問: 丙出南門直行一百三十五步而立, 甲出東門東行一十六步見之, 問城徑幾里?].

Li Ye designed five different approaches to the problem. In the fifth one he finally reduced to the sixth degree equation:

$$-2x^6 - 714x^5 - 62156x^4 - 2220302x^3 + 82926x^2 + 1725602816x + 51336683776 = 0.$$

In the same way, Li Ye constructed 160 problems which consisted of the main contents of the Sea-Mirror of Circle Measurement. Basically speaking, the figure was used rather as a model for constructing and designing mathematical problems than as a practical model. If we would like to identify any relations between this figure and practice, there are only some apparent terms such as town, gates, and the name of the diagram.

We would like to point out two facts:

1. By means of the model, Li Ye constructed his mathematical background, his mathematical contexts and mathematical problems which were necessary for his study and his construction of mathematical theory.
2. The figure was not very practical although it was connected with some practical terms. In practice, those kinds of calculations are unnecessary. It was, therefore, rather a mathematical model than a practical model.

He designed such a model not for the need of solving problems from living but for the need of construction of mathematical theory.

3 Zhu Shijie's Algebra and the Five Sums and Five Differences of Right-angled Triangle

3.1 From the Heavenly Element to the Four Elements

Since the time when Li Ye formulized the theory of heavenly Element method algebra , and especially after Sea-Mirror of Circle Measurement was published in 1282, the idea of this theory was developed rapidly. There used to be only one unknown in the algebra of heavenly element. Then the number of unknowns was added one by one, from one to two, then to three. Finally, Zhu Shijie added the fourth unknown to the algebraic systems. We know little about the course of the development between Li Ye and Zhu Shijie. The only material we can refer to is the preface by Mo Ruo [莫若] to the Jade Mirror of Four Elements in 1303.

The expressions of unknowns and equations in the four elements algebra were similar to the former. An equation with only one unknown was expressed by putting its coefficients vertically from the lowest term to higher term one by one. With the increase of the numbers of unknowns, the four directions of a plane were engaged in the expression of the equation with more unknowns. It is obvious that the spirit of the expressions for different unknowns is consistent. Because there are only four directions in a plane, it is obvious that only four unknowns could be expressed in a plane in the above method. When more than four unknowns were needed, Zhu Shijie had to set new unknowns for the second time [9]. The details of the four elements method are narrated in almost all treatises on the history of mathematics in China. We will not talk about it any more.

Before saying too much, we would like to remark that Zhu Shijie was at the same situation as Li Ye. He wanted a context under which he could construct more mathematical problems to explain his algebraic method. He also found resource in the traditional mathematical works. The resource is the operations of the three lines of a right-angled triangle which is traditionally called arithmetic for right-angled triangles [勾股算術 *gou-gu suan-shu*] [6, pp. 391–396].

3.2 The Model of the five Sums and five Differences of a Right-angled Triangle

Chinese mathematicians had studied the right-angled triangle long before Zhu Shijie. It was one of the objects of the Mathematics of Gnomon in Zhou [周髀算経 *Zhou Bi Suan-jing*] which was an astronomical work written about 2nd century B.C. and Nine Chapters of the Mathematical Arts. The right-angled triangle theorem [勾股之法 *gou-gu zhi fa*], or Pythagorean theorem, was proved and applied in Mathematics of Gnomon in Zhou. According to the record of Gnomon in Zhou, Shang Gao [商高] taught the theorem to Zhou-Gong [周公], the brother of the first king of the Western Zhou Dynasty in the ninth century B.C.

In the third century, mathematician Zhao Shuang [趙爽] studied Mathematics of Gnomon in Zhou and added his comments to the book. He wrote a paragraph titled Comments to the Diagram of Right-angled Triangle with its Inscribed Circle and Square [勾股方圓圖注 *Gou-gu fang-yuan tu-zhu*] and drew a diagram of hypotenuse [弦圖 *xian-tu*] to discuss the calculations and operations of the two legs and hypotenuse. There are also similar discussions in the Nine Chapters of the Mathematical Arts. Li Ye also engaged the Five Sums and Five Differences in his Sea-Mirror of Circle Measurement.

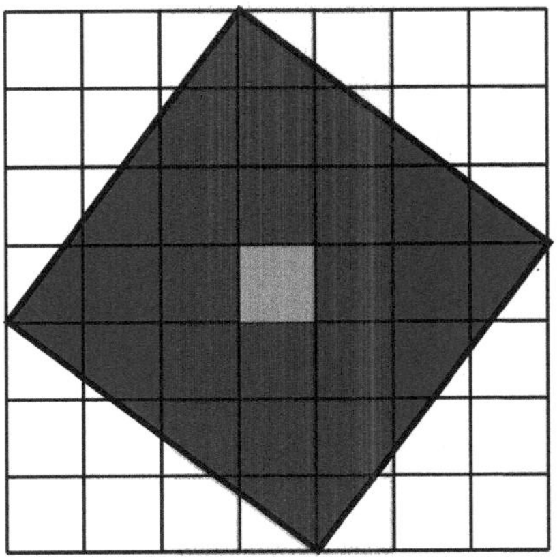

Fig. 2 Diagram of Hypotenuse

Based on the above studies, Zhu Shijie formularized his model:
Let the two legs be a and b, and the hypotenuse be c. The Five Sums are

$$a+b,\ \ a+c,\ \ b+c,\ \ a+b+c \text{ and } c+(b-a)$$

and the Five Differences

$$b-a,\ \ c-a,\ \ c-b,\ \ (b+a)-c \text{ and } c-(b-a).$$

Zhu Shijie constructed his mathematical problems by combining the Five Sums and Five Differences with their operations, addition, subtraction and multiplication, and then established higher degree equations which he needed. It is obvious that the problems constructed in this way are of no practical meanings. The Five Sums and Five Differences are totally a mathematical model rather than a practical model. Because of their importance in the Jade Mirror of Four Elements, Zhu put the "diagram of self-multiplying of the five sums" [五和自乘演段之圖 *wu-he zi-cheng yan-duan zhi-tu*] (see Fig.3) and the "diagram of self-multiplying of the five differences" [五較自乘演段之圖 *wu-jiao zi-cheng yan-duan zhi-tu*] (see Fig.4) at the beginning of the treatise. In another important "diagram of self-multiplying of the four elements"

[四元自乘演段之圖 *si-yuan zi-cheng yan-duan zhi-tu*] (see Fig.5) the four elements are also represented by the three sides of right-angled triangle a, b, c and the yellow square side [黄方] $(a+b-c)$, that is, the diameter of the inscribed circle.

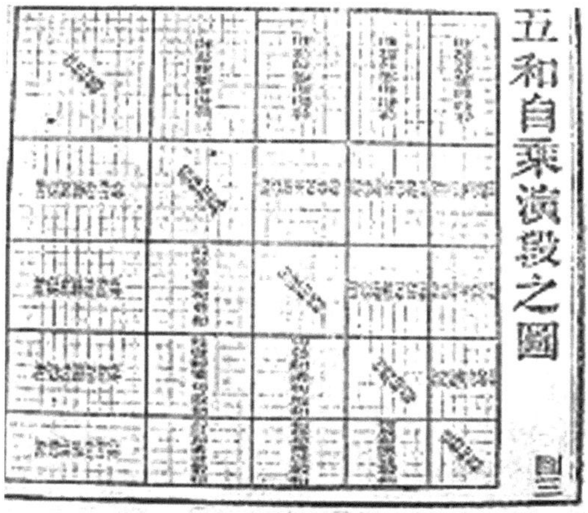

Fig. 3 Diagrams of Self-Multiplying of the Five Sums

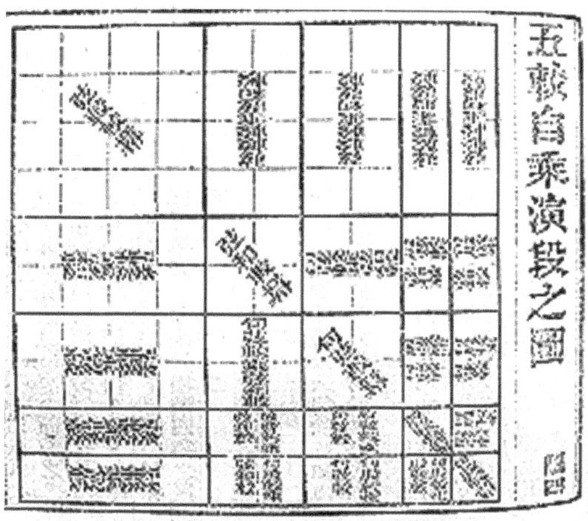

Fig. 4 Diagrams of Self-Multiplying of the Five Differences

Making use the sums and differences, Zhu could construct many problems which lead to higher degree equations. Here is the last problem among 288 problems in his book:

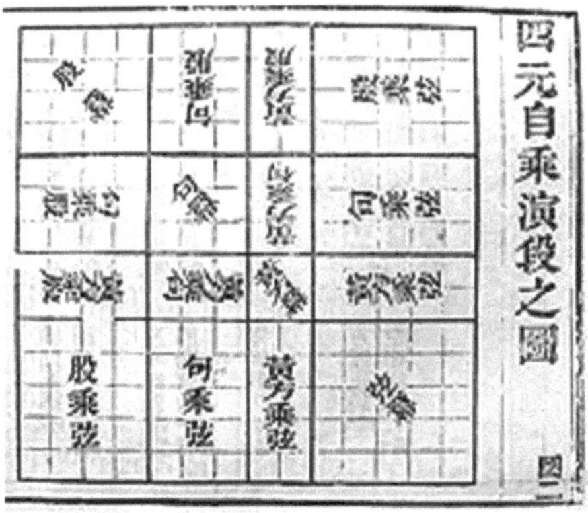

Fig. 5 Diagrams of Self-Multiplying of the Four Elements

There is an unknown number. We know that it is a root of the equation of the fourth degree in which the coefficient of the constant term is equal to the sum of the squares of the five differences, that of the first degree -3, that of the second degree 1, that of the third degree 1, and that of the fourth degree -2. Also know that subtracting twice the product of the leg gou and the leg gu from the square of the sum of gou and gu and then adding the sum of the three sides of the triangle, we get a number equal to the square of the self-multiplying of the unknown number adding to the square of hypotenuse and subtracting the leg gu. Again, adding half of the sum of the three sides of the triangle to the diameter of the inscribed circle in the triangle we have the cubic of the number. Ask: what is the number? [今有一数, 不知多少, 但言五較各自乘并之, 為正實, 以三為益方, 一為縱上廉, 一為縱下廉, 二為益隅, 三乘方開之, 与其数相等. 只云: 勾股和冪減二直積, 加三相和; 与其数冪自乘并弦冪減股相同. 又云: 半之三相和加黄方, 与其数再自乘亦等. 問元数幾何?]

The solution: let x be gou (shorter leg, a), y be gu (longer leg, b), z be xian (hypotenuse, c), w be the wanted number, the known conditions are:

$$-2w^4+w^3+w^2-3w+(y-x)^2+(z-x)^2+(z-y)^2 +(y+x-z)^2+(z-y+x)^2=0, \quad (1)$$

$$(x+y)^2-2xy+(x+y+z)=w^4+z^2-y, \quad (2)$$

$$\frac{1}{2}(x+y+z)+(x+y-z)=w^3, \quad (3)$$

$$x^2+y^2-z^2=0. \quad (4)$$

Reducing the system of simultaneous equations, Zhu obtained:

$$2006w^{14}-11112w^{13}+22292w^{12}-19168w^{11}+2030w^{10}+12637w^9-8795w^8 -8799w^7+19112w^6-9008w^5-384w^4+1792w^3-640w^2-768w+1152=0.$$

Extracting its root, he gets the answer: $w = 2$.

This is one example of many problems in the Jade Mirror of Four Elements. From the example it can be understood how Zhu Shijie made use of the five sums and differences in construction of mathematical problems. Again, the model is not related to the application in practices. Therefore, Zhu created his mathematical contexts and problems by means of his model rather than practices. In this way he designed problems in his book which derived 124 equations of fourth or higher than fourth degree [4, p. 307]. It should be pointed out that Zhu also engaged other models in his book.

4 Re-evaluation of Qin Jiushao's Algebra

Qin Jiushao was a mathematician in the thirteenth century. We have remarked that Qin also wanted equations of higher degree, like Li Ye and Zhu Shijie, to demonstrate his method of solving numerical equations, and it was also difficult for him to construct appropriate mathematical problems, although there appeared several equations of higher degree in his treatise. Because of the lack of suitable mathematical models, he had to increase the degree of his equation by squaring two sides of an equation or other operations. For example, from a problem named Circle-town measurement from remoteness [遥度圆城 *yao-du yuan-cheng*] he managed to establish an equation of the 10th degree in his book, meanwhile Li Ye derived an equation of only third degree. Most historians didn't understand Qin Jiushao and criticized him seriously.

The above analysis of Li Ye's and Zhu Shijie's models help us very much for understanding Qin Jiushao. In fact, in my opinion, Qin had to do that, otherwise, he could not derive appropriate method to get equations of higher degree, for all problems which appeared in his book were based on living practices. This is why he complicated his operations in establishment of equations.

5 Concluding Remarks

Both models discussed above originated from the traditional studies on right-angled triangles and belonged to the field of traditional arithmetic for right-angled triangles. What both Li Ye and Zhu Shijie were concerned with were not the models themselves but their functions of generating mathematical problems. The mathematicians afterward accepted the two mathematical models and studied them very much. Many mathematicians were influenced. They studied both models very much, especially during the 18th and 19th centuries. Mathematicians of Qing [清] dynasty (1644–1911) added two lines to the Illustration of a Circle Town, one is the perpendicular to the hypotenuse starting from the center of the circle, the other the parallel line to the hypotenuse and passing through the center of the circle (see Fig.6), which enlarge the function of the model. Korean mathematicians also did the same. The model of Five Sums and Five Differences was also engaged by mathematicians such as Gu Yingxiang [顧應祥] (1483–1565), Mei Wending [梅文鼎] (1633–1721) and others.

From above discussion we come to conclusion as follows:

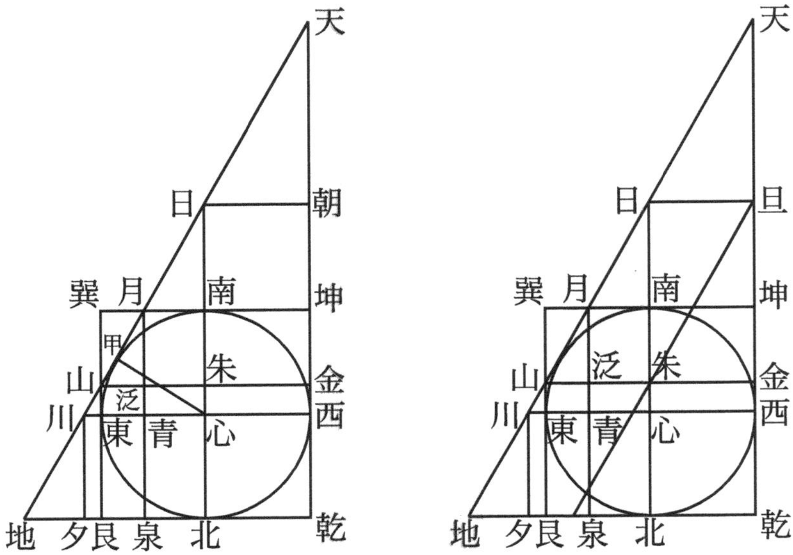

Fig. 6 Revised illustration of a circle town

Firstly, it is one of the characteristics of traditional mathematics to pay more attention to practice and abstract mathematical problems from practices, but this is not the whole of the traditional Chinese mathematics. The establishment of the models of Li Ye and Zhu Shijie makes it out that the theoretical aspect has not been studied sufficiently and should be paid more attention by historians of mathematics. Chinese mathematics had changed its style in some way during the Song and Yuan dynasties and strengthened its theoretical aspect.

Secondly, during the 13th and 14th centuries, Chinese mathematicians paid great attentions to practical problems, but practical problems did not satisfy the needs of their research works and their thinking. They turned their attentions to construct new mathematical models different from living practices. Discovery of new models provided mathematicians with new problems and new thought and, therefore, promoted the development of mathematics.

Lastly, Li Ye's and Zhu Shijie's models are based on geometry. The program of problem-solving in the mathematics during the 13th and 14th centuries coincides with the spirit of Descartes' program.

References

1. K. Chemla [林力娜], translated by Guo Shirong [郭世栄]: The Textual Structure and Expression of Mathematical Knowledge of Li Ye's "Sea-Mirror of Circle Measurement" [李冶《測円海鏡》的版本結構与数学知識的表述], *Collected Papers on History of Mathematics* edited by Li Di [李迪], vol. 5 [数学史研究文集, 第五輯], pp. 122–140 (1993).
2. Hikosaburo Komatsu [小松彦三郎]: Zhu Shijie, the Teacher of Seki and Takebe, *Book of Abstracts, XXII International Congress of History of Science, Beijing* [第 22 届国際科学史大会論文摘要集 北京], p. 290 (2005).
3. Li Di [李迪]: Changes of Mathematical Structure in the Song and Yuan Periods [宋元時期数学形式的転変], *Selected Papers on History of Science and Technology in China*, vol. 1 [中国科学技術史論文集 (一)], Inner Mongolia Education Press [内蒙古教育出版社], pp. 219–233 (1991).
4. Li Di [李迪]: *General History of Mathematics in China, vol. Song and Yuan Dynasties* [《中国数学通史》宋元卷], Jiang-su Education Press [江蘇教育出版社], Nanjing (1999).
5. Li Wenlin [李文林], Mechanization Characteristics of Descartes' *La Geometry* [笛卡儿《幾何学》的機械化特征], *Studies in History of Natural Sciences* [自然科学史研究], 12, pp. 225–234 (1993).
6. Liu Dun [劉鈍]: *Great Discussion on Mathematics* [大哉言数], Liao Ning Education Press [遼寧教育出版社], Shenyang (1993).
7. G. Polya, *Mathematical Discovery* [数学發現], vol. 1, John Wiley & Sons, 1962.
8. Qian Baocong [錢宝琮]: Study of Qin Jiushao's "Treatise of Mathematics in Nine Chapters" [秦九韶《数書九章》研究], *Collected Papers on History of Mathematics of Song and Yuan Dynasties* [宋元数学史論文集], Science Press [科学出版社], pp. 60–103 (1966).
9. Wang Yanyu [王艶玉]: Zhu Shijie's method of multiple setting of the heavenly elements [朱世杰的 "多次立天元法"], *Collected Papers on History of Mathematics*, vol. 1 [数学史研究文集, 第一輯], Inner Mongolia University Press [内蒙古大学出版社与九章出版社], pp. 83–93 (1990).
10. Wu Wen-tsun [呉文俊]: *Mathematics Mechanization: Mechanical Geometry Theorem-Proving, Mechanical Geometry Problem-Solving and Polynomial Equation-Solving*, Science Press, Beijing / Kluwer Academic Publishers, Dordrecht (1997).

The *Suanxue Qimeng* and Its Influence on Japanese Mathematics

Mitsuo Morimoto

Abstract The *Suanxue Qimeng* (the Introduction of Mathematics) was written by Zhu Shijie in 1299 during the *Yuan* dynasty. The book had been lost in the *Ming* dynasty but was conserved in Korea and reprinted several times as mathematical textbook for the education of mathematical experts of the Korean dynasty. In the last decade of the 16th century a copy of the book was transferred to Japan. In the *Edo* period in Japan, the book was reprinted several times with annotation. The last chapter of the book explained the procedure of celestial element (*tianyuanshu* or *tengen jutsu*), a method for representing a polynomial of one variable with numerical coefficients by a column vector of its coefficients. Having mastered this procedure, Seki Takakazu generalized it in such a way that polynomials of several variables could be handled with.

1 Chinese sources of Japanese mathematics

The *wasan*, or Japanese traditional mathematics, was flourishing in Japan during the *Edo* period (1603–1867), on the basis of the Chinese traditional mathematics stemmed from the Nine Chapters on the Mathematical Art [九章算術 Kyūshō Sanjutsu] [11], which dated back at least to the first century, the *Han* Dynasty.

The three Chinese monographs helped Japanese mathematicians to develop their own mathematics: the Systematic Treatise on Arithmetic [算法統宗 Sanpō Tōsō], the Yang Hui's Methods of Mathematics [楊輝算法 Yōki Sanpō], and the Introduction to Mathematics [算学啓蒙 Suanxue Qimeng, Sangaku Keimō]. Among these books the most influential on Japanese mathematicians was the Introduction to Mathematics, from which they learned the procedure of celestial element [天元術 tianyuanshu, tengen jutsu].

Mitsuo Morimoto
1-64-5, Wada, Suginami-ku, Tokyo 166-0012, Japan
e-mail: mormormor@jcom.home.ne.jp

1.1 Early years of the wasan

In the *Edo* period, Japan was secluded from the world by her own foreign policy. The *wasan* was investigated in the island almost independently of foreign influences. Of course, the main source of the Japanese mathematics was Chinese, but there is some speculation that there might be a Western influence through Christian missionaries.

The Portuguese drifted to Tanegashima island in 1543. This incident was the first contact of Japan with European powers. Then F. Xavier, a Jesuit missionary, arrived in Kagoshima in 1549. In 1580's Jesuits were authorized to organize a "collegio" in Azuchi, capital of Japan of the time, for a few years, where mathematics was one of courses. Mōri Shigeyoshi [毛利重能] published the Book on Division [割算書 Warizan-sho] (1622), the preface of which contained a distorted story of biblical subjects, although the book treated the traditional mathematics of every day life. Yoshida Mitsuyoshi [吉田光由], a disciple of Mōri, published the Book on Things Small and Large [塵劫記 Jinkōki] (1627), more than 300 versions of which were repeatedly reproduced during the *Edo* period. The main source of this Japanese best-seller on merchant mathematics was the Systematic Treatise on Arithmetic, which was written by Cheng Dawei [程大位] (1533–1606) in 1592 during the *Ming* dynasty. The Systematic Treatise on Arithmetic explained the calculation on abacus, which was invented for commercial use at the time and spread quickly in East Asia.

1.2 Zhu Shijie and his two books

Zhu Shijie [朱世傑] was a Chinese mathematician of the *Yuan* dynasty (1206–1368). He published two books, the Introduction to Mathematics in 1299, and the Jade Mirror of the Four Elements [四元玉鑑 Shigen Gyokukan] in 1303.

The Introduction to Mathematics consists of three parts divided into twenty chapters forwarded by Preface and Introduction. In China, this book is not appreciated so much as the Jade Mirror of the Four Elements; the former is introductory and elementary while the latter is considered as the culmination of the development of Chinese mathematics during the *Song* and the *Yuan* dynasties.

The last chapter of the Introduction to Mathematics is devoted to the procedure of celestial element, on which we shall discuss later in this paper. This is a way to handle polynomials and algebraic equations of one variable with integer coefficients, while the Jade Mirror of the Four Elements developed a method to handle certain kinds of algebraic equations of four variables. Therefore, the latter book is usually evaluated higher than the former in the history of Chinese Mathematics. Both books disappeared in China during the *Ming* dynasty (1368–1644) and were re-found around 1800 in the *Qing* dynasty.

Because the Introduction to Mathematics was a systematic treatise of mathematics starting with the four rules of arithmetic, it was chosen as an important textbook for mathematics students in the Korean *Yi* dynasty (1392–1910). King Seajong the Great [世宗] (1418–1450) himself learned the book. The first Korean print of the book was published around 1450 with the copper font of the *Geng-wu* [庚午] year. One copy of this edition survived until now and was discovered recently. Then the second print was published during the reign of King Seongjong [成宗] (1470–1494) and the third print during the reign of King Jungjong [中宗] (1506–1544), both with the copper font of the *Yi-hai* [乙亥] year. Only the middle volume of these editions survived in Korea, while one reprint of the third edition was transferred to Japan in the late 16th century, possibly during the Japanese expeditions to Korea, 1592–1598. This reprint has been conserved in Tsukuba University, Japan. Later around 1660 the Introduction to Mathematics was reprinted in Korea with Preface by Jin Shizhen [金始振]. This edition was later revised in the *Yi-wei* [乙未] year and published in the *Geng-wu* [庚午] year, that is, 1810. This last edition was the basis of the Chinese edition of 1839 by Luo Shilin [羅士琳] and all other later Chinese editions. For the details, we refer the reader to [8].

1.3 Acceptance of the Introduction to Mathematics in Japan

In the *Edo* period, Haji Dōun [土師道雲] and Hisada Gentetsu [久田玄哲] reprinted the original book with Japanese reading signs in 1658. At first, Japanese mathematicians could not understand the procedure of celestial element. For example, Satō Masaoki [佐藤正興], the author of the Origin of Mathematical Methods [算法根源記 Sanpō Kongen-ki] (1669) and Sugiyama Sadaharu [杉山貞治], the author of the Elementary Introduction to Mathematical Methods [算法発蒙集 Sanpō Hatsumō-shū] (1670) took the procedure of celestial element as one of the most advanced method imported from China and used the terminology to give their books some authority without knowing its real meaning. Later, Hoshino Sanenobu [星野実宣] published the New Commentary on the Introduction to Mathematics [新編算学啓蒙註解 Shinpen Sangaku Keimō Chūkai] (1672). In this commentary, Hoshino's annotations were obscure and it seems to me that he could not understand the real meaning of the procedure of celestial element.

The first Japanese mathematician who could understand the procedure of celestial element was Hashimoto Masakazu [橋本正数] of Osaka. His disciple Sawaguchi Kazuyuki [沢口一之] published a book called the Mathematical Methods Old and New [古今算法記 Kokon Sanpō-ki] (1671). In this book, Sawaguchi applied the procedure of celestial element correctly to many problems and successfully solved them. At the end of the book, he gave fifteen challenge problems [遺題 idai] which could not be solved by the procedure of celestial element.

Having read the Introduction to Mathematics, Seki Takakazu [関孝和] (ca. 1642–1708) also mastered the procedure of celestial element, generalized it, applied his generalized method to Sawaguchi's fifteen problems, and published his solutions in the Mathematical Methods of Exploring Subtle Points [発微算法 Hatsubi Sanpō] (1674). As this book gave only the final solution of the fifteen problems, there arose some doubt about the correctness of the solutions. Then Takebe Katahiro [建部賢弘] (1664–1739), a disciple of Seki Takakazu, published the Colloquial Commentary on

Operations (in the Mathematical Methods for Exploring Subtle Points) [発微算法演段諺解 Hatsubi Sanpō Endan Genkai] (1685) (See [12] for an English translation.) in four parts and revealed all the secrets of Seki's new algebraic method.

Takebe Katahiro was one of the greatest Japanese mathematicians of this epoch. He published the Complete Colloquial Commentary (on the Introduction to Mathematics) [算学啓蒙諺解大成 Sangaku Keimō Genkai Taisei] (1690) in seven parts, which was almost two times larger than the original book and contained detailed explanation on the procedure of celestial element. With Takebe's Complete Colloquial Commentary Japanese mathematicians could understand Zhu Shijie's Introduction to Mathematics fully and obtained the algebraic foundation for the *wasan*.

For lives of Seki and Takebe and their mathematics, see Horiuchi [3]. We also refer the reader to [12], which contains English translation of Takebe's three mathematical monographs and his short biography in English.

1.4 The Complete Colloquial Commentary on the Introduction to Mathematics

While the Introduction to Mathematics consists of four parts, Summary, Upper volume, Middle volume, and Lower volume with total 137 sheets, Takebe's Complete Colloquial Commentary consists of seven parts; Summary, Upper first, Upper second, Middle first, Middle second, Lower first, and Lower second volumes with total 219 sheets. In the Complete Colloquial Commentary, Takebe's annotation is printed with half-sized characters, Takebe's Complete Colloquial Commentary is more than two times of the original book. (Note that one sheet consists of two pages.)

	Introduction to Math.	C. C. Commentary
Preface	2 sheets	2 sheets
Table of Contents	1 sheets	1 sheets
Summary	7 sheets	13 sheets
Upper Volume	35 sheets	54 sheets
Middle Volume	44 sheets	62 sheets
Lower Volume	48 sheets	87 sheets
Total	137 sheets	219 sheets

The Japanese name of the Complete Colloquial Commentary is *the Genkai Taisei*, where *genkai* means the colloquial explanation and *taisei* the great accomplishment. In Japan, the Chinese classics used to be read literally word by word, but Takebe tried to annotate the original text in colloquial Japanese. Note that the Japanese in Takebe's commentary was written using *kata-kana* and Chinese characters. This means Takebe's readers were supposed to belong to the "warrior" (*samurai*) class, while the Book on Things Small and Large was written using *hira-kana* and Chinese characters and widely used in private primary schools (*terakoya*) for the merchant class.

The last chapter of the Introduction to Mathematics named Chapter for Extracting the Root [開方积鎖門 Kaihō Sekisamon] is composed of 34 problems. The first seven problems deal with the procedure for extracting the root [開方術 kaihō-jutsu] and the other 27 problems concern with the procedure of celestial element.

In order to explain the Takebe's understanding of the procedure of celestial element, first we have to explain the counting tools of the *wasan*, i.e., the counting-rods and the counting board.

2 Counting board algebra

2.1 Counting-rods

The *wasan* is in the stream of Chinese traditional mathematics, where the numbers are, basically, natural numbers represented decimally by counting-rods. There are two ways to represent numbers; in the orders of 1, 100, 10^4, etc. the counting-rods are placed vertically on the counting board, while in the order of 10, 10^3, 10^5, etc., they are placed horizontally.

	1	2	3	4	5	6	7	8	9
Order of 1, 100, 10^4, ⋯	𝍠	𝍡	𝍢	𝍣	𝍤	𝍥	𝍦	𝍧	𝍨
Order of 10, 10^3, 10^5, ⋯	𝍩	𝍪	𝍫	𝍬	𝍭	𝍮	𝍯	𝍰	𝍱

There are two kinds of counting-rods, red and black. The red rods represent positive numbers and the black rods negative numbers. If one has to write numbers on paper with black ink, negative numbers are written with oblique line.

1 2 3 4 5 6 7 8 9

Negative numbers 

If there is no counting-rods on the counting board, it means the digit is 0. On the paper, the empty digit is represented by the sign ○. For example, 𝍫𝍠○𝍡𝍮𝍧 represents 310268.

The notion of positive and negative numbers had been clearly established. But their operation was complicated and required some techniques because the numbers were closely related with counting tools like counting board, counting-rods, or abacus.

For example, the addition of integers was called if same, add; if different, subtract [同加異減 dōka-igen]. This means that if two numbers are represented by the rods of the same color, we add and that if two numbers are of different color, we subtract. As in the case of abacus, the addition and the subtraction of natural numbers were fundamental operation; the notion of addition of integers were secondary operation and thus required some explanation.

Similarly, the subtraction of integers was called if same, subtract; if different, add [同減異加 dōgen-ika].

This rule had been known, since the Nine Chapters on the Mathematical Art, as the sign rule [正負術 seifu-jutsu] (p.404, [11]) and was repeated by Zhu Shijie in the Introduction to Mathematics. Takebe recognized its importance and, in Summary of the Complete Colloquial Commentary, he stated the rule of addition and subtraction of positive and negative numbers and zero with great care.

2.2 Counting Board

The counting board looks like the following: The rows of the counting board are named, from top to bottom, Quotient [商 shō], Reality [実 jitsu], Square [方 hō], Side [廉 ren], and Corner [隅 gū].

10^3	10^2	10	1	10^{-1}	10^{-2}	10^{-3}	
							Quotient
							Reality
							Square
							Side
							Corner

The counting board was used in many kinds of calculation, the most important of which was the extraction of root. For example, an algebraic equation of order 3 with numerical coefficients

$$a_0 + a_1x + a_2x^2 + a_3x^3 = 0 \tag{1}$$

was represented on the counting board in the *wasan*. The constant term a_0 was placed in the Reality row, a_1 in the Square row, a_2 in the Side row, and a_3 in the Corner row; that is, algebraic equation (1) was represented by the configuration of counting-rods on the counting board:

$$\begin{bmatrix} a_0 \\ a_1 \\ a_2 \\ a_3 \end{bmatrix} \tag{2}$$

Problem No. 1 in Chapter for Extracting the Root of the Introduction to Mathematics reads as follows:

> There is a square shaped area of 4096 *bu* [squared]. (*bu* is a unit for length.) Question: how much is one side? Answer: 64 *bu*.

The equation to be extracted [開方式 kaihō-shiki]

$$4096 - x^2 = 0 \tag{3}$$

was represented on the counting board with counting-rods as follows:

	10^3	10^2	10	1	10^{-1}	10^{-2}	10^{-3}	
								Quotient
Red rods	𝍬		𝍱	𝍥				Reality
								Square
Black rods				𝍠				Side
								Corner

An equation of any order could be solved numerically by the procedure for extracting the root. In the sequel, we treat only a cubic equation for simplicity.

We regard the Reality row A_0, the Square row A_1, the Side row A_2, and the Corner row A_3 as memories in a computer and the coefficients of (1) a_0, a_1, a_2, and a_3 as values of A_0, A_1, A_2, and A_3. If we place a value q in the Quotient row Q, then the calculation is done on the counting board as follows: (we use here the BASIC-like language)

$$A_2 = A_2 + A_3 \times Q, \quad A_1 = A_1 + A_2 \times Q, \quad A_0 = A_0 + A_1 \times Q,$$
$$A_2 = A_2 + A_3 \times Q, \quad A_1 = A_1 + A_2 \times Q,$$
$$A_2 = A_2 + A_3 \times Q.$$

Let us denote by a_0', a_1', a_2', and a_3' the values of A_0, A_1, A_2 and A_3 after these operations. Then we have

$$a_0 + a_1 x + a_2 x^2 + a_3 x^3 = a_0' + a_1'(x-q) + a_2'(x-q)^2 + a_3'(x-q)^3.$$

Further, if we add q' to q in Q, then the values a_0'', a_1'', a_2'', and a_3'' in

$$a_0 + a_1 x + a_2 x^2 + a_3 x^3 = a_0'' + a_1''(x-q-q') + a_2''(x-q-q')^2 + a_3''(x-q-q')^3.$$

can be calculated from a_0', a_1', a_2', and a_3' by the same program. If we can make the value in the Reality row empty, i.e., zero, after several operations, the value $q + q' + \cdots$ in the Quotient row becomes a root of the equation. Usually, the root is sought in this way, digit by digit from the top digit. This is the principle of the procedure for extracting the root. This principle was well known in Chinese traditional mathematics since the age of the Nine Chapters on the Mathematical Art. In Summary of the Introduction to Mathematics Zhu Shijie described succinctly the procedure for extracting the root saying

> "Place the product in the Reality row and operate in Square, Side, Corner rows adding if same and subtracting if different."

Zhu Shijie explained also this procedure in Chapter for Extracting the Root and Takebe, in the Complete Colloquial Commentary, commented further how to manipulate counting-rods in the procedure for extracting the root.

2.3 Counting Board Algebra

The procedure of celestial element can be said, in today's terminology, a method to represent a polynomial

$$a_0 + a_1 x + a_2 x^2 + a_3 x^3 \tag{4}$$

by a configuration of the counting board $\begin{bmatrix} a_0 \\ a_1 \\ a_2 \\ a_3 \end{bmatrix}$. Because, in the procedure for extracting the root, the same configuration represents algebraic equation (1), it was difficult, psychologically, to admit this ambiguity of the meaning of a configuration. For example, making the Reality row empty and placing one rod in the Square row, we form the following configuration:

$$\begin{bmatrix} \bigcirc \\ | \end{bmatrix}. \tag{5}$$

In the procedure for extracting the root, configuration (5) represents the equation $x = 0$ but in the procedure of celestial element same configuration (5) represent a virtual number x. To make configuration (5) on the counting board was called "to place the celestial element unit" in the *wasan*.

Let us examine how the argument goes in the procedure of celestial element. Problem No. 8 in Chapter for Extracting the Root of the Introduction to Mathematics reads as follows:

> There is a rectangular rice field of area 8 *mu* 5 *fen* 5 *li* [squared]. Given: the sum of the length and width is 92 *bu*. Question: how much is the length and width respectively? Answer: width 38 *bu*, length 54 *bu*.

Because 1 *mu* is equal to 240 *bu* [squared], 8 *mu* 5 *fen* 5 *li* $= 8.55 \times 240 = 2052$ *bu* [squared].

Zhu wrote as follows:

> Method of Solving: Place the celestial element unit $\begin{bmatrix} \bigcirc \\ | \end{bmatrix}$ as the width. Take this and subtract from the given sum, and let this be the length. Multiply this with the width, and we get the area: $\begin{bmatrix} \bigcirc \\ \text{𝍱𝍡} \\ \not| \end{bmatrix}$. Move this aside to the left. Take the area and convert the unit from *mu*

to *bu*. We subtract this from the area and we obtain the algebraic equation 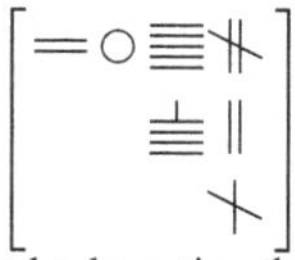.

Extracting the root from this, we obtain the width. Taking the given sum and subtracting the width, we obtain the length. End of Problem.

Zhu's method can be translated into today's terminology almost literally as follows: Let x be the width. Then $92-x$ is the length and

$$x(92-x)=92x-x^2$$

is the area. As the area is equal to the given 2052, we obtain the equation

$$x(92-x)-2052=0.$$

Solving this equation, we find the width.

Takebe interpreted Zhu's method as follows: Place the celestial element unit and consider the configuration 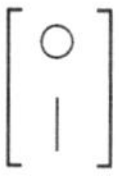 as the virtual width. Subtracting this configuration from the sum 92, we obtain the configuration , which is considered as the virtual length.

Here Takebe inserted a long explanation on addition of configurations on the counting board. In today's terminology this amounts to the addition of column vectors.

Now multiplying the two configurations, the virtual width and the virtual length, we obtain the configuration 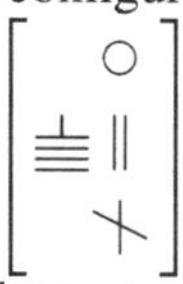, which is considered as the virtual area. Canceling the virtual area with the true area 2052, we find the equation to be extracted 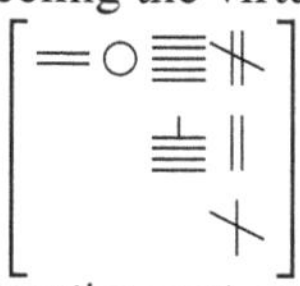. Extracting the root from this equation by the procedure for extracting the root, we find the width.

Takebe recognized the configuration of counting-rods on the counting board as the virtual number [仮の数, *kari no kazu*] and formulated the three rules of arithmetic, i.e., addition, self-multiplication, and mutual multiplication. As we mentioned earlier, the addition was defined as vector addition. The rule of powers was formulated as follows:

Method of self-multiplication and mutual multiplication

If the configuration with 2 rows $\begin{bmatrix} | & \text{Reality} \\ | & \text{Square} \end{bmatrix}$ is to be multiplied by itself, the Reality multiplied by itself is placed in the Reality, the doubled product of the Reality and the Square is placed in the Square, and the Square multiplied by itself is placed in the row below, thus we obtain the configuration with 3 rows. For example, if we multiply $\begin{bmatrix} \text{𝍦} \\ \text{𝍡} \end{bmatrix}$ by itself, we obtain $\begin{bmatrix} \text{𝍬𝍨} \\ \text{𝍪𝍧} \\ \text{𝍣} \end{bmatrix}$.

If the configuration with 3 rows $\begin{bmatrix} | & \text{Reality} \\ | & \text{Square} \\ | & \text{Side} \end{bmatrix}$ is to be multiplied by itself, the Reality multiplied by itself is placed in the Reality, the doubled product of the Reality and the Square is placed in the next row, the doubled product of the Reality and the Side, added by the squared Square, is placed in the third row, the doubled product of the Square and the Side is placed in the fourth row, and the squared Side is placed in the fifth row. For example, if we multiply $\begin{bmatrix} \text{𝍡̸} \\ \text{𝍢} \\ \text{𝍠} \end{bmatrix}$ by itself, we obtain $\begin{bmatrix} \text{𝍣} \\ \text{𝍩𝍡̸} \\ \text{𝍤} \\ \text{𝍥} \\ \text{𝍠} \end{bmatrix}$.

In today's terminology, the method of self-multiplication described above can be stated as follows:

$$(7+2x)^2 = 49+28x+4x^2,$$
$$(-2+3x+x^2)^2 = 4-12x+5x^2+6x^3+x^4,$$

or more generally

$$(a+bx+cx^2)^2 = a^2+2abx+(2ac+b^2)x^2+2bcx^3+c^2x^4.$$

Takebe also stated the rule of mutual multiplication of configurations and gave the following examples (in today's terminology)

$$(-7+2x)(3+x) = -21-x+2x^2,$$
$$(1-6x+2x^2)(2-3x+x^2) = 2-15x+23x^2-12x^3+2x^4.$$

Thus, Takebe knew that the configurations on the counting board could be regarded as virtual numbers and could be operated addition, self-multiplication and mutual multiplication in the same way as the true numbers [真の数 makoto no kazu]. In today's terminology, Takebe recognized that the procedure of celestial element was a way of manipulating polynomials. In this sense, I would like to say that the configurations on the counting board form the "counting board algebra", which is canonically isomorphic to the ring of polynomials of one variable with numerical coefficients.

Seki wrote the Methods of Solving Implicit Problems [解隠題之法 Kai indai no hō] around 1683 and developed the "counting board algebra" in a systematic way. A traditional Chinese book on mathematics followed the style of the Nine Chapters on the Mathematical Art and looked like a problem book. But in the Methods of Solving Implicit Problems Seki stated the rule of operations on configurations without introducing any problem. Horiuchi [2] argues that this book was a separating point of the *wasan* from the tradition of Chinese mathematics.

3 Method of side writing

3.1 Seki Takakazu and Takebe Katahiro

Seki Takakazu is considered to be one of founders of the *wasan*. He studied Chinese mathematics reading the Yang Hui's Methods of Mathematics of Yang Hui [楊輝], a Chinese mathematician of Southern *Song* in the late 13th century, and the Introduction to Mathematics. In 1974, Seki's existent 27 mathematical works were compiled in [1] with explanations in Japanese as well as in English. During his life time Seki had only one publication, the Mathematical Methods for Exploring Subtle Points, which we cited earlier.

Seki's another book, the Concise Collection of Mathematical Methods [括要算法 Katsuyō Sanpō] was published posthumously in 1712. (See [1].) In this book we find, among others, Seki's calculation of the circular constant $\pi = 3.141592...$ with twelve digits accuracy. This is one of his remarkable results on the circular principle [円理 enri], i.e., the study on the circle.

Seki also discovered the theory of resultants and determinants. (See the Methods of Solving Concealed Problems [解伏題之法 Kai fukudai no hō] [1].)

Takebe Katahiro entered Seki's school in 1676 when he was 13 years old. His first monograph, the Mathematical Methods for Clarifying Slight Signs [研幾算法 Kenki Sanpō] (See [12] for an English translation.) was published in 1683. Then he published in 1685 the Colloquial Commentary on Operations, and in 1690 the Complete Colloquial Commentary. Takebe completed all these three monographs in his twenties.

In Takebe's Mathematical Methods for Clarifying Slight Signs, as well as in Seki's Mathematical Methods for Exploring Subtle Points, a final solution to each problem was given in a form of algebraic equation of one variable, which was written in Chinese, without any explanation how to derive the equation. However, in the Colloquial Commentary on Operations, Takebe explained how the equations were derived employing the method of side writing [傍書法 bōsho hō], which generalizes the procedure of celestial element, and in the Complete Colloquial Commentary, Takebe explained the procedure of celestial element as the "counting board algebra" in the same way as Seki Takakazu did in the Methods of Solving Implicit Problems. Through these monographs, Takebe showed the method of side writing is a method to handle polynomials with several unknowns and applied them to various kinds of problems.

He collaborated with his master in many mathematical researches and in editing the Complete Book of Mathematics [大成算経 Taisei Sankei]. In his thirties and forties, Takebe was busy as a government officer but he resumed his mathematics in his fifties and wrote, in 1722, the Mathematical Treatise on the Technique of Linkage [綴術算経 Tetsujutsu Sankei] and the Fukyū's Technique of Linkage [不休綴術 Fukyū Tetsujutsu] (See [12] for an English translation.) on the technique of linkage [綴術 zhuishu, tetsu jutsu]. In these books, Takebe described, among others, his calculation of the circular constant π with more than forty digits accuracy and three formulas to represent the length of the arc when the sagitta (the length of the arrow) is given. One of the formulas coincides with the Taylor expansion of the square of the inverse trigonometric function $(\arcsin x)^2$. We can say that with these results he could compete with his European contemporaries (For the details, see Morimoto [6]).

3.2 Method of side writing

Seki introduced the method of side writing in the Methods of Solving Explicit Problems [解見題之法 Kai kendai no hō] and then combined it with the procedure of celestial element in the Methods of Solving Concealed Problems (see [1]). The method of side writing can be considered as a generalization of the procedure of celestial element and allowed Seki and Takebe to obtain the equation to be extracted even if the data were not given numerically. Note that Seki's Trilogy [三部抄 Sanbushō], i.e., the Methods of Solving Explicit Problems, the Methods of Solving Implicit Problems, and the Methods of Solving Concealed Problems, was completed around 1685 as manuscripts but was being kept secretly in Seki's school. It was Takebe who first published results relying on the method of side writing.

If we state Problem No. 8 above in a manner of Takebe's Mathematical Methods for Clarifying Slight Signs, it reads as follows:

> There is a rectangular rice field of given area A. The sum of the length and width is given to be B. Question: how much is the length and width respectively?

In the method of side writing, we place the celestial element unit $\begin{bmatrix} \bigcirc \\ \mid \end{bmatrix}$ and consider it as the virtual width, and let the configuration $\begin{bmatrix} \mid \; B \\ \not{\mid} \end{bmatrix}$ be the virtual length, and the configuration $\begin{bmatrix} \bigcirc \\ \mid B \\ \nmid \end{bmatrix}$ the virtual area. Then the equation to be extracted can be represented as $\begin{bmatrix} \nmid A \\ \mid B \\ \nmid \end{bmatrix}$.

In Takebe's Mathematical Methods for Clarifying Slight Signs many problems were given in this form and the final equations were described in Chinese. But in the Colloquial Commentary on Operations, Takebe showed how the equations were derived with the method of side writing. In this way, in the *wasan* polynomials with polynomial coefficients could be manipulated easily although the notation was cumbersome. Because of this, in the Meiji period Japanese abandoned the *wasan* and could switch to the western mathematics with almost no difficulties.

K. Sato concluded his article [10] stating "While *tengen jutsu* experienced a rigorous change in Japan, we have another question whether this technique became the counter part of "algebra" or not. Straightforwardly, the answer is negative". But as we explained above, Seki and Takebe recognized the configurations on the counting board can be calculated in the same manner as true numbers, and thus form an algebra. As the "counting board algebra" is canonically isomorphic to the algebra of polynomials of one unknown with numerical coefficients, they can be identified naturally.

Note

We cite mostly articles and monographs in western languages. The Bibliography of Ogawa [9] contains almost all treatises on Japanese mathematics written in European languages. As general references for the history of Chinese mathematics, we refer the reader to Martzloff [5] and Li-Du [4].

Some part of this article was read at an international conference in Chiang Mai, Thailand ([7]).

References

1. Akira Hirayama, Kazuo Shimodaira and Hideo Hirose, editors: *Takakazu Seki's Collected Works Edited with Explanations* [関孝和全集 *Seki Takakazu Zenshū*], Osaka Kyōiku Tosho, Osaka (1974).
2. Annick Horiuchi: Sur un point de rupture entre les traditions chinoise et japonaise des mathématiques, *Rev. Hist. Sci.*, XLII/4, 375–390 (1989).
3. Annick Horiuchi: *Les mathématiques japonaises à l'époque d'Edo*, J.VRIN (1994).
4. Li Yan and Du Shiran (translated by John N. Crossley and Anthony W.-C. Lun): *Chinese Mathematics, A Concise History*, Clarendon Press, Oxford (1987).

5. H.-C. Martzloff: *A History of Chinese Mathematics*, Springer (1997); The French original edition: *Histoire des mathématiques chinoises*, Masson, Paris (1987).
6. Mitsuo Morimoto: Japanese mathematics in the 18th century, Manila (2001); reproduced in Science Research Reports, International Christian University (http://science.icu.ac.jp/srr).
7. Mitsuo Morimoto: A Chinese root of Japanese traditional mathematics — wasan, Chiang Mai, Thailand (2003); reproduced in Science Research Reports, International Christian University (http://science.icu.ac.jp/srr).
8. Nam Kwon-Heui [南権熙]: A bibliographical Study of the 'Sinpyun-Sanhakkyemong' which was printed by KyungOja and other editions (in Korean), *The Bibliographical Studies* [書誌学研究], vol. 16, 336–360 (1998).
9. Tsukane Ogawa: A review of the history of Japanese mathematics, *Revue d'Histoire des Mathematiques*, vol. 7, 137–155 (2001).
10. Ken'ichi Sato: Reevaluation of *tengen jutsu* or *tianyuanshu*; in the context of comparison between China and Japan, *Historia Scientiarum*, vol. 5, 57–67 (1995).
11. Kanshen Shen, John N. Crossley and Athoney W.-C. Lun: *The Nine Chapters on the Mathematical Art, Companion and Commentary*, Oxford University Press, and Science Press, Beijing (1999).
12. Osamu Takenouchi and Mitsuo Morimoto: *Selected Mathematical Works of Takebe Katahiro (1664–1739)*, Wasan Institute (2004).

Power Series Expansions in India Around A. D. 1400

Setsuro Ikeyama

Abstract Mādhava of Saṅgamagrāma was a mathematical astronomer who flourished around 1400 A.D. in South India. Not a few mathematical formulas attributed to him have been transmitted by scholars of his school to these days. The most important among them are power series expansions of trigonometrical functions sine, cosine, arctangent, and so on. Because these series are not found in his extant astronomical treatises it is not always clear what parts of them were found by Mādhava himself, though some, the expansion of the circumference of a circle for example, can be attributed to him with a certainty.

After taking a glance at Mādhava himself and his school, I will explain in this paper how Mādhava derived the power series of the circumference C of a circle with diameter d together with the last corrective term:

$$C = \frac{4d}{1} - \frac{4d}{3} + \frac{4d}{5} - \frac{4d}{7} + \cdots + (-1)^{n-1}\frac{4d}{2n-1} + (-1)^n \cdot 4d \cdot \frac{n}{(2n)^2+1},$$

according to the commentary *Kriyākramakarī* (*ca.* AD 1550), composed by Śaṅkara, who was a scholar situated near the end of the Mādhava school.

1 Introduction

The purpose of this paper is to explain the fruit of a school of mathematicians that flourished in the 14th to 17th century in South India. Almost all the materials are from Chapter 1 of Studies in Indian Mathematics [インド数学研究, Indo Sūgaku Kenkyū] , by Takao Hayashi, Takanori Kusuba, and Michio Yano [1]. It is a pity

Setsuro Ikeyama
Kyoto Sangyo University, Kamikamo-motoyama, Kita-ku, Kyoto 603-8555, Japan
e-mail: QWN11071@nifty.com

that this is written in Japanese. So I am very glad to introduce the achievements of this book in English.

2 Mādhava and his school

A school of mathematicians, which we call Mādhava school, was active in Kerala, a state located in the southernmost part of the Indian subcontinent.

Generally speaking, there remains little information about individuals in India, and mathematicians are no exceptions. Mādhava, the founder of the school, flourished between 1380 and 1420. He belonged to a Brāhmaṇa, Hindu priest family, and lived in Saṅgamagrāma, the modern Irinjalakuda, a village about 50km north of Cochin.

As for his writings, some astronomical treatises exist now but mathematical works are not available. Fortunately parts of them are cited or summarized in his successors' treatises and we can reconstruct his theories from them.

After Mādhava his school continued for about five generations. Śaṅkara, a scholar situated near the end of the school tree and a pupil of Dāmodara and also of Nīlakhaṇṭha, composed a commentary which we deal with here.

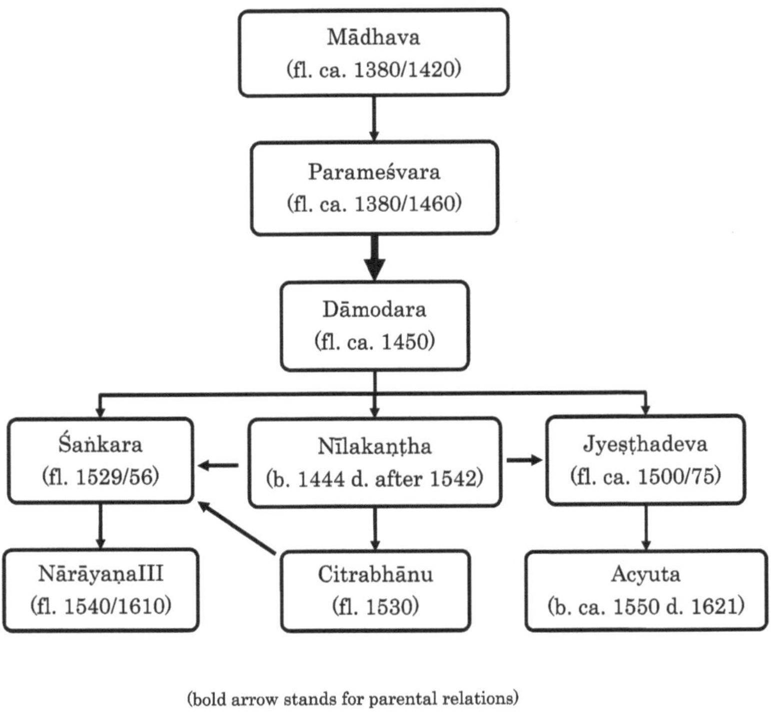

Fig. 1 Tree of the Mādhava School

3 Mādhava's series for calculating circumferences

3.1 Source of the Text

Śaṅkara wrote a commentary titled *Kiryākramakarī* on a mathematical textbook named *Līlāvatī* composed by Bhāskara in the 12th century. *Līlāvatī* is one of the most famous books on mathematics which is written in plain but beautiful verse and was widely used as a textbook of mathematics. The *Kiryākramakarī* is one of many commentaries on it.

Generally in India when one learns something, first he memorizes basic treatises. Because the treatises were versified it must be easy, at least for Indian people, to memorize them but was very hard to understand what the verse meant before teachers explained them. And this explanation was compiled as a commentary by successors.

The stanza 199 of the *Līlāvatī* is placed at the top of verse about geometrical problems concerning circles and gives two approximate values of π:

$$\frac{22}{7}, \ \frac{3927}{1250}.$$

Śaṅkara begins his commentary on this stanza with word by word explanation of the the verbal expression, then summarizes variety of πs known in India by his time, and finally explains the theories for calculation of more accurate circumferences which were figured out by Mādhava.

First, he demonstrates a method for calculation of circumferences using circumscribed regular polygons beginning with a square. This procedure attracts our interest but is not a current topic. Then Śaṅkara starts to demonstrate the second method for calculation of circumferences using a power series, the so called "Mādhava series".

3.2 Original Expression of the Series

Śaṅkara quotes four stanzas which, he affirms, were composed by "him", that is, Mādhava.

An easier way to get the circumference is mentioned by him (Mādhava). That is to say:

1. *Subtract or add alternately the diameter multiplied by four and divided in order by the odd numbers like three, five, etc., to or from the diameter multiplied by four and divided by one.*
2. *Assuming that division is completed by dividing by an odd number, whatever is the even number above (next to) that (odd number), half of that is the multiplier of the last [term].*

3. *The square of that (even number) increased by 1 is the divisor of the diameter multiplied by 4 as before. The result from these two (the multiplier and the divisor) is added when [the previous term is] negative, when positive subtracted.*
4. *The result is an accurate circumference. If division is repeated many times it will become very accurate.*

The first stanza gives the Mādhava series and the other three stanzas express a corrective term.

Let C be circumference of a circle whose diameter is d and $C(n)$ be a corresponding circumference calculated using the Mādhava series and the corrective term added after the nth term of the series.

$$C(n) = \frac{4d}{1} - \frac{4d}{3} + \frac{4d}{5} - \cdots + (-1)^{n-1}\frac{4d}{2n-1} + (-1)^n \cdot 4d \cdot \frac{n}{(2n)^2+1}.$$

3.3 Derivation of the Series

Now, let us trace how Mādhava derived his series expressed in the first stanza.

Step 1

Śaṅkara begins by drawing a figure for it. Figure 2 is a fourth part of the drawing instructed by Śaṅkara but is the only part which is used in the following demonstration.

Let O be the center of a circle. Sides OE and EA are equal to the radius of the circle r, which is equal to a half of d.

Having placed as many dots as you wish with equal intervals on line EA, draw lines from the center of the circle O to the dots.

In figure 2, we place m dots on EA and name them A_0 (A sub 0), which is identical to E, A_1, A_2, up to A_m which is equal to A. Thus EA is divided into m parts. Each part has the same length of a. Only two lines from O to dots A_n and A_{n+1}, the nth and $n+1$th dots, are shown and the length of these two lines are denoted by k_n and k_{n+1} respectively. The intersections of the lines and the circle are denoted by B_n and B_{n+1} and the arc between them is c_n. Finally G_{n+1} denotes the foot of the perpendicular from A_n to OA_{n+1}.

Here let us draw one more line from O to A or A_m. This is a diagonal line of the square. The intersection of this line OA_m and the arc of the circle is denoted by B_m.

Mādhava's idea is that because the arc B_0B_m, that is, an eighth of circumference C, is divided into small arcs $c_0, c_1, \cdots c_n, \cdots c_{m-1}$ by lines $OA_1, OA_2, \cdots OA_n, OA_{n+1}, \cdots, OA_m$, c_n should be calculated from the given length.

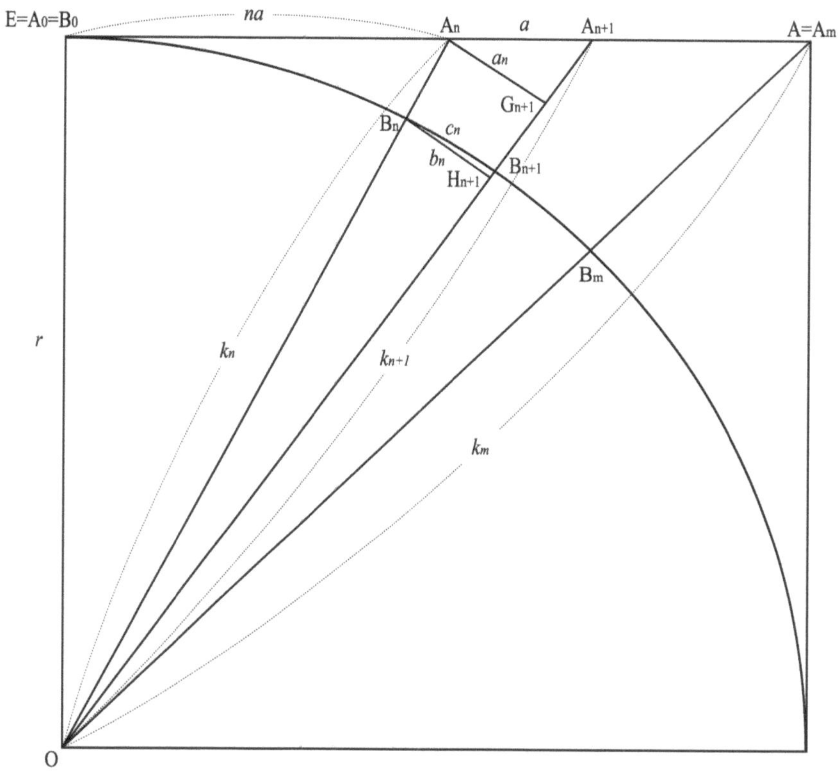

Fig. 2 Derivation of the Series

$$\frac{C}{8} = \sum_{n=0}^{m-1} c_n \,.$$

(An eighth of C equals the summation from 0 to $m-1$ of c sub n.)

Because $\mathrm{EA}_n = na$,

$$k_n^2 = r^2 + (na)^2 \,.$$

(k sub n squared equals to r squared plus na squared.)

Because $\triangle \mathrm{A_n A_{n+1} G_{n+1}}$ is similar to $\triangle \mathrm{OA_{n+1}E}$, a_n, the length of side $\mathrm{A}_n\mathrm{G}_{n+1}$, can be calculated as:

$$a_n = \mathrm{A_n G_{n+1}} = \frac{\mathrm{A_n A_{n+1}} \cdot \mathrm{OE}}{\mathrm{OA_{n+1}}} = \frac{ar}{k_{n+1}} \,.$$

Here let b_n be the $r\sin$ of the arc c_n. When the interval of the dots placed on EA is very small, that is, when $a \approx 0$, b_n is approximately equal to c_n. Therefore Mādhava derives b_n from a_n.

Because $\triangle \mathrm{B_n H_{n+1} O}$ is similar to $\triangle \mathrm{A_n G_{n+1} O}$,

$$b_n = \mathrm{B_nH_{n+1}} = \frac{\mathrm{OB_n} \cdot \mathrm{A_nG_{n+1}}}{\mathrm{OA_n}} = \frac{r \cdot a_n}{k_n} = \frac{ar^2}{k_n k_{n+1}}\,.$$

Also because when a is approximately 0:

$$k_n \approx< k_{n+1},$$

(k sub n is approximately equal to but less than k sub n plus 1,)

$$\frac{ar^2}{k_{n+1}^2} \approx< b_n \approx< \frac{ar^2}{k_n^2}\,.$$

(ar squared over k_{n+1} squared is approximately equal to but less than b_n and b_n is approximately equal to but less than ar squared over k_n squared.)

Therefore the summation of b_n when n is from 0 to m minus 1 is:

$$\sum_{n=0}^{m-1} \frac{ar^2}{k_{n+1}^2} = \sum_{n=1}^{m} \frac{ar^2}{k_n^2} \approx< \sum_{n=0}^{m-1} b_n \approx< \sum_{n=0}^{m-1} \frac{ar^2}{k_n^2}\,.$$

From this, Mādhava or Śaṅkara seems to have thought:

$$\begin{aligned}\sum_{n=0}^{m-1} b_n &\approx \sum_{n=1}^{m} \frac{ar^2}{k_n^2} + \frac{\sum_{n=0}^{m-1} \frac{ar^2}{k_n^2} - \sum_{n=1}^{m} \frac{ar^2}{k_n^2}}{2} \\ &= \sum_{n=1}^{m} \frac{ar^2}{k_n^2} + \frac{\frac{ar^2}{k_0^2} - \frac{ar^2}{k_m^2}}{2} \\ &= \sum_{n=1}^{m} \frac{ar^2}{k_n^2} + \frac{\frac{ar^2}{r^2} - \frac{ar^2}{2r^2}}{2} = \sum_{n=1}^{m} \frac{ar^2}{k_n^2} + \frac{a}{4}\,,\end{aligned}$$

and because one fourth of a approximately equals to 0 when a is approximately 0, he finally gets the relation:

$$\frac{C}{8} = \sum_{n=0}^{m-1} c_n \approx \sum_{n=0}^{m-1} b_n \approx \sum_{n=1}^{m} \frac{ar^2}{k_n^2} \left(= \frac{ar^2}{k_1^2} + \frac{ar^2}{k_2^2} + \frac{ar^2}{k_3^2} + \cdots + \frac{ar^2}{k_{m-1}^2} + \frac{ar^2}{k_m^2} \right)\,.$$

Step 2

Here the denominator of this formula, k_n, varies from term to term. For avoiding the variation, Mādhava-Śaṅkara says "a method should be sought for in order to derive divisors of one kind" (hārāṇām ekavidhatvaṃ netum upāyo 'trānveṣyaḥ) and transform the denominators.

Here they introduce a general relation:

when $0 < y < x$, (x is greater than y and y is greater than 0,)

$$\frac{ay}{x} = a - \frac{a(x-y)}{y} + \frac{a(x-y)^2}{y^2} - \frac{a(x-y)^3}{y^3} + \cdots$$

(the product of a and y over x is a minus the product of a and the sum of x and y over y plus the product of a and the sum of x and y squared over y squared minus the product of a and the sum of x and y cubed over y cubed $\cdots$)

without any proof or demonstration here.

Then they convert the divisors using it. (They demonstrate it after getting the series for calculation of circumferences.)

That is, Applying this expansion for ar^2/k_n^2, that is, thinking k_n^2 as x and r^2 as y, and replacing numerators by

$$k_n^2 - r^2 = (na)^2 ,$$

they get:

$$\begin{aligned}\frac{ar^2}{k_n^2} &= a - \frac{a(k_n^2-r^2)}{r^2} + \frac{a(k_n^2-r^2)^2}{r^4} - \cdots + (-1)^p \frac{a(k_n^2-r^2)^p}{r^{2p}} + \cdots \\ &= a - \frac{a(na)^2}{r^2} + \frac{a(na)^4}{r^4} - \cdots + (-1)^p \frac{a(na)^{2p}}{r^{2p}} + \cdots \\ &= a + \sum_{p=1}^{\infty} (-1)^p \frac{a(na)^{2p}}{r^{2p}} .\end{aligned}$$

Therefore an eighth part of C is:

$$\begin{aligned}\frac{C}{8} \approx \sum_{n=1}^{m} \frac{ar^2}{k_n^2} &= \sum_{n=1}^{m} \left\{ a + \sum_{p=1}^{\infty} (-1)^p \frac{a(na)^{2p}}{r^{2p}} \right\} \\ &= ma + \sum_{p=1}^{\infty} \left\{ (-1)^p \sum_{n=1}^{m} \frac{a(na)^{2p}}{r^{2p}} \right\} \\ &= r + \sum_{p=1}^{\infty} (-1)^p \frac{a \cdot \sum_{n=1}^{m} (na)^{2p}}{r^{2p}} . \qquad (1)\end{aligned}$$

Step 3

Next, Mādhava-Śaṅkara transforms the numerator of formula (1) above into

$$a \cdot \sum_{n=1}^{m} (na)^{2p} = a \cdot a^{2p} + a \cdot (2a)^{2p} + a \cdot (3a)^{2p} + \cdots + a \cdot \{(m-1)a\}^{2p} + a \cdot (ma)^{2p} .$$

First of all, they deal with the simplest form;

$$\sum_{n=1}^{m} na\,,$$

which is the sum of the partial sides ($\mathrm{EA_n}; n = 1, 2, \cdots, m$) and we denote it by $A(m)$.

The method of the calculation of $A(m)$ adopted by Mādhava-Śaṅkara is as follows.

They break every term of the right side into "r", which is equal to ma, and "decreased amount":

$$\begin{aligned}\sum_{n=1}^{m} na = A(m) &= a + 2a + 3a + \cdots + (m-1)a + ma \\ &= \{r-(m-1)a\} + \{r-(m-2)a\} + \cdots + (r-a) + r\,.\end{aligned}$$

From this it is easily obtained that:

$$A(m) = mr - \{A(m) - r\}\,.$$

Therefore,

$$2A(m) - r = mr\,,$$

that is,

$$A(m) = \frac{(m+1)r}{2}\,.$$

And when $a \approx 0$,

$$a \cdot A(m) = \frac{a(m+1)r}{2} = \frac{(ma+a)r}{2} = \frac{(r+a)r}{2} \approx \frac{r^2}{2}\,, \tag{2}$$

where $ma = r$. And generally when $a \approx 0$,

$$a \cdot A(n) = \frac{a(n+1)na}{2} = \frac{(na+a)na}{2} \approx \frac{(na)^2}{2}\,. \tag{3}$$

Then using these two relations, Mādhava-Śaṅkara examines

$$\sum_{n=1}^{m} (na)^2\,,$$

that is,

$$A^{(2)}(m).$$

They break the right side in the same way as in the case of $A(m)$. That is,

$$\begin{aligned}A^{(2)}(m) &= a^2 + (2a)^2 + \cdots + \{(m-1)a\}^2 + (ma)^2 \\ &= a\{r-(m-1)a\} + (2a)\{r-(m-2)a\} + \cdots + \{(m-1)a\}(r-a) + ma \cdot r \\ &= rA(m) - \{a \cdot (m-1)a + (2a) \cdot (m-2)a + \cdots + (m-1)a \cdot a\}\,.\end{aligned}$$

Furthermore, because

$$
\begin{aligned}
A(m-1) &= a+2a+3a+\cdots+(m-2)a+(m-1)a,\\
A(m-2) &= a+2a+3a+\cdots+(m-2)a,\\
\cdots \quad & \cdots,\\
A(3) &= a+2a+3a,\\
A(2) &= a+2a,\\
A(1) &= a\,,
\end{aligned}
$$

Mādhava-Śaṅkara transforms the second term:

$$
\begin{aligned}
&\{(m-1)a\}\cdot(1a)+\{(m-2)a\}\cdot(2a)+\cdots\\
&\quad +(2a)\cdot\{(m-2)a\}+(1a)\cdot\{(m-1)a\} = a\cdot\textstyle\sum_{n=1}^{m-1}A(n)\,.
\end{aligned}
$$

Thus, they get:

$$A^{(2)}(m) = rA(m) - a\cdot\sum_{n=1}^{m-1}A(n)\,. \tag{4}$$

Applying relation (3), they get:

$$a\cdot\sum_{n=1}^{m-1}A(n) = \sum_{n=1}^{m-1}a\cdot A(n) \approx \sum_{n=1}^{m-1}\frac{(na)^2}{2} = \frac{1}{2}A^{(2)}(m-1)\,,$$

and hence,

$$
\begin{aligned}
A^{(2)}(m) &\approx rA(m) - \frac{1}{2}A^{(2)}(m-1)\\
&\approx rA(m) - \frac{1}{2}A^{(2)}(m)\,,\\
A^{(2)}(m) &\approx \frac{2}{3}rA(m)\,.
\end{aligned}
$$

Therefore, using relation (2),

$$a\cdot A^{(2)}(m) \approx \frac{2}{3}arA(m) \approx \frac{2}{3}r\cdot\frac{r^2}{2} = \frac{r^3}{3}\,, \tag{5}$$

and generally

$$a\cdot A^{(2)}(n) \approx \frac{2}{3}a\cdot(na)A(n) \approx \frac{2}{3}(na)\cdot\frac{(na)^2}{2} = \frac{(na)^3}{3}\,, \tag{6}$$

when $a \approx 0$.

In the same way, using relations (5) and (6), they demonstrate the case of $A^{(3)}(m)$:

$$\begin{aligned}A^{(3)}(m) &= a^3+(2a)^3+\cdots+\{(m-1)a\}^3+(ma)^3\\&= a^2\{r-(m-1)a\}+(2a)^2\{r-(m-2)a\}+\cdots\\&\quad+\{(m-1)a\}^2(r-a)+(ma)^2r\\&= rA^{(2)}(m)-a\cdot\sum_{n=1}^{m-1}A^{(2)}(n)\,.\end{aligned}$$

Therefore,

$$\begin{aligned}A^{(3)}(m) &\approx rA^{(2)}(m)-\sum_{n=1}^{m-1}\frac{(na)^3}{3},\\&\approx rA^{(2)}(m)-\frac{1}{3}A^{(3)}(m)\,,\\A^{(3)}(m) &\approx \frac{3}{4}\cdot rA^{(2)}(m)\,.\end{aligned}$$

Hence,

$$a\cdot A^{(3)}(m)\approx\frac{3}{4}\cdot arA^{(2)}(m)=\frac{3}{4}\cdot r\cdot\frac{r^3}{3}=\frac{r^4}{4}\,,$$

and generally

$$a\cdot A^{(3)}(n)\approx\frac{3}{4}\cdot(na)\cdot\frac{(na)^3}{3}=\frac{(na)^4}{4}\,.$$

Having demonstrated up to $A^{(3)}(m)$, Śaṅkara gives a general relation, saying "each sum should be multiplied by the half-diameter and diminished by its own part divided by the number [of the power] increased by 1", that is,

$$A^{(p)}(m)\approx r\cdot A^{(p-1)}(m)-\frac{r\cdot A^{(p-1)}(m)}{p+1}=\frac{p}{p+1}rA^{(p-1)}(m)\,.$$

Then starting from

$$a\cdot A(m)\approx\frac{r^2}{2}\,,$$

using this recursive calculation, he finally get

$$a\cdot A^{(p)}(m)\approx\frac{r^{p+1}}{p+1}\,. \tag{7}$$

Step 4

Applying this relation (7) to formula (1) in Step 2, Mādhava-Śaṅkara get

$$a\cdot A^{(2p)}(m)=a\cdot\sum_{n=1}^{m}(na)^{2p}\approx\frac{r^{2p+1}}{2p+1}\,,$$

and then

$$\frac{C}{8} \approx r + \sum_{p=1}^{\infty} (-1)^p \frac{a \cdot \sum_{n=1}^{m} (na)^{2p}}{r^{2p}},$$
$$= r + \sum_{p=1}^{\infty} (-1)^p \frac{aA^{(2p)}(m)}{r^{2p}} \approx r + \sum_{p=1}^{\infty} (-1)^p \frac{r^{2p+1}}{(2p+1)r^{2p}},$$
$$= r + \sum_{p=1}^{\infty} (-1)^p \frac{r}{2p+1}.$$

Here applying $d = 2r$, we finally get the Mādhava's series:

$$C \approx 8r + \sum_{p=1}^{\infty} (-1)^p \frac{8r}{2p+1},$$
$$= 4d + \sum_{p=1}^{\infty} (-1)^p \frac{4d}{2p+1},$$
$$= \frac{4d}{1} - \frac{4d}{3} + \frac{4d}{5} - \cdots + (-1)^{n-1} \frac{4d}{2n-1} + \cdots.$$

3.4 *Other Formulas Based on the Same Principle*

Then Mādhava-Śaṅkara gives two formulas for calculating arcs or circumferences.

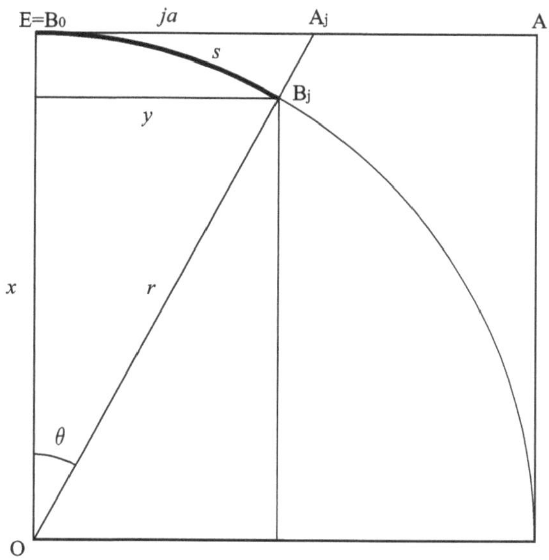

Fig. 3 Arc (Arctangent)

Arc (Arctangent)

The first formula is for "calculating an arc from a given sine" (iṣṭajyāyāḥ cāpānayanam).

Please refer the reader to Figure 3. The formula gives the arc s corresponding to a central angle θ of a circle using y ($r$sine corresponding to $theta$), x (its $r$cosine), and radius r of the circle.

$$s\left[=r\cdot\arctan\frac{y}{x}\right]=\frac{ry}{1x}-\frac{ry^3}{3x^3}+\frac{ry^5}{5x^5}-\cdots+(-1)^{n-1}\frac{ry^{2n-1}}{(2n-1)x^{2n-1}}+\cdots. \qquad (8)$$

This is a general form of Mādhava series. Śaṅkara gives no explanation about how to derive it. We infer the way as follows. In figure 3, concerning an arc s, that is B_0B_j, we can use the same way as in the case of calculation of the circumferences, dividing EA_j into j parts which have the equal length a:

$$\begin{aligned} s &= \sum_{n=0}^{j-1} c_n \approx \sum_{n=0}^{j-1} b_n = \sum_{n=0}^{j-1}\frac{ar^2}{k_nk_{n+1}} \approx \sum_{n=0}^{j-1}\frac{ar^2}{k_{n+1}^2} = \sum_{n=1}^{j}\frac{ar^2}{k_n^2},\\ &= \sum_{n=1}^{j}\sum_{p=0}^{\infty}(-1)^p\frac{a(na)^{2p}}{r^{2p}} = \sum_{p=0}^{\infty}(-1)^p\sum_{n=0}^{j}\frac{a(na)^{2p}}{r^{2p}},\\ &= \sum_{p=0}^{\infty}(-1)^p\frac{aA^{(2p)}(j)}{r^{2p}}. \end{aligned}$$

And the relation

$$aA^{(p)}(j)\approx\frac{(ja)^{p+1}}{p+1}$$

is also concluded.

Then using the simple proportional relation:

$$ja=\frac{ry}{x},$$

we get:

$$\begin{aligned} s &\approx \sum_{p=0}^{\infty}(-1)^p\frac{(ja)^{2p+1}}{(2p+1)r^{2p}} = \sum_{p=0}^{\infty}(-1)^p\frac{(\frac{ry}{x})^{2p+1}}{(2p+1)r^{2p}},\\ &= \sum_{p=0}^{\infty}(-1)^p\frac{ry^{2p+1}}{(2p+1)x^{2p+1}},\\ &= \frac{ry}{1x}-\frac{ry^3}{3x^3}+\frac{ry^5}{5x^5}-\cdots+(-1)^{n-1}\frac{ry^{2n-1}}{(2n-1)x^{2n-1}}+\cdots. \end{aligned}$$

It is clear that $s=C/8$ when $x=y$.

Circumference

Now let us think of the second formula:

$$C = \frac{\sqrt{12d^2}}{1} - \frac{\sqrt{12d^2}}{3\cdot 3} + \frac{\sqrt{12d^2}}{5\cdot 3^2} - \cdots + (-1)^{n-1}\frac{\sqrt{12d^2}}{(2n-1)3^{n-1}} + \cdots .$$

If formula (8) is applied in the case of $\theta = \frac{\pi}{6}(=30^\circ)$, because

$$\frac{y}{x} = \frac{1}{\sqrt{3}},$$

$$\begin{aligned} s\left(\frac{\pi}{6}\right) &= \frac{r}{\sqrt{3}} - \frac{r}{3}\left(\frac{1}{\sqrt{3}}\right)^3 + \cdots + (-1)^{n-1}\frac{r}{2n-1}\cdot\left(\frac{1}{\sqrt{3}}\right)^{2n-1} + \cdots, \\ &= \frac{r}{\sqrt{3}} - \frac{r}{3\cdot 3\sqrt{3}} + \cdots + (-1)^{n-1}\frac{r}{(2n-1)3^{n-1}\sqrt{3}} + \cdots . \end{aligned}$$

Because the obtained arc $s\left(\frac{\pi}{6}\right)$ is one twelfth of the circumference C,

$$C = 12\cdot s\left(\frac{\pi}{6}\right) .$$

Furthermore because $2r = d$,

$$\frac{12r}{\sqrt{3}} = \sqrt{12d^2}$$

. Thus the formula

$$C = \frac{\sqrt{12d^2}}{1} - \frac{\sqrt{12d^2}}{3\cdot 3} + \frac{\sqrt{12d^2}}{5\cdot 3^2} - \cdots + (-1)^{n-1}\frac{\sqrt{12d^2}}{(2n-1)3^{n-1}} + \cdots$$

seems to be obtained.

3.5 Corrective Term

Then Śaṅkara demonstrates in detail the corrective term added to Mādhava series:

$$(-1)^n \cdot 4d \cdot \frac{n}{(2n)^2+1} . \tag{9}$$

Concerning this correction, the authors of Studies in Indian Mathematics [1] published an article [2] in *Centaurus*. Please refer the reader to it for detailed explanations and discussions. Now I just show the outline of Śaṅkara's demonstration.

He mentions his plan for correction as follows:

> When division for that purpose was made by a certain odd number, one should make a correction separately. Then one should make [another] correction separately immediately after dividing by the next odd number. When it has been done in this way, if the two circumferences obtained are equal, then the correction is ascertained to be accurate.

Let C be the true circumference, $C(n)$ be a partial sum of n terms of the Mādhava series increased by the corrective term.

Śaṅkara's plan is that if

$$C(N) = C(N+1)\,,$$

where N is a natural number, then

$$C(n) = C$$

where n is a natural number greater than N.

Let $C(n)$ be:

$$C(n) = \frac{4d}{1} - \frac{4d}{3} + \frac{4d}{5} + \cdots + (-1)^{n-1}\frac{4d}{2n-1} + (-1)^n \cdot 4d \cdot F(n)\,.$$

When

$$C(n) = C(n+1)\,,$$

$$F(n) = \frac{1}{2n+1} - F(n+1)\,,$$

that is,

$$F(n) + F(n+1) = \frac{1}{2n+1}\,. \tag{10}$$

Therefore he seeks a function $F(n)$ which satisfies this condition.

If

$$F(n) = F(n+1) = \frac{1}{2(2n+1)}\,,$$

the condition mentioned above would be satisfied. But such a function $F(n)$ never exists. When

$$F(n) = \frac{1}{2(2n+1)}\,, \tag{11}$$

then $F(n+1)$ must be:

$$F(n+1) = \frac{1}{2\{2(n+1)+1\}} = \frac{1}{2(2n+1)+4}\,. \tag{12}$$

The difference between denominators of formula (11) and formula (12) is just 4.

Observing this difference Śaṅkara first of all think a function, $F_1(n)$:

$$F_1(n) = \frac{1}{2(2n+1)-2} = \frac{1}{4n}\,. \tag{13}$$

Here,

$$F_1(n+1) = \frac{1}{2\{2(n+1)+1\}-2} = \frac{1}{2(2n+1)+2} = \frac{1}{4n+4} \tag{14}$$

where the difference of denominators remains 4. However,

$$F(n) + F(n+1) = \frac{2n+1}{2n(2n+2)}$$

which does not satisfy condition (10).

Mādhava-Śaṅkara, who are not satisfied with $F_1(n)$, start to seek better function by trial and error and find another function $F_2(n)$:

$$F_2(n) = \frac{1}{4n + \frac{4}{4n}} = \frac{n}{(2n)^2+1},$$

which is the formula included as the corrective term (9).

References

1. Takao Hayashi [林隆夫], Takanori Kusuba [楠葉隆徳] and Michio Yano [矢野道雄]: *Studies in Indian Mathematics: Series, Pi and Trigonometry* [インド数学研究: 数列・円周率・三角法 *Indo Sugaku Kenkyu: Suretsu, Enshuritsu, Sankakuhou*], Kōseishakōseikaku [恒星社厚生閣], Tokyo (1997).
2. Takao Hayashi, Takanori Kusuba and Michio Yano: The Correction of the Mādhava Series for the Circumference of a Circle, *Centaurus*, vol. 33, pp. 149–174 (1990).

An Early Japanese Work on Chinese Mathematics in Vietnam: Yoshio Mikami's Study of the Vietnamese Mathematical Treatise *Chi Minh Toan Phap*

Alexei Volkov

Abstract In 1934 Yoshio Mikami (1875–1950) published a paper devoted to the Vietnamese mathematical treatise Guide [towards] Understanding of Calculational Methods *Chi Minh Toan Phap*. His analysis of several topics discussed in the treatise (representation of numbers with counting rods, format of multiplication table, generic problems of various categories, etc.) allowed him to advance hypotheses concerning the origin and the time of compilation of the treatise. The book studied by Mikami nowadays is not available. In the present paper the author examines Mikami's work and provides a description of the Vietnamese mathematical treatise *Chi Minh Lap Thanh Toan Phap* by Phan Huy Khuong (preface 1820) textually close to that investigated by Mikami.

1 Introduction

The earliest attempt to investigate the extant materials on Vietnamese mathematics was made by the outstanding Japanese historian of mathematics Yoshio Mikami [三上義夫] (1875–1950) who provided an analysis [12] of the Vietnamese mathematical treatise *Chi Minh Toan Phap*[1] [指明算法], yet did not have access to other extant Vietnamese mathematical books.[2] Unfortunately, the current whereabouts of the book explored by Mikami are unknown. On the basis of Mikami's description the author of the present paper was able to identify a treatise with a slightly dif-

Alexei Volkov
Center for General Education and Institute of History, National Tsing Hua University, Hsinchu 30013, Taiwan
e-mail: alexei.volkov@gmail.com

[1] Due to technical reasons, in this paper I omit the diacritics adopted in the transliteration system Quoc Ngu; the interested reader will find the correct transliteration of all the used Vietnamese terms in the Glossary at the end of the paper.

[2] I am grateful to Professor Yukio Ōhashi [大橋由紀夫] who drew my attention to the work of Mikami and kindly sent me a copy of his 1934 paper [12].

ferent title which seems matching, at least partly, with Mikami's description; two manuscript copies of it are currently preserved in Vietnam and a microfilm copy of one of them, in Paris.

The motivation for writing this paper is threefold: firstly, to describe in greater detail an extant mathematical treatise arguably close to that studied by Mikami; secondly, on the basis of the information concerning Vietnamese mathematics accumulated during the 75 years that passed since the publication of Mikami's seminal article, to answer, at least partly, the questions he posed; thirdly, to discuss the methodology Mikami used to study the isolated document he had at his disposal.

2 The book Mikami studied

Mikami's paper appeared in the journal School Mathematics [學校數學 Gakkō Sūgaku] in 1934; he begins his relatively short (nine pages) article with a mention of the ethnologist Nobuhiro Matsumoto [松本信廣] (1897–1981),[3] who, after a field trip to Vietnam conducted in 1933, brought to Japan a mathematical book titled Guide [towards] Understanding of Computational Methods [指明算法 *Chi Minh Toan Phap*]. It is not known whether the book still exists physically, and if it does, where it is to be found.[4] Mikami did not specify whether the book was handwritten or block-printed. The name of the author and the date of compilation were not mentioned either, and this might mean that the book he obtained from Matsumoto lacked one or several opening pages.

An inspection of the titles of the extant mathematical treatises does not reveal any treatise bearing the title 指明算法.[5] According to the catalogue *Noi Cac Thu Muc* [内閣書目] of the department Noi Cac [内閣] of the Imperial Library in Hue compiled in 1908 and preserved in Han-Nom Institute (call number A113/1–2), a book with this title was once stored in the Imperial Library, yet it is impossible to know whether it was the book explored by Mikami or another (presumably, Chinese) treatise.[6] However, Mikami's cursory description of the treatise he investigated allows to identify a Vietnamese book which is the closest to it among the extant treatises: this is the *Chi Minh Lap Thanh Toan Phap* [指明立成筭法] authored by Phan Huy Khuong [潘輝框] (preface 1820). Even though there are reasons to believe that the latter book is not textually identical with the book found by Matsumoto Nobuhiro

[3] On the life and activities of Matsumoto see, for example, Ito [6].

[4] The collection of the late Prof. Matsumoto is currently preserved in the library of Keiō University, Mita Campus, Tokyo, yet the book is not found there, and Professor Ōhashi was unable to locate the book in other Japanese libraries either (Professor Yukio Ōhashi's private communication, September 2009).

[5] The number of the mathematical treatises so far located by the author of this paper equals to 22; for a list of the 19 treatises see Volkov [18] (three more books were located only recently).

[6] In the aforementioned catalogue the *Chi Minh Toan Phap* [指明算法 Zhi Ming Suan Fa] is listed together with Chinese mathematical treatises such as the 周髀算經, 海島算經, 孫子算經, 綺古算經, 詳解九章算法, and 算法統尊 (=算法統宗). There existed several Chinese books bearing the title Zhi Ming Suan Fa [指明算法] compiled during the Ming [明] (1368–1644) and Qing [清] (1644–1911) dynasties, see Li Di [9, pp. 366–367].

(see below), it has the same structure (four chapters) and contains, as far as Mikami's description allows to judge, numerous elements common with the former.

There exist two manuscript copies of the *Chi Minh Lap Thanh Toan Phap* preserved in the Han-Nom Institute in Hanoi; their call numbers are A.1240 and VHv.1185. The catalogue of Tran and Gros [14] (entry 433) also mentions a microfilm (call number MF. 2391) of the edition with the call number A.1290. However, the book with the call number A.1290 is not the mathematical treatise under consideration but a chronology of the kings of Cambodia entitled *Cao Mien The Thu* [高綿世次]. The call number of the microfilm of the latter book (MF.476) differs from that of the microfilm of the mathematical treatise mentioned above, and thus the entry in [14] apparently contains a misprint.[7] In the present study I used a microfilm copy of the manuscript A.1240 preserved in the library of EFEO in Paris and a partial copy of the manuscript VHv.1185.[8]

In his paper Mikami made several quotations from the Vietnamese books he studied; these quotations allow one to claim that his book was close enough to the manuscript A.1240. More specifically, Mikami mentions 11 subtitles and names of methods he found in the four chapters of his book; as Table 1 shows, the same (or almost the same, in the cases of items 3 and 6) items are found in the corresponding chapters of the *Chi Minh Lap Thanh Toan Phap*:

Item no.	Items in the *Chi Minh Toan Phap* [指明算法] inspected by Mikami	The corresponding items in the *Chi Minh Lap Thanh Toan Phap* [指明立成筭法] by Phan Huy Khuong [潘輝框] (1820), MS A.1240
1	1 算法綱領詩 in chap. 1	chap. 1; first item in the table of contents
2	九章算數法式 in chap. 1	chap. 1; third item in the table of contents
3	開平方法 in chap. 2	Table of contents (chap. 2) contains several items whose names include the term 開平方法
4	量倉窖歌 in chap. 3	chap. 3: first item.
5	平分法詩 in chap. 4	chap. 4: first item
6	異乘同除 in chap. 4	chap. 4: 異乘同除法 is item 12 in the table of contents
7	盈不足詩法 in chap. 4	chap. 4: item 17
8	兩盈兩不足詩法 in chap. 4	chap. 4: item 18
9	盈適足不足適足詩法 in chap. 4	chap. 4: item 19
10	望木高求路遠法 in chap. 4	chap. 4: item 20
11	算題試文格式 in chap. 4	chap. 4: item 21

Table 1

[7] It is possible that the microfilm MF.2391 is that of the manuscript A.1240 (the digit "4" may have been misprinted as "9" in [14] ; however, I was unable to check this hypothesis, since the microfilm MF.2391 currently remains unavailable, according to a private communication of Nguyen Thi Duong (September 2009).

[8] I would like to express my gratitude to Nguyen Thi Duong who kindly sent me copies of a number of pages of the manuscript VHv.1185.

As the reader can see, the book inspected by Mikami contained the titles of subsections almost completely identical with those found in one of the extant copies of the *Chi Minh Lap Thanh Toan Phap*. Certainly, one cannot exclude the possibility that the *contents* of these subsections may have been different. Indeed, there are strong reasons to believe that the two books were not entirely identical. For instance, Mikami speaks about the table of contents as if the treatise was opening with it (p. 3), while both the manuscripts A.1240 and VHv.1185 contain a preface and a section preceding the table of contents. The latter section is similar to that found in the Chinese treatise Summarized Fundamentals of Computational Methods [算法統宗 Suan Fa Tong Zong] [2] by Cheng Dawei [程大位] (1533–1606) completed in 1592 and contains a diagram of the Chinese abacus [算盤 suan pan]; Mikami, in turn, states that the treatise he investigated does not contain any mentions of the abacus (p. 4). Moreover, a comparison of five excerpts quoted by Mikami and their counterparts found in *Chi Minh Lap Thanh Toan Phap* presented below shows a number of discrepancies:

(1) Mikami quotes "一例算學直看、先認詩歌、次 。。。" (p. 4). In A.1240 the opening phrase of a paragraph following the table of contents (p. 6a) reads 一例學直看、先認詩歌、次 。。。; the character 算 is missing.
(2) Mikami mentions the subtitle "學習算例詩歌" (p. 4), while in A.1240 the subtitle reads 學習算例詩 (with the missing character 歌) is item 5 in table of contents (chapter 1); in the text, the section titled 學習算例詩歌 is found on p. 10b.
(3) Mikami mentions the expression "得算子" in the multiplication table 九九 (p. 4). In A.1240 the section 九章算數法式 of chapter 1 contains systematic computations of the number of 算子 expressed as 得算 N 子 with various numerals N (pp. 8b–9b).
(4) Mikami quotes "六不聚兮五不員" (p. 4), while in A.1240 the section 學習算例詩歌 (p. 10b) contains the phrase 六不聚兮五不單, the last character in Mikami's copy is apparently a mistake of the copyist (or the carver, if the book inspected by Mikami was block-printed).
(5) Mikami provides a long quotation from the model examination paper which reads as follows: 對 士 謂筭法。中來因除。不越衰分。上 看 多少有差。此執事 算[9] 問。而 士 所以復之也。茲見題中所問。惟照奉 恨 惠及本屬略說。平分而主用差分之法。諒知 算 無 究 之妙用矣。士 請 算 而 。。 (p. 7; I keep the punctuation of Mikami). The parallel excerpt found in the fourth chapter of A.1240 and VHv.1185 reads (the punctuation is mine):[10] 對。愚 謂筭法中來因除不越衰分。上 有 多少。有差。此執事 筭 ⟨ 河 ⟩ [問] 而 愚 所以復之也。茲

[9] The characters 筭 and 算, historically, had different meanings: according to the dictionary [說文解字 Shuo wen jie zi] by Xu Shen [許慎] (AD 55?–149?), the character suan 筭 meant the counting rods while the suan 算 meant the operations performed with the instrument. Interestingly, the manuscript copy of the text preserves the form 筭, while Mikami systematically uses 算; however, it remains unknown whether the copy he inspected indeed contained the form 算, or it also contained 筭 which was later changed to 算 by the publishers of the journal.

[10] I underline the diverging or missing characters in both excerpts, use angular brackets to mark the characters to be removed, and put in square brackets the characters to be inserted.

見題中所〈河〉[問] 惟照奉 銀 惠及本屬。略説平分而主用差分之法。諒知 筭法 無 窮 之妙用矣。愚 請 筭 而排陳之。(Chapter 4, p. 30b)[11]

Even though the manuscript A.1240 has a "running header" (placed at the unattached edge of each double leaf) that reads 指明算法,[12] on the basis of the above-mentioned comparison one can conclude that the book inspected by Mikami and the manuscript copy titled *Chi Minh Lap Thanh Toan Phap* are not textually identical. Moreover, Mikami's copy contained two elements lacking in the extant copy of *Chi Minh Lap Thanh Toan Phap*: (1) the pictures of the numerals represented with the counting rods and (2) an appendix. Below I will discuss the elements missing in Mikami's copy, in particular, a preface authored by the compiler of the treatise, but it seems appropriate to begin with a description of the received copies of *Chi Minh Lap Thanh Toan Phap*.

3 The Chi Minh Lap Thanh Toan Phap

The manuscripts A.1240 and VHv.1185 preserved in the library of the Institute for Han-Nom Studies (Hanoi) are listed in the catalogue of Han-Nom books by Tran and Gros [14, vol. 1, p. 258], yet the catalogue does not provide much information about the treatise and its author; in particular, it does not mention that the book contains an appendix (or rather an independent treatise bound together with *Chi Minh Lap Thanh Toan Phap*) entitled *Cuu Chuong Lap Thanh Toan Phap* [九章立成算法] that can be provisionally rendered in English as Ready-Made Computational Methods of Nine Categories. Both manuscripts A.1240 and VHv.1185 contain four chapters [卷] and a short introductory section.

The introductory section contains:
– a front page featuring the title of the treatise (the author's name is not mentioned) (p. 1a);[13]
– a Preface signed by the presumed compiler of the treatise, Phan Huy Khuong [潘輝框] containing the date of its compilation (pp. 1b-2b);
– a diagram explaining the construction of, and the operations with, the abacus (p. 3a) which is an exact reproduction of the picture found in the Chinese mathematical treatise Summarized Fundamentals of Computational Methods [2, p. 113];
– a table of correspondences between powers of 10, monetary units, units of length, weight, and volume (p. 3b);

---

[11] For a translation and discussion of this excerpt see Volkov [20] .

[12] I did not see this header in the partial copy of VHv.1185 I had at disposal.

[13] I refer to the double leaves of the manuscript A.1240 using letters a and b, for recto and verso, respectively. Chapters 1, 2, and 4 have individual pagination beginning with page 1, while the pagination of chapter 3 continues the pagination of Chapter 2 (thus the first page of Chapter 3 has number 44a, and the last one, 54a).

– diagrams of 32 plane figures (referred to as shapes of fields [田勢] (pp. 4a-b) the areas of which are calculated in Chapter 2 of the treatise (some of the figures are identical with those discussed in [2, p. 113]);
– the table of contents of the treatise (pp. 5a–6a) (the Appendix is not mentioned);
– the rules for studies of computations [筭學條例] (pp. 6a–6b).

Chapter 1 contains eleven subsections devoted to arithmetical operations; it includes, in particular, the multiplication table 9×9 entitled Table for the Method of Computing with Numbers [classified according to] Nine Categories [九章筭數法式] (pp. 8a–9b) discussed by Mikami. It also contains descriptions of the methods of multiplication and division accompanied by eleven arithmetical problems apparently designed to illustrate the introduced operations. A large section of Chapter 2 (pp. 1a–42a) is devoted to the calculation of areas of plane figures (a square, a rectangle, a right-angle triangle, a trapezium, a circle, etc.). This section does not include mathematical problems written down in the traditional format "condition – answer – algorithm"; instead, this section features computational procedures described with the numerical values written in smaller characters. A short concluding section of the Chapter (pp. 42b–43b) is devoted to the computation of square roots and contains two problems provided with detailed numerical solutions.

Chapter 3 contains algorithms of calculation of volumes accompanied by 14 problems illustrating the application of these algorithms; some algorithms are presented in a rhymed form and placed before the corresponding problems.

Chapter 4 of Manuscript A.1240 contains 38 problems devoted to flat-rate and weighted distribution (problems 1–20),[14] the Rule of Three and its modifications (problems 21–28), the method of Double False Position (problems 29–34), and to the proportionality of sides of triangles (problems 35–37).[15] The last problem is a model examination problem on weighted distribution.[16]

4 The Preface by Phan Huy Khuong

Some information concerning the history of the treatise can be found in the Preface authored by Phan Huy Khuong found in both manuscript copies A.1240 and VHv.1185; it reads as follows:[17]

[14] The original number of the problems in this chapter is unknown, since the manuscript copy A.1240 was made from a damaged original, or the copyist eye-skipped one or several pages: page 16a begins with the solution of an unstated problem, while page 15b does not contain the condition of the problem solved on page 16a. The numeration of the problems in this chapter is therefore conventional: I assign number 14 to the problem solution of which begins on page 16a.

[15] I avoid using the term "similar triangles" which might impose a modernizing interpretation.

[16] This problem, its solution, and the context are discussed in Volkov [20] .

[17] The original text found in the manuscript copy A.1240 is not punctuated, while the copy VHv.1185 contains punctuation. The punctuation I use in this paper is that found in the latter copy. I am thankful to Lin Hung-Chun [林虹君] for discussions of the preface.

潘家筭法指明序。
夫筭者併別也。多少併其數。平差別其分。莫非筭法。推之爲修齊之道。[18] 措之為平治之規。包數元義。固如是哉。予姓潘字框。學習古人。探求遺法。參今酌古。曾已有年。念全書繁衍。初學難通。倘一毫之差。[有] 千里之謬。[19] 如予力學筭辨。粗得統宗。可不立成法訓。以示後人。使易精識。爲自淺入深之學者乎。遂於庚辰清明節始生明良辰。自樂園中。日記故事。激發乎九經之奧旨。盤查乎六藝之淵微。惟知筭者所存。足爲近用。豈可以小技觀。而輕廢乎大典。以是神會尖微。乎提綱領。集成筭書。永垂家訓。庶使學者易通。方合正宗之妙術。觀此指明筭法。吾非敢作古以術籠人。蓋吾思先聖之功無窮。他日所用不外今日所存。後來學者。當以類推。遂書為序。
時。
明命元年庚辰清明節始生明。
山西鎮國威府慈廉縣東鄂社老圃潘輝框撰。
樂天窩藏書。

Preface of the *Toan Phap Chi Minh* [transmitted in] the Phan family.[20]

As far as the computations (筭) are concerned, [there are the procedures of] joining together and separating apart. [When performing the procedure of] the "excessive [amount] and deficient [amount]," [one] joins their quantities together; [when doing] the "equal [distribution] and proportional [distribution]" [of an amount], [one] separates it apart. There is nothing [here] that would not amount to the counting methods. "Pushing them [the methods?] forward" [= developing them by analogy] is the universal way of "perfection" and "equalizing" [them]; grouping them [together in an appropriate way] is the standard tool for "pacification" and "ordering".[21] The original meaning of the "embracing numbers" is certainly like that!

My name is Phan, style Khuong. I studied and exercised [the knowledge] of the people of the past, explored and looked for the methods [they] left behind, was involved in the [activities] of the present while taking the ancient [things into consideration]. For quite some years I have already studied the complexity and vastness of the entire [collection of mathematical?] books.[22] [They] are difficult to comprehend for the beginners. If [the

[18] The manuscript VHv. 1185 is damaged and the four characters 推之為修 are missing.

[19] The manuscript A.1240 does not contain the character 有.

[20] Technically, the title of the treatise mentioned here is different from the one on the front page.

[21] The four terms mentioned here 修, 齊, 治, 平 are quoted from Chapter 2 of the classic Great Learning [大學 Da Xue] : 古之欲明明德於天下者, 先治其國；欲治其國者, 先齊其家；欲齊其家者, 先修其身；欲修其身者, 先正其心；欲正其心者, 先誠其意；欲誠其意者, 先致其知；致知在格物；物格而後知至, 知至而後意誠, 意誠而後心正, 心正而後身修, 身修而後家齊, 家齊而後國治, 國治而後天下平. The four clauses 修身, 齊家, 治國, 平天下 were rendered by J. Legge [8, p. 357] as "to cultivate their persons," "to regulate their families," "to order well their states," "and the whole kingdom was made tranquil and happy," respectively. I am thankful to Hsu Yu-Fang [許瑜芳] who drew my attention to this quotation from the Great Learning. Phan Huy Khuong's remark can be understood in the sense that the four Confucian terms can be interpreted mathematically in the context of two algorithms he mentioned. For a discussion of the possible mathematical interpretation of this excerpt, see below.

[22] Even though the expression Quan Shu [全書] may have referred to the Chinese Compendium Complete Library in Four Branches of Literature [四庫全書 Si Ku Quan Shu] completed in 1782, it looks rather unlikely that the Vietnamese author was ever able to consult this collection. However, the possibility that he had access to certain editions of mathematical books based upon the Chinese Compendium cannot be ruled out, especially given that editions of the year 41 of Qian Long [乾隆] era (1775) of several Chinese mathematical treatises are mentioned in the catalogue of another section of the Imperial library in Hue, *Co hoc vien thu tich thu sach* [古學院書籍守冊] (Han-Nom Institute, call number A2601/6).

student] diverts [in understanding?] for [only] one hao [毫][23], [it will result] in an error of [one] thousand li [里][24]. As I was studying vigorously the [correct] understanding of computations, I comprehended [only] roughly the major principles of the system [or: the Summarized Fundamentals of Computational Methods [算法統宗 Suan Fa Tong Zong]]. How could it be possible not to establish [these] ready-made [computational] methods [to provide] instructions?! [I did it] in order to expose [the methods] to the people of the future generations [of my family] and to make it easy [to acquire] refined knowledge for those of them who begin their study from the "shallow waters" [in aspiring] to "enter into the depth"! Thus, on the Thanh Minh Festival [清明節] of the [year] Canh-thin [庚辰], early at the dawn in auspicious hour, in [my] "Garden of Bliss" I am writing [this] story today. Being inspired by the hidden indices of the Nine Canons,[25] I examined the origins and the subtleties of the Six Arts [of Antiquity, of which the sixth one was mathematics]. [If one relies upon] only that what is held by those who comprehend computations, [it] would suffice [only] for a short-range application; [yet] could it be possible to use [only] minor skills and thus to consider the Great Models [of Antiquity?] unimportant and useless?! [In applying] the sharpness and subtlety of their divine knowledge to [my] "grasping the main rope and collar" [= the central point], [I] assembled and completed a book on counting [which] will be forever transmitted [in our] family to instruct multitudes of learners [making them] comprehend [the subject] easily and then to gather [all] the subtle techniques of the truthful origin. When contemplating this *Chi Minh Toan Phap* [指明算法], [one can see that] I did not dare to "manufacture the antiquity" in order to "put others in a basket" [= to restrain them] with the algorithms [or: technicalities]. [This is because] I think that the merits of the sages of the past were endless, [but] the [things they] used in their days do not go beyond those existing nowadays. The future students should take [these methods] to develop [them by analogy]. This is the preface for the book that follows.

On the Thanh Minh Festival of the [year] Canh-thin, the first year of the [era] Minh mang, early at the dawn in auspicious hour,

Compiled by "Old Peasant" Phan Huy Khuong from the Commune Dong Ngac in District Tu Liem of the Administrative Region Quoc Oai [under jurisdiction] of the Administrative Center Son Tay.

[Done] in the Library "Retreat of Blissful Heaven".

The text of this Preface contains a number of elements relevant to the present study. To begin with, it provides the family name of the author, Phan, and his style (literary name), Khuong. According to the front page of the manuscript A.1240, the name of the author is Phan Huy Khuong [潘輝框]; this name is also mentioned in the beginning of chapters 3 and 4. The identity of the author, elements of his biography and even his lifetime remain unknown.[26] The year Canh-thin coinciding with the

---

[23] *Hao* is a Chinese unit of length equal to 10^{-4} of the unit *chi*; during the Qing dynasty (1644–1911) 1 *chi* was approximately equal to 32 cm.

[24] *Li* is the largest unit in the traditional Chinese system of linear measures; during the Qing dynasty it was approximately equal to 576 m.

[25] This may be a reference to the "Nine Categories", i.e., the Computational Algorithms of Nine Categories [九章算術 Jiuzhang Suanshu], the Chinese mathematical classic of the Han dynasty; see [1].

[26] According to an anonymous author, Khuong was one of the names of Phan Huy On (1754–1786) see http://www.vietgle.vn/trithucviet/detail.aspx?key=Phan+Huy+%C3%94n&type=A0 (retrieved on August 31, 2009); yet this identification cannot match well with the Preface: its author suggests that the Preface was compiled on the Thanh Minh festival of the year Canh-thin (Chinese gengchen), but in the year Canh-thin 1760 Phan Huy On was only 6 years old, and by the next

first year of the era Minh mang [明命元年庚辰] mentioned as the date of compilation of the Preface in both extant manuscripts provides the date of the publication of the treatise (1820) found in the catalogue of [14].

An interesting aspect of the Preface is that the author quotes and mentions a number of treatises while incorporating their titles in his text as parts of sentences, for example, the obscure phrase "The original meaning of the 'embracing numbers' is certainly like that!" in the opening part of the Preface contains an almost explicit reference to the *Yuan Bao Shu Yi* [元包數義] chapter of the numerological treatise *Yuan Bao Shu Zong Yi* [元包數總義] by Zhang Xingcheng [張行成] (進士 jinshi 1132). Another example is offered by the phrase 粗得統宗。可不立成法 which can be understood as "··· I comprehended [only] roughly the major principles of the system. How could it be possible not to establish [these] ready-made [computational] methods [to provide] instructions?!" as well as a reference to two mathematical books: "I comprehended [only] roughly the Summarized Fundamentals [統宗 Tong Zong] [of Computational Methods [算法 Suan Fa]]. How could it be possible not to establish [these] Ready-made methods of [computations] [立成筭法] [to provide] instructions?!" The former title could be referring to the aforementioned Summarized Fundamentals of Computational Methods [算法統宗 Suan Fa Tong Zong] (1592) by Cheng Dawei [程大位], as well as to an abridgment of it available in Vietnam, for example, to the Systematic Treatise on Computational Methods [統宗算法 *Thong Tong Toan Phap*], a treatise of unknown date authored by one Ta Huu Thuong [謝有常].[27] The title of this treatise makes an obvious allusion to the Summarized Fundamentals of Computational Methods. Indeed, certain parts of the Chinese treatise are quoted verbatim, as, for example, the versified rules of calculation of areas of plane figures,[28] the problem of two walkers,[29] and so on. Nevertheless, the compiler of the *Thong Tong Toan Phap* considerably modified sections of the Chinese book in inserting a number of problems not found in the original Chinese treatise, adapting the Chinese original to the Vietnamese measure units, and providing his explanations in Nom (the writing system used to transcribe Vietnamese language with Chinese characters and newly created characters based on Chinese ones). The reader will find below an example of a mathematical problem in Phan Huy Khuong's treatise quoted practically verbatim from the treatise of Cheng Dawei.

The other title "hidden" in the preface by Phan Huy Khuong, the Ready-made methods of [computations] [立成筭法], may have referred to the *Chi Minh Lap Thanh Toan Phap* itself as well as to the mathematical treatise *Cuu Chuong Lap Thanh Toan Phap* [九章立成算法] probably edited by Phan and appended to the former one (see below). But the most intriguing among these hidden references is the obscure mention of the Great Model(s) [大典 Da Dian]; this term, on the

Canh-thin year, 1820, he had been long dead. According to Gaspardone 1934, Phan Huy On's original style was Trong-Duong [仲洋], which he changed to Hoa-Phu [和甫] in 1780 (p. 83, n.1).

[27] One manuscript copy of the latter treatise is preserved in the National Library of Vietnam (Hanoi). Call number R.1194; this treatise is not listed in Tran and Gros [14].

[28] *Thong Tong Toan Phap*, pp. 27–29, Cheng [2, pp. 226–227].

[29] *Thong Tong Toan Phap*, pp. 207–208, Cheng [2, pp. 895–896].

one hand, may have been referring to the great mathematical books of antiquity in general, while, on the other hand, suggesting that the author was familiar with the mathematical sections of the collection Great Canon of the Yongle Era [永樂大典 Yong Le Da Dian] compiled in China 1404–1408, or with some mathematical treatises based upon them.[30]

In the opening part of the Preface the author mentions four actions coming from the Great Learning: self-improvement, family regulation, ordering of a state, and pacification of the civilized world [修, 齊, 治, 平]; he states that these terms can be given mathematical interpretations. The exact meaning of this claim remains unknown, yet one can suggest a tentative explanation of the rather obscure opening paragraph. Phan mentions the procedures of "excessive and deficient [amounts]" and "equal and proportional [distribution]," that is, the Rule of Double False position and the method of weighted distribution. The first problem on the Rule of Double False position in his treatise (Problem 29 of Chapter 4) reads as follows: a certain number of people buy a thing or things; if each of them pays 5 *van* [文 wen], then the total amount will exceed the price by 6 *van*; if each of them pays 3 *van*, the total amount will be less than the price by 4 *van*. The number of people and the price of the things are to be found. The answer is found according to the algorithm for the first time recorded in the Book on Computations with Counting Rods [算数書 Suan Shu Shu][3, pp. 81-88] and Computational Algorithms of Nine Categories [九章算術 Jiu Zhang Suan Shu] [1, p. 558 ff.]: 5 *van* is to be multiplied by 4 *van*, 3 *van* by 6 *van*, the products added to obtain the "dividend of things" [物寔], 38 ; one should also sum up the excess and deficit to obtain the "dividend of people" [人寔], 10; the difference of the amounts paid by each individual $(5-3=2)$ is simply called "Divisor" [法]. To calculate the amounts of the people, 5, and the price of the "thing(s)", 19, one has to divide each respective "dividend" by the "Divisor."

In modern notation, the problem and the method can be rendered as follows: if $Nm_1 = X+E, Nm_2 = X-D$, then $X = (m_1D+m_2E)/(m_1-m_2), N = (E+D)/(m_1-m_2)$; here m_1 and m_2 are amounts of money paid by each individual $(m_1 > m_2)$, E and D are excess and deficit, respectively, and X is the price of the purchased object.[31] An interesting detail is that the choice of the numerical values in the Vietnamese treatise is rather particular: the author uses the consecutive natural numbers 3,4,5,6: $m_2 = 3, D = 4, m_1 = 5, E = 6$. However, the numerical parameters of this problem were not invented by the Vietnamese author himself; they are the same as those of Problem 1 of Chapter 10 of the Summarized Fundamentals of Computational Methods [2, p. 675]. The texts of the two problems differ only slightly: in the condition of his problem Cheng Dawei uses different monetary units, and the technical terms employed in the algorithms are not the same.

---

[30] A large part of the mathematical section of the Great Canon of the Yongle Era is now lost; in particular, lost are the volumes of the collection containing fragments of the ancient mathematical treatise Computational Algorithms of Nine Categories [九章算術 Jiu Zhang Suan Shu], except for small portions of the text, see Chemla and Guo [1, pp.72–73].

[31] The formulas were discussed by numerous historians; for one of the most recent works presenting the origins of the method in ancient China the reader is referred to Chemla and Guo [1, pp. 549–555 and 849–860].

Among the operations performed to obtain the result one can see the multiplication of the values D and E by m_1 and m_2, respectively. Among the four terms mentioned by the author, 修, 齊, 治 and 平, the term 齊 was used by the Chinese commentator Liu Hui [劉徽] (fl. AD 263) in the context of reduction of fractions p_1/q_1 and p_2/q_2 to common denominator as referring to the multiplication of the numerators p_1 and p_2 by denominators q_2 and q_1, respectively. One can notice a certain parallelism between the mathematical meaning of the term 齊 used by Liu Hui and by Phan Huy Khuong, while the mathematical interpretation of the three other terms, 修, 治 and 平, meant by Phan remains to be investigated.

One more obscure statement of Phan found in the very beginning of his preface, that the method of false position is related to "joining together" while the method of weighted distribution, to "separating apart", may have a rather straightforward interpretation: the final result of the operations performed to find the weighted distribution of a given amount corresponding to given weighting coefficients $k_1, k_2, \cdots$ consists of calculating two or more magnitudes proportional to the coefficients; the given amount is thus "separated apart". On the contrary, the algorithm of solution of problems on "double false position", in very general terms, consists of constructing the sole amount $(m_1D + m_2E)/(m_1 - m_2)$ on the basis of the given "elements" E, D, m_1, and m_2.

One more difficult question to answer is: what was the book Phan wrote his Preface for? The Preface itself refers to a *Toan Phap Chi Minh* [筭法指明], this title, formally speaking, differs from both *Chi Minh Toan Phap* and *Chi Minh Lap Thanh Toan Phap*. It is possible that *Toan Phap Chi Minh* and *Chi Minh Toan Phap* may have been perceived by Vietnamese *literati* as one and the same title written down according to the norms of literary Chinese and of Nom (in which the order of the adjectives and nouns is reversed comparing to Chinese). However, the contents of the Preface does not contain any specific reference to the *Chi Minh Lap Thanh Toan Phap* and to its contents; the author mentions, although not explicitly, only two topics, the Rule of the False Position and the method of weighted distribution. Both topics are treated in Chapter 4 of the *Chi Minh Lap Thanh Toan Phap*, and if the Preface originally accompanied the extant version of the book, it remains unclear why other topics discussed in the treatise remained unmentioned by the author.[32]

5 The Appendix of Mikami's book

In his paper Mikami mentions that the book he investigated contained an (apparently, untitled) Appendix in one chapter devoted, as he reports, to the "extraction of roots and other [topics]" [12, p. 8]; he provides a cursory analysis of this chapter

[32] Since Mikami does not mention that the treatise he investigated had a preface, one cannot completely rule out the possibility that the Preface was originally written by Phan Huy Khuong for another mathematical treatise and was combined together with the rest of the treatise by later editors.

and quotes three excerpts from it (pp. 8–9).[33] One contains a problem of division of a truncated cone, the other is a problem on production of gunpowder, and the third one is a problem of conversion of monetary units.

The first problem, to my knowledge, is not found in the extant Vietnamese treatises and deserves to be reproduced here. It reads as follows:

今有木一段。長十尺。上徑七寸九分。下徑一尺一寸九分。秤量一百斤。茲鋸為二段。每重五十斤。問。每段長並徑各若干。

答曰。上段長五尺五寸。原上徑七寸九分。今下徑一尺一分。下段長四尺五寸。原下徑一尺一寸九分。今上徑一尺一分。

法曰。置上、下徑為一。折半。以長因之。得木九百九十寸。[······]

Now [let us assume that] there is a piece of wood, [its] length is 10 *xich* [尺],[34] the upper diameter is 7 *thon* [寸] 9 *phan* [分], lower diameter is 1 *xich* 1 *thon* 9 *phan*. The weight is 100 *can* [斤]. And now [one] cuts it into two pieces. Each piece weighs 50 *can*. The question is: what is the length and also the diameter of each piece?

The answer is: the length of the upper piece is 5 *xich* 5 *thon*, the original upper diameter is 7 *thon* 9 *phan*, the lower diameter newly [obtained] is 1 *xich* 1 *phan*. The length of the lower piece is 4 *xich* 5 *thon*, the original lower diameter is 1 *xich* 1 *thon* 9 *phan*, the upper diameter newly [obtained] is 1 *xich* 1 *phan*.

The method: set [on the counting device] upper and lower diameters and make them one. Divide [the obtained value] in halves. Multiply it by the length. Obtain that the wood is 990 *xich*. [······]

Unfortunately, Mikami stops quoting here the solution of the problem while reporting that the entire algorithm occupies about ten lines. The wording of the problem makes the reader think that the piece of tree having the shape of a truncated cone is supposed to be visualized in horizontal position (hence the "length" instead of "height"). One has to find a position for a cross-section going parallel to the two bases of the truncated cone such that the upper and the lower sections would have equal volumes. Since the area of the cross-section is a quadratic function of the distance from the upper (or lower) base, the correct solution of the problem would have involved dealing with a cubic equation. However, as the numerical answer shows, the solution was mathematically incorrect and instead of the division of a truncated cone the anonymous author solved a plane problem of division of a trapezium with the upper and lower bases $a = 0.79$ and $c = 1.19$ and the height $H = 10$ into two sections with equal areas (see Fig. 1).

[33] Technically, Mikami explicitly states that only the first excerpt (the problem on the dissection of a truncated cone) is found in the Appendix, yet the layout of his paper makes the reader understand that the two other excerpts quoted in sections 11 and 12 of his paper (pp. 8–9) were also found in the Appendix. Moreover, the two latter excerpts quoted by Mikami are not found in the manuscript A.1240.

[34] In this translation one encounters the units of measure *phan*, *thon* and *xich* (1 *xich* = 10 *thon* = 100 *phan*) and the unit of weight *can*; since the date of compilation of the text quoted by Mikami is unknown, their actual values cannot be specified.

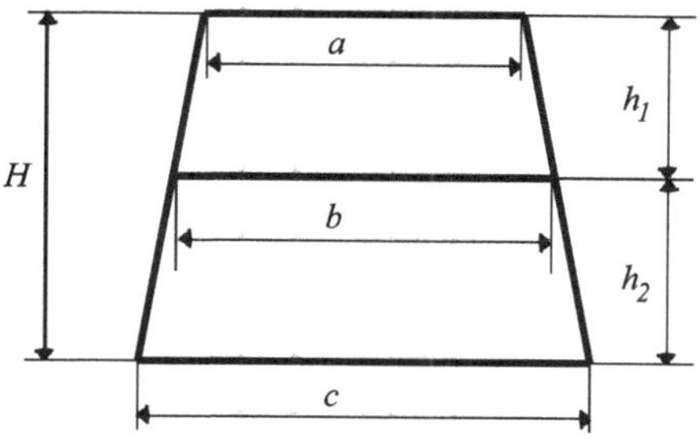

Fig. 1. Subdivision of a trapezium into two equiareal parts

According to the solution, $h_1 = 5.5, h_2 = 4.5$, and $b = 1.01$; for these values the area of the upper part of the trapezium is indeed equal to the area of the lower part. Conversely, as one might expect, the values h_1 and h_2 do not yield the correct solution of the original problem, since the ratio of the volumes of the upper and the lower parts for these values of h_1 and h_2 approximately equals to $1 : 1.22$. The beginning of the solution quoted by Mikami contains the calculation of the area of the trapezium; this operation may be necessary for solving the planar problem. However, Mikami does not quote the entire algorithm and thus does not leave any clue as to what was the method used to solve the planar problem (which would have involved solution of a quadratic equation). It remains entirely unknown whether the original cubic problem was supposed to be solved with an algebraic method originating from China, Europe or elsewhere.[35]

The second excerpt from the Appendix quoted by Mikami is directly related to his attempt to find the date of compilation of the Appendix. The problems he quotes from the Appendix mentions the composition of gunpowder. Mikami does not provide any definitive dates for the introduction of firearms and gunpowder in Vietnam (which in itself is an interesting and difficult question which certainly could not be answered satisfactorily in 1930s), yet cautiously suggests that a problem focusing on such topic most likely was not composed earlier than the late Song – early Yuan dynasty (p. 9). According to modern scholarship, the use of artillery in Vietnam is documented from as early as the late 14th century. Since then, there are various accounts concerning the use of cannons by the Vietnamese: by 1631 a state foundry was established in the area of Hue (Central Vietnam), and since the late 1650s the cannons were produced massively under the direction of an obscure half-Portuguese military expert João da Cruz (ca. 1610–1682); in 1740s a Western visitor reported that he saw some 1200 cannons in Central Vietnam.[36] The production of good quality gunpowder was one of essential activities of the trade, and it would be quite natural that mathematical problems of this type may have become rather popular by the late 17th – early 18th century, or even earlier. The mention of gunpowder

[35] A legend about Seki Takakazu [關孝和] preserved in the Burin Inkenroku [武林隱見録] mentions a problem of division of a trunk of a tree (most likely, a cone or a truncated cone) into parts according to a given ratio of volumes, Horiuchi [5, p. 130] ; if this interpretation of the legend is correct, the problem found in the Appendix is a special case of Seki's problem for the ratio 1:1.

[36] On the history of firearms in Vietnam see, for example, Li Tana [10, pp. 43-46], Sun [13] and Volkov [21] .

thus could provide but a very approximate date for the time of compilation of the Appendix while suggesting that in Vietnam the weaponry production might have required particular mathematical training. To my knowledge, a problem on composition of gunpowder (with different numerical values) can be also found in the Great Compendium of Mathematical Methods [算法大成 *Toan Phap Dai Thanh*],[37] another book of unknown origin and date, conventionally credited to the authorship of Luong The Vinh [梁世榮] (1441–1496?) and preserved in two manuscript copies in Han-Nom Institute in Hanoi.[38]

On p. 10 Mikami concludes that the Appendix is compiled later than the treatise, basing this conclusion on the mention of gunpowder. Even though Mikami avoids suggesting any decisive dates for the *Chi Minh Toan Phap*, his detailed discussion of the multiplication table gives the impression that Mikami was inclined to think of the treatise he had at disposal as of a rather early work, probably of the late Yuan or early Ming time. His detailed discussion of the multiplication table found in the treatise (pp. 5–7), if read in the context of Mikami [11], is directly related to the search for the date of the treatise. In his 1921 paper [11] he showed that the order of products in the multiplication tables found in Chinese treatises was reversed by the late Song – early Yuan period: in the earliest extant books the tables always began with 9×9, while in the later treatises they began with 1×1. The table in the Vietnamese treatise under investigation begins with the product 9×9 and thus could have been considered by Mikami a piece of evidence supporting a hypothesis of the early origin of the Vietnamese text (or at least of the multiplication table found in it). Together with certain archaic mathematical terms and the counting instruments (counting rods) presumed to be used to solve the problems, the particular order of the multiplication table convinced him that the treatise he explored was anterior to the late Ming while the mention of gunpowder in the Appendix made him believe that the Appendix was written later than the treatise. I will return to his conclusion in the last part of this article.

6 The Appendix of A.1240

Interestingly enough, the manuscript copy of the *Chi Minh Lap Thanh Toan Phap* with the call number A.1240 also contains an Appendix in one chapter. However, unlike the text seen by Mikami, this Appendix has an individual title *Cuu Chuong Lap Thanh Toan Phap* [九章立成算法], a short preface, and a table of contents. The table of contents contains editorial remarks indicating which subjects have already been discussed in the *Chi Minh Lap Thanh Toan Phap* and therefore were intentionally removed from this text. Moreover, the title *Chi Minh Toan Phap* is systematically featured on pages of the Appendix as a running head in the manuscript A.1240. This suggests that this treatise was reworked before being appended to the previ-

[37] Problem 68, MS A.2931, pp. 62a–b.

[38] For a preliminary analysis of this treatise see Volkov [15]; on Luong and his treatise, see below.

ous one, yet the extent of the editorial work is difficult to evaluate. The Preface to the *Cuu Chuong Lap Thanh Toan Phap* contains two parts. The first part is a rather long excerpt extracted, as a commentary suggest, from the biography of Luong The Vinh[39] [梁世榮] (1441–1496?), a famous *literatus* and a high-rank official of the Le [黎] dynasty (1428–1789) conventionally considered to be good at mathematics. His biography is indeed found in the collection of biographies Records of Successful Examinees [登科録 *Dang Khoa Luc*] by Nguyen Hoan [阮俒] (1712–1791);[40] the microfilm copy preserved in EFEO (Paris) does contain the biography on p. 10b of chapter 1. It is mentioned that Luong compiled (established [立]) the *Dai Thanh Toan Phap* [大成算法]. The title of the book differs from the title of the extant book *Toan Phap Dai Thanh* [算法大成] credited to his authorship and preserved, in two manuscript copies, in the Han-Nom Institute in Hanoi.[41] The name of the presumed author ("Doctor Luong The Vinh") is written only on the first page of each manuscript next to its title; however, a close inspection of the original manuscript A.2931 shows this page was added later, and, since the manuscript VHv.1152 is a copy of A.2931, the authorship of Luong The Vinh cannot be considered proven.

The second part of the preface is a brief note stating that one Pham Huu Chung [范有鍾] "gathered the most important elements [of the book of Luong?] and compiled [the *Cuu Chuong Lap Thanh Toan Phap*] in the 'phonetic [script] of the country' (that is, in Nom, the phonetic script using Chinese characters or their derivatives) in order to make it easier for the beginners (後學范有鍾。。。撮要撰為國音。以便初學。" and then block-printed his work in the second year of the Bao Thai [保泰] era (1720–1729), that is, 1721. The treatise is, indeed, written in Nom. We shall meet Pham Huu Chung later when discussing the use of the counting rods in the treatise investigated by Mikami. To conclude this brief investigation of the Preface, one can argue that its author(s) suggested a new hypothesis as to what the original title of the book authored by Luong The Vinh was: according to the preface, its original title was *Dai Thanh Toan Phap*, and it was reworked by Pham Huu Chung in 1721 to produce a book in Nom (the original book of Luong thus must have been written in classical Chinese) titled differently, namely, *Cuu Chuong Lap Thanh Toan Phap* or possibly simply *Cuu Chuong Toan Phap* [九章算法]. It is interesting that a book with the latter title was indeed also credited to the authorship of Luong The Vinh.[42]

The *Cuu Chuong Lap Thanh Toan Phap* contains a number of rhymed sections devoted to general matters related to mathematics as well as to some particular methods such as the calculation of the areas of plane figures and the problems on weighted distribution. However, I was unable to locate in the *Cuu Chuong Lap*

---

[39] For biographies of Luong The Vinh see Volkov [17].

[40] See [4, p. 68, n. 1], for a biography of Nguyen Hoan.

[41] The call numbers are VHv.1152 and A.2931. It is unknown when manuscript A.2931 was produced (it certainly happened prior to 1934), while manuscript VHv.1152 is a copy of the manuscript A.2931 made in 1944.

[42] The book with this title as authored by Luong The Vinh is mentioned in the *Nam Su Tap Bien* [南史輯編] (1896), while an inscription in his temple located in his native village mentions a book titled *Cuu Chuong Toan Hoc* [九章算學], Volkov [16].

Thanh Toan Phap the three excerpts quoted by Mikami from the Appendix of the book he explored. One can therefore conclude that the Appendices attached to the Mikami version and to the manuscript A.1240 were different.

7 The counting rods in Vietnam: Mikami's evidence

Mikami reports that the *Chi Minh Toan Phap* contains pictures of configurations of counting rods representing numbers in the multiplication table 9×9 (p. 4). It remains unknown whether the counting rods represented all the numbers in the table or only the products, and whether they represented all products, from 9×9 to 1×1, or only some of them. Mikami provides the configurations of counting rods used in China and stresses that the way of representing numbers in the Vietnamese treatise differed from the Chinese method (pp. 4–5); the Vietnamese configurations of rods representing the numbers from 1 to 9 looked as follows (Fig. 2):

Fig. 2. The counting rods numerals as found in the *Chi Minh Toan Phap*

It is known that the mathematicians (and, in particular, astronomers) of Tonkin (Northern Vietnam) still used counting rods in the mid-17th century [19]. Moreover, the solution of the model examination problem found in the *Chi Minh Lap Thanh Toan Phap* (Problem 38 of Chapter 4) contains a reference to the instrument *toan* [筭 suan in Chinese] which well may have been the counting rods.[43] However, the received manuscript copies A.1240 and VHv.1185 of the *Chi Minh Lap Thanh Toan Phap* do not contain any pictures of configurations of counting rods. My cursory inspection of the available Vietnamese mathematical texts revealed only one mathematical treatise containing the images of counting rods, the *Cuu Chuong Lap Thanh Tinh Phap* [九章立成併法] Ready-made Methods of Addition of Nine Categories.

The catalogue [14] lists this book under the title *Cuu Chuong Lap Thanh Toan Phap*, even though the actual titles of the two listed block-printed editions read *Cuu Chuong Lap Thanh Tinh Phap*, that is, Ready-made Methods of Addition of Nine Categories.[44] One more copy of it is preserved in the Bibliothèque Nationale (Paris, France). This relatively short (the block-printed book with the call number AB.53 contains only 22 double pages) treatise featuring problems and algorithms in classical Chinese with long explanations in Nom is authored by the aforementioned

---

[43] As states the commentary on the *Cuu Chuong Lap Thanh Toan Phap* appended to the manuscript A.1240 and probably also written by Phan Huy Khuong, the "round counters" [圓算子] referred to the abacus [算盤], while the "square counters" [方算子], to the counting rods.

[44] The catalogue mentions two editions with the call numbers AB.53, AB.173; for the descriptions in [14], see items 638 and 3503. There exists one more block-printed copy (dated 1721) of the text preserved in the Library of the Han-Nom Institute entitled 九章立成 and 九章立成併法; it has the call number VHb.374 and is slightly different from the editions AB.53 and AB.173; it is not mentioned in [14].

Pham Huu Chung [范有鍾]. One of the three block-printed editions found in the Institute of Han-Nom Studies (AB.173) was printed in 1713, while the dates of printing of the other editions remain unknown. The treatise bearing the same title *Cuu Chuong Lap Thanh Tinh Phap* preserved in the National Library of Vietnam (Hanoi) bearing the call number R.1649, is identical with the block-print AB.53 from the Han-Nom Institute.

The treatise is not subdivided into chapters and consists of short sections devoted to discussions (often in versified form) of various topics such as multiplication table, calculation of areas of plane figures, weighted distribution, and operations with common fractions. The treatise also contains a number of mathematical problems presented in the traditional format "question – numerical answer – algorithm". A cursory analysis shows that the block-printed book differs considerably from the *Cuu Chuong Lap Thanh Toan Phap* appended to the *Chi Minh Lap Thanh Toan Phap*, as far as its composition and the contents are concerned.

The counting rods numerals printed in the *Cuu Chuong Lap Thanh Tinh Phap* (call number A.53), however, are quite particular. To begin with, the counting rods numerals are provided only for the products of 9 (that is, for the numbers 81, 72, 63,$\cdots$, 9) and for the table of division by 9 which, textually, does not differ from the multiplication table (pp. 3a–4a and 4a–5a, respectively).[45] Some of the configurations are easily recognizable, while others appear rather unconventional:

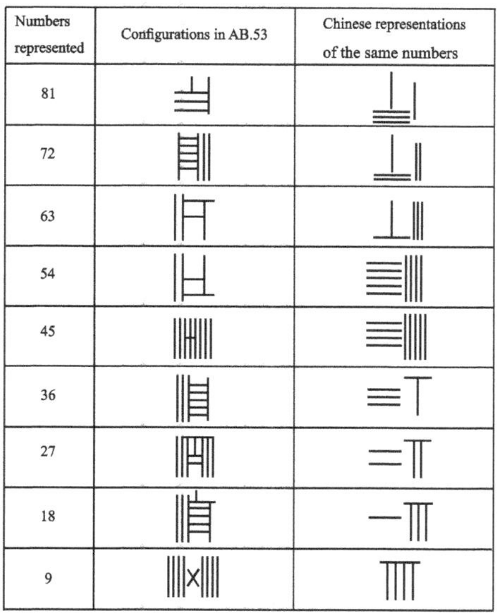

Table 2. Configurations of counting rods used to represent the numbers $n\times 9, n=9,8,\cdots,1$ in the edition AB.53 of the *Cuu Chuong Lap Thanh Tinh Phap*

[45] The block print AB.173 provides the counting rods configurations for other parts of the multiplication table.

As the reader can see, some configurations are identical with their traditional Chinese counterparts (such as the one representing the number 81), while some others look rather different. They remain difficult to interpret even if one adopts the modified method of representation of digits described by Mikami. It is not impossible that the carvers distorted considerably the original configurations and even placed them in wrong positions,[46] yet the available data does not allow to a sensible reconstruction of the "counting rods numerals" found in this Vietnamese treatise.[47]

The information provided by Mikami can be corroborated, yet only partly, by a commentary on the *Cuu Chuong Lap Thanh Toan Phap* bound together with the manuscript A.1240: a commentator (most probably, Pham Huu Chung himself) describes the representation of numbers with both instruments, the counting rods and the abacus. In particular, he states that the numbers 6 and 9 are to be represented with the counting rods as shown in Table 3. The representation of 6 differs from the Chinese but is identical with that described by Mikami, while the representation of 9 differs from both the traditional Chinese one and from the one provided by Mikami (p. 4).

Number	Representation according to Pham Cuu Chung	Representation as reconstructed by Mikami	Standard Chinese representation
6			
9			

Table 3. Representations of numbers 6 and 9.

The discrepancies between the Vietnamese configurations of counting rods described by Mikami and those found in the extant materials, on the one hand, and the differences between the representations of numbers with counting rods in China and in Vietnam, on the other, deserve a special attention and will be discussed in a further publication.

[46] The configurations for 72 and 36 are symmetrical and, therefore, at least one of them is placed in a wrong position since the symmetrical configurations must represent the numbers ab and ba composed of the same digits.

[47] The configurations from the Vietnamese treatise resemble the pictures used to represent the outcomes of mantic procedures in traditional China, see Kalinowski [7, pp. 43–44, 59].

8 Discussion and Conclusions

The nowadays lost Vietnamese mathematical treatise briefly presented by Yoshio Mikami in 1934 and the extant manuscripts A.1240 and VHv.1185 were apparently related to each other, even though the exact nature of this relationship remains to be investigated. However, on the basis of the known elements one can advance a number of hypotheses concerning the prototype of the treatise and its later versions, and pose several questions related to Mikami's findings and methodology.

1. It appears plausible to ask whether the hypothetical original version of the treatise titled *Chi Minh Toan Phap* (which I shall refer to as "prototype" below) contained the diagrams representing configurations of counting rods in the multiplication table mentioned by Mikami. Technically, there are two possibilities: (a) the prototype did not contain these images, and they were added later (in this case Mikami's copy featured these later additions, unlike the two manuscripts stored in Han-Nom Institute); (b) the prototype originally contained the configurations preserved in Mikami's copy but later editors (or copyists) removed them from the treatise, and this is the reason why they were no longer found in A.1240 and VHv.1185.

The second option looks to me more probable for the following reasons:

(1) The above-mentioned use of the counting rods reported in the mid-17th century and the (distorted?) images of the counting rods found in the extant printed copies of the early 18th century mathematical text *Cuu Chuong Lap Thanh Tinh Phap* suggest that the counting rods were used continuously until the early 18th century or even later.

(2) The text of the model examination paper in Chapter 4 of the *Chi Minh Lap Thanh Toan Phap* mentions a counting instrument *toan* [筭] which most likely was meant by the author to be the counting rods and not the abacus, given that the algorithms described in all the four chapters of the treatise do not mention the operations with latter, at least explicitly. This piece of evidence thus suggests that the counting rods may have been used even as late as the early 19th century.

(3) The manuscripts A.1240 and VHv.1185 do not contain images of the counting rods yet do contain a picture of the abacus apparently copied from the [算法統宗 Suan Fa Tong Zong] (or its Chinese or Vietnamese derivatives). However, this picture of the abacus is found on a page placed in the very beginning of the manuscripts between the title page and the table of contents. The manuscripts A.1240 and VHv.1185 do not contain other references to the abacus or pictures of it. Mikami reports that no mentions of the abacus were found in his copy. It is therefore possible that the picture of the abacus was added to the original text at some point; if it is true, it still remains unknown whether the pictures of the counting rods originally found in the multiplication table were removed at the same time. At any rate, this assumption implies that the copy of Mikami was of an earlier date than the two manuscript copies stored in Han-Nom Institute.

2. The preface of the treatise found in the manuscript copies is dated of the year 1820, while the version studied by Mikami did not contain a preface; this fact could be explained in two ways:

(1) Mikami obtained a damaged copy of the treatise without several front pages but otherwise textually almost completely identical (except for the pictures of the counting rods and abacus) with the A.1240 and VHv.1185, or

(2) the copy of Mikami was complete and originally did not contain a preface. The first option seems to be less probable since the two extant copies contain the name of the presumed compiler in the beginning of the chapters 3 and 4, while Mikami's copy did not contain any mention of the author. The second option, however, still does not rule out the possibility that the book was indeed compiled by Phan Huy Khuong and completed in 1820.

3. If the mathematical problems found in Mikami's version were identical with those of the manuscripts A.1240 and VHv.1185, the mathematical book he investigated could not be compiled earlier than the late 16th century; this conclusion follows from the fact that the first problem on the Rule of Double False Position discussed above is identical with a problem from the Summarized Fundamentals of Computational Methods [算法統宗 Suan Fa Tong Zong] printed in 1592.

4. The observations listed above suggest various scenarios of the production and textual evolution of the treatise in which Phan Huy Khuong would have played different roles, from the author of the entire treatise to the editor and/or the author of detailed computational procedures he added to an anonymous treatise; yet at any rate the time when the treatise was compiled cannot be anterior to the late 16th century for the reasons mentioned above. The question which naturally emerges under these circumstances is: how the latter fact can be reconciled with the observations of Mikami related to the archaic structure of the multiplication table?[48] Theoretically, if the multiplication table was copied from an early Chinese or Vietnamese treatise, while the rest of the treatises was partly written anew and partly borrowed from the Summarized Fundamentals of Computational Methods [算法統宗 Suan Fa Tong Zong], it may explain the discrepancy, yet this reconstruction remains disturbingly conjectural.

5. The above-mentioned problem related to the multiplication table challenges the very methodology of the search of datable elements in the mathematical treatises under investigation suggested by Mikami. If various elements of a treatise can be dated of different time periods, what could be the sensible hypotheses concerning the time of its compilation? More broadly, what precisely the "time of compilation" would mean in the context of production of mathematical textbooks in which a number of elements might have been intentionally preserved in their archaic forms?

[48] The idea of Mikami (later used by Li Yan) related to the structure of the multiplication table was applied to the investigation of another Vietnamese mathematical treatise in [15, pp. 383–384].

Glossary of Vietnamese terms

This Glossary contains the Vietnamese terms used in the article. The list is organized alphabetically, it provides

(1) the term as found in the article (without diacritics),

(2) the same term written in Hán-Nôm [漢喃] script (Vietnamese written language), and

(3) the term written in the Quoc Ngu transliteration system used in Vietnam nowadays.

B

Bao Thai	保泰	Bảo Thái

C

can	斤	*cân*
Canh-thin	庚辰	Canh-thìn
Cao Mien The Thu	高綿世次	*Cao Miên Thế Thự*
Chi minh lap thanh toan phap	指明立成筭法	*Chỉ minh lập thành toán pháp*
Chi minh toan phap	指明算法	*Chỉ minh toán pháp*
Cuu chuong lap thanh tinh phap	九章立成併法	*Cửu chương lập thành tính pháp*
Cuu chuong lap thanh toan phap	九章立成算法	*Cửu chương lập thành toán pháp*
Cuu chuong toan hoc	九章算學	*Cửu chương toán học*
Cuu chuong toan phap	九章算法	*Cửu chương toán pháp*

D

Dai thanh toan phap	大成算法	*Đại thành toán pháp*
Dang khoa luc	登科錄	*Đăng khoa lục*
Dong Ngac	東鄂	Đông Ngạc

H

Hoa-Phu	和甫	Hoà-Phủ

L

Le	黎	Lê
Luong The Vinh	梁世榮	Lương Thế Vinh

N

Nam su tap bien	南史輯編	*Nam sử ttập biên*
Nguyen Hoan	阮俒	Nguyễn Hoản
Nguyen Thi Duong	阮氏楊	Nguyễn Thị Dương
Noi cac	內閣	Nội các
Noi cac thu muc	內閣書目	*Nội các thư mục*
Nom	喃	Nôm

P

Pham Huu Chung	范有鍾	Phạm Hữu Chung
phan	分	*phân*
Phan Huy Khuong	潘輝框	Phan Huy Khuông
Phan Huy On	潘輝溫	Phan Huy Ôn

Q

Quoc Ngu	國語	Quốc Ngữ
Quoc Oai	國威	Quốc Oai

S

Son Tay	山西	Sơn Tây

T

Ta Huu Thuong	謝有常	Tạ Hữu Thường
thon	寸	*thõn*
Thong tong toan phap	統宗算法	*Thống tông toán pháp*
Toan	筭/算	Toán
Toan phap chi minh	筭法指明	*Toán pháp chỉ minh*
Toan phap dai thanh	算法大成	*Toán pháp đại thành*
Trong-Duong	仲洋	Trọng-Dương
Tu Liem	慈廉	Từ Liêm

V

van	文	*văn*

X

xich	尺	*xích*

Acknowledgements This work was partly supported by the research grant allocated to the Project 近代東西文明的遭遇與衝撞 (Modern Civilizations of West and East: Encounter and Mutual Impact) in the framework of the Project of Promoting Academic Excellence & Developing World Class Research Centers of the National Tsing Hua University, Hsinchu, Taiwan.

References

1. Karine Chemla et Guo Shuchun [郭書春] : *Les Neuf chapitres. Le classique mathématique de la Chine ancienne et ses commentaires*, Dunod, Paris (2004).
2. Cheng Dawei [程大位] : *Summarized Fundamentals of Computational Methods, with Emendations and Explanations* [算法統宗校釋 Suan Fa Tong Zong Jiaoshi], edited by Mei Rongzhao [梅榮照] and Li Zhaohua [李兆華], Anhui Education Publ. [安徽教育出版], Hefei (1990).
3. Christopher Cullen : The Suan Shu Shu 'Writing on reckoning' : A Translation of a Chinese Mathematical Collection of the Second Century BC, With Explanatory Commentary, *Needham research institute working papers 1*, Needham Research Institute, Cambridge (2004).
4. Emile Gaspardone : Bibliographie Annamite, *Bulletin de l'École Française d'Extrême-Orient*, vol. 34, no. 3, pp. 1–172 (1934).
5. Annick Horiuchi : *Les mathématiques japonaises à l'époque d'Edo*, Vrin, Paris (1994).
6. Seiji Ito [伊藤清司] : In memorial of Master Matsumoto Nobuhiro [松本信廣先生を偲んで Matsumoto Nobuhiro sensei wo shinonde], *The Japanese journal of ethnology* [民族學研究] vol. 46(1), pp. 125–127 (1981).
7. Marc Kalinowski : La divination par les nombres dans le manuscrits de Dunhuang, *I. Ang and P.-E. Will (eds.) : Nombres, astres, plantes et viscères : Sept essais sur l'histoire des sciences et des techniques en Asie orientale*, pp. 37–87, Collège de France, Paris (1994).
8. James Legge : *Confucius*, Dover, New York (1971).
9. Li Di [李迪] (vol. ed.) : *Integrated Catalogue of Chinese Mathematical Books* [中国算学書目匯編 Zhongguo Suanxue Shumu Huibian], Supplementary volume 2, *Wu Wenjun* [吳文俊] *(gen. ed.) : Grand Encyclopedia on the History of Chinese Mathematics* [中国数学史大系 Zhongguo Suanxueshi Daxi], Beijing Shifan Daxue Publishers [北京師範大学出版社], Beijing (2000).
10. Li Tana : *Nguyen Cochinchina, Southern Vietnam in the Seventeenth and Eighteenth Centuries*, Cornell University Press, Ithaca (N.Y.), SEAP (1998).
11. Yoshio Mikami [三上義夫] : On the [multiplication table] 'Nine times nine' [九九に就て Kuku-ni tsuite], *Tōyō gakuhō* [東洋學報], vol. 11, pp. 102–118 (1921).
12. Yoshio Mikami [三上義夫] : On one mathematical book from Annam (= Vietnam) [安南の一算書に就て Annan-no ichi sansho-ni tsuite], *Gakkō sūgaku* [學校數學], vol. 14, pp. 3–11 (1934).
13. Sun Laichen : Military Technology Transfers from Ming China and the Emergence of Northern Mainland Southeast Asia, *Journal of Southeast Asian Studies*, vol. 34, pp. 495–517 (2003).
14. Tran Nghia et François Gros : *Catalogue des livres en Han-Nom*, Edition des sciences sociales, Hanoi (1993).
15. Alexei Volkov : On the origins of the Toan phap dai thanh (Great Compendium of Mathematical Methods), *Y. Dold-Samplonius et al. (eds.) : From China to Paris: 2000 years transmission of mathematical ideas*, Franz Steiner, Stuttgart pp. 369–410 (2002).
16. Alexei Volkov : Traditional Vietnamese Mathematics: The case of Lương Thế Vinh (1441–1496?) and his treatise Toan phap dai thanh (Great Compendium of Mathematical Methods), *U Kyi Win (ed.) : Traditions of Knowledge in Southeast Asia*, Myanmar Historical Commission, Yangon, vol. 3, pp. 156–177 (2005).
17. Alexei Volkov : State mathematics education in traditional China and Vietnam: formation of mathematical hagiography of Luong The Vinh (1441-1496?), *Trinh Khac Manh and Phan Van Cac (eds.) : Nho Giao o Viet Nam / Confucianism in Vietnam*, Social Sciences Publishing House, Hanoi, pp. 272–309 (2006).

18. Alexei Volkov : Mathematics in Vietnam, *Helaine Selin (ed.) : Encyclopedia of the History of Science, Technology, and Medicine in Non-Western Cultures*, (2nd ed.), Springer-Verlag, Dordrecht etc., pp. 1425–1432 (2008).
19. Alexei Volkov : Mathematics and Mathematics Education in Traditional Vietnam, *E. Robson and J. Stedall (eds.) : Oxford Handbook of the History of Mathematics*, Oxford University Press, pp. 153–176 (2009).
20. Alexei Volkov : Demonstration in traditional Chinese and Vietnamese Mathematics: Argumentation for State Examinations, *K. Chemla (ed.) : History and historiography of mathematical proof in ancient traditions*, Cambridge UP, Cambridge (forthcoming).
21. Alexei Volkov : Evangelization, politics, and technology transfer in the 17th century Cochinchina: the case of João da Cruz, *L. Saraiva and C. Jami (eds.) : Portugal and East Asia IV: Portuguese Missionaries in late Ming and early Qing China* (forthcoming).

The *Jinkōki* of Yoshida Mitsuyoshi

Ken'ichi Sato

Abstract Yoshida Mitsuyoshi (1598–1672) published the Jinkōki first in 1627. This was a problem book of elementary mathematics for everyday use but it also contained many interesting problems which attracted readers. This book became so popular that there have been more than 300 versions published during the *Edo* era (1603–1868) in Japan. In these notes, we shall survey the first edition of the Jinkōki, and the problems which were added in later editions.

1 Introduction

Japanese mathematics has been developed on the basis of the mathematics that came over from China and Korea, first in the 6th century. In the Asuka Era (593–710) and the Nara Era (710–794), Japan gradually established systems as a nation. In order to bring up government officials they started the University [大学寮 daigakuryō]. As for the Mathematics education Yōrō Legal Codes [養老律令 Yōrō Ritsuryō] (718) regulates that the University have 2 doctors in mathematics who taught mathematical arts [算術 sanjutsu] to 30 students. The textbooks used in this school were not written in Japan, but instead, these were from China and Korea. The 30 students were divided into two groups of 15 members and they were educated separately. Each group used different textbooks.

One group learned mainly the actual practical use of mathematics, while the other group specialized in theoretical mathematics. When the students had completed their lessons, they had to take an examination; the contents of the examination depended on groups. In the first group, the knowledge of Nine Chapters on Arithmetic Arts [九章算術 Kyūshō Sanjutsu], a mathematics textbook from China, was regarded as the most important. The examinations of the other group were

Ken'ichi Sato
Wasan Institute, 48-23 Sakae-cho, Kita-ku, Tokyo 114-0005, Japan
e-mail: kensatoh@tt.rim.or.jp

mainly based on the Technique of Linkage [綴術 tetsu jutsu] and Six Chapters [六章 rokushō]. The contents of the Technique of Linkage are not known.

The abacus was brought to Japan in the Muromachi Era (1333–1568). The tally trade [勘合貿易 kangō bōeki] with the Ming dynasty China started from the shōgunate of the Ashikaga Yoshimitsu [足利義満] regime and the abacuses used by Chinese were brought into Japan. The exact date of transfer is unknown, but it is believed to be sometime around the middle of the Muromachi Era. In the course of time, the Japanese started to make abacuses of their own; and it started to get widely used by merchants and samurais. The oldest abacus remaining in Japan is the one that is believed to have been used by Maeda Toshiie [前田利家] at Nagoya castle in Kyūshu at the time when he was dispatched for the action in Korea. It is a known fact that in the Age of Civil Wars (1467–1568) the samurais used the abacus. There were also a large number of samurais who were familiar with mathematics. It is written in many samurai family's house laws [家訓 kakun] that mathematics is important. In the Muromachi Era, the government was unstable and it became a period where the person's abilities were more important than their family lines, so that everyone was required to work using all his potentials at all times. Because there were many people like these, they were able to construct castles and other civil construction works.

After the battle of Sekigahara in 1600, there were no more people who battled with swords and guns and the world became a peaceful place. The number of samurais, that is, soldiers was widely reduced and those people lost their jobs. Still, there were people who tried to work for their clans, while others changed their jobs wanting to make a better use of their talents.

Around this time, abacus was becoming popular in the Kansai area like Kyoto; but most people did not know well how to use it. Naturally, there were people who opened schools that taught the use of abacuses. The number of students increased and a book called the Notes on Arithmetic [算用記 Sanyōki] was written to be used as a textbook. There was a teacher named Mōri Shigeyoshi [毛利重能] in one of these abacus schools. Mōri used to be a samurai who served Ikeda Terumasa. After the battle of Sekigahara, he resigned his position as a samurai. Then, he moved to the area near Kyōgoku in Kyoto and opened an abacus school. Mōri also used the Notes on Arithmetic which was popular at that moment in his school. The author of this book was unknown but it was published. The book was well organized and the edition was one that was published after many revisions of the original manuscripts. Mōri's school became popular and it had many students. His name was not only known in Kyoto, but also in nearby villages and towns, all the way to the area currently known as Hyōgo Prefecture. He became so popular that there was even rumors saying that he had learned the mathematics of Ming dynasty China. Even after his death, he is described as the person who created the Japanese mathematics in numerous mathematics books.

Mōri had been thinking about remaking the Notes on Arithmetic. He finally finished revising the book and published it in the year 1622. An introduction was added to the book, and the publication year was written on it. Mōri's book that still remains today, has a book cover but the title of the book is missing. In the book, after the introduction, there is an index. The words Index for Division [割算目録之次第 Warisan Mokuroku no Shidai] is written on the index. That is the reason why the book was named Notes on Division [割算書 Warisansho]. Since then, it has been called the Notes on Division up to the present date.

In the same year 1622 as the Notes on Division was published, Momokawa Jihē [百川治兵衛] from Sado wrote a book called the Notes of Various Calculations [諸勘分物 Shokanbumono]. This was also a mathematics book and its quality was equal to that of Mōri, but the words used in it were unique.

A little before entering the Edo Era (1603–1868), Imamura Tomoaki [今村知商] was born in Kawachi, near Osaka. There was no one who could teach him around the area of Kawachi; but he heard that a teacher familiar even with the mathematics of the Ming dynasty named Mōri Shigeyoshi had a school in Kyōgoku, Kyoto. He decided to become his disciple and traveled to Kyoto right away. Imamura learned everything Mōri knew in a short period of time. Afterwards, he kept studying on his own and opened a school in Edo. Right after this, his ability was recognized and he was admitted into the Iwaki clan where he achieved great success.

Quite obviously Mōri's fame was also well known in his home city of Kyoto. Later, Yoshida Mitsuyoshi [吉田光由], a boy about 10 years old, entered Mōri's school. Yoshida Mitsuyoshi was born as the second son of Yoshida Shūan [吉田周庵] in Saga, Kyoto in the year 1598, the same year Toyotomi Hideyoshi [豊臣秀吉] died. He knew that Mōri's mathematics school was a very prestigious mathematics school located in Nishi-Kyōgoku, Kyoto; so he entered the school without any doubt. The Yoshida is a famous family in Kyoto also known as the Suminokura [角倉]. The Suminokura was a family dedicated to the medical business and their real family name was Yoshida. Since Mitsuyoshi was the second son, he did not have to take over the family business, so he was able to devote himself to the mathematics he loved. Since he studied eagerly at Mōri's school, he was able to learn everything Mōri knew in a short period of time. Mōri must have been an excellent mathematics teacher with an exceptional ability to teach, since in both Imamura's and Yoshida's cases, he was able to teach them all he knew in a short period of time.

After Yoshida left Mōri's school, he went to the house of his main family, Suminokura Ryōi [角倉了以]. Ryōi was known as a lord of rivers [河川大名 kasen daimyōu] and he could conduct difficult river works, at the same time had a great knowledge of mathematics. In addition, his son, Soan [素庵], was involved in the foreign trades with Annam [安南] (the present North and Middle Vietnam), who was also known as an intellectual, a calligrapher, and a master of tea ceremonies. He had also started a publisher known for its luxurious printed books Saga editions [嵯峨本 sagabon] in collaboration with Hon-ami Kō'etsu [本阿弥光悦]. Yoshida learned from this Suminokura father and son the mathematics from the Systematic Treatise on Arithmetic [算法統宗 Sanpō Tōsō] (1593), a problem book by Cheng Dawei [程大位 Tē Taii] brought from China. By learning the mathematics of the Systematic Treatise on Arithmetic, Yoshida gained great knowledge. It is said that even after Yoshida left Mōri's school, they met each other in various occasions. In these occasions, Yoshida would teach mathematics to Mōri instead, and the master and disciple relation was reversed.

Yoshida was writing a mathematics book that would fit to the Japanese society of that era based on the information in the Systematic Treatise on Arithmetic. In the Kan'ē period (1624–1644), Yoshida finally finished his book and went to the famous temple of Tenryūji in Saga to visit Reverend Priest Genkō [玄光] to ask him to name the book and write the introduction for it. Genkō named the book the Jinkōki [塵劫記]. The first edition of the Jinkōki was published in 1627. After its publication, there were people who would publish it without Yoshida's permission. As a countermeasure, Yoshida changed the book's contents and published new editions. The editions of the Jinkōki published by Yoshida were only the ones published in the years 1627, 1629, 1631, 1634, and 1641.

There are many research papers and monographs on the Jinkōki in Japanese but few in European languages. The English translation published by the Wasan Institute [2] is one of the rare examples.

2 Contents of the *Jinkōki*

The first edition of the book published in 1627 consisted of 26 chapters as explained below.

Chapter 1. The Naming of Large Numbers

First, Yoshida Mitsuyoshi listed the names of the basic numbers. This was because at the beginning of the Edo Era, the decimal notation system [十進名数法 jusshin mēsūhō] was utilized but the names of large numbers were changing. In the first edition of Jinkōki the units of large numbers changed at each time they were multiplied by ten:

One [一 ichi], two [二 ni], three [三 san], four [四 shi], five [五 go], six [六 roku], seven [七 shichi], eight [八 hachi], nine [九 ku], ten [十 jū],
100 [百 hyaku], $1000 = 10^3$ [千 sen], $10000 = 10^4$ [万 man], 10^5 [億 oku], 10^6 [兆 choō], which is called the method of small multiplication [小乗法 shōjōhō].

In the later editions, however, he switched to the method of large multiplication [大乗法 daijōhō]. in which
10^4 [万 man], 10^5 [十万 jūman], 10^6 [百万 hyakuman], 10^7 [千万 senman],
10^8 [億 oku], 10^9 [十億 jūoku], 10^{10} [百億 hyakuoku], 10^{11} [千億 senoku],
10^{12} [兆 chō], 10^{13} [十兆 jutchō], 10^{14} [百兆 hyakuchō], 10^{15} [千兆 senchō],
10^{16} [京 kei], ...

Soon the method of large multiplication became the standard in Japan which lasts until today, including his wrong choice of the character 秭 for 𥝱 meaning 10^{24}.

Chapter 2. The Naming of Numbers Smaller than 1

The naming system for these numbers are the same as the system used today and the name changes every one tenth of a number.

Chapter 3. Units of Capacity

The most representative example of the name of units of capacity is the units used for counting the capacity of rice.
1 *koku* [石] = 10 *to* [斗]
1 *to* = 10 *shō* [升]
1 *shō* = 10 *gō* [合]. 70000 grains of rice
1 *gō* = 10 *shaku* [勺]. 7000 grains of rice
1 *shaku* = 10 *sai* [抄]. 700 grains of rice
1 *sai* = 10 *satsu* [撮]. 70 grains of rice
1 *satsu* = 10 *kē* [圭]. 7 grains of rice
1 *kē* = 10 *zoku* [粟].

Chapter 4. Units of Area

Ho [歩] is the standard unit of an area, also known as *tsubo* [坪]. It is the area of 1 *ken* [間] square (2 *jō* [畳]). 30 *ho* is equal to 1 *se* [畝], and 300 *ho* is equal to 1 *tan* [反].

Chapter 5. Multiplication Table

The multiplication table is lined up starting from the row of 1 just like it is taught at current elementary schools. In the ancient times, the multiplication table started from the row of 9 both in China and in Japan, so it was hard to memorize it. In the Jinkōki, however, the numbers were lined up in an easier to memorize way.
In ichi ga ichi (1×1=1), in ni ga ni (1×2=2), in san ga san (1×3=3),
in shi ga shi (1×4=4), in go ga go (1×5=5), in roku ga roku (1×6=6)
in nana ga nana (1×7=7), in hachi ga hachi (1×8=8), in ku ga ku (1×9=9)
ni nin ga shi (2×2=4), ni san ga roku (2×3=6), ni shi ga hachi (2×4=8),
ni go jū (2×5=10), ni roku jūni (2×6=12), ni shichi jūshi (2×7=14),
ni hachi jūroku (2×8=16), ni ku jūhachi (2×9=18),
sa zan ga ku (3×3=9), san shi jūni (3×4=12), san go jūgo (3×5=15),
sabu roku jūhachi (3×6=18), san shichi nijūichi (3×7=21),
san pa nijushi (3×8=24), san ku nijūshichi (3×9=27),
shi shi jūroku (4×4=16), shi go nijū (4×5=20), shi roku nijūsi (4×6=24),
shi shichi nijūhachi (4×7=28), si ha sanjūni (4×8=32), si ku sajūroku (4×9=36),
go go nijūgo (5×5=25), go roku sanjū(5×6=30), go shichi sanjūgo (5×7=35),
go ha yonjū (5×8=40), gokku shijūgo (5×9=45),
roku roku sajūroku (6×6=36), roku shichi shijūni (6×7=42),
roku ha shijūhachi (6×8=48), roku ku gojūshi (6×9=54),
shichi shichi shijūku (7×7=49), shichi ha gojūroku (7×8=56),
shichi ku rokujūsan (7×9=63),
hachi ha rokujūshi (8×8=64), hachi ku shichijūni (8×9=72),
ku ku hachijūichi (9×9=81).

Chapter 6. Names of the Division Table (1)

The division system written on the Jinkōki was called eight calculations [八算 hassan]. This division table omits the divisions by 1 and starts from the divisions by 2. When 10 is divided by 2, it is called "ni ichi ten saku no go." The numbers are lined up in the order of divisor, dividend, quotient, and remainder; however, it does not simply line up the numbers, but it places a letter between the numbers to make it easier to memorize. In this way, this chapter shows examples of how to divide even up to numbers that can be divided by 9.

Chapter 7. Names of the Division Table (2)

This division table is called ken-ichizan [見一算]. The division table when dividing with one digit numbers is called eight calculations, while the division table when dividing with two-digit numbers is called ken-ichizan.

Chapter 8. Division and Multiplication

It explains how to calculate by multiplication instead of division.
Multiplying by 0.2 is equivalent to dividing by 5.
Multiplying by 0.04 is equivalent to dividing by 25.
Multiplying by 0.5 is equivalent to dividing by 2.
Multiplying by 0.008 is equivalent to dividing by 125.

Chapter 9. Problem on Rice Trading

(Unit price) × (Amount) = (Price)

Rice trade was used as an example to represent a simple problem that uses the formula shown above. From this Chapter on, the format of how the problems on the book are written becomes apparent. First, the problem is written and then the answer is written on the next line, after indenting 2 characters. The calculation procedure is written at the beginning of the next line as the process. This format became the basic format for mathematical problems of the Edo Era. The numbers in the problems comes with units, so the units written on Chapters 1 through 4 becomes necessary.

Here is an example problem.
The price of 1 *koku* of rice is 26 *monme* [匁] 5 *bu* [分]. How much would be the price of 123 *koku* of rice?

Answer: 3 *kan* [貫] 259 *monme* 5 *bu*.

Chapter 10. Calculation for Exchanging Gold and Silver

Three types of currencies, gold, silver and *zeni* [銭] (copper coins) were used in the Edo Era. Silver was a currency by weight so its value depended on its purity and weight. The types of gold were 1 *Ōban* [大判] = 10 *ryō* [両], 1 *Koban* [小判] = 1 *ryō*, 1 *bu* = one fourth of a *ryō*, and 1 *shu* [朱] = one fourth of a *bu*. In Japan's Edo Era there was a manufacturing technique to make pure silver [灰吹銀 haifuki gin]. The silvers used as currency were the *chō gin* [丁銀] and *mameita gin* [豆板銀] which had silver contents of 80%. But the *zeni* was the most commonly used for shopping in everyday life, so exchanging currencies was constantly necessary.

Example:
A man has an ingot of silver weighting 976 *monme*. He wants to change this with an ingot of refined pure silver. When refining silver, 20% is lost during the process.
How much silver can he get after the refinement?

Chapter 11. Buying and Selling the *zeni*

It was common to exchange gold coins and silver into *zeni*, but it was called "buying *zeni* with silver" and it was not considered as an exchange.

Chapter 12. Calculation of Interest

This chapter explains how to calculate the interest of borrowing silver or rice. The interest rates at that time period were between 20 to 26% and very different from the present rate.

Chapter 13. Buying and Selling Silk

In the Japanese society of the past, the measuring scale differed by the product it was measuring. The basic measuring scale was called the carpenter's square [曲尺 kanejaku] because it was mainly used by carpenters. The measuring scale used to measure textile fabrics was called tailor's scale [呉服尺 gofuku shaku]. The relation between these two scales is : 1 *shaku* [尺] in the tailor's scale = 1 *shaku* 2 *sun* [寸] in the carpenter's square.

For example:
The price of cotton was in silver 30 *monme* per *tan*. The length of 1 *tan* was 2 *jō* 5 *shaku*. How much will 1 *shaku* of cotton cost?

All fabric made in Japan have the same width. The amount of fabric needed for one man is considered as one *tan*. The size of fabric imported from China was not uniform.

Chapter 14. Purchasing Foreign Products

Japan imported silk and thread from foreign countries. *Kin* [斤] was used as the unit for measuring weight. The weight of one *kin* was considered as 250 *monme* for Japanese products, but it was considered as 160 *monme* for foreign products. For this reason, it was necessary to make problems for these kinds of calculations. When the Jinkōki was first published in 1627, foreign ships were coming in and out not only in Nagasaki but also in Hirado. For this reason, the title of this problem was originally "Trading with the Black Ships." However later on, since Hirado was closed, the title was changed to "Trading in Nagasaki." Only one problem was written in this chapter and it was on proportional distribution.

Example:
Three merchants were purchasing imported goods. The first merchant brought 64 *kan* 800 *monme* of silver. The second merchant brought 52 *kan* 300 *monme*. The third merchant brought 42 *kan* 900 *monme*. So the total amount of silver of the three merchants was 160 *kan*. The total products bought between the three of them were 250 *kin* of ginsengs, 70 *kin* of agallochs, 280 scrolls, and 8400 *kin* of thread. If the

total cost of the products were divided proportionately according to the amount of silver brought by each merchant, how much would each merchant have to pay?

Chapter 15. Expenses to Transport by Ship

Because Kyoto is in a basin, there are many rivers running through it. Suminokura Ryōi was also famous as an engineer for civil construction works. He modified these rivers so that ships could travel through them to transport goods. The Ooi River flows from the neighboring country of Tamba into Saga, Kyoto. This river was modified and lumber cut upstream was bound together into "rafts" and sent adrift downstream to Kyoto. These lumber were collected and traded in Saga. In addition, rice was also transported by ships through the river. The freight was paid with the transported rice, so problems using these scenarios could be created such as the following one.

Example:
When transporting 250 *koku* of rice on a ship, the charge for transporting every 100 *koku* is 7 *koku*. If the charge is paid from the 250 *koku* of rice transported, how much would the freight charge be?

Chapter 16. The Size of a *Masu* (measurement device)

The measure [升 masu] was established by the government as a device to measure the volume of objects. At the beginning of the Edo Era, the volume of an 1 *shō* measure (1.804 liter) was 62.5 cubic *sun*. However, the Edo shōgunate [江戸幕府] changed it to 64.827 cubic *sun* in the year 1627. Because the amount of rice was commonly measured using this *masu*, people were very confused. For this reason, in the Jinkōki, the usual measure is still used by changing its name to old measure.

Example:
There is a cylindrical shaped container. Both its diameter and height are 9 *sun* 3 *bu*. What is the volume of this container?

Calculation method
$9.37 \times 9.37 \times 9.37 \times 0.79 \div 64.827 = 10.025128$

Chapter 17. Land Surveying

Land surveying is to measure the area of a land. Practically most agricultural fields are rectangular. The shapes mentioned in this chapter are a triangle, an equilateral triangle, a rectangle, a trapezoid, a rhombus, a sector, and a circle. 3.16 is used as the value equivalent to the circumference ratio or pi. In addition, this chapter also mentions complicated shapes made by the combination of these previously mentioned shapes; however there are no problems with complex calculations.

Chapter 18. Amount of Yield and Tax

The amount of yield of rice was proportional to the area of the rice field. However, the amount of the sun light and the temperature of the water also influence the yield, hence the amount of expected yield was ranked into 4 levels from A to D. The tax also varied depending on the level. Most of the tax rates that come up in the problems in the Jinkōki is set as about 65%. The tax on crop yield was called mononari [物成]. When the mononari was decided, taxes called kuchimai [口米] and bumai [夫米] were added to this. These were similar to the present regional tax.

Chapter 19. Trading of Gold and Silver Foils

It was common to coat folding screens and the family altars with gold and silver foils. These objects were coated with small squared foils, so the cost was proportional to the surface area of the object.

Chapter 20. Problem on the Volume of Lumber

The prices of the log-shaped lumber cut from the mountains were decided by their volume. In actuality, it cannot be used as a log so the actual volume was measured after its shape was changed.

Chapter 21. River Construction

The problems in this chapter are about the volume of stone baskets called snake basket [蛇籠 ja kago] and cubic frame [角枠 kaku waku]. These baskets are used in constructions as container of stones.

Chapter 22. Various Constructions

This chapter deals with problems for calculating the number of workers and amount of money needed for the construction of objects like tile roofed mud-wall [築地 tsuiji], fences, and moats, etc.

Chapter 23. Measuring the Height of a Tree

This chapter has a problem about measuring the height of a tree by using square paper.

Chapter 24. Calculating the Distance

This chapter has an example for measuring the distance to a destination by using a simple method.

Chapter 25. Obtaining the Square Root

This chapter explains how to obtain the length of a side of a square-shaped land with an area of 15129 *tsubo* square by using an abacus.

Chapter 26. Obtaining the Cube Root

This chapter explains how to obtain a cube root by using an abacus.

The contents stated so far are from the first edition published in 1627. The book was revised many times after this, and there were some changes made on each edition. The most significant changes are stated below.

3 Revised Edition of the *Jinkōki*

2 years after the first edition, Yoshida Mitsuyoshi republished the Jinkōki with some additional contents. The additional contents are stated below.

One of the additional contents was the calculation of nesting boxes [入れ子算 ireko-san]. This was a problem about placing different sized measure or pots sequentially inside each other. The sizes of the objects, etc. were in arithmetic or geometric progression.

Another additional content was the Joseph's problem [継子立 mamako date]. This new content has a big illustration, which is followed by an explanation.

Problem:
A man has 30 children. 15 of them are the children he had with his first wife and the other 15 are the children he had with his current wife. He decides to choose one child from all of them to leave all his belongings. His current wife, lines up all 30 children around a pond in the same way as it is shown in the illustration. This man decides to count the children clockwise starting from a random child and remove every 10th child. The one child who remains in the end would receive his belongings. The wife pointed out one child and started to count. When 14 children were removed, all 14 of them were the child of the former wife. The only remaining child of the former wife complained that it was unfair that only the children of the former wife have been removed and that he wished that this time the counting started from him. The man accepted this opinion and every 10th child from this child started to be removed. In the end, the remaining child were former wife's children. In the illustration, the former wife's children are dressed in a white kimono and the current wife's children in a black kimono.

Fig. 1 In the illustration, the former wife's children are dressed in a white kimono and the current wife's children in a black kimono.

This problem also appears in a book written in the Kamakura Era (1185–1333). A very similar problem also exists in Europe.

Another additional problem was called the calculation of rats [鼠算 nezumi-san] (geometrical growth). Rats have a high reproductive power and they are known to give birth several times a year. Since rats eat rice, which is the Japanese principal food, this problem represents how much damage rats can cause.

Problem:

There is a couple of rats at the beginning of January. This couple gives birth to 12 rats by the end of January. At the beginning of February, these 14 rats make 7 couples, and by the end of the month, each couple gives birth to 12 rats. If these rats keep giving birth at the same pace, how many rats will there be by the end of December?

The answer for this problem is 27682574402 rats.

The next problem is, if one rat eats 0.5 *gō* of rice a day, how much rice would 27682574402 rats eat a day?

- Another content is the calculation of crows [烏算 karasu-san].

Problem:

If there are 999 crows on 999 beaches and if each crow caws 999 times, how many caws would there be in total?

The source of this problem is an old Chinese mathematics book called the Sunzi Mathematical Canon [孫子算経 Sonshi Sankē].

-The next problem is about "finding out how many people stole silk." This problem was also taken from the Sunzi Mathematical Canon.

Problem:

Silk was stolen from a person's house. The voices of the thieves splitting the silk under a bridge near the house can be heard.

If the thieves split the silk into 8 *tan* each, there would be 7 *tan* lacking. If they split it by 7 *tan* each, there would be 8 *tan* left. How many thieves are there and how much *tan* of silk was stolen?

-Another content is the calculation of sharing oil [油はかり分け算 abura hakari wake san].

Problem:

A 1 *to* (10 *shō*) barrel is filled with 10 *shō* of oil. By using a 3 *shō* measure and a 7 *shō* measure, divide the oil so that there is 5 *shō* of oil in the 1 *to* barrel and 5 *shō* of oil in the 7 *shō* measure.

This is possible by following the next procedure.

Use the 3 *shō* measure to take out 3 *shō* of oil from the 1 *to* barrel and pour it in the 7 *shō* measure. Take out 3 *shō* of oil again from the 1 *to* barrel and fill it in the 7 *shō* measure. Take out 3 *shō* of oil from the 1 *to* barrel again and fill the 7 *shō* measure. There is 2 *shō* of oil remaining in the 3 *shō* measure. Drain all the oil in the 7 *shō* measure back into the 1 *to* barrel, and pour the 7 *shō* measure with the 2 *shō* of oil in the 3 *shō* measure. Then take out another 3 *shō* of oil from the 1 *to*

barrel and pour it inside the 7 *shō* measure.
The source of this problem is unknown. Very similar problems exist also in Europe.

The Jinkōki was later revised and republished many times. The last edition Yoshida Mitsuyoshi published was in 1641. This edition contains an illustration of what is now called the Pascal's triangle. This illustration was taken from the Chinese Systematic Treatise of Arithmetic, but since it was not in the former editions of the Jinkōki. In the end page of book, there are 12 problems without their answers written on the book.

Challenge problems [遺題 idai]
Yoshida Mitsuyoshi included 12 challenge problems without answers at the end of the book. These types of challenge problems without answers became very popular in the Edo era.

4 The Influence of the *Jinkōki* on Japanese Mathematicians

The first 3 challenge problems are written below.

Challenge Problem 1. Right-angled Triangle
The sum of the length of the east side and the northwest side is 81 *ken*. The sum of the length of the east side and the southwest side is 72 *ken*. How long is the east side? How long is the northwest side? How long is the southwest side?

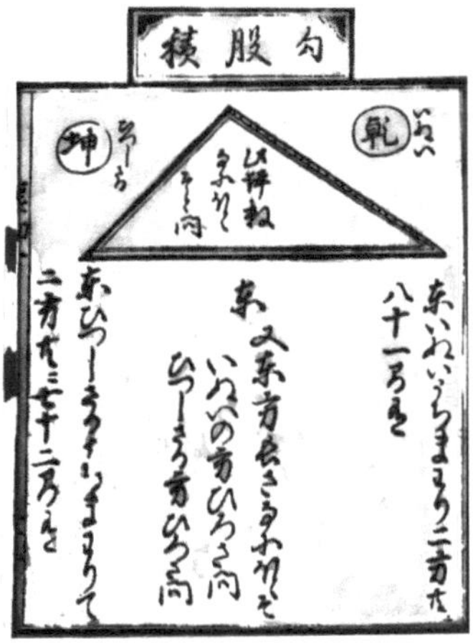

Fig. 2 Right-angled Triangle

Challenge Problem 2. Cutting Fragrant Wood
There is a piece of Fragrant Wood imported from China. The length is 3 *ken*, the circumference of the bottom part is 5 *shaku*, and the circumference of the top part

is 2 *shaku* 5 *sun*. Its cost is 10 silver coins. Three people buy this piece jointly. They want to cut it so that each part would have the same price. Where is the piece cut from the bottom and from the top?

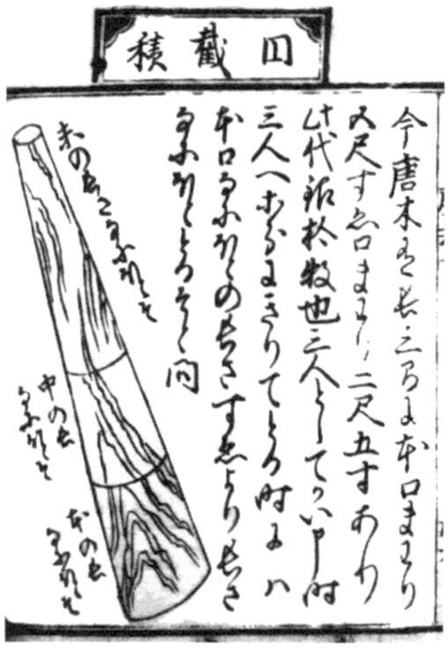

Fig. 3 Cutting Fragrant Wood

Challenge Problem 3. Different Pairs of Four Objects

Pine trees	80^{hon}	
Japanese cypresses	50^{hon}	The total price is 2 *kan* 790 *monme* of silver.
Pine trees	120^{hon}	The pine tree has the same price as above.
Cedars	40^{hon}	The total price is 2 *kan* 322 *monme* of silver.
Cedars	90^{hon}	The cedar has the same price as above.
Chestnut trees	150^{hon}	The total price is 1 *kan* 932 *monme* of silver.
Chestnut trees	120^{hon}	The chestnut tree has the same price as above.
Pine tree	7^{hon}	The total price is 419 *monme* of silver.

What is the price of each pine tree, Japanese cypress, cedar, and chestnut tree?

Challenge Problems 4 through 12 are left out from this report. See [1].

Conclusion

After the Jinkōki was published, since it was revised several times and published on several occasions, it had a great influence on later Japanese mathematics. The mathematics that is useful in everyday life included in the Jinkōki was praised even in the Meiji Era (1868–1912). By the time period of Seki Takakazu [関孝和], the mathematical games were explained mathematically by many mathematicians including himself. The theoretical mathematics of the Edo Era developed triggered by the Idai in the Jinkōki; and its development lead to the completion of the solution methods of equations and to the discovery of determinants.

References

1. Ken'ichi Sato [佐藤健一] (ed.): *The First Edition of the Jinkōki* [塵劫記初版本], Kenseisha [研成社] (2006).
2. Osamu Takenouchi [竹之内脩] et al. (ed.): *Jinkōki*, Wasan Institute, Tokyo (2000).

Résumé of Works on Mathematics of Seki Takakazu

Osamu Takenouchi

Abstract We shall describe briefly the contents of the Résumé of Works on Mathematics (*Katsuyō Sanpō*) of Seki Takakazu (ca.1640 - 1708). This monograph is the posthumous publication of the great mathematician of the *Edo* period and contains most of his representative works on mathematics.

Introduction

In his lifetime Seki Takakazu [関孝和] (ca.1640–1708) had only one publication, the Mathematical Method for Clarifying Subtle Points [発微算法, *Hatsubi Sanpō*], but his manuscripts were kept in the hands of his disciples. After his death, they gathered together to make a publication to celebrate their master's achievements. It was completed after 3 years and the Résumé of Works on Mathematics [括要算法, *Katsuyō Sanpō*] was published in 1712. (See [1].) Though there are yet many important results which are not taken in this monograph, the works contained in it are representative ones. The monograph consists of four books, which we will describe briefly in this paper. For the details, we refer the reader to the author's recent book [2].

Book 1

Book 1 contains the research on the sum of powers of natural numbers, among others, the formula to calculate the sum of powers of natural numbers

Osamu Takenouchi
Professor Emeritus of Osaka University, 1-3, Machikaneyama, Toyonaka, Osaka 560-8531, Japan
e-mail: osamu-ta@mxs.mesh.ne.jp

$$1^p + 2^p + \cdots + n^p.$$

He presented the result in a table:

	Φ											
0	12	Φ										
Multiply 5/66 and Add	66	11	Φ									
0	220	55	10	Φ								
Multiply 1/30 and Subtract	495	165	45	9	Φ							
0	792	330	120	36	8	Φ						
Multiply 1/42 and Add	924	462	210	84	28	7	Φ					
0	792	462	252	126	56	21	6	Φ				
Multiply 1/30 and Subtract	495	330	210	126	70	35	15	5	Φ			
0	220	165	120	84	56	35	20	10	4	Φ		
Multiply 1/6 and Add	66	55	45	36	28	21	15	10	6	3	Φ	
Multiply 1/2 and Add	12	11	10	9	8	7	6	5	4	3	2	1
Whole	1	1	1	1	1	1	1	1	1	1	1	1
Divisor	12	11	10	9	8	7	6	5	4	3	2	
p	11	10	9	8	7	6	5	4	3	2	1	Base

This table is based on the table of binomial coefficients. This is used as follows. As an example, we take up the case when $p = 6$. Look at the column from the bottom. The numbers are 1, 7, 21, 35, 35, 21, 7. These numbers are to be multiplied with n^7, n^6, $,n^5$, $\cdots$ and processed according to the instruction given on the left part. After all, divide the result by the number 7 written underneath. The final result is

$$\frac{1}{7}\left(n^7 + \frac{1}{2}\times 7n^6 + \frac{1}{6}\times 21n^5 + 0 - \frac{1}{30}\times 35n^3 + 0 + \frac{1}{42}\times 7n\right).$$

This just corresponds to the expression which comes from the Bernoulli polynomial:

$$\frac{1}{p+1}n^{p+1} + \frac{1}{2}n^p + \frac{p}{2}B_1n^{p-1} - \frac{p(p-1)(p-2)}{2\cdot 3\cdot 4}B_2n^{p-3} + \frac{p(p-1)(p-2)(p-3)(p-4)}{2\cdot 3\cdot 4\cdot 5\cdot 6}B_3n^{p-5} - \cdots.$$

Bernoulli was able to use the writing method of binomial coefficients due to Pascal, while Seki could not. So Seki gave these numbers as explicit numerals.

The numbers which appear on the left part, *i.e.*, $\frac{1}{6}, \frac{1}{30}, \frac{1}{42}, \frac{1}{30}, \frac{5}{66}, \cdots$, are the so-called Bernoulli numbers. Seki Takakazu also mentioned the way to get these numbers which is quite the same as Bernoulli gave.

Book 2

Book 2 is devoted to the research of the solution of the indeterminate equation of natural numbers.

The original problem is what is usually known as "Chinese remainder theorem." It appears in the Sunzi Mathematical Canon [孫子算経 *Sunzi Suanjing*] of the 4th century. It is a problem to obtain the number of objects. It reads as follows.

> There are a number of objects. The number of objects is not known. We want to know how many they are.
>
> Counting 3 by 3, 2 leave. Counting 5 by 5, 1 leaves. Counting 7 by 7, 5 leave. Then we know the total number can be calculated as follows.
>
> > Remainder by $3 \times 70 +$ remainder by $5 \times 21 +$ remainder by 7×15. If the number thus obtained exceeds 105, then subtract 105 and 105, till it becomes less than 105. That is the total number.

This is, one sees, a problem to obtain the solution of residual systems:

$$x \equiv a_1 \pmod{m_1}, \quad x \equiv a_2 \pmod{m_2}, \quad \ldots, \quad x \equiv a_n \pmod{m_n}.$$

He made an extensive study of this problem. He generalized the problem to more number of moduli and to the cases where the moduli are not relatively prime.

Book 3

Book 3 studies systematically the regular polygons from the triangle up to the polygon with 20 sides.

Fix one of the vertices as A_0, and name successively the adjacent vertex as A_1, the second vertex as A_2, the third adjacent vertex as A_3, and so on.

The length of the side is denoted by a, the length of the radius of the inscribed circle by r, and the length of the radius of the circumscribed circle are denoted by R.

Starting with one of the vertices, the line segment connecting it to the adjacent vertex and its length is denoted by $a_1 (= a)$, the one to the second adjacent vertex by a_2, the one to the third adjacent vertex by a_3, and so on. Namely, $A_0A_1 = a$, $A_0A_2 = a_2$, $A_0A_3 = a_3$, etc.

The distance from the center to the mid-point of the chord joining a vertex with its second next vertex, e.g. A_0A_2, is denoted by r_2, the distance from the center to the mid-point of the chord joining a vertex with its third next vertex, e.g. A_0A_3, is denoted by r_3, and so on.

Moreover, he uses b_3 b_4, b_5, b_6, etc. as the lengths of the segments connecting the center O with the following points:

b_3: the intersection point of OA_1 and A_0A_3,

b_4: the intersection point of the segment joining O with the mid-point of A_0A_3 and A_0A_4,

b_5: the intersection point of OA_2 and A_0A_5,

b_6: the intersection point of the segment joining O with the mid point of A_0A_5 and A_0A_6.

It seems to us that he prepared several lemmas, though not explicitly given, to make up general settings.

[Formula 1] $2a_kr_k = Ra_{2k}$

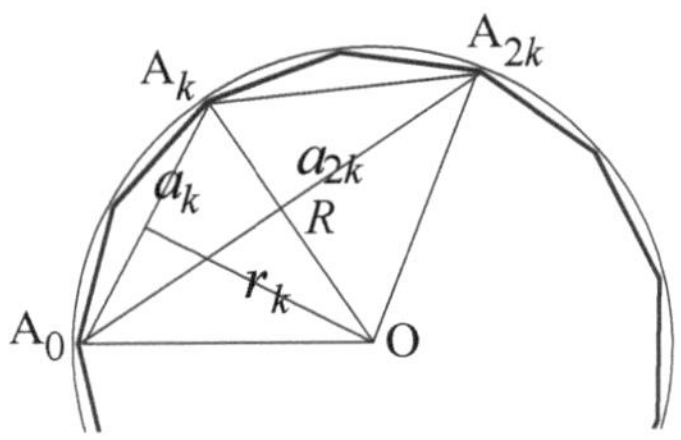

[Formula 2] $2Rr_{2k} = 2R^2 - a_k^2$

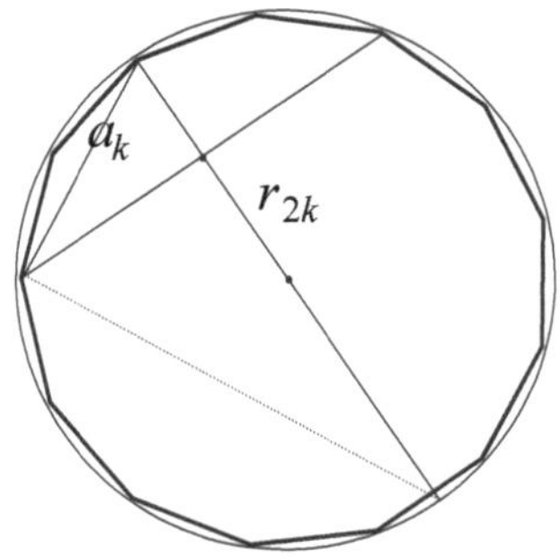

[Formula 3] $R + b_3 = 2r_2$, $R^2 - a^2 = Rb_3$

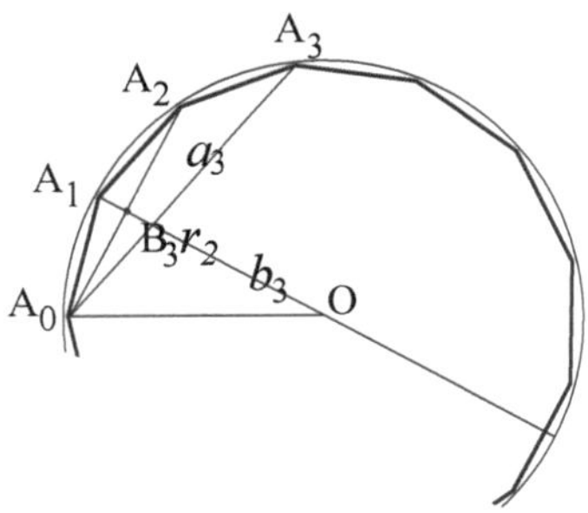

[Formula 4] $2r_k = b_{k-1} + b_{k+1}$

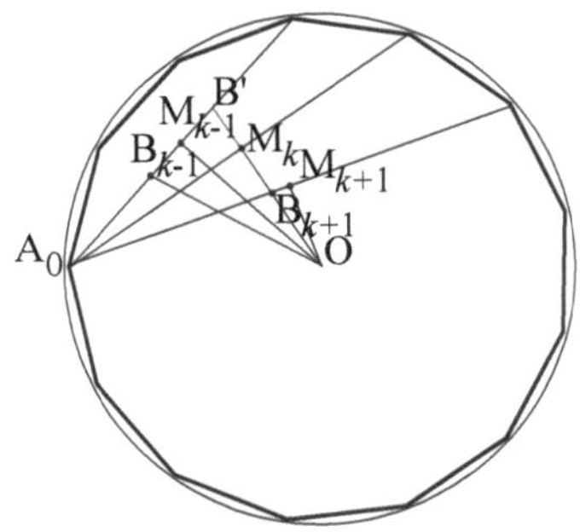

[Formula 5] $2at = Ra_3$, $t = R + b_3/2$

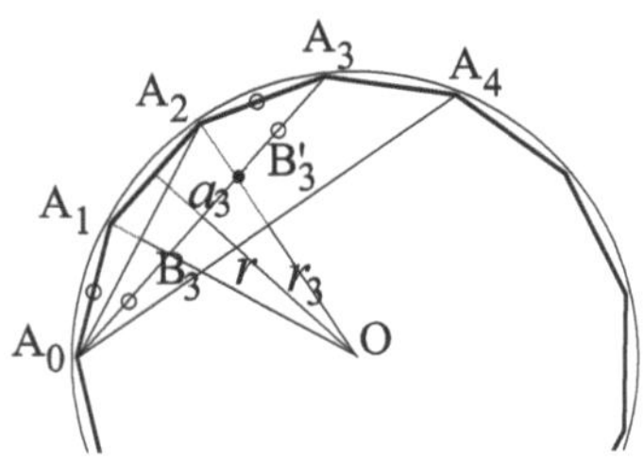

$$A_0B_3 = A_3B'_3 = A_0B_1 = a$$

$$B_3B'_3 = A_1A_2 \times \frac{OB_3}{OA_1} = a \cdot \frac{b_3}{R}$$

$$Ra_3 = 2aR + ab_1 = 2at$$

We show how he made use of these lemmas to establish his result in the case of the regular polygon with 11 sides. By above formulas, we get the followings:

$$R^2 - a^2 = Rb_3 \quad \text{by [Formula 3];}$$
$$2R^2 + Rb_3 = 2Rt \quad \text{by [Formula 5];}$$
$$4r^2a^2 = R^2{a_2}^2 \quad \text{by [Formula 1];}$$
$$2R^4 - R^2{a_2}^2 = 2R^3r_4 \quad \text{by [Formula 2];}$$
$$2R^3r_4 - R^2 \cdot Rb_3 = R^3b_5 \quad \text{by [Formula 4].}$$

Therefore, the basic relation $R^5 = 32rr_2r_3r_4r_5$, which is obtained by [Formula 1], implies

$$R^3b_5 \cdot Rb_3 \cdot 2Rt \cdot 4r^2 = R^{10} = 32R^5rr_2r_3r_4r_5.$$

And we get the following equation.

$$\left(2R^4 - (4R^2 - a^2)a^2 - R^2(R^2 - a^2)\right)(R^2 - a^2)\left(2R^2 + (R^2 - a^2)\right)(4R^2 - a^2) = R^{10}.$$

Seki's intention was to establish numerical relations among the length of the side, radius of the circumcircle, and radius of the inscribed circle.

Book 4

In Book 4 is devoted to the rigorous and accelerated determination of the circle number π, the formula for the length of the circular arc, and the volume of the sphere.

The classical method to obtain the length of the circumference of a circle was to approximate it by the length of the perimeter of the inscribed regular polygon. Seki calculated the length of the perimeter, beginning from the square, and making the double of the number of sides up to 131,072. After having made these calculations, he proceeded as follows.

Let a, b, c be the following values, the length of the diameter being put equal to 1.

$a =$ the perimeter of the regular polygon with 32,768 sides,
$b =$ the perimeter of the regular polygon with 65,636 sides,
$c =$ the perimeter of the regular polygon with 131,072 sides.

He calculated $b + \dfrac{(b-a)(c-b)}{(b-a)-(c-b)}$ and got 3.14159265359. He declared then that this is the exact number of the ratio of the length of the circumference of a circle against its diameter. The method used here is the first acceleration method we ever had.

The next problem he concerned with was to get the fractional approximation of π. It was known that the fraction $\frac{355}{113}$ was a good approximation of π. But it was not clear how to obtain this fraction. Starting with $\frac{3}{1}$, he successively made fractions $\frac{7}{2}, \frac{10}{3}, \frac{13}{4}, \frac{16}{5}, \frac{19}{6}, \frac{22}{7}, \cdots$. At each stage, he examined how close it was to the value 3.14159265359 obtained in the above. Arriving at the fraction $\frac{355}{113}$, he recognized that the degree of approximation was extremely good. From this point after, he always used this fraction when the value of π was needed.

In the last place, he determined the volume of a sphere. He took a sphere of diameter 10 and sliced it to 50, 100, 200 thin disks. Then he added the square of the diameters of these disks. Let them be a, b, c. To these, he applied the same accelerating process $b + \frac{(b-a)(c-b)}{(b-a)-(c-b)}$ as in the above. He got thus $666\frac{2}{3}$. Multiply to this the circle number π and divide by 4. As already noted, as the value of π, he used the fractional approximation $\frac{355}{113}$. So the result he got was $523\frac{203}{339}$. Divide this by 1000. Then he got $\frac{355}{678}$ and he called this value as the volume rate of the sphere. When one wants to have the volume of a sphere of diameter r, one will make the product of this rate with r^3.

References

1. Hirayama, Shimodaira and Hirose (ed): *Takakazu Seki's Collected Works Edited with Explanations* [関孝和全集], Osaka Kyōiku Tosho [大阪教育図書] (1974).
2. Osamu Takenouchi [竹之内脩]: *Mathematics of Seki Takakazu* [関孝和の数学], Kyōritsu Shuppan [共立出版] (2008).

Seki Takakazu's Measuring Process of the Volume of Solids Derived from Spheres

Toshio Sugimoto

Abstract In this paper, I shall explain a crucial point of Seki Takakazu's argument which claims that there is a close relationship between the solid of revolution of an arc-figure (called here a finger ring) and an obliquely cut segment of a cylinder (called an *onglet*). I think his argument is the most brilliant; no one has ever attained his level. But his explanation is too concise; no one has ever found its proper interpretation. Several years ago I succeeded in deciphering his arguments in seeking the volume of solids of revolution for the first time and published it in a series of papers in Japanese. Here I shall show some of my interpretations with many figures for illustration.

1 Volume of a sphere

Problems. Seki Takakazu wrote an important book called Measurements [求積 Kyūseki] [4], in which he considered many problems of measuring the area and the volume of figures. There are 15 problems of area and 34 problems of volume. Each problem consists of four parts: a question, an answer with numerical values, a method for calculations, and the solution by formulas. In Measurements the circular constant π is assumed to be $355/113$. For example, the area of a disc with diameter 1 is written to be $355/452$, i.e., $\pi/4$ times that of the circumscribing square (Fig.1).

Pyramid. The volume of a pyramid with square base 1×1 and height $1/2$ is one third of that of a half cube of the same dimensions. Although Seki gave general explanation, which seems to me complicated, I prefer a solid model (Fig.2), which shows three pyramids coming out of the half cube.

Toshio Sugimoto
2-40-3, Kikunodai, Chōfu, Tokyo 182-0007, Japan

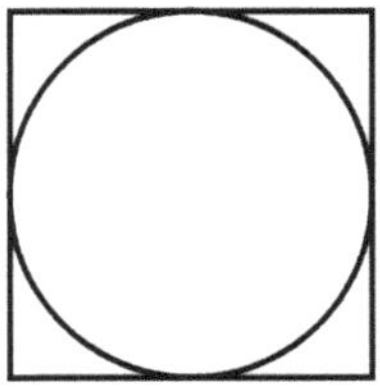

Fig. 1

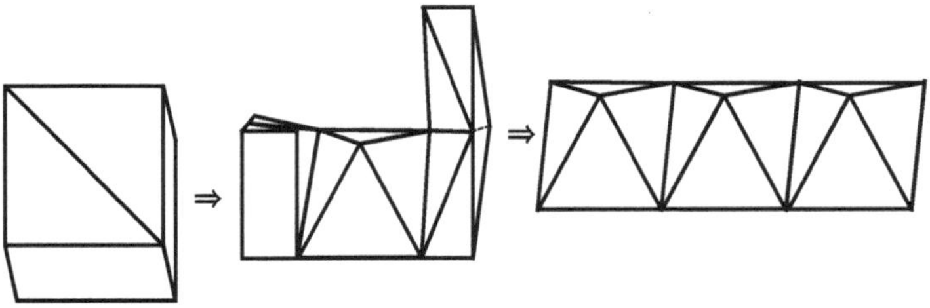

Fig. 2

Half sphere. In place of Seki's difficult explanations, it is convenient to use Liu Hui's [劉徽] solid, or the half matching cover [合蓋 hegai][1] (square cover of dishes) (Fig.3), which looks like a square from the upper side and a half circle from the lateral side. We obtain the volume of the half matching cover as $2/3$ of the half cube. By the same reasoning, a half sphere has also $2/3$ volume of the cylinder with the same height, so we multiply the half matching cover with $355/452$, i.e., $\pi/4$, to obtain the half sphere. According to Seki's explanation in [4, p.237], a half sphere is composed of a main cone [円錐 ensui] and a subsidiary cone [旁錐 bōsui], but the latter is not discernable in his explanatory figure.

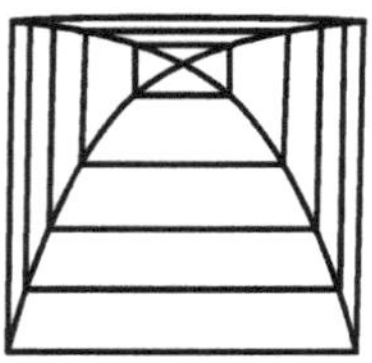

Fig. 3

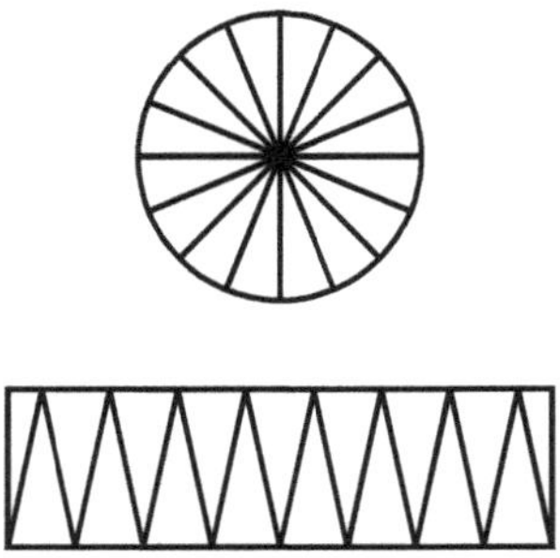

Fig. 4

[1] [牟合方蓋 mohe fangai] employed in Chapter 3 of Nine Chapters of Mathematics [九章算術 Kyūshōsanjutu]. See these proceedings, p. 71.

In another article Method of seeking the volume of a sphere [求立円積術 Kyū Ritsuenseki Jutsu] in the Concise Collection of Mathematical Science [括要算法 Katsuyō Sanpō] [3, pp.361–366] Seki claimed that the volume of a sphere is composed of many thin discs. The thinner discs become, the closer to the sphere the total volume becomes.

After this consideration of regarding as discs, he attained the above synthetic explanation. There are two standpoints, the analytic one (gathering small parts) and the synthetic one (modifying into a combination of well-known solids).

Volume of sphere. We conclude that the volume V of a sphere of diameter d is given by

$$V = (2/3)(355/452)d^3 = (355/678)d^3 = (\pi/6)d^3.$$

For example,

$$\text{if } d = 1, \text{ then } V = 0.52359882;$$
$$\text{if } d = 0.8, \text{ then } V = 0.26808260;$$
$$\text{if } d = 0.6, \text{ then } V = 0.11309734.$$

2 Area of spheric surfaces

There is a close relationship between the surface area S and the volume V of a sphere, — the V can be obtained from the S, and vice versa. The relations can be explained in two ways:

(1) The volume V can be regarded as a collection of triangular cones, and V is obtained from S;

(2) The volume V can be regarded as a collection of thin spheric surfaces S, which are differences of two close spheres of different sizes, and S is obtained from V.

See [2, pp. 73–74 of the August 1982 issue] for details.

The area S of the spheric surface with diameter d is given by

$$S = (355/113)d^2 = \pi d^2.$$

These ideas are well known in a simple figure. The area of a disc can be obtained from a collection of triangles, which shows that the area is the product of the radius and a half of the circumstance (Fig.4).

3 Solids made by revolution

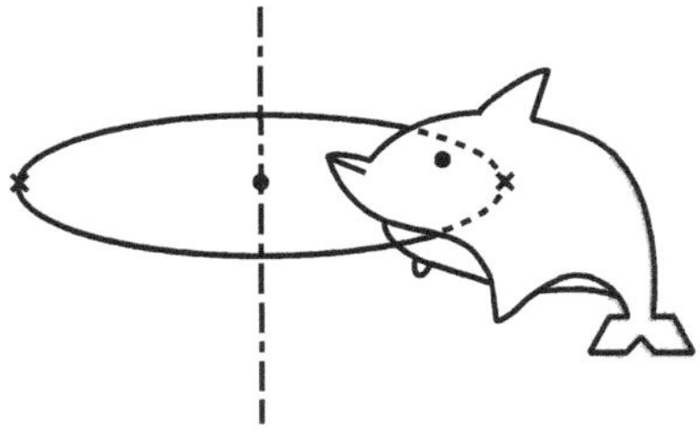

Fig.5

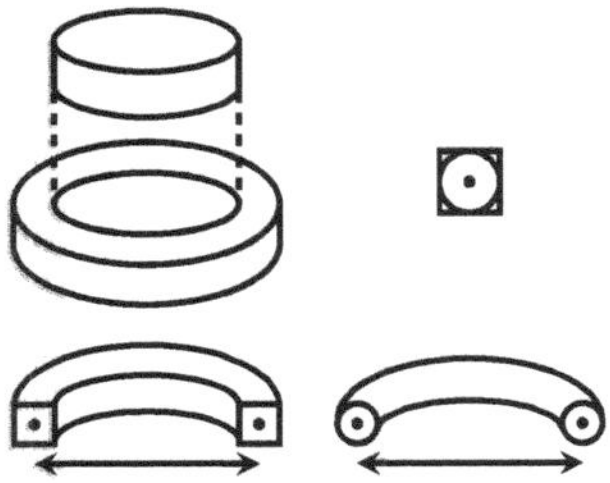

Fig.6

If we revolve a plane figure Q (for example a dolphin) around a straight line called the center axis, we shall get a dolphin-shaped ring (Fig.5). The volume is the product of the area of Q and the circumference C of locus of the center of gravity of Q. Seki called the center of gravity simply the "center," as he had no concept of dynamics and regarded it as a purely geometrical notion of a figure. A diameter of the circumference C is the line segment which binds two centers $\times$ on the opposite sides of the figure Q in Fig.5.

In this paper, we shall use the word "barycenter" for mathematical clarity. In the Western tradition, this fact is called the Pappos-Guldin formula. It is very easy to calculate the volume of a torus, because the barycenter coincides with the center of the circle. But in a general case of different figure, it is not always easy to find the barycenter.

In the square segment case, I noted already that a square-shaped ring can be found as the difference of two discs with different sizes $d+e$ and $d-e$ (Fig.6, left-hand side).

$$V = (355/452)[(d+e)^2 - (d-e)^2]e = (355/113)de^2.$$

In this case, the section is the square $e \times e$. The ring of a circular section is obtained by analogy to Fig.1, as shown in Fig.6, right-hand side. If we investigate the general case, we are confronted with a calculus of definite integrals, as Western mathematicians have done.

4 Sphere segments and finger rings

We now enter into the main theme of this paper – figures derived from a sphere. A sphere segment E is a solid in appearance of a flat cap, which is obtained by cutting a sphere of diameter 1 with a flat plane (Fig.7). After cutting two segments Es from the sphere (one in the upper side and the other in the lower), we take off a cylinder F as shown in Fig.7. Then we get a ring G (G' being the other section), which is a ring having an arc shaped segment. Seki called it a "regular arc ring," but I shall call it a finger ring based on my intuition. After these operations the sphere is divided into 4 parts: $2E+F+G$. The diameter a of the small circle, the height f of the cylinder, and the diameter d of the sphere satisfy the well-known relation (Fig.8):

$$a^2+f^2=d^2, \tag{1}$$

which is Pythagorean theorem, and was called in the Japanese traditional mathematics [和算 Wasan] the law of right triangles [勾股弦の法 kōkogen no hō]. Seki called f a "sub-diameter," which played an important role in his theory.

The diameter d is assumed to be 1 in the calculation. The volume F of the cylinder is given by

$$F=(355/452)a^2f=(\pi/4)a^2f.$$

The height c of a sphere segment E is given by

$$c=(1-f)/2,$$

and the diameter is a. Then the volume E of the sphere segment is given by

$$E=(355/2712)(3a^2+4c^2)c=(\pi/24)(3a^2+4c^2)c.$$

This is the formula of the sphere segment in Measurements. (See [1, p.238], and also Sugimoto [6, p.467].)

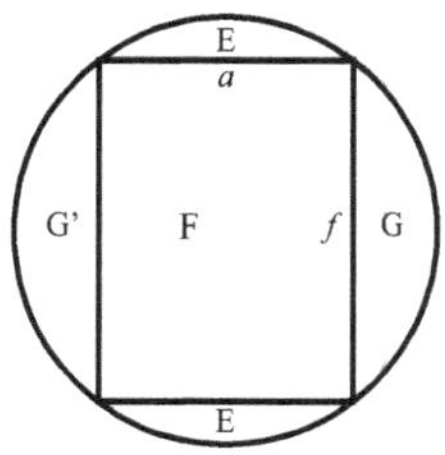

Fig.7

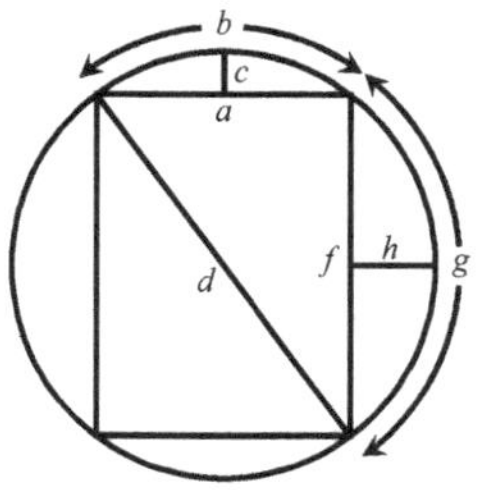

Fig.8

Finally the volume of the finger ring G is given by

$$G=V-2E-F=(355/678)1^3-2E-F,$$

where V is the volume of the sphere and is equal to $(4\pi/3)(1/2)^3$.

5 Thinking patterns hereditary in *Wasan*

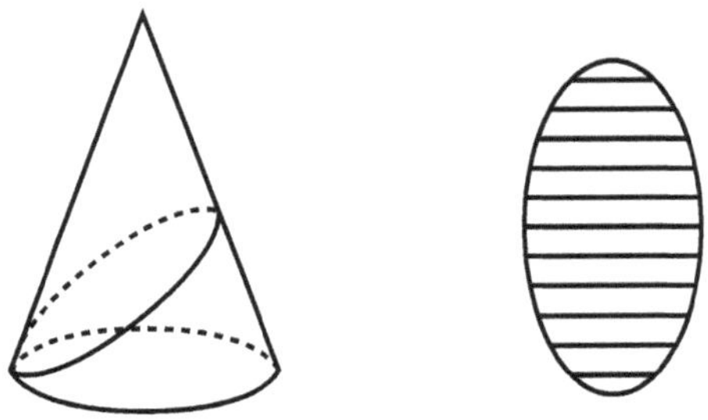

Fig.9

Ellipse. In the Western tradition, an ellipse means a curve obtained as a section of a cone, or a conic section. Seki also obtained this figure by the same operation, but an ellipse meant for him a full figure with its interior included. He was much concerned with its area and was indifferent to the nature of quadratic curves (Fig.9). I once remarked that an ellipse looks like *waraji* (sandals made of rice-stalks in old days in Japan). It looks like an ellipse with thick lateral lines. In fact, Saki calculated the area by adding all the length of lateral lines.

(Recently I published a book [6] in which all of my papers on Seki Takakazu and his mathematics are collected. For details of ellipses for Seki, we refer the reader to [6, p.550].)

Arc-figures. Similarly, a circle in Wasan is a disc, and an arc is the plane figure surrounded by the arc and the cord connecting two end points of the arc.

Let us consider a bow, an arrow, a chord and a figure surrounded by the bow and the chord. We shall call the bow as an arc, the arrow as a sagitta, and the figure as an arc shaped figure, or simply an arc-figure. As I noted earlier, Seki considered figures filled with interior and he was very much concerned with the area. Bows and arrows used to be weapons for warriors. Although Seki worked at the accountant department of the *shōgun* government, he was a *samurai* and should have been familiar with these figures.

The plane section of the area E with arc b, chord a, sagitta c, and diameter d, in Fig.8 is calculated as follows :

$$E = (1/4)(bd - af),$$

where the sub-diameter f can be calculated from a and d by (1).

Arcsine formula. To calculate the area of an arc-figure, we need to express the arc b by the other variables. Nowadays we calculate b by the arcsine formula, but in his times Seki calculated it in a fractional expression, which was very inconvenient for him to use. (See [5].) It is one of the reasons why Seki used only two numerical examples:

$$\text{(Case 1)}\ \ a = 0.6,\ c = 0.1,\ b = 0.6435011;\quad (E \text{ in Fig.7\&8}),$$
$$\text{(Case 2)}\ \ f = 0.8,\ h = 0.2,\ g = 0.92729522,\ (G \text{ in Fig.7\&8}),$$

with $d = 1$. See Fig.8.

The area of an arc-figure (in this context, E and G mean plane figures) is calculated by the formula:

$$E = (1/4)(bd - af) = 0.040875275, \qquad \text{(Case 1)},$$
$$G = (1/4)(gd - fa) = 0.111823805, \qquad \text{(Case 2)}.$$

Using these results, we can get the volume of the finger ring in Fig.7 as follows : using $a = 0.6$ and $c = 0.1$ in Fig.8, we find that the volume of the sphere segment E in Fig.7 is given by

$$E = (355/2712)(3 \times 0.6^2 + 4 \times 0.1^2) \times 0.1 = 0.014660767$$

and the volume of the sphere ($d = 1$) is a double of the semi-sphere

$$2 \times [1^2 \times 0.5 \times (2/3) \times (355/452)] = 2 \times (355/1356) = 355/678 = 0.52359882.$$

The volume F of the cylinder is given by

$$F = 0.6^2 \times (355/452) \times 0.8 = 0.22619469,$$

and finally the volume G of the finger ring is given by

$$G = 355/678 - 2E - F = 0.26808260.$$

We notice an interesting fact that this value is equal to the volume of a sphere with diameter 0.8. See Section 1.

6 Barycenters of arc-figures

In the case of a straight cylinder, we get the height by dividing the volume by the area of the circle. If we divide the volume of a finger ring by the area of its section (i.e., an arc-figure), we get a track of the barycenter of the arc-figure. You may consult the Pappos-Guldin formula. Dividing 0.26808260 by 0.111823805, you get 2.3973661. This value means a track of the barycenter of the arc-figure, or circumference of revolution. If we divide this by 355/113, i.e. π, then we get $j = 0.76310527$. This distance means the interval between two centers of the arc-figure Gs. It can be calculated as follows :

$$i = (j - a)/2 = (0.76310527 - 0.6)/2 = 0.081552635.$$

The value i is the distance of the barycenter of the arc-figure from the chord (Fig.10).

Although the concept of barycenter or center of balance is relatively natural, it is difficult to determine the barycenter of a plane figure of various shape and in a general position.

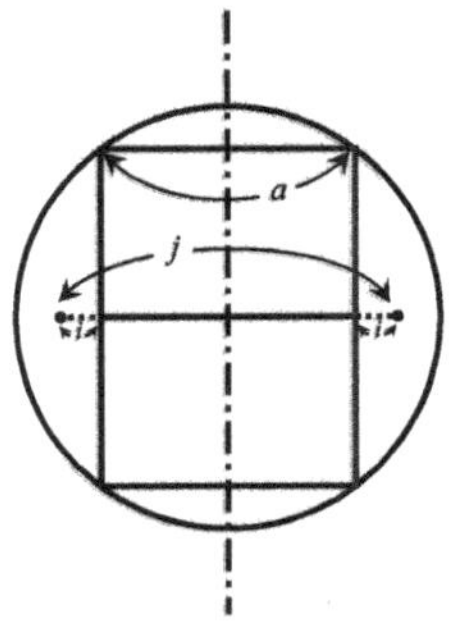

Fig.10

The second example. We assume the G and G' of Fig.7 as sphere segments. Then the upper E and the lower E would be sections of the identical finger ring. After calculation, we get 0.113097345 as the volume of the new finger ring, and 0.040875275 as the section of the new finger ring. We must notice that the value 0.113097345 is coincident with the volume of a ball of diameter 0.6. See Section 1. By dividing the former by the latter, we get 2.766889 (as a track of barycenter of the arc-figure), and further dividing by $355/113$, we get 0.88072802 (as the interval between two barycenters). By a similar calculation (in this case, using 0.8), we get

$$(0.88072802 - 0.8)/2 = 0.04036401.$$

This is nothing but the position of the barycenter in the second example.

A miracle Theorem. Suppose that you take off a cylinder from a sphere and that the height of the intersection of the cylinder and the sphere is equal to k. (We have already two examples, $k = 0.8$ and $k = 0.6$.) If the diameter of the sphere changes, the remaining finger ring has holes with changing diameter but the height of the hole remains the same k. The larger the sphere becomes, the thinner the section of the finger ring becomes. The smaller the sphere becomes, the thicker the section of the finger ring becomes (Fig.11).

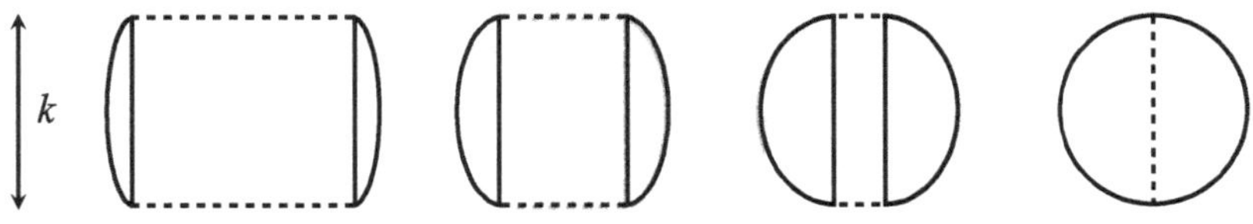

Fig.11

Now you may discover a miracle having noted that the remaining finger rings have always the same volume! Once you admit this miracle, you can find the volume as the limiting case; i.e. if the height k coincides with the diameter of the sphere, then the remainder is nothing but the sphere itself and the volume is $(355/678)k^3$. You may ascertain this miracle by several calculations, then by demonstration. I suppose that our Seki should have also noticed this theorem numerically, although in his manuscript, there is no mention about it. I believe that he must have noticed this fact.

7 *Onglets* and the barycenters of arc-figures

Seki called the solid of Fig.12 a "cylinder cut by an oblique plane." Pascal called this solid an *onglet*, which means a hollow into which we can put a nail to look a word up in a thick dictionary (in French, *ongle* means nail, *onglet* nail-putting hollow). We shall use Pascal's terminology in this paper. Fig.13 shows a half *onglet*. How can we calculate the volume of this solid ? CGBA means a half arc-figure, triangle HAB is equal to an isosceles right triangle, and also CIHA is similar to CGBA, with the relation of EI : EG = AH : AB = $\sqrt{2}$: 1. To calculate the half *onglet* solid CHAB, we collect all triangles IEG (isosceles right triangles) from point C to point A. To calculate it as a definite integral, we have to know $(1/2)\Sigma EG^2\delta$, where δ means the thickness of a triangular board.

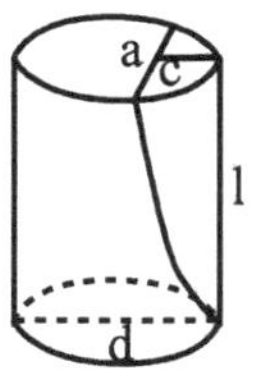

Fig.12

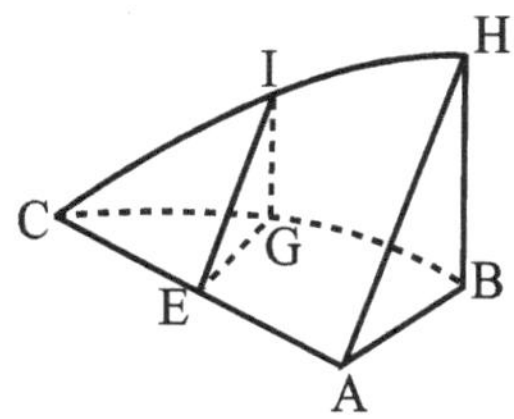

Fig.13

Seki did not write anything about its value. How could he then obtain the volume of this solid ? I suppose that he obtained it by a thinking similar to Pascal's. We slice off an *onglet* from *a* chord of arc-figure (with sagitta c on the top of a cylinder with diameter d) obliquely downward toward a point beneath l from the top. Seki gave the volume P of this solid by

$$P = (l/12c)(a^3 - 6lE),$$

where E is the area of the arc-figure. When the sagitta c is equal to $d/2$, then

$$P = (1/6)d^2 l.$$

Barycenters of arc-figures, again. As I said above, the barycenter of the arc-figure is a point which situates inner with distance i from the chord a. Seki claimed that if the area of the arc-figure is E and the volume of the *onglet* is P, then there is the relation :

$$P/E = (l/c)i.$$

The left-hand side means a kind of "average", dividing P by E. When we cut the cylinder obliquely (half right angle) by a plane, then the coefficient (l/c) will be 1 and the right-hand side will become i.

8 A wonderful relation

(1) If we regard the finger ring as a body of revolution, then we can get the diameter between barycenters of each section (arc-figure), and we can obtain the barycenter naturally.

(2) If we calculate the i in the above discussion about *onglet*, we can smoothly get the exactly same barycenter.

This is a wonderful relation between far distant things; one is the finger ring and the other is the *onglet*. In the Western culture, a dynamics has been developed for seeking the barycenter of a plane-figure and of a solid, or looking for the center of rotation of a wheel. But Seki sought the barycenter, which he called the center, by purely geometrical relations. I admire his keen intuition about geometrical figures. In the end of this paper, I shall give an imagination about his root of intuition.

Decipherment of a sentence of Seki. We have arrived at the most important point of this paper. Seki solved a riddle which binds the finger ring with the *onglet*. Japanese mathematicians in the Edo period wrote scientific works mostly in Chinese, not in formal Chinese, but with somewhat Japanese flavor. I shall translate his sentence 是伸弧環, 而去中之弧壔, 則兩旁適作此形. into English with my comments. (See [6, p.515–519].)

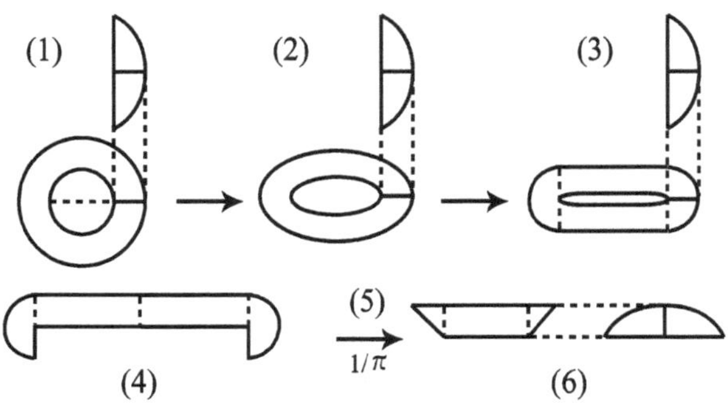

Fig.14

Two *onglet* solids are made from the following processes (Fig.14).

(1) There is a finger ring with arc-section.
(2) If we stretch the finger ring from both sides,
(3) it will be flat.
(4) We reconnect two solids after cutting in several parts.
(6) We get a straight open rainwater pipe.
(7) We remove from this solid the straight open rainwater pipe, then we have two *onglets* on both sides (Fig.15).

I input one sentence (5) between (4) and (6) for better understanding.
(5) We divide this solid by 355/113, i.e., π.

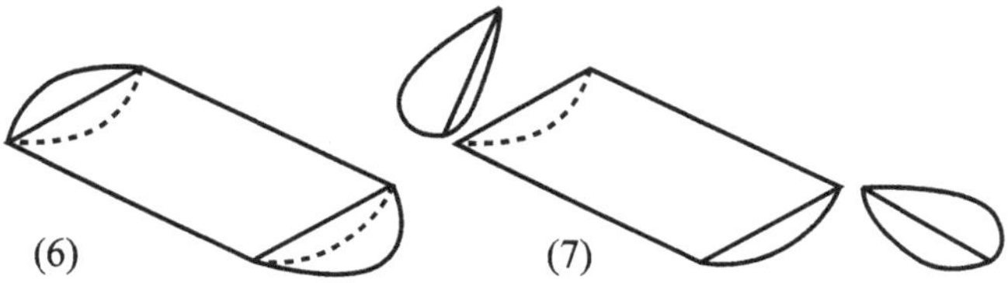

Fig.15

By this transformation, from a finger ring we get two *onglet* solids. As a result, the finger ring and the *onglet* both have the same arc-section, and the same barycenter. Seki showed this fact by a splendid magic.

We have obtain the theorem: "the volume of a finger ring is the product of (a) the area of an arc-figure section and (b) the circumference drawn by a diameter between two barycenters".

You may compare this with the straight cylinder case, where the volume is the product of (a) the area of an arc-figure section and (b) the height of the cylinder.

We obtained this theorem when the finger ring is an actual part of a sphere. But this theorem is still valid for more general cases where a diameter is longer or shorter than the original case. The so-called Pappos-Guldin formula is usually stated in this general form. Of course Seki should also have noticed this general form.

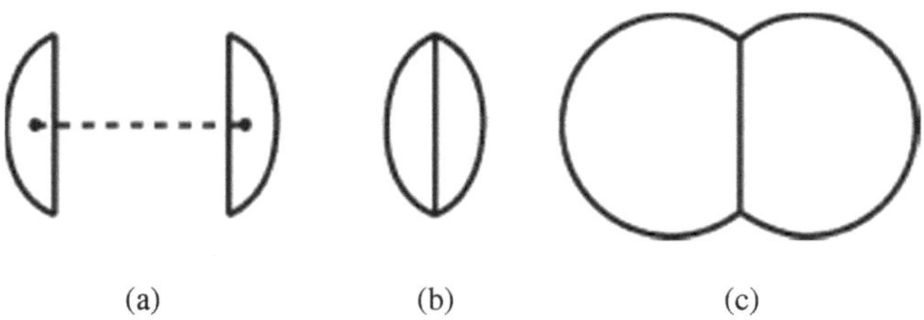

Fig.16

Fig.16 shows various cases:
(a) far separated,
(b) two chords coincide,
(c) two circles cut off by the other-side of the center, and two chords coincide.

Kepler [7, pp.192–197] once studied various kinds of wine barrel and gave various names to various rotating figures. He named (b) as "lemon" and (c) as "apple". If we put together two segments in Fig.14, (4), which are obtained by cutting off the intervening center open rainwater pipe, then we get a solid "lemon".

9 An umbrella-shaped solid

We can see one of the exhaustive characters of Seki's study in the case where two arc-figures are placed in oblique situation. Fig.17 shows a special case in which the tops of the two arc-figures coincide with one another. Based on my intuition, I called it "umbrella-shaped", [6, p.530]. In Fig.17, there are two arc-shaped segments Es. You may read each size in the figure.

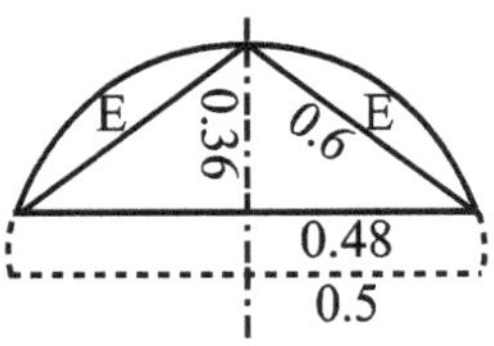

Fig.17

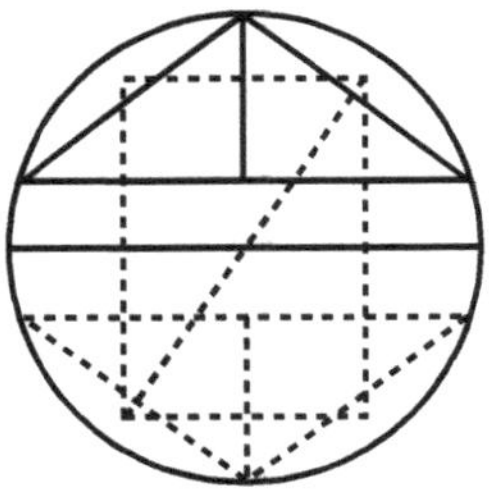

Fig.18

As you see, rotating this figure around the center axis, we get an umbrella-shaped solid. The volume U of the umbrella-shaped solid was measured, in the same procedure as in the finger ring case, by subtraction of a cone C from a sphere section A:

$$A = (355/2712)(3\times 0.96^2 + 4\times 0.36^2)\times 0.36 = 0.15471717,$$
$$C = (1/3)(355/452)\times 0.96^2\times 0.36 = (355/1356)\times 0.96^2\times 0.36 = 0.08685876,$$
$$U = A - C = 0.06785841.$$

Seki solved the problem of measuring the volume as shown in Fig.18. We draw the same figure in the lower side, then we connect four barycenters with dotted lines each other in rectangle. Then we bind two centers of upper right-hand figure and lower left-hand figure by an oblique line as shown in the figure. The diameter binding two arc-figures obliquely is nothing but that of the preceding case in Fig.7, which bind two centers of Es.

Even though the arc-figure is placed vertically, horizontally or obliquely, the barycenter always represents all the weights of figure, no matter what shape of the figure is. This is a very nature of the barycenter. Whether Seki thought of this nature seriously or not, I can not say, because he wrote nothing about it. Probably he should have recognized it by intuition.

In Fig.8 and in Section 6, especially in the second example, we calculated the diameter between two centers, which was 0.88072802 in the second example. This is the very distance between two barycenters obliquely in Fig.18. We can see the figure of a *Kōkogen* in Fig.18, similar to the right triangle 0.36, 0.48 and 0.6 in Fig.17. Therefore, we can calculate other two line segments in Fig.18 proportionally:

$$\text{horizontal line} = (0.36/0.6) \times 0.88072802 = 0.528436812,$$
$$\text{vertical line} \quad = (0.48/0.6) \times 0.88072802 = 0.704582416.$$

In the case of an umbrella-shaped solid U in Fig.17, the distance between two barycenters is nothing but 0.528436812. As a result of this, and with the area 0.040875275 (in the case 1) of the arc-figure, the volume of this solid was shown to be

$$U = 0.040875275 \times (355/113) \times 0.528436812 = 0.06785841,$$

where $(355/113) \times 0.528436812$ is the circumference of the circle as a track of the center of arc-figure. This result coincides with U calculated through another root $A-C$, as we did before. This is the case where the arc of two arc-figures is placed along a same circle. We can generalize this result further for the case where the distance between centers is either longer or shorter. This is what Seki did in various examples in Measurements [4].

The marvelous intuition of Seki. Everywhere in Measurements, we can find Seki's intuitive grasp of geometrical objects. I suppose that his idea came from concrete models made with something available for him like pasta-material. Seki also named many solids in his book after commodities, which have been used by people in the Edo period. I suppose that he recognized mathematical figures in familiar things in everyday life and that he contemplated in those things the inner structure composed of basic geometrical figures.

References

1. Akira Hirayama [平山諦], Kazuo Shimodaira [下平和夫] and Hideo Hirose [広瀬秀雄] eds.: *Takakazu Seki's Collected Works Edited with Explanations* [關孝和全集 Seki Takakazu Zenshū], Ōsaka Kyōiku Tosho [大阪教育図書], Ōsaka (1974).
2. Tamotsu Murata [村田全]: *Takebe Katahiro's mathematics and its philosophy* [建部賢弘の数学とその思想 Takebe Katahiro no sūgaku to sono shisō], *Seminar in Mathematics* [数学セミナー Sūgaku Seminā] (August 1982–January 1983).

3. Seki Takakazu [關孝和]: *Concise Collection of Mathematical Methods* [括要算法 Katsuyō Sanpō] (1712); in *Takakazu Seki's Collected Works edited with explanations*, pp. 269–370 (1974).
4. Seki Takakazu [關孝和]: *Measurements* [求積 Kyūseki]; in *Takakazu Seki's Collected Works edited with explanations*, pp. 219–250 (1974).
5. Toshio Sugimoto [杉本敏夫]: Seki Takakazu's construction process for arcsine formula — A tendency of separate treatment among the mathematical branches, Tatsuhiko Kobayashi [小林龍彦] et al. eds., *Proceedings of the Fourth International Symposium on the History of Mathematics and Mathematical Education using Chinese Characters*, Maebashi Institute of Technology [前橋工科大学], pp.189–197 (2001).
6. Toshio Sugimoto [杉本敏夫]: *Commentary on Seki Takakazu's Mathematics, Thinking process of a genius* [解読 · 關孝和, 天才の思考過程 Kaidoku · Seki Takakazu, Tensai no Shikōkatei], Kaimeisha [海鳴社], Tokyo (2008).
7. D. J. Struik (ed.) : *A Source Book in Mathematics 1200–1800*, Harvard U. P., 1st ed., (1969); 2nd ed., (1986).

Seki Takakazu's Method on the Remainder Problem

Sumie Tanabe

Abstract It is known that a system of simultaneous congruences of first degree which will be henceforth called a remainder problem, first appeared in the Sunzi's Arithmetical Canon (c.AD400). Later in China, remainder problems were discussed in many books, and some of these Chinese books were introduced into Japan by the seventeenth century. The eminent Japanese mathematician Seki Takakazu (c.1642-1708) investigated remainder problems adopting the term the art of cutting bamboo [jianguan shu] which is found in the Chinese book, Yang Hui's Arts on Arithmetic [Yanghui Suanfa] (1275) by Yang Hui. Seki is supposed to have consulted the Chinese book, but Seki's method is much more advanced than Yang Hui's. Seki generalized the theory on the remainder problem and showed the procedure for the solution systematically. The aim of this paper is to analyze Seki's method on the remainder problem in comparison with Chinese books, especially with the Mathematical Treatise in Nine Chapters [Shushu Jiuzhang] (1247) by Qin Jiushao.

1 The Remainder Problem

Let us consider a system of simultaneous congruences of first degree

$$a_i x \equiv r_i \pmod{m_i} \text{ for } i = 1, 2, \cdots, n, \tag{1}$$

which will be called a remainder problem in this paper. It is known that a remainder problem first appeared in the Sunzi's Arithmetical Canon [孫子算経 Sunzi Suanjing] (C. AD 400, Fig.1). The problem in the book is as follows:

$$x \equiv 2 \pmod 3, \quad x \equiv 3 \pmod 5, \quad x \equiv 2 \pmod 7. \tag{2}$$

Sumie Tanabe
Sacred Heart Senior High School, 4-11-1, Shirokane, Minato-ku, Tokyo 108-0072, Japan
e-mail: tstanabe@da.airnet.ne.jp

The answer given in it was only "$x = 23$", and was not the general solution "$x = 23 + 105k$ ($k \in Z$)".

It is supposed that remainder problems were investigated in ancient China in connection with the calendrical astronomy and the divination lore. After the Sunzi's Arithmetical Canon, remainder problems were discussed in many books such as the Mathematical Treatise in Nine Chapters [数書九章 Shushu Jiuzhang] (1247) by Qin Jiushao [秦九韶], the Yang Hui's Arts on Arithmetic [楊輝算法 Yanghui Suanfa] (1275, Fig.2) by Yang Hui [楊輝], and the Systematic Treatise on Arithmetic [算法統宗 Suanfa Tongzong] (1592, Fig.3) by Cheng Dawei [程大位]. The remainder problem and its solution are identified by many terms such as the unknown total [物不知其総数 wu bu zhi qi zong shu] in the Sunzi's Arithmetical Canon, the great art to seek the total from remainders [大衍総数術 dayan zongshu shu] in the Mathematical Treatise in Nine Chapters, the method of calling the roll by General Hanxin [韓信點兵 hang xin dian bing] in the Systematic Treatise on Arithmetic, the art of cutting bamboo [翦管術 jianguan shu] in the Yang Hui's Arts on Arithmetic and so forth. Nowadays, the solution of the remainder problem is commonly called "Sunzi's theorem" or "the Chinese remainder theorem".

The eminent Japanese mathematician Seki Takakazu [関孝和] (c.1642-1708) investigated remainder problems adopting Yang Hui's use of the art of cutting bamboo in his posthumous work with the title Compendium of Mathematics [括要算法 *Katsuyō Sanpō*] (1712, Fig.4) and others. Seki is supposed to have consulted the Chinese book, Yang Hui's Arts on Arithmetic. But Seki's method is much more advanced than Yang Hui's.

2 Traditional Chinese remainder problem and its solution

In almost all Chinese books except for the Mathematical Treatise in Nine Chapters, remainder problems (1) are given under the condition that positive integers m_1, m_2, $\cdots$, m_n are pairwise co-prime, and also coefficients a_i are equal to one, that is "$a_i = 1$ for all i".

Let us examine the traditional solution in such a situation.

1. First solve each indeterminate equation for $i = 1, \cdots, n$

$$\frac{M}{m_i}x - m_i y = 1 \text{ where } M = m_1 \cdot m_2 \cdot \cdots \cdot m_n. \tag{3}$$

   And set the solution $\frac{M}{m_i}x = x_i$.
2. Then each x_i fulfills the conditions

$$x_i \equiv 1 \pmod{m_i}, \quad x_i \equiv 0 \pmod{m_j} \text{ for all } i, j \text{ where } i \neq j. \tag{4}$$

3. Therefore, we can see that the solution of a remainder problem (1) is a remainder class

$$x \equiv \sum_{i=1}^{n} r_i x_i \pmod{M}.$$

4. The last stage of the solution of a traditional Chinese remainder problem is to subtract L.C.M. of m_i for $i = 1, \cdots, n$ several times as the need arises for the purpose of getting the minimum positive answer. We should mention that in ancient China and also in Japan before the mid-nineteenth century, mathematicians did not seek the general solution, but only sought for the minimum positive answer. This shows us the difference of concept on a system of number between that era and nowadays.

Now let us see the solution of the problem (2) in the Sunzi's Arithmetical Canon. The book says that the answer is obtained by the following calculation:

$$70 \times 2 + 21 \times 3 + 15 \times 2 - 105 \times 2 = 23.$$

23 is truly the minimum positive answer of this problem. The next part of the book explains briefly to obtain the answer in the general case of arbitrary remainder r_1, r_2, r_3 but under the particular condition with moduli 3,5,7. The calculation indicated is as follows :

$$x = 70r_1 + 21r_2 + 15r_3 - 105n.$$

But there is no mention of procedures to get key numbers 70, 21, 15. How to seek the key numbers is the core of the solution, and that is to solve equations (3), the first part of the solution above-mentioned. All the more, we should pay attention that the necessary and sufficient condition to solve indeterminate equation (3) is that $\frac{M}{m_i}$ and m_i are pairwise co-prime, that is to say moduli m_i are pairwise co-prime for all i. Almost all remainder problems in Chinese books except for the Mathematical Treatise in Nine Chapters are under the condition that all moduli are pairwise co-prime.

3 The spread of remainder problems in Japan

In the beginning of the seventh century, at first through Korea, Chinese civilization of Sui and Tang dynasties arrived in Japan along with the Ten Classics of Mathematics [算経十書 Suanjing shishu], a part of which was Sunzi's Arithmetical Canon. Therefore, the remainder problem and its simple solution might be known to some people of specially educated officers in the Nara period (710-784) in Japan.

In the Edo period (1603-1867), the remainder problem had become well-known by the publication of Inalterable Treatise [塵劫記 Jinkōki] in 1627 by Yoshida Mitsuyoshi [吉田光由] (1598-1672). The book was revised several times by Yoshida Mitsuyoshi himself and furthermore, there were hundreds of pirated editions of the book. It was a best and long seller over a period of two hundred years. The author Yoshida Mitsuyoshi consulted the Chinese book Systematic Treatise on Arithmetic, and introduced a remainder problem with the same condition on moduli as in the Sunzi's Arithmetical Canon and the Systematic Treatise on Arithmetic. He named the remainder problem the subtraction by one hundred and five [百五減算 hyakugo genzan], because at the last stage for seeking the minimum positive answer one should subtract by 105.

After several decades, Seki Takakazu investigated remainder problems in a more general framework than Yang Hui. Seki's results can be found in the Compendium of Mathematics and also in the Complete Book of Mathematics [大成算経 *Taisei Sankei*]. After that, Seki's method to solve remainder problems used in general by Japanese mathematicians in the Edo period. For example the Clever Man's Booklet on Mathematics [勘者御伽雙紙 Kanja Otogi Zōsi] (1743) written by Nakane Genjun [中根彦循] (1701-1761) discussed not only a problem called the subtraction by one hundred and five but also a problem called the subtraction by three hundred and fifteen [三百十五減 sanbyakujūgo gen] of which moduli are five, seven and furthermore a problem the subtraction by sixty three [六十三減 rokujūsan gen] of which moduli are seven and nine. The details of indeterminate equations in the Clever Man's Booklet on Mathematics are discussed in [17].

4 The Compendium of Mathematics

The Compendium of Mathematics was compiled by Seki's disciples and published for memorial to the anniversary of Seki's death. It consists of Seki Takakazu's works from 1680 to 1683 and is divided into four volumes named as *gen* [元], *kō* [亨], *ri* [利], and *tei* [貞]. The second volume *kō* treats the elementary number theory and is divided into two parts. The first part deals with basic problems on number theory, for example the greatest common divisor, the least common multiple and indeterminate equation like:

$$Ax - By = 1, \text{ where positive integers } A, B \text{ are given and } (A,B) = 1. \tag{5}$$

Seki called the procedure to solve these indeterminate equations (5) the art to leave remainder one [剰一術 jōichi jutsu]. Seki's method on indeterminate equations, the art to leave remainder one, is mutual division. It is the same as Euclid's algorithm. Almost all Chinese books do not mention the procedure to solve indeterminate equations (5). Among all the Chinese books, only the Mathematical Treatise in Nine Chapters states the procedure with the term the great art to seek remainder one [大衍求一術 dayan quyi shu/daien kyūichi jutsu].

The second part of the volume *kō* deals systematically with the art of cutting bamboo, a general procedure to solve remainder problems, by means of the results in the first part of the volume *kō*.

The following are 9 problems treated in the second part:

No. 1. $\begin{cases} x \equiv 1 \pmod{5} \\ x \equiv 2 \pmod{7} \end{cases}$

Ans. $x = 16$

No. 2. $\begin{cases} x \equiv 2 \pmod{36} \\ x \equiv 14 \pmod{48} \end{cases}$

Ans. $x = 110$

No. 3. $\begin{cases} x \equiv 2 \pmod{3} \\ x \equiv 1 \pmod{5} \\ x \equiv 5 \pmod{7} \end{cases}$

Ans. $x = 26$

No. 4. $\begin{cases} x \equiv 3 \pmod{6} \\ x \equiv 3 \pmod{8} \\ x \equiv 5 \pmod{10} \end{cases}$

Ans. $x = 75$

No. 5. $\begin{cases} x \equiv 3 \pmod{5} \\ x \equiv 2 \pmod{7} \\ x \equiv 2 \pmod{9} \\ x \equiv 7 \pmod{11} \end{cases}$

Ans. $x = 128$

No. 6. $\begin{cases} 35x \equiv 35 \pmod{42} \\ 44x \equiv 28 \pmod{32} \\ 45x \equiv 35 \pmod{50} \end{cases}$

Ans. $x = 13$

No. 7. $\begin{cases} 8x \equiv 2 \pmod{3} \\ 7x \equiv 3 \pmod{4} \\ 6x \equiv 3 \pmod{5} \end{cases}$

Ans. $x = 13$

No. 8. $\begin{cases} 34x \equiv 6 \pmod{8} \\ 34x \equiv 14 \pmod{20} \\ 34x \equiv 23 \pmod{27} \end{cases}$

Ans. $x = 11$

No. 9. $\begin{cases} 13x \equiv 3 \pmod{7} \\ 13x \equiv 8 \pmod{9} \end{cases}$

Ans. $x = 11$

As we can see, the nine distinctive problems are given with simple numbers. Problems No.1 to No.5 are with condition "$a_i = 1$" and problems No.6 to No.9 are with condition "$a_i \neq 1$". Furthermore, problems of odd number are under the condition that moduli are pairwise co-prime, and problems of even number are under the condition that moduli are not pairwise co-prime. Because the term, the art of cutting bamboo, was used in the Yang Hui's Arts on Arithmetic, Seki certainly consulted the Chinese book. But Seki's method to solve remainder problems is more general than Yang Hui's. Seki generalized the theory of the Chinese remainder problem to the cases in which moduli are not pairwise co-prime and coefficients are not necessarily equal to one, and presented systematically the procedure for the solution.

As an example, let us examine the problem No. 6.

$$35x \equiv 35 \pmod{42}, \quad 44x \equiv 28 \pmod{32}, \quad 45x \equiv 35 \pmod{50}.$$

In this problem, $a_1 = 35$, $a_2 = 44$, $a_3 = 45$, that is "$a_i \neq 1$ for $i = 1,2,3$" and moduli $m_1 = 42, m_2 = 32, m_3 = 50$, that is, m_1, m_2, m_3 are not pairwise co-prime.

Let us see Seki's method for solving it. First, by reducing each equation he got the following expression:

$$5x \equiv 5 \pmod 6, \quad 11x \equiv 7 \pmod 8, \quad 9x \equiv 7 \pmod{10}.$$

But moduli 6, 8 and 10 are still not pairwise co-prime. So, he changed moduli 6, 8 and 10 by new moduli 3, 8 and 5 by paying attention to L.C.M. of 3, 8 and 5 is equal to those of 6, 8 and 10, where L.C.M. denotes the least common multiple. Seki called this procedure on reduction respective reduction [逐約 chikuyaku]. Thus he arrived at the following congruences:

$$5x \equiv 5 \pmod 3, \quad 11x \equiv 7 \pmod 8, \quad 9x \equiv 7 \pmod 5.$$

Let us compare the two expressions (6) and (7) under the following conditions:

- m'_i is a divisor of m_i for i =1,2,3;
- m_1, m_2 and m_3 are not pairwise co-prime;
- m'_1, m'_2 and m'_3 are pairwise co-prime;
- L.C.M. of m_1, m_2 and m_3 is equivalent to that of m'_1, m'_2 and m'_3.

$$a_1x \equiv r_1 \pmod{m_1}, \quad a_2x \equiv r_2 \pmod{m_2}, \quad a_3x \equiv r_3 \pmod{m_3}. \tag{6}$$

$$a_1x \equiv r_1 \pmod{m'_1}, \quad a_2x \equiv r_2 \pmod{m'_2}, \quad a_3x \equiv r_3 \pmod{m'_3}. \tag{7}$$

Generally speaking, simultaneous congruences expressions (6) and (7) are not always equivalent. The necessary and sufficient condition for the equivalence between expressions (6) and (7) is as follows:

$$\begin{aligned} a_1r_2 &\equiv a_2r_1 \pmod{(m_1, m_2)}, \\ a_2r_3 &\equiv a_3r_2 \pmod{(m_2, m_3)}, \\ a_3r_1 &\equiv a_1r_3 \pmod{(m_3, m_1)}. \end{aligned} \tag{8}$$

Suppose that the condition (8) is satisfied, then the remainder problem (6) is solvable and also has a solution uniquely to modulus L.C.M. of m_1, m_2 and m_3. Proofs of these theories are described in many books on the elementary number theory, for example, [14], [19]. The details of procedure for solving problem No. 6 and others were stated in [15] and [16].

All remainder problems Seki presented in his works satisfied the condition (8), but there is no mention of this fact. Why and how did Seki only take up problems which satisfy the condition (8) in his work? Moreover, did he ever recognize the condition (8) by calculating so many equations? Those are the questions that should be considered carefully.

5 The Complete Book of Mathematics

The Complete Book of Mathematics is a treatise which Seki wrote in collaboration with Takebe brothers, Kataakira [建部賢明] and Katahiro [建部賢弘]. It took twenty-eight years from 1683 to 1711 for them to complete the treatise in twenty volumes. Because Seki passed away before the completion and Katahiro was very busy with his official business at that time, Kataakira, elder brother of Katahiro, completed it alone applying the finishing touches.

The Complete Book of Mathematics is a manuscript and it hadn't been published after all. So it had never been adopted as a textbook of Seki's school. Seki's theory had been taught after his death by using some treatise which Seki wrote and one of the core was the Compendium of Mathematics published in 1712.

The sixth volume of the Complete Book of Mathematics deals with remainder problems; the sixth chapter of the sixth volume is named *Cutting Bamboo* [翦管 Jianguan/Senkan]. Furthermore, the sixth chapter is divided into two parts. The first part titled *Seeking the Total* [求総数 kyūsōsū] discusses eight remainder problems and the second part takes up further technical nine problems.

The following are the eight problems in the first part, *Seeking the Total* :

No. 1 $\begin{cases} x \equiv 1 \pmod 5 \\ x \equiv 2 \pmod 7 \end{cases}$ Ans. $x = 16$

No. 2 $\begin{cases} x \equiv 3 \pmod 6 \\ x \equiv 3 \pmod 8 \\ x \equiv 5 \pmod{10} \end{cases}$ Ans. $x = 75$

No. 3 $\begin{cases} x+6 \equiv 3 \pmod 3 \\ x-9 \equiv 6 \pmod 7 \end{cases}$ Ans. $x = 22$

No. 4 $\begin{cases} \dfrac{x}{2} \equiv 3 \pmod 5 \\ \dfrac{x}{3} \equiv 4 \pmod 7 \\ \dfrac{x}{4} \equiv 6 \pmod{19} \end{cases}$ Ans. $x = 96$

No. 5 $\begin{cases} 35x \equiv 35 \pmod{42} \\ 44x \equiv 28 \pmod{32} \end{cases}$ Ans. $x = 13$

No. 6 $\begin{cases} 24x \equiv 12 \pmod{30} \\ 35x \equiv 7 \pmod{42} \\ 44x \equiv 28 \pmod{32} \end{cases}$ Ans. $x = 53$

No. 7 $\begin{cases} \dfrac{2}{3}x \equiv 4 \pmod 7 \\ \dfrac{3}{4}x \equiv 4 \pmod 8 \end{cases}$ Ans. $x = 48$

No. 8 $\begin{cases} \dfrac{x+5}{2} \equiv 3 \pmod 6 \\ 3(x-4) \equiv 4 \pmod 7 \\ \dfrac{3}{5}(x+2) \equiv 5 \pmod 8 \end{cases}$ Ans. $x = 73$

We can see that problems No.1, No2, No.5 and No.6 are the same kind as those of the Compendium of Mathematics. Especially as far as problems No.1 and No.2, even coefficients are exactly the same as those of problems in the Compendium of Mathematics, while problems like No.3, No.4, No.7 and No.8, which need process of calculation such as addition, subtraction or division of the total x, were not found in the Compendium of Mathematics.

So, the remainder problems in the Complete Book of Mathematics surpassed those of the Compendium of Mathematics, both in variation and in quantity. But the most important part of the art of cutting bamboo is just *Seeking the total* which had been expounded in detail in the Compendium of Mathematics. We can say that the Compendium of Mathematics is a book which states Seki's theory on the remainder problem narrowing it down to its essential.

6 The Mathematical Treatise in Nine Chapters

The Mathematical Treatise in Nine Chapters by Chinese mathematician Qin Jiushao was published in 1247. All other Chinese books before and after this book only dealt with remainder problems of which moduli were pairwise co-prime and did not mention the details of solution of the indeterminate equation (5). Even under such circumstances, Qin Jiushao discussed in this book remainder problems in the case of moduli which were not pairwise co-prime and furthermore, expounded on the solution of these remainder problems, including details of the procedure for solving the indeterminate equation (5). In this book, he named the procedure for solving the indeterminate equation (5) the great art to seek remainder one [大衍求一術 dayan quyi shu/daien kyūichi jutsu].

As the title indicates, this book consists of nine chapters and is divided into eighteen volumes. The first two volumes are for the first chapter Problems on Remainder [大衍類 dayan lei/daien rui] and present nine remainder problems, eight problems out of which are concerned with moduli not being pairwise co-prime.

The following are some typical problems from the first and second volumes:

$$\text{No.2.}\begin{cases} x \equiv 4 & \pmod{365\frac{1}{4}} \\ x \equiv 8 & \pmod{29\frac{499}{940}} \\ x \equiv 0 & \pmod{60} \end{cases}$$

$$\text{No.5.}\begin{cases} x \equiv 0.32 & \pmod{0.83} \\ x \equiv 0.70 & \pmod{1.10} \\ x \equiv 0.30 & \pmod{1.35} \end{cases}$$

$$\text{No.8.}\begin{cases} x \equiv 60 & \pmod{130} \\ x \equiv 30 & \pmod{120} \\ x \equiv 20 & \pmod{110} \\ x \equiv 30 & \pmod{100} \\ x \equiv 30 & \pmod{60} \\ x \equiv 30 & \pmod{50} \\ x \equiv 5 & \pmod{25} \\ x \equiv 10 & \pmod{20} \end{cases}$$

All the remainder problems of the Mathematical Treatise in Nine Chapters are given in practical situations such as the calendrical astronomy, trade of rice, tactics, divination lore and so forth. Consequently those are intricate with complicated moduli such as fractions, decimals and big figures. This point is a remarkable contrast with the simplicity of problems Seki presented. The details of problems of the Mathematical Treatise in Nine Chapters are stated in [4], [7], [8], [10] and [12].

7 Seki's method on the remainder problem

It is well-known that some Chinese books on mathematics, for example the Sunzi's Arithmetical Canon, the Yang Hui's Arts on Arithmetic and the Systematic Treatise on Arithmetic, were imported into Japan and Seki Takakazu or other Japanese mathematicians studied and consulted these Chinese books. But these Chinese books did not deal with problems in the case of moduli not being co-prime.

In Japan, before Seki, Hoshino Sanenobu [星野実宣] (1638–1699) described a remainder problem of which moduli were not co-prime in the Works on Right Triangles [股勾弦鈔 Kōkogen Shō] which was published in 1672. The problem is as follows:

$$x \equiv 5 \pmod 6,\ x \equiv 7 \pmod 8,\ x \equiv 5 \pmod{10}.$$

There is only the answer "95" on the heels of the problem, but no mention of its solution in the book.

In such a context in the Edo period, Seki discussed remainder problems in the case of moduli were not co-prime and furthermore showed the details of the solution clearly and systematically. On the other hand, more than three hundred years previously in an adjacent country China, Qin Jiushao dealt with remainder problems in the case of moduli being not co-prime. Between Qin Jiushao's method on remainder problems and those of Seki, there are some points of similarity. But there is no evidence that the Mathematical Treatise in Nine Chapters was imported into Japan. Had Seki a chance to know Qin Jiushao's work by oral instruction or any other way? Or else, was the similarity between method of Qin Jiushao and those of Seki caused by the mathematical inevitability?

Even if Seki knew Qin Jiushao's work in a rare possibility, it is obvious that there are clear differences between Qin jiushao's work and those of Seki. First, Qin Jiushao considered problems in which coefficients a_i are all equal to one, while Seki considered the more general case where coefficients a_i are not necessarily equal to one. Second, moduli in problems of Qin Jiushao are complicated, on the other hand those of Seki are quite simple. Finally, Qin Jiushao's problems are practical, in contrast, Seki's problems are completely abstract. Consequently, in Seki's work, the core of the theory on remainder problems was singled out in strong relief. This is just the point which shows Seki's advantage. Seki had certain sensitivity on abstraction and generalization which is the most important and indispensable for mathematical investigation. We may say that Seki's penetration is far ahead of his times, the Edo period.

今有物不知其數三三數之賸二五五數之賸三七七
數之賸二問物幾何答曰二十三
孫子算經 卷下 七
術曰三三數之賸二置一百四十五五數之賸三置六
十三七七數之賸二置三十并之得二百三十三以二
百一十減之即得凡三三數之賸一則置七十五五數
之賸一則置二十一七七數之賸一則置十五一百六
以上以一百五減之即得

Fig. 1 The Sunzi's Arithmetical Canon
[孫子算経 Sunzi Suanjing] (AD400?)

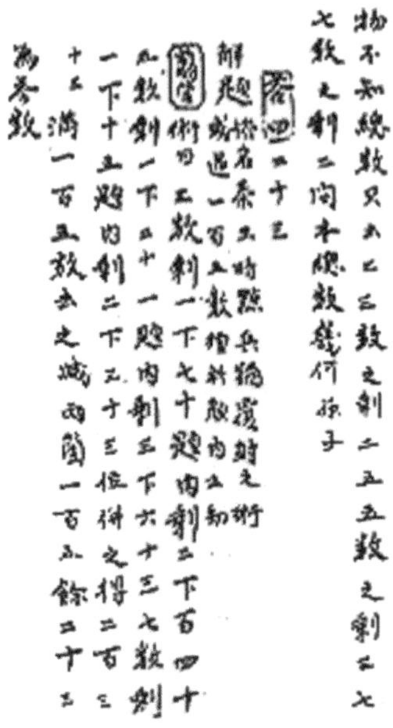

Fig. 2 The Yang Hui's Arts on Arithmetic
[楊輝算法 Yanghui Suanfa] (1275) by Yang Hui [楊輝]

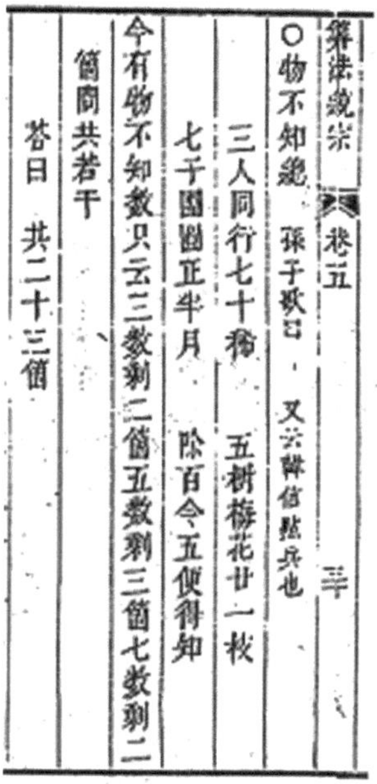

算法統宗 卷五

○物不知總 孫子歌曰 又云韓信點兵也

三人同行七十稀 五樹梅花廿一枝

七子團圓正半月 除百令五便得知

今有物不知數只云三數剩二箇五數剩三箇七數剩二

箇問共若干

荅曰 共二十三箇

Fig. 3 The Systematic Treatise on Arithmetic [算法統宗 Suanfa Tongzong] (1592) by Cheng Dawei [程大位]

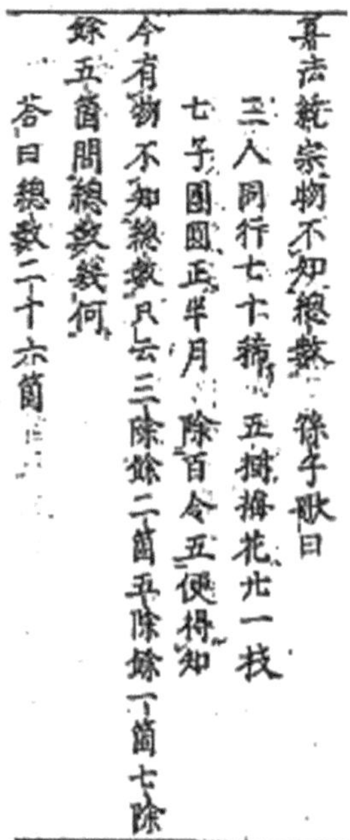

算法統宗物不知總數 孫子歌曰

三人同行七十稀 五樹梅花廿一枝

七子團圓正半月 除百令五便得知

今有物不知總數只云三除餘二箇五除餘一箇七除

餘五箇問總數幾何

荅曰總數二十六箇

Fig. 4 The Compendium of Mathematics [括要算法 Katsuyō Sanpō] (1712) by Seki Takakazu [関孝和]

References

1. Akira Hirayama, Kazuo Shimodaira and Hideo Hirose: *Seki Takakazu Zenshū* (The Collected Works of *Seki Takakazu*, in Japanese with English summary), Ōsaka Kyōiku Tosho [大阪教育図書] (1974).
2. Akira Hirayama: The time when "Yang-Hui-Suanfa" was copied by Seki Takakazu (in Japanese), *Journal of History of Mathematics Japan*, **68**, pp. 1–2 (1976).
3. Akira Hirayama: *The Birth of Wasan* [和算の誕生 *Wasan no Tanjō*], Kōseisha Kōseikaku [恒星社厚生閣] (1993).
4. Shuntaro Ito: *History of Mathematics, 2; Mathematics of the Middle Ages* [数学の歴史 2, 中世の数学 *Sugaku no Rekisi 2, Cyusei no Sugaku*], Kyōritu Shuppan [共立出版] (1987).
5. Shigeru Jochi: Another "Celestial Element Method"; "The Technique of Acquiring One in Dayan" (in Japanese), *Journal of History of Mathematics Japan*, **148**, pp. 1–12 (1996).
6. Shigeru Jochi: On the Yanhui's Arts of Arithmetic, copied by Nakada Takahiro and conserved by Ishiguro Nobuyoshi (in Japanese), *RIMS Kôkyûroku*, **1392**, Kyoto Univ., pp. 46–59 (2004).
7. Ulrich Libbrecht: *Chinese Mathematics in the Thirteenth Century*, The MIT Press (1973); Dover edition (2005).
8. Jean-Claude Martzloff: *A History of Chinese Mathematics*, Springer (1997).
9. Nihon Gakushiin [日本学士院]: *History of Japanese Mathematics before the Meiji Restoration* (in Japanese), Iwanami Shoten [岩波書店] (1954–1960).
10. Qian Baocong: *History of Chinese Mathematics* (in Japanese), Misuzu Shobō [みすず書房] (1990).
11. Ren Jiyu: *Comprehensive collection of Chinese scientific and technical classical works - mathematics* (in Chinese), Henan Jiaoyu Chubanshe [華南教育出版社] (1993).
12. Shen Kangshen: "*Dai'ensosujutsu*" studied by Qin Jiushao and "*Shoyakujutsu*" by Seki Takakazu (in Japanese), *Journal of History of Mathematics, Japan*, **109**, pp. 1–23 (1986).
13. Ken'ichi Sato: *History of Mathematics in Modern Japan, A Study of Seki Takakazu* (in Japanese), University of Tokyo Press (2005).
14. Teiji Takagi: *Lecture on Elementary Number Theory* (in Japanese), Kyōritu Shuppan (1931).
15. Sumie Tanabe: Seki Takakazu's Art of Cutting Bamboo (in Japanese), *RIMS Kôkyûroku*, **1257**, Kyoto Univ., pp. 114–124 (2003).
16. Sumie Tanabe: Seki Takakazu's Art of Cutting Bamboo, Part 2 (in Japanese), *RIMS Kôkyûroku*, **1513**, Kyoto Univ., pp. 91–103 (2006).
17. Sumie Tanabe: The Art of Cutting Bamboo in the Clever Man's Booklet on Mathematics (in Japanese), *RIMS Kôkyûroku*, **1583**, Kyoto Univ., pp. 40–50 (2008).
18. Sumie Tanabe: A Brief History of the Art of Cutting Bamboo (in Japanese), *Reports of Institute for Mathematics and Computer Science*, **29**, Tsuda College, pp. 272–285 (2008).
19. I. M. Vinogradoff: *Elements of Number Theory*, Dover (1953).

Seki Takakazu's Method of Calculating the Volume of Solids of Revolution and His Mathematical Object

Fumiaki Ozaki

Abstract It is usually said that Seki Takakazu (ca. 1642 – 1708) calculated the volume of a solid of revolution using the Pappus-Guldinus theorem. However, it is difficult to say that Seki was influenced by Mechanics. In fact, Seki calculated the volumes of all parts which constitute a figure and added them. We may claim that Seki found the Pappus-Guldinus theorem from these calculations and will explain Seki's method of construction for mathematical objects.

1 Introduction

The manuscript book Draft for Deformations of Sphere Segments [毬闕変形草解 Kyūketsu Henkei Sōkai] [1] is contained in Seven Documents [七部書 Shichibu sho] which are the Seki's typical works. This manuscript treats only the volume of solids of revolution and has fewer problems and pages than other books in Seven Documents. For instance, Measurements [求積 Kyūseki] [2], which is one of the Seven Documents, has 59 problems on areas and volumes in 39 pages, whereas Draft for Deformations of Sphere Segments has 9 problems in 7 pages. Moreover, the same four problems are dealt with in both manuscripts. Hence, the Collected Works [3] writes:

> This Draft for Deformations of Sphere Segments was one of Seki's incomplete manuscripts.

But it is not discussed why this manuscript is incomplete. We explain Seki's method of measuring the volume of solids of revolution and show that how this manuscript was written along with his modification theory of the figure.

Fumiaki Ozaki
Faculty of Mathematics, Kyushu University, 744, Motooka Nishi-ku, Fukuoka 812-0395, Japan
e-mail: f-ozaki@math.kyushu-u.ac.jp

2 Method of calculating the volume of solids of revolution

2.1 *Torus*

The discussion on solids of revolution in Measurements begins with a torus .
<Problem 47 of Measurements>

Now given a torus, whose inner and outer circumferences are 6 *shaku* 1 *sun* and 8 *shaku* 1 *sun* respectively.[1] What is the volume?
Answer: 565 *sun*.

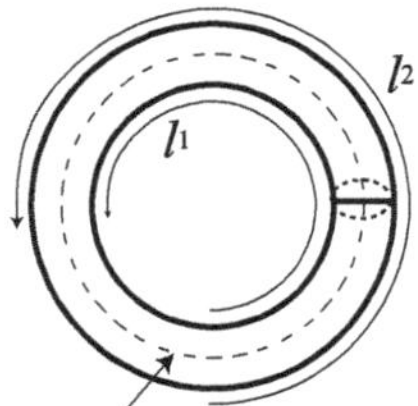

Fig. 1

Let l_2, l_1 denote the circumferences of outer and inner circles respectively (Figure 1). Seki calculated the volume of torus as follows:

$$V = \frac{1}{32\pi}(l_2 - l_1)^2(l_2 + l_1). \tag{1}$$

Earlier Seki calculated the width between outer and inner circumferences of a ring in Problem 16 of Measurements.

$$\text{Width} = \frac{l_2 - l_1}{2\pi}.$$

This is the diameter of a section of the torus.

Seki stated as follows at the end of explanation for this problems.

We calculated in the volume of torus as the area of the section of torus times center circumference.

$$\text{Volume of torus} = \text{area of section} \times \text{center circumference}. \tag{2}$$

This is almost the same as the Pappus-Guldinus theorem. In the sequel, we will call this Theorem (2).

Hence, we predict that the ring volume is given by

$$V = \pi\left(\frac{l_2 - l_1}{4\pi}\right)^2 \cdot \frac{1}{2}(l_2 + l_1). \tag{3}$$

How he found Theorem (2)? He drew a development of ring which has a trapezoidal shape in Problem 16 of Measurements. Therefore, he knew that a development of the torus was a hose which pointed at both ends.

In the Seki's period, a pail was made for sets of wood blocks in the shape of flat-headed stack. They took the pail apart and put together the pieces of these wood blocks. They could create a wide flat-headed stack.

---

[1] *shaku* [尺] is a unit of length, which is about 1/3 m. *sun* [寸] is a unit of length, 1 *shaku* = 10 *sun*.

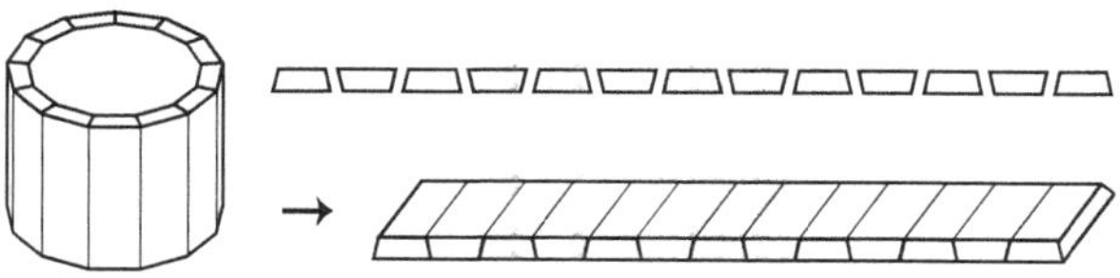

It was popular to calculate the volume of a flat-headed stack. Therefore, we assume that Seki applied these things to the torus and perceived Theorem (2).

In this problem two circumferences were given as condition. This was the influence of Nine Chapters on the Mathematical Art [九章算術 Kyūshō Sanjutsu]. Furthermore, Sugimoto [6] analyzed two numerical values 81, 61 of this problem and gave consideration: In formula (1), $(81+61)/2 = 71$, and $71 \times 1/\pi = 113/5$ (π : $355/113$). It becomes easy to calculate by this reduction. Therefore, Seki adopted these values.

2.2 Arc Ring

On the other hand, in Draft for Deformations of Sphere Segments the discussion begins with an outer arc ring [外弧環 gaikokan].

<Problem 1 of Draft for Deformations of Sphere Segments>

Now given an outer arc ring, whose height [高 kō] is 8 *sun*, sagitta [矢 shi] is 2 *sun*, and imaginary diameter [虚徑 kyokei] is 6 *sun*. What is the volume?

Answer: 268 *sun* 0832.

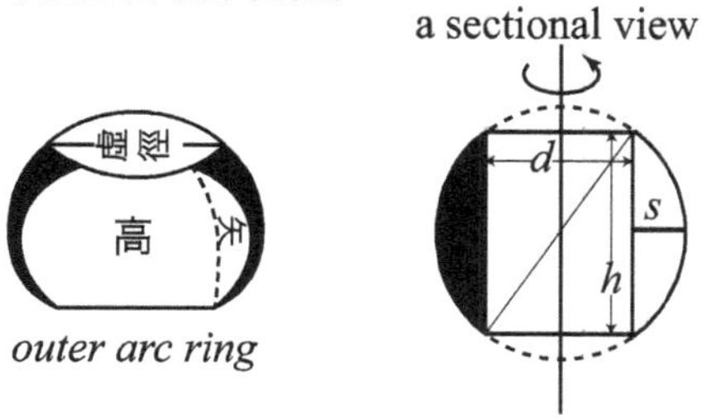

d = imaginary diameter,
s = sagitta,
h = height.

Fig. 2

Seki did not use Theorem (2) to solve this problem. Instead he calculated the volumes of all parts which constitute the figure and then added them.

The sphere was divided into three parts: outer arc ring, cylinder, and sphere segment. Earlier the volumes of these parts had already been calculated. Therefore, he wrote down only a result for this problem:

$$
\begin{aligned}
V^{\mathrm{I}} &= \frac{\pi}{6}(\sqrt{d^2+h^2})^3[\text{sphere}] - \frac{\pi}{4}d^2h[\text{cylinder}] \\
&\quad - 2 \times \frac{\pi}{24}(3d^2+4s'^2)s'[\text{sphere segment}] = \frac{\pi}{6}h^3. \qquad (4) \\
&\quad (2s' = \sqrt{d^2+h^2} - h).
\end{aligned}
$$

where the volume of a sphere segment is calculated in other manuscript Methods of Solving Explicit Problems [解見題之法 Kaikendai no Hō], which are reproduced in this proceedings.

Then problems of outer arc ring continue.

< Problem 2 of Draft for Deformations of Sphere Segment>

Now given an outer arc ring, whose height is 8 *sun*, sagitta is 2 *sun*, imaginary diameter is 1 *shaku* 1 *sun*, what is the volume?

Answer: 443 *sun* 73791

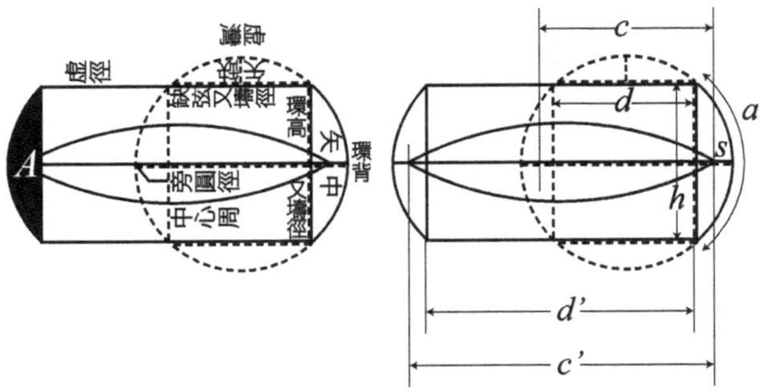

Fig. 3

This problem is same as Problem 48 of Measurements. Seki explained the calculation by drawing a figure (Figure 3) in Measurements.

He assumed that length of arc is $a = 9.273$ in this problem, and area of arc is as $A = 11.1825$. From these values, we can verify what calculation Seki did. See below for further details.

His method of calculation for the volume is as follows:

$$V^{\mathrm{II}} = \frac{\pi}{6}\left(h^3 + (d' - d) \times A \times 6\right). \tag{5}$$

Making use of two relations (2) and (4) we can prove Formula (5).

$$V^{\mathrm{II}} = \pi \times c' \times A, \tag{6}$$

$$V^{\mathrm{I}} = \pi \times c \times A = \frac{\pi h^3}{6}, \tag{7}$$

and $c' = d' - d + c$, $c = \dfrac{h^3}{6A}$. Then we obtain formula (5).

In Draft for Deformations of Sphere Segments, Seki drew a figure (Figure 4) instead of Figure 3. In Figure 4, this figure shows that problem 1 compare with problem 2. and we also see that he used two relations (2) and (4).

Fig. 4

Seki solved the following problems based on this procedure.

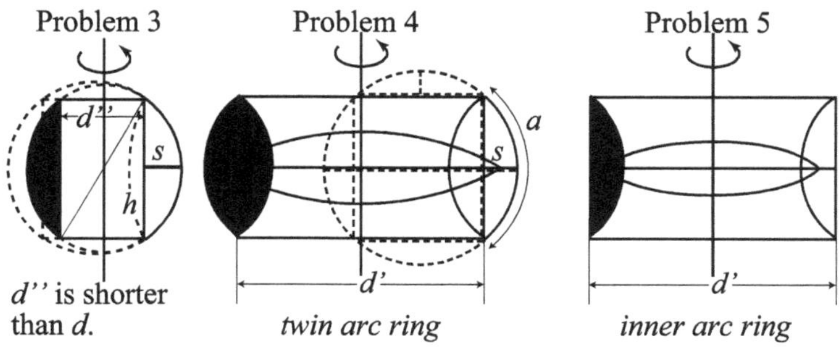

Fig. 5

Solutions are respectively as follows:

<Problem 3>

$$V^{\mathrm{III}} = \frac{\pi}{6}\left(h^3 + (d'' - d) \times A \times 6\right), \tag{8}$$

<Problem 4>

$$V^{\mathrm{IV}} = \frac{\pi}{6} \times d' \times 2A \times 6, \tag{9}$$

<Problem 5>

$$V^{\mathrm{V}} = \frac{\pi}{6}\left((d' + d) \times A \times 6 - h^3\right). \tag{10}$$

Seki got this formula (10) by subtracting (5) from (9).

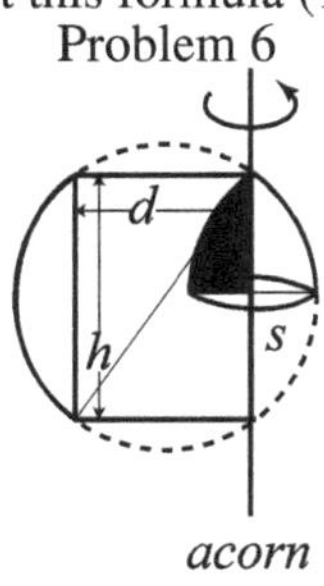

Fig. 6

<Problem 6>

$$V^{\mathrm{VI}} = \frac{\pi}{6}\left(4 \times \left(\frac{1}{2}h\right)^3 - d \times \frac{1}{2}A \times 6\right). \tag{11}$$

Only the formula (11) was written in the book. We can get this formula as follows:

$$V^{\mathrm{VI}} = \frac{1}{2}\left(\frac{\pi}{6}\left(h^3 - d \times A \times 6\right)\right).$$

At first, Seki improved Problem 1 varying the length $d = 6$. Then he operated on the parameter $s = 2$, then on $h = 8$. We focused on the final statement of explanation for Problem 2.

> h and s form a pair [相対 aitai]. So that if h equals $2s$, the sphere segment becomes the half-circle, and d and h form a pair. So that if $d = 0$, the cylinder is disappeared.

He tried to vary the values of the parameters h, s and d respectively. Seki's notion of the pair was to indicate the pair in which two parameters were correlated in magnitude. In this case, he remarked that, when h equaled $2s$, an arc became a half-circle. He also noted that a half-circle was against the definition of the arc ring. Therefore, he dealt with the problems which brought forth figures in this process.

The quoted phrase showed us Seki's thought on the figure. He transformed the basic figure to compose different figures. Whenever he found a new figure, he considered the procedure of composition for new one. In these problems, Problem 1 has two ways to calculate the volume. One is to divide the sphere. The other is to use the Pappus-Guldinus theorem. In his manuscript Seki did not write down the latter calculation for Problem 1. But we claim that he recognized the possible application of the Pappus-Guldinus theorem on this problem and that he based his recognition on these things. He began to apply the new procedure on new figures. Moreover he constructed new mathematical objects using this operation on the 'pair', which is explained in detail in Volume 4 of the Complete Book of Mathematics [大成算経 Taisei sankei].

3 Volume 4 of the Complete Book of Mathematics

The Complete Book of Mathematics was written by Seki and Takebe brothers (Katahiro and Kataakira) from 1683 to 1711. They spent 28 years to compile all the 20 volumes. Each volume has the title which indicates its contents. Volume 4 is named Three Essentials [三要 Sanyō] and consists of three sections: Phenomenon and Figure [象形 shōkei], Flow and Ebb [满干 mankan], and Number [數 sū].

This volume is very abstruse. Fujiwara Matsusaburo, the editor of [4, volume2, p.385], writes as follows:

> Volume 4 is very strange and meaningless as mathematical theory. But as it contains the notion which is the basis of the organization of the Complete Book of Mathematics, we deal with this briefly.

Xu Zelin [7] explains that Volume 4 is related to the art of divination of ancient China.

We comment on Volume 4. The three essentials are as follows:

- 'Phenomenon and Figure': Classification of mathematical problems; a mathematical object in geometrical problems is called a figure [形 kei] , while that of numerical problems (e.g., trade, tax, modular equality, measure, specific gravity, magic squares, magic circles, etc.) a phenomenon [象 shō].
- 'Flow and Ebb': Phases of varying parameters of mathematical objects.
- 'Number': Classification of numbers in a problem. A source number is called static [静 sei] and a number derived from static [静 sei] numbers is called dynamic [動 dō]. For instance, the base and the height of a triangle are static; the

area and the length of the oblique side are dynamic. In construction of a problem, it is important to know how to select these two kinds of numbers.

In the last part of the Number section, numbers are classified into four classes; integer [全 zen], rational number [繁 han], algebraic number [畸 ki], and non-algebraic number [零 rei].

The concept of the 'pair' was explained in 'Flow and Ebb' section. The authors stated as follows:

> Flow and Ebb are fundamentally associated with Phenomenon and Figure and have the three phases; *zen* [全], *kyoku* [極] and *hai* [背]. In short, Flow is to increase. It reaches no limit as it increases. Ebb is to decrease. It reaches an end as it decreases. The phase *zen* indicates the general or ordinary position, the phase *kyoku* the extreme position, and the phase *hai* is opposite to the phase *zen*. The phenomenon and figure have the paired properties like long and short, large and small, high and low, heavy and light. If there is no paired properties, there exits only one total number.
>
> Nevertheless, the paired properties can be distinguished between new [新 sin] and old [旧 kyū]. In a word, if the parameter is given originally, it is called old. On the other hand, if the parameter is not given originally but is a result of subtraction, it is recognized to be new.

The authors recognized the paired properties, that is, the paired parameters with magnitude relationship. They fixed one of the pair and varied the other. They made parameter to increase or to decrease. The recognized three phases in the variation of a parameter. the phase *zen* is ordinary, the phase *kyoku* is extreme, and the phase *hai* is excessive.

In the sequel we call *zen* ordinary, *kyoku* extreme, and *hai* excessive.

For instance, in Problem 22 of Three Essentials, when we exchange gold for silver, the unit price of gold must be higher than the unit price of silver. I mean, gold and silver are in the pair. Because relation of gold and silver had been decided originally, this relation is old.

Let x denote the gold, a the unit price of gold, and b the unit price of silver. The obtained silver is denoted by S. We note that x, a and b are known parameters and we have

$$S = x \times \frac{b}{a}. \tag{12}$$

x is no paired properties. Consider first $x > 0$. x varies according to 'flow and ebb', that is, x increases and decreases.

ordinary ebb: $x \searrow 0$	extreme ebb: $x = 0$	excessive ebb: $x < 0$
ordinary flow: $x \nearrow \infty$	extreme flow: none	excessive flow: none

Parameters a and b are also in pair.

Consider $a < b$, b being fixed. a increases and decreases.

ordinary ebb: $a \searrow b$	extreme ebb: $a = b$	excessive ebb: $a > b$
ordinary flow: $b \nearrow \infty$	extreme flow: none	excessive flow: none

Problem 35 in Three Essentials is a problem of geometry. A square hut has a top square and a bottom. We know top square's edge y is less than bottom one x. These parameters are in 'pair' and this relation of these parameters is old. The authors varied the parameter y as in Figure 7.

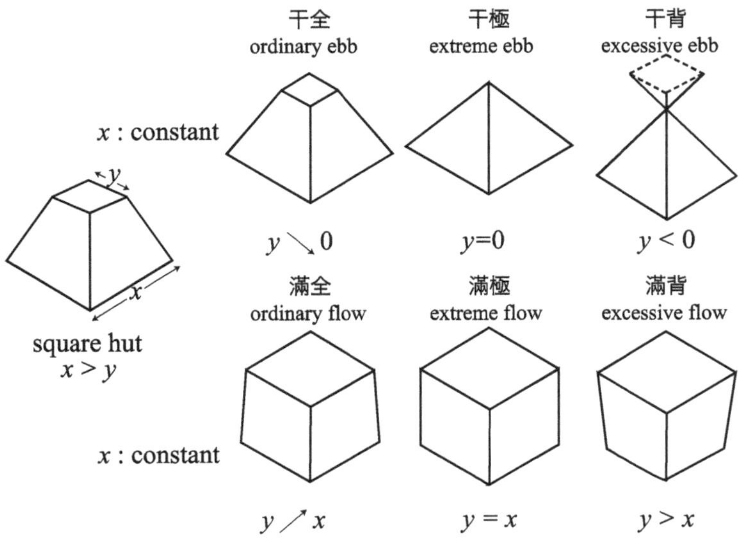

Fig. 7

We omit the detail here. The authors gave a relation between bottom square's edge and height and varied the parameters. This relation of these values is new. Thus, they observed other parameters of the square hut.

They applied this operation to various figures and extended his mathematical objects. They wanted to formulate their objects of study something like "schemes" in the recent mathematics.

But figures related with the sphere (sphere, a circular corn, a solids of revolution, etc.) were not included in Volume 4.

4 length of arc

Seki calculated a square of arc in the problem 22 of Measurements.

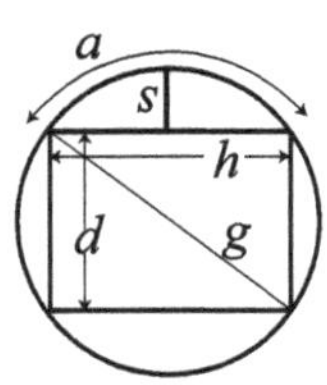

Fig. 8

$$\text{square of arc} = \frac{1}{4}\big(ag - (g-2s)h\big) = 11.18238. \tag{13}$$

This a is given as 9.272952 *kyō* [強], where *kyō* means that it was rounded off to the seventh decimal place.

In the Concise Collection of Mathematical Methods [括要算法 Katsuyō Sanpō], Seki got $a = 9.272953$ solving the equation.

$$-41841459200a^2 + 3597849073280 = 0. \tag{14}$$

In [8], Seki and his pupil Takebe Katahiro calculated the length of arc as follows:

$$a^2 = \frac{Asg^4 + Bs^2g^3 + Cs^3g^2 + Ds^4g + Es^5}{\alpha g^3 + \beta sg^2 + \gamma s^2g + \delta s^3}. \tag{15}$$

$$\begin{array}{ll}
\alpha = 975503374, & A = 39020125496, \\
\beta = -18610356125, & B = -61434714678, \\
\gamma = 10948798854, & C = 25918266069, \\
\delta = -1913138432. & D = -1828448393, \\
 & E = -102756994.
\end{array}$$

Hence, when $s = 2$, $g = 10$, we have

$$a = 9.27295218.$$

The value 9.27295218 is 9.272952 *kyō*. Seki may have used this value.

5 Summary

Why Seki performed operation of parameters in Volume 4 of the Complete Book of Mathematics? Earlier in [9], we pointed out that Seki constructed various figures from basic ones in Volume 4 and concluded that this volume was written for finding clue to the measuring areas and volumes. Observing that he always divided a figure into some basic parts to find the area or the volume of a figure in his manuscripts,

we may think that Volume 4 was written in order not to raise a wrong problem.

But we think that the authors' real intention was to discover new mathematical objects. These new objects must be accompanied by a new procedure. Actually Seki varied various parameters of a mathematical object into extreme positions and succeeded to find a new object in the problem of solids of revolution. We believe that this was the aim of his writing down the incomplete manuscript Draft for Deformations of Sphere Segments for the posterior generations.

References

1. Seki Takakazu [関孝和]: Draft for Deformations of Sphere Segments [毬闕変形草解 Kyūketsu Henkei Sōkai], Hand copies preserved at Kyūshū University [九州大学図書館], Chiba Prefectural [千葉県立図書館] and National Diet [国立国会図書館] Libraries, available at digital collection of Tohoku University [東北大学デジタルコレクション] (http://dbr.library.tohoku.ac.jp/infolib/meta_pub/G9200001CROSS).
2. Seki Takakazu [関孝和]: Measurments [求積 Kyūseki], Hand copies preserved at Kyūshū University [九州大学図書館] and University of Tokyo [東京大学付属図書館] Libraries, available at digital collection of Tohoku University [東北大学デジタルコレクション] (http://dbr.library.tohoku.ac.jp/infolib/meta_pub/G9200001CROSS).
3. A. Hirayama, K. Shimodaira and H. Hirose [平山諦, 下平和夫, 広瀬秀雄]: *Takakazu Seki's Collected Works edited with Explanations* [関孝和全集], Ōsaka Kyōiku Tosho [大阪教育図書], Osaka (1974).
4. Japan Academy [日本学士院]: *History of Japanese Mathematics before the Meiji Restoration* [明治前日本数学史], 5 volumes, Iwanami Shoten [岩波書店], Tokyo (1954-1960).
5. K. Chemla et Guo Shuchun : *Les Neuf Chapitres, Le Classique mathématique de la Chine ancienne et ses commentaires*, Dunod, Paris (2005).
6. Toshio Sugimoto [杉本敏夫]: *Decipherment of Seki Takakazu* [解読関孝和 Kaidoku Seki Takakazu], Kaimeisha [海鳴社], Tokyo (2008).
7. Xu Zelin [徐澤林]: Takebe Katahiro's Epistemology of Mathematics [建部賢弘的数学認識論—論『大成算経』中 "三要"], *Studies in the History of Natural Sciences* [自然科學史研究], Vol. 21, No. 3, pp. 232–243 (2002).
8. Keishi Iwashita [岩下 啓史]: On the volume 12 of the Complete Book of Mathematics and the route of transmission of manuscripts [大成算経巻之十二と写本の系統について], *RIMS kôkyûroku*, **1317**, pp. 125–133 (2003).
9. Fumiaki Ozaki [尾崎 文秋] : Volume 4 of the Complete Book of Mathematics and the mystery of the Three Essentials (Phenomenon and Figure, Flow and Ebb, and Number) [『大成算經』巻之四 三要 (象形, 満干, 数) の謎], *RIMS kôkyûroku*, **1392**, pp. 186–196 (2004).

Leibniz's Theory of Elimination and Determinants

Eberhard Knobloch

Abstract As late as 1938, people doubted whether Gottfried Wilhelm Leibniz ever dealt with determinants. Thus, Gerhard Kowalewski wrote in 1938: "Strangely, nothing relating to determinants and their application has been found in his (viz. Leibniz's) manuscripts until now" [19, p. 125]. Later on, Morris Kline, hinting at Leibniz's often cited letter dating from 1693 to L'Hospital, erroneously wrote in 1972: "The solutions of simultaneous linear equations in two, three, and four unknowns by the method of determinants was created by Maclaurin, probably in 1729, and published in his posthumous Treatise of Algebra (1748)" [8, p. 606].

In 1972, the most important Leibnizian treatise on systems of linear equations appeared as Knobloch [9]. A long sequence of papers [10, 11, 12, 13, 14, 15, 16, 18] by Knobloch followed dealing with Leibniz's theory of elimination and determinants. Yet, all those papers remained partly unknown. In 2000 a historical survey of the evolution of algebra [1] appeared that again knew only Leibniz's letter to L'Hospital that was published in 1850 for the first time and added on 149f further false information about Leibniz's index notation. The booklet represented the state of the art of 1850.

In the following paper I would like to summarize Leibniz's main ideas and results regarding determinants and elimination theory in order to demonstrate that Leibniz laid the foundation of the theory of determinants in Europe between 1678 and 1713, in other words, at the same time as his famous Japanese contemporary Seki.

Eberhard Knobloch
Technische Universität Berlin, Berlin-Brandenburgische Akademie der Wissenschaften, Strasse des 17. Juni, 135, 10623 Berlin, Germany
e-mail: eberhard.knobloch@mailbox.tu-berlin.de

1 The Fundamental Ideas

Leibniz's great interest in determinants was directly connected with his paramount interest in three disciplines which he spent his entire life trying to perfect and organize: the *ars characteristica*, that is the art of inventing suitable characters, signs; the *ars combinatoria*, that is the art of combination; the *ars inveniendi*, that is the art of inventing new theorems, new results, new methods.

These arts are strongly correlated with each other. The art of combination, being by far more than combinatorics in the strict mathematical sense of the word, teaches how to combine signs or characters. The art of inventing suitable signs supports the art of inventing that was the main aim of all Leibnizian scientific, especially mathematical, activities.

There are two important, famous examples for the usefulness and success of Leibniz's invention of suitable signs in order to foster the mathematical development:

1) the differential and integral calculus;
2) determinant theory.

The second example is the subject that will be dealt with in the following sections. What did the three arts contribute to the determinant theory? The art of inventing suitable characters led to numerical double indices. The art of combination helped to represent a determinant as a sum.

In order to avoid any misunderstanding it might be useful to remember the modern definition of a determinant: Let A be a square matrix of order n, $d(A)$ its determinant:

$$d(A) := \sum_{s \in S_n} \operatorname{sign}(s) a_{1s(1)} \cdots a_{ns(n)}.$$

Thanks to the art of inventing Leibniz became the founder of determinant theory. In fact, the literature on the history of science mentions a variety of different authors as being the founders of that theory, depending on what criteria were employed to justify the choice: Gauss (who coined the name determinans (viz. numerus)), Cauchy (who derived a system of theorems to form a new mathematical discipline), Cayley (who introduced symbols, the vertical determinant lines or rules), Weierstrass (who established the axiomatic approach: the determinant is a linear, alternating, normal mapping), and so on.

Apart from those of Weierstrass, Leibniz had already met all these criteria in full [15, p. 58]:

1. he talked about resultans (viz. aequatio),
2. he formulated a series of general theorems concerning the combinatorial aggregates, which he called "resultants", without however proving these.
3. he invented symbols for the resultants and used a very clever subscript notation that was capable of being generalized.
4. he discovered important results in the theory of systems of linear equations and elimination theory which he expressed in the language of determinants.

In a deliberate departure from the tradition of Viète, Leibniz used fictitious numbers as simple, double, and multiple subscripts, since numbers have a threefold advantage:

1. they can be checked at any stage of the calculation.
2. they can express the various arrangements, orders, positions and relationships between the quantities and characters.
3. they serve mathematical progress since they allow us to recognize laws of formation and harmonics.

In January 1675, he explicitly emphasized the greatest advantage of his method that is, to use numbers without using a numerical calculation [24, VII, 1, p.530]. He invented well over fifty different notations for algebraic and differential equations. A larger selection of these notations is explained in [13]. Only three particularly important descriptions can be included here.

From June 1678 at the latest he used double indices for systems of linear equations:

$$10 + 11x + 12y = 0,$$
$$20 + 21x + 22y = 0,$$

where today we write

$$a_{10} + a_{11}x + a_{12}y = 0,$$
$$a_{20} + a_{21}x + a_{22}y = 0.$$

The numbers on the left indicate that they relate to the equation, those on the right to the variables. This index notation was used by him in his letter to the Marquis de l'Hospital dating from April 28, 1693 (old style[1]) that was published in 1850 for the first time [22, II, pp. 236–241]. Leibniz himself never published this index notation during his lifetime. This is not true of the way of writing several polynomials with a common unknown:

$$\begin{matrix} 10x^2 + 11x + 12 \\ 20x^2 + 21x + 22 \end{matrix} \quad \text{or} \quad \begin{matrix} 10 + 11y + 12y^2 \\ 20 + 21y + 22y^2, \end{matrix}$$

where today we write

$$\begin{matrix} a_{10}x^2 + a_{11}x + a_{12} \\ a_{20}x^2 + a_{21}x + a_{22} \end{matrix} \quad \text{or} \quad \begin{matrix} a_{10} + a_{11}y + a_{12}y^2 \\ a_{20} + a_{21}y + a_{22}y^2. \end{matrix}$$

The numbers on the left again refer to the equation, while those on the right either, together with the exponents of the related variable power, give the degree of the polynomial or agree with the exponent of the variable power. The first possibility

[1] i.e. in Julian calendar

was published by him in 1700 and in 1710 [20, 21]. Apart from Charles Reyneau in 1708 and by Karl Friedrich Hindenburg from 1779 onward [15, p. 57] nobody paid attention to Leibniz's explanations in this respect.

In the most important study on systems of linear equations of January 12 (22[2]), 1684, Leibniz developed a shorthand on the basis of the solutions or results he had obtained [9]. His procedure might be described as follows:

$$10+11a=0,$$
$$20+21a=0.$$

In order to eliminate a, the first equation is multiplied by 21, the second equation by -11. The sum of the two multiplied equations is

1) $$\begin{aligned}&10\cdot 21+11\cdot 21a\\ &-11\cdot 20-11\cdot 21a=0\end{aligned}$$

or

2) $$\begin{aligned}&10\cdot 21\\ &-11\cdot 20=0.\end{aligned}$$

The rising (equation) numbers on the left do not change their order. They can be written smaller:

3) $$\begin{aligned}&{}_{1}0\cdot{}_{2}1\\ &-{}_{1}1\cdot{}_{2}0=0\end{aligned}$$

or they can be even left out since they can be added again at any time:

4) $$\begin{aligned}&+0\cdot 1\\ &-1\cdot 0=0\end{aligned}$$

and eventually

5) $$\overline{0\cdot 1}=0.$$

Leibniz's notation $\overline{0\cdot 1}$ is equivalent to the modern term $\begin{vmatrix}10 & 20\\ 11 & 21\end{vmatrix}$. In other words, Leibniz represents a determinant by the product of the elements of its main diagonal and that for up to five-row determinants. In principle $\overline{1\cdot 2\cdot 3\cdots n}$ denotes the same system of coefficients as $\begin{vmatrix}a_{11} & \cdots & a_{1n}\\ \vdots & & \vdots\\ a_{n1} & \cdots & a_{nn}\end{vmatrix}$.

When Leibniz calculated the resultant of $n+1$ equations with n unknowns, his horizontal lines had to be modified accordingly. His equation

$$\overline{0\cdot 1}\cdot 22-\overline{0\cdot 2}\cdot 21+\overline{1\cdot 2}\cdot 20=0$$

[2] In parentheses is the date in Gregorian calendar.

results from a development of a three-row determinant $\overline{0\cdot 1\cdot 2}$ after the second row. In the two-row determinants $\overline{0\cdot 1}, \overline{0\cdot 2}$ and $\overline{1\cdot 2}$ those numbers on the left which do not appear in the right-hand factor 22, 21 and 20 need to be added in their natural order in the solution. Now $\overline{0\cdot 1}$, for example, stands for $10\cdot 31 - 30\cdot 11$.

Leibniz talked about an "aequatio resultativa," "aequatio resultans" or simply "resultans," just as Etienne Bézout in 1764 spoke of "équations résultantes," terminology which Laplace, Binet, Cauchy followed later on.

2 Inhomogeneous Systems of Linear Equations: motivation, rules of signs, general theorems

2.1 Motivation

Why was Leibniz so interested in determinants? The reason can be found in his many manuscripts dealing with the algorithmic solution of the quintic equation. In order to solve this problem Leibniz generalized Cardano's approach by using a substitution of the form

$$x = a_1 + a_2 + a_3 + a_4 \ .$$

His calculations resulted in essential results in the theory of symmetric functions and additive number theory and led him to believe that the problem could be reduced to the solution of systems of linear equations. In May 1678 he wrote [17, p. 115]:

> Yet I shall demonstrate that the labor of calculating is not difficult, because what is important is the fact that the quantities looked for are not multiplied by each other or by themselves. Hence every calculation arising from the equations used for elimination (destructitiae) is done exclusively by addition and subtraction of arbitrary quantities leaving only signs and known (sufficiently simple) numerical coefficients. This is neither laborious, nor difficult, nor prolix.
>
> Indeed, these tiny equations used for elimination cannot lead to confusion because they do not ascend either to rectangles or to powers, even if their number is large. For example, in the case of the [equation of the] fifth degree there are at most 284 equations serving for elimination or—if one uses an abbreviation—about 160 such equations. Yet, they are written down without any calculation and afterward the calculation derived from them is carried out by addition and subtraction alone.
>
> Hence to carry out the elimination calculation is no more difficult than to diligently write down 160 small lines, that is, the values of the unknowns. The value of every unknown can be written down at once without any calculation by the estimation of a glance.
>
> The writing down of the equations used for elimination does not require a great deal of attention. If somebody should follow the method prescribed by me, a mere description would do. In order to write down the values deduced from the equations no calculation is necessary. One who is not distracted could carry out the whole calculation for an equation of the fifth degree, within, I think, the space of one day, provided that everything has been rightly prepared [23, p. 113].

One month later Leibniz erroneously believed to have found the correct law of formation of such determinants and that in the case of three linear equations with three unknowns [12, pp. 9–11].

$$12,2+13,3+14,4-119 \text{ aequ. } 0,$$
$$22,2+23,3+24,4-209 \text{ aequ. } 0,$$
$$32,2+33,3+34,4-299 \text{ aequ. } 0.$$

$$4 \text{ aequ. } \frac{\begin{array}{c}-12,23,299+12,33,209-22,33,119\\+13,22,299-12,32,209+23,32,119\end{array}}{\begin{array}{c}-12,23,34+12,33,24-22,33,14\\+13,22,34-13,32,24+23,32,14\end{array}}.$$

2.2 *Rules of sign*

At present we know six sign rules formulated by Leibniz for the terms of determinants. In June 1678 he formulated his first sign rule:

(1) The terms have to be arranged according to the largeness of the factors. The signs plus or minus are alternatively assigned to them.

This rule cannot be generalized. What is to be done in the case of four equations? Neither the left nor right figures of the double indices keep their order.

A second trial dates from 1683 or a bit later [12, p. 80]:

The resultant of the linear equation system

$$10a+11x+12y=0,$$
$$20a+21x+22y=0,$$
$$30a+31x+32y=0$$

reads:

$$+10.21.32-10.22.31-11.20.32+11.22.30+12.20.31-12.21.30=0.$$

Leibniz formulated his second rule as follows:

(2) Let an arbitrary term be positive or negative. Those terms that differ from this term by an even number of coefficients obtain the opposite sign.

Again this rule cannot be generalized. It is only valid in the case of two or three different coefficients. While the second rule is based on the number of different coefficients, the third rule, written down at more or less the same time as the foregoing is based on the number of common coefficients [12, p. 89]. It refers to the same equation as before:

(3) Terms have opposite signs if they have a single common coefficient or an odd number of such coefficients. Terms have two or an even number of common coefficients have the same sign.

The rule is only valid in the case of three factors that have one common factor. Moreover it contradicts to the second rule.

Leibniz took the crucial step once he had recognized that the coefficients themselves have to be replaced by the right or left figures of the double indices.

Shortly before he discovered the correct general sign rule on January 12 (22), 1684, he relied on cyclic permutations of the right figures of the double indices [12, p. 44]. Considering the same equation system as before he observed that:

(4) Permutations of the right figures have the same sign if they result from each other by a cyclic permutation. The (three) other permutations have the opposite sign.

Indeed (leaving aside the left figures that retain their order) the products 0.1.2, 1.2.0, 2.0.1 are positive, the other three products are negative.

Yet, this rule cannot be generalized because in general two cyclic permutations do not differ by an even number of transpositions:

$+0.1.2.3$ implies $-1.2.3.0$ because three transpositions are needed to carry out this permutation.

A bit later Leibniz solved the sign problem in the study "De sublatione literarum ex aequationibus seu de reductione plurimum aequationum ad unam," "On the elimination of letters from equations or on the reduction of several equations to one equation" [9]. On the margin of the manuscript he added the remark:

> In this attempt I solved the problem, whereas earlier I always got stuck at some point. What is done here is an eminent example of the combinatorial art.

First of all he formulated a recursion rule [9, p. 176]:

(5) Two terms that differ from each other only by two corresponding coefficients (so that *nm kl* is replaced by *nl km*) of the same two equations have opposite signs.

It is worth mentioning that in 1729 MacLaurin gave exactly the same rule [25, pp. 81–85]:

> Opposite signs are assigned to combinations which contain the products of two opposite coefficients.

Moreover Leibniz himself communicated a complicated version of this recursion rule to L'Hospital on April 28 (May 8), 1693 [22, II, p. 240]:

> Eae combinationes opposita habent signa, si in eodem aequationis prodeuntis latere ponantur, quae habent tot coefficientes communes, quot sunt unitates in numero quantitatum tollendarum unitate minuto; caetera habent eadem signa.

Those combinations have opposite signs provided that they are set on the same side of the resulting equation, which have as many common coefficients as there are units in the number of quantities that have to be eliminated, whereby this number was diminished by one. The others have the same signs.

If there are $(n+1)$ equations, n unknowns, it is a matter of $(n-1)$ common coefficients or of two different factors. In other words, Leibniz consciously concealed his knowledge by communicating an unnecessarily complicated version of the recursion rule. Nearly immediately after the recursion rule Leibniz writes down the general, correct sign rule [9, pp. 176f.].

(6) Two terms that differ from each other only by an odd number of transpositions of left or right figures have opposite signs. Those that differ from each other by an even number have the same sign.

This sign rule is indeed equivalent with the rule given by Gabriel Cramer in 1750 [4, p. 658]:

> Qu'on compte, pour chaque terme, le nombre des dérangements: s'il est pair ou nul, le terme aura le même signe +; s'il est impair, le terme aura le signe −.

While Leibniz used the notion of transposition, Cramer used the notion of inversion. A permutation is called even or odd if it contains an even or odd number of inversions, respectively. Now, a permutation is even or odd if and only if it can be generated by an even or odd number of transpositions that in general is smaller than that of its inversions. In other words, Leibniz's sixth sign rule and Cramer's rule are equivalent.

2.3 General theorems

Without demonstrating them, Leibniz added three general theorems in this decisive paper dating from the end of May 1684.

(1) Diagonalising around the main diagonal:
The same laws of arrangement and signs result if one uses the left-hand numbers 1, 2, 3 etc. instead of the right-hand numbers, and instead of 0, 1, 2, 3 the numbers 1, 2, 3, 4. In modern terms: For the formation of the determinant the rows and columns of the matrix A are interchangeable.

(2) Interchanging rows or columns:
If we interchange two rows (columns), then the value of the determinant changes its sign. If we bring the rows $1,2,\dots,n$ (columns) of A into a sequence $k_1,k_2,\dots,k_n$, then the determinant is multiplied by the sign of this permutation:

$$\overline{2\cdot 1\cdot 3} = -\overline{1\cdot 2\cdot 3}, \quad \overline{3\cdot 1\cdot 2} = \overline{1\cdot 2\cdot 3}.$$

(3) Expanding determinants according to Laplace:
The equation $\overline{0\cdot 1\cdot 2\cdot 3} = 0$ can be written in the following for ways:

$$\overline{0\cdot 1\cdot 2}\cdot^4 3 - \overline{0\cdot 1\cdot 3}\cdot^4 2 + \overline{0\cdot 2\cdot 3}\cdot^4 1 - \overline{1\cdot 2\cdot 3}\cdot^4 0 = 0,$$
$$\overline{0\cdot 1\cdot 2}\cdot^3 3 - \overline{0\cdot 1\cdot 3}\cdot^3 2 + \overline{0\cdot 2\cdot 3}\cdot^3 1 - \overline{1\cdot 2\cdot 3}\cdot^3 0 = 0,$$
$$\overline{0\cdot 1\cdot 2}\cdot^2 3 - \overline{0\cdot 1\cdot 3}\cdot^2 2 + \overline{0\cdot 2\cdot 3}\cdot^2 1 - \overline{1\cdot 2\cdot 3}\cdot^2 0 = 0,$$
$$\overline{0\cdot 1\cdot 2}\cdot^1 3 - \overline{0\cdot 1\cdot 3}\cdot^1 2 + \overline{0\cdot 2\cdot 3}\cdot^1 1 - \overline{1\cdot 2\cdot 3}\cdot^1 0 = 0.$$

Leibniz started with the system of four (n) linear equations:

$$
\begin{aligned}
10+11a+12b+13c&=0,\\
20+21a+22b+23c&=0,\\
30+31a+32b+33c&=0,\\
40+41a+42b+43c&=0.
\end{aligned}
$$

He expanded the resultant according to the coefficients of the fourth, third, second, first row. His determinants $\overline{0\cdot1\cdot2}$, $\overline{0\cdot1\cdot3}$, etc. are minors $d_{ik}(A)$ of $(n-1)$th order, but no adjoints. In modern terms we get:

$$\det(A)=d_{i3}(A)a_{i3}-d_{i2}(A)a_{i2}+d_{i1}(A)a_{i1}-d_{i0}(A)a_{i0},\ i=1,2,3,4.$$

The minors $d_{ik}(A)$ are not multiplied by $(-1)^{i+k-1}$, that is, Leibniz did not introduce adjoints. The exponent reads $(i+k-1)$ instead of $i+k$ because Leibniz began the indexation with 0. Provided that Leibniz expanded the determinant according to the $(4-i)$-th row, $i=0,1,2,3$, the value of the determinant is multiplied by $(-1)^i$. This difference between his own expansion and that according to Laplace's expansion does not matter in the case of resultants because the resultant must be equal to zero.

3 Elimination of a Common Variable

Another important application of Leibniz's determinant theory concerned the resultant of two polynomials from the ring $R[x]$ [of polynomials with coefficients in an integral domain R] and the elimination of a common unknown from algebraic equations with two and more variables. He thus anticipated methods and results that were attributed to James Joseph Sylvester, Etienne Bézout and Leonhard Euler later on.

3.1 Sylvester's dialytic method

In the year 1679–1681 Leibniz explained in the essay "De tollendis incognitis" (On the elimination of unknowns) [12, n. 36, p. 159] how the two initial equations

$$
\begin{aligned}
a+bx+cx^2+dx^3+ex^4\,\text{etc.}&=0,\\
l+mx+nx^2+px^3+qx^4\,\text{etc.}&=0
\end{aligned}
$$

can be multiplied by gradually increasing powers of x. If we set the coefficient determinant

$$\begin{vmatrix} a & b & c & d & e & \cdot & \cdot & \cdot \\ \cdot & a & b & c & d & e & \cdot & \cdot \\ \cdot & \cdot & a & b & c & d & e & \cdot \\ \cdot & \cdot & \cdot & a & b & c & d & e \\ l & m & n & p & q & \cdot & \cdot & \cdot \\ \cdot & l & m & n & p & q & \cdot & \cdot \\ \cdot & \cdot & l & m & n & p & q & \cdot \\ \cdot & \cdot & \cdot & l & m & n & p & q \end{vmatrix}$$

to zero, then we obtain the so-called Sylvester determinant and thus the resultant of the two polynomials. The English mathematician Sylvester published this solution in 1840 [26]. Later on, it became known under the name of the "dialytic method."

3.2 *Bézout's method and Euler's second method*

Around 1683/84, Leibniz in several studies developed a method of reducing the problem of calculating resultants to the solution of a system of algebraic equations which he was already able to determine in an elegant fashion at that time [12, ns. 16, 50, 54]:

> Let two polynomials of e-th and f-th degree be given, then these are multiplied together with auxiliary polynomials of the $(f-1)$-th and $(e-1)$-th degree, respectively. The product polynomials are added and the coefficients of each power of the unknowns are set equal to zero. One gets a system of linear equations that is a sufficient number of equations to determine the auxiliary variables.

This rule was formulated by Leibniz in its complete general form. The procedure appears in Euler [7] in 1766 and Bézout [2] in 1767.[3]

3.3 *Euler's resultants*

Except for Newton [and Leibniz], Euler [5, 6] was the first in Europe to develop the general theory of elimination. He was interested in the number of intersection points of two algebraic curves on a plane and wanted to prove that it is less than or equal to nm if one is of order n and the other of order m,

This is known today as Bézout's theorem because Bézout [2] proved it in the general case. Bézout's paper was submitted to the Academy in Paris in 1764. In the same year Euler submitted his third article [7] to the Academy in Berlin and claimed that he also proved it. Both made use of resultants and Euler had already computed resultants when $m \leq n \leq 3$ in [5, 6]. However, Euler didn't know determinants so

[3] These two methods of Euler and Bézout always give the same result but their theoretical backgrounds are different. The following subsection is added by the editor to ease the reader's understanding.

that he had no means to express the resultants in the general case. Much later in 1840 Cauchy [3] showed that Euler's resultants were the same as Sylvester's in the following way.

Suppose that

$$f(x) = a_0 + a_1 x + \cdots + a_n x^n = 0, \tag{1}$$
$$g(x) = b_0 + b_1 x + \cdots + b_m x^m = 0, \tag{2}$$

are two equations with coefficients in an integral domain R. We assume that $n \geq m$, $a_n \neq 0$ and $b_m \neq 0$.

Sylvester's resultant, which is equal to Bézout's [and Seki's, may also be written

$$\mathscr{R}_{\text{Sylvester}}(f,g) = \begin{vmatrix} f(x) & a_1 & \cdots\cdots\cdots & a_n & 0 & \cdots & 0 \\ xf(x) & a_0 & a_1 & \cdots\cdots\cdots & a_n & 0 & 0 \\ & & \ddots & & & \ddots & \\ x^{m-1}f(x) & \cdots & 0 & a_0 & a_1 & \cdots\cdots\cdots & a_n \\ g(x) & b_1 & \cdots\cdots & b_m & 0 & \cdots & 0 \\ xg(x) & b_0 & b_1 & \cdots\cdots & b_m & 0 \cdots & 0 \\ & & \ddots & & \ddots & & \\ & & & & & & 0 \\ x^{n-1}g(x) & \cdots\cdots & 0 & b_0 & b_1 & \cdots\cdots & b_m \end{vmatrix} \tag{3}$$

$$= p(x)f(x) + q(x)g(x) \tag{4}$$

with polynomials $p(x)$ and $q(x)$ of degree $\leq m-1$ and $n-1$, respectively.

On the other hand, Euler [7] starts with the assumption that (1) and (2) have a common root ω in an extended field K of R, which was the complex number field C for Euler. Then, we have the factorizations

$$f(x) = (x-\omega)\delta(x) = (x-\omega)(d_0 + d_1 x + \cdots + d_{n-1}x^{n-1}), \tag{5}$$
$$g(x) = (x-\omega)\gamma(x) = (x-\omega)(c_0 + c_1 x + \cdots + c_{m-1}x^{m-1}), \tag{6}$$

as polynomials with coefficients in K. Hence we have

$$f(x)\gamma(x) - g(x)\delta(x) = 0, \tag{7}$$

that is,

$$
\begin{pmatrix}
a_0 & 0 \cdots 0 & & b_0 & 0 & & \cdots & 0 \\
a_1 & a_0 & 0\ 0 & b_1 & b_0 & 0 & 0 & 0 \\
\vdots & \ddots & 0 & \vdots & & \ddots & & \\
a_{m-1} & \cdots & a_1\ a_0 & b_{m-1} & \ddots & \ddots & \ddots & 0 \\
a_m & \cdots & a_1 & b_m & b_{m-1} & & & 0 \\
\vdots & a_m & \vdots & 0 & b_m & \ddots & & b_0 \\
a_n & & \vdots\ \vdots & 0 & 0 & \ddots & & \vdots \\
0 & \ddots & \vdots & 0 & & \ddots & & \vdots \\
 & a_n & & & & & & b_{m-1} \\
0 & \cdots & 0\ a_n & 0 & 0 & \cdots\cdots & & b_m
\end{pmatrix}
\begin{pmatrix}
c_0 \\ c_1 \\ \vdots \\ \vdots \\ c_{m-1} \\ -d_0 \\ \vdots \\ \vdots \\ \vdots \\ -d_{n-1}
\end{pmatrix}
= 0. \qquad (8)
$$

Thus it follows from Bézout's Lemme 1 [2] that the determinant of the matrix, which is the transpose of Sylvester's, should vanish.

3.4 Rules of formation and sign rule of the resultant

Around 1692/93 Leibniz had found the most important dimensional and homogeneity properties of the resultant. Two essays are especially interesting in this respect: "De tollendis literis" (On the elimination of unknowns) and "De tollenda litera ex duabus aequationibus" (On the elimination of an unknown from two equations) [12, ns. 54, 56].

Let two polynomials of e-th and f-th degree be given. The resultant will have the following properties:

(1) It is homogeneous of degree f in the coefficients of the first polynomial and of degree e in the coefficients of the second polynomial (law of homogeneity).
(2) Each of its terms consists of $e+f$ factors.
(3) The sum of the right-hand subscripts is ef.

3.5 Leibniz's explication theory—Euler's first method

Around 1693/94, Leibniz developed a type of "explication theory" on resultant calculation [10], that was explained and used by Euler in three publications [5, vol. 2, p. 269] in 1748, [6] in 1750 and [7] in 1766.

Let two polynomial equations be given:

$$10x^3+11x^2+12x+13=0,$$
$$20x^3+21x^2+22x+23=0.$$

Using twice a cross wise multiplication and subtraction procedure, Leibniz produced two equations of second degree:

$$\begin{matrix}10\cdot23\\-13\cdot20\end{matrix}\,x^2\begin{matrix}+11\cdot23\\-13\cdot21\end{matrix}\,x\begin{matrix}+12\cdot23\\-13\cdot22\end{matrix}=0\,,$$

$$\begin{matrix}10\cdot21\\-11\cdot20\end{matrix}\,x^2\begin{matrix}+10\cdot22\\-12\cdot20\end{matrix}\,x\begin{matrix}+10\cdot23\\-13\cdot20\end{matrix}=0\,,$$

or

$$(10)x^2+(11)x+(12)=0,$$
$$(20)x^2+(21)x+(22)=0.$$

The coefficients within brackets are "explicated" by the differences between the products, for example

20 is explicated by $10\cdot21-11\cdot20$,
21 is explicated by $10\cdot22-12\cdot20$ etc.

This procedure has to be repeated until the elimination of x from two linear equations yields the resultant. The degree of the resultant is of course too high this way. Yet, Leibniz became interested in the law of formation of the terms produced by repeated explications. By concentrating his efforts on the explication of the coefficients of the second equation he elaborated a "dichotomic tree," as he called it:

$$\begin{array}{c}20\\ \overbrace{\hspace{22em}}\\ 10\cdot21 \qquad\qquad\qquad - \qquad\qquad\qquad 11\cdot20\\ \overbrace{10\cdot22-12\cdot20}\qquad\qquad\qquad\overbrace{10\cdot21-11\cdot20}\\ \overbrace{10\cdot23-13\cdot20}\quad\overbrace{10\cdot21-11\cdot20}\quad\overbrace{10\cdot22-12\cdot20}\quad\overbrace{10\cdot21-11\cdot20}\end{array}$$

etc.

If these substitutions (explications) are really carried out, one gets:

$$20,$$
$$10\cdot21-11\cdot20,$$
$$10^2\cdot22-10\cdot12\cdot20-10\cdot11\cdot21+11^2\cdot20,$$
$$10^3\cdot23-10^2\cdot13\cdot20-10^2\cdot12\cdot21+10\cdot11\cdot12\cdot20$$
$$-10^2\cdot11\cdot22+10\cdot11\cdot12\cdot20+10\cdot11^2\cdot21-11^3\cdot20.$$

In order to inquire into the structure of these rows Leibniz denoted certain domains, that is the second half of all terms of a row by A, the second quarter of all terms of a row by B, the second eighth of all terms of a row by C etc.

The capitals within brackets like (A), (B), (C) denote the sequence of first terms of a sequence of further bisections:

(A) means: we have to consider the second half (the domain). Then we have to take the first term of the second half, the first term of the second quarter, the first term of the eighth etc.

(B) AB means: we have to consider the second quarter of the second half of the second quarter (the domain). Then we have to take the first term of the second half, the first term of the second quarter etc.

Let us consider the above mentioned fourth row as an example. We are looking for the sequence (A):

A denotes the second half of the row that is

$$-10^2 \cdot 11 \cdot 22 + 10 \cdot 11 \cdot 12 \cdot 20 + 10 \cdot 11^2 \cdot 21 - 11^3 \cdot 20.$$

The first term of the second half of this partial row is

$$10 \cdot 11^2 \cdot 21.$$

The first term of the second quarter of this partial row is $10 \cdot 11 \cdot 12 \cdot 20$. There is only one term. There are no further possible bisections. Hence we get:

$$(A) = 10 \cdot 11^2 \cdot 21 + 10 \cdot 11 \cdot 12 \cdot 20.$$

Leibniz did not mention a non-trivial theorem that is obviously true though I was not able to demonstrate it up to now:

Theorem. *Any permutation of the letters of such a combination of letters leads to the same sum.*

For example: $(A)BB = (B)AB = (B)BA$ etc. The brackets do not matter.

It is worth mentioning, however, that Leibniz added 34 corollaries regarding the formation of the terms of a row and their relations with each other.

Epilogue

None of all these fascinating results were published by Leibniz during his lifetime. They are documents of a restless scholar. In October 1674 he himself said [24, VII, 3, p. 539].

> "Malo enim idem facere, quam semel nihil." "For I prefer to do the same twice instead of doing nothing once."

References

1. Izabella G. Bashmakova and G. S. Smirnova: *The Beginnings and Evolution of Algebra*, translated from the Russian by A. Shenitzer with the editorial assistance of D. A. Cox, Mathematical Association of America,Washington, D.C. (2000); The Russian original "The origin and development of algebra," *Outline of the History of Mathematics*, pp. 94–246, Izdat. Moskov. Univ., Moscow (1997).
2. Etienne Bézout: Recherches sur le degré des équations résultantes de l'évanouissement des inconnues, et sur les moyens qu'il convient d'employer pour trouver ces équations, *Histoire de l'académie royale des sciences*, année 1764, Paris, pp. 288–338 (1767).
3. Augustin Louis Cauchy: Mémoire sur l'élimination d'une variable entre deux équations algébriques, *Exercices d'analyse et de physique mathématique*, vol. 1, pp. 91–104 (1840); *Œuvres complètes d'Augustin Cauchy*, sér. II, tom. XI, pp. 466–509, Gauthier-Villars, Paris (1913).
4. Gabriel Cramer: *Introduction à l'analyse des lignes courbes algébriques*, Genève (1750).
5. Leonhard Euler: *Introductio in analysin infinitorum* in 2 vols., Lausanne (1748); I cite the reprint in: Leonhard Euler, *Opera omnia*, ser. I, vols. 8 and 9 (= EO I, 8 and 9), Basel (1922 and 1945).
6. Leonhard Euler: Démonstration sur le nombre des points où deux lignes des ordres quelconques peuvent se couper, *Mémoires de l'académie des sciences de Berlin*, IV année, 1748, Berlin, pp. 234–248 (1750); = EO ser. I, vol. 26, pp. 46–59 (1953).
7. Leonhard Euler: Nouvelle méthode d'éliminer les quantités inconnues des équations, *Mémoires de l'académie des sciences de Berlin*, XX année, 1764, Berlin, pp. 91–104 (1766); = EO ser. I, vol. 6, pp. 197–211 (1921).
8. Morris Kline: *Mathematical Thought from Ancient to Modern Times*, New York (1972).
9. Eberhard Knobloch: Die entscheidende Abhandlung von Leibniz zur Theorie linearer Gleichungssysteme, *Studia Leibnitiana*, vol. 4, pp. 163–180 (1972).
10. Eberhard Knobloch: Unbekannte Studien von Leibniz zur Eliminations- und Explikationstheorie, *Archive for History of Exact Sciences*, vol. 12, pp. 142–173 (1974).
11. Eberhard Knobloch: Studien von Leibniz zum Determinantenkalkül, *Studia Leibnitiana*, Suppl. 13, pp. 37–45 (1974).
12. Eberhard Knobloch: *Der Beginn der Determinantentheorie, Leibnizens nachgelassene Studien zum Determinantenkalkül, Textband*. Gerstenberg Verlag, Hildesheim (1980).
13. Eberhard Knobloch: Zur Vorgeschichte der Determinantentheorie, *Studia Leibnitiana*, Suppl. 22, pp. 96–118 (1982).
14. Eberhard Knobloch: Erste europäische Determinantentheorie, *G. W. Leibniz, Das Wirken des großen Philosophen und Universalgelehrten als Mathematiker, Physiker, Techniker*, published by Erwin Stein und Albert Heinekamp, Hannover, pp. 32–41 (1990); 2nd. ed., pp. 55–64 (2000).
15. Eberhard Knobloch: First European theory of determinants, *G.W. Leibniz, The work of the great universal scholar as philosopher, mathematician, physicist, engineer*, ed. by Karl Popp, Erwin Stein. Hannover, pp. 55–64 (2000).
16. Eberhard Knobloch: Déterminants et élimination chez Leibniz, *Revue d'histoire des sciences*, vol. 54, pp. 143–164 (2001).
17. Eberhard Knobloch: Beyond Cartesian limits: Leibniz's passage from algebraic to "transcendental" mathematics, *Historia Mathematica* vol. 33, pp. 113–131 (2006).
18. Eberhard Knobloch: Mettre au jour de nouveaux corpus: Les œuvres inconnues de Leibniz, *Archives Internationales d'Histoire des Sciences*, vol. 57, pp. 89–104 (2007).
19. Gerhard Kowalewski: *Große Mathematiker*. München-Berlin (1938).
20. Gottfried Wilhelm Leibniz: Responsio ad Dn. Nic. Fatii Duillerii imputationes. Accessit nova artis analyticae promotio specimine indicata, dum designatione per numeros assumtitios loco literarum, algebra ex combinatoria arte lucem capit, *Acta Eruditorum*, pp. 198–208 (May 1700); (= Leibniz 1849–1863, V, pp. 340–349).

21. Gottfried Wilhelm Leibniz: Monitum de characteribus algebraicis, *Miscellanea Berolinensia*, 1, pp. 155–160 (1710); (= Leibniz 1849–1863, VII, pp. 218–223).
22. Gottfried Wilhelm Leibniz: *Mathematische Schriften*, published by Carl Immanuel Gerhardt, Berlin–Halle (1849–1863).
23. Gottfried Wilhelm Leibniz: *Die mathematischen Studien zur Kombinatorik, Textband*, published by Eberhard Knobloch, Wiesbaden (1976).
24. LSB = Gottfried Wilhelm Leibniz: *Sämtliche Schriften und Briefe*, published by the Berlin-Brandenburgische Akademie der Wissenschaften and the Akademie der Wissenschaften zu Göttingen, Berlin, since 1923.
25. Colin MacLaurin: *Treatise of Algebra* in three parts, London (1748).
26. James Joseph Sylvester: A method of determining by mere inspection the derivations from two equations of any degree, *Philosophical Magazine*, vol. 16, pp. 132–135 (1840); *The Collected Mathematical Papers of James Joseph Sylvester*, vol. I, Cambridge, pp. 54–57 (1904).

Algebra, Elimination and the Complete Book of Mathematics

Hikosaburo Komatsu*

Abstract Seki Takakazu (1642?–1708) is a mathematician in the Edo Period (1603–1868) of Japan. He was distinguished far from the other mathematicians in Japan at that time. We have so far failed to find any name of person who taught him mathematics in spite of all our efforts of investigations at this occasion of 300 years after his death. His disciples are few and his monumental treatise *Complete Book of Mathematics* (1683–1711) of approximately 1800 pages has practically been ignored by mathematicians who claimed themselves to be in Seki's school of mathematicians and also by later historians of mathematics until these days. Yet he was not isolated not only in Japan but also in the world. We will show the evidence in what follows.

1 Historical Background

1.1 Chinese mathematics

There used to be two types of mathematics. One was mathematics for government officials of centralized and powerful countries, and the other for common people participating in market economy under less centralized administrations. Japan (and the united Korea) imported the first type of mathematics in the seventh century from China of the Tang dynasty [唐朝] (618–904). Nine Chapters of Mathematics [九章算術] [1] was the main textbook in schools for government officials majoring in mathematics. There remain records that Zhuì Shù [綴術] by Zǔ Chōng-Zhī [祖沖之] (ca. 500) was also one of the textbooks. But it disappeared quite soon from three

Hikosaburo Komatsu
Graduate School of Mathematical Sciences, the University of Tokyo,
3-8-1 Komaba, Meguro-ku, Tokyo 153-9814, Japan
e-mail: komatsu@ms.u-tokyo.ac.jp

* Supported in part by JSPS Grant-in-aide for scientific research (C) 20540107.

countries including the Tang China probably because eight digits of the ratio π of circumferences were of no use for administrative works.

Japan also imported in the seventh century calendar systems from China often through Korea. The last imported calendar Xuān Míng Lì [宣明暦 Senmyō reki] was adopted in 862 and was used in Japan until 1685 when it was replaced by Jōkyō Calendar [貞享暦 Jōkyō reki] by Shibukawa Harumi [澁川春海] (1639–1715). This means that the minimal knowledge was maintained to compile a new calendar every year but no more for approximately a thousand years. The same was true with mathematics.

On the other hand, Chinese mathematics and calendrical science made a remarkable progress in the Sòng [宋] (960–1279) and Yuán [元] (1279–1368) dynasties. As the famous picture Riverside Scene at Qingming Festival [清明上河圖] shows the urban life in China at that time was much modernized. By the middle of the eleventh century mathematicians established the root-extraction method [開方術 kāifāng shù] of any algebraic equation of any degree in a single unknown with numerical coefficients, and Qín Jiǔ-Sháo [秦九韶 Shin Kyū Shō] refined it in the Book of Mathematics in Nine Chapters [数書九章 Shùshū Jiǔzhāng] (1247) [2] to the same algorithm as the so-called Horner's Method in the nineteenth century.

The way to construct an algebraic equation for the root-extraction method is called the heavenly element method [天元術 Tiānyuán shù], because it starts with fixing an unknown called the heavenly element [天元]. The Song-Yuan mathematics needed to solve higher degree equations because, as their problems became complicated, they had to eliminate auxiliary unknowns, and during these processes the degrees of resulting equations got high.

Zhū Shì-Jié [朱世傑 Shu Seiketsu] invented algebraic expressions with up to four variables and gave many methods to eliminate a common variable from two such equations in the Jade Mirror of Four Elements [四元玉鑑 Siyuán Yùjiàn] (1303).

In China many books written in the Song–Yuan period were lost during the Ming [明] dynasty (1368–1644). However, some of them were reprinted in Korea of the Li [李] dynasty (1392–1910) with copper types. Among these, Introduction to Mathematics [算学啓蒙 Suàn-Xué Qǐ-Méng] (1299) [4] by Zhū Shì-Jié [朱世傑] and Yang Hui's Methods of Mathematics [楊輝算法 Yáng Huī Suàn-Fǎ] (1275) [3] by Yáng Huī [楊輝] were brought in Japan at the end of the sixteenth century and became very important source books for Japanese mathematicians in the Edo period.

Traditional computing tools in China and its vicinities were counting rods [算木 suàn mù] on counting boards [算盤 suàn pán]. In the Ming period abacuses [珠算] took their places in China and Japan. Summary of Mathematical Methods [算法統宗 Suànfǎ Tǒngzōng] (1592) [6] by Chéng Dà-Wèi [程大位] was widely used as the standard textbook of mathematics using abacuses. It modeled after Nine Chapters of Mathematics and in a sense took over the position.

Lineage of Mathematicians in Kyoto and Osaka

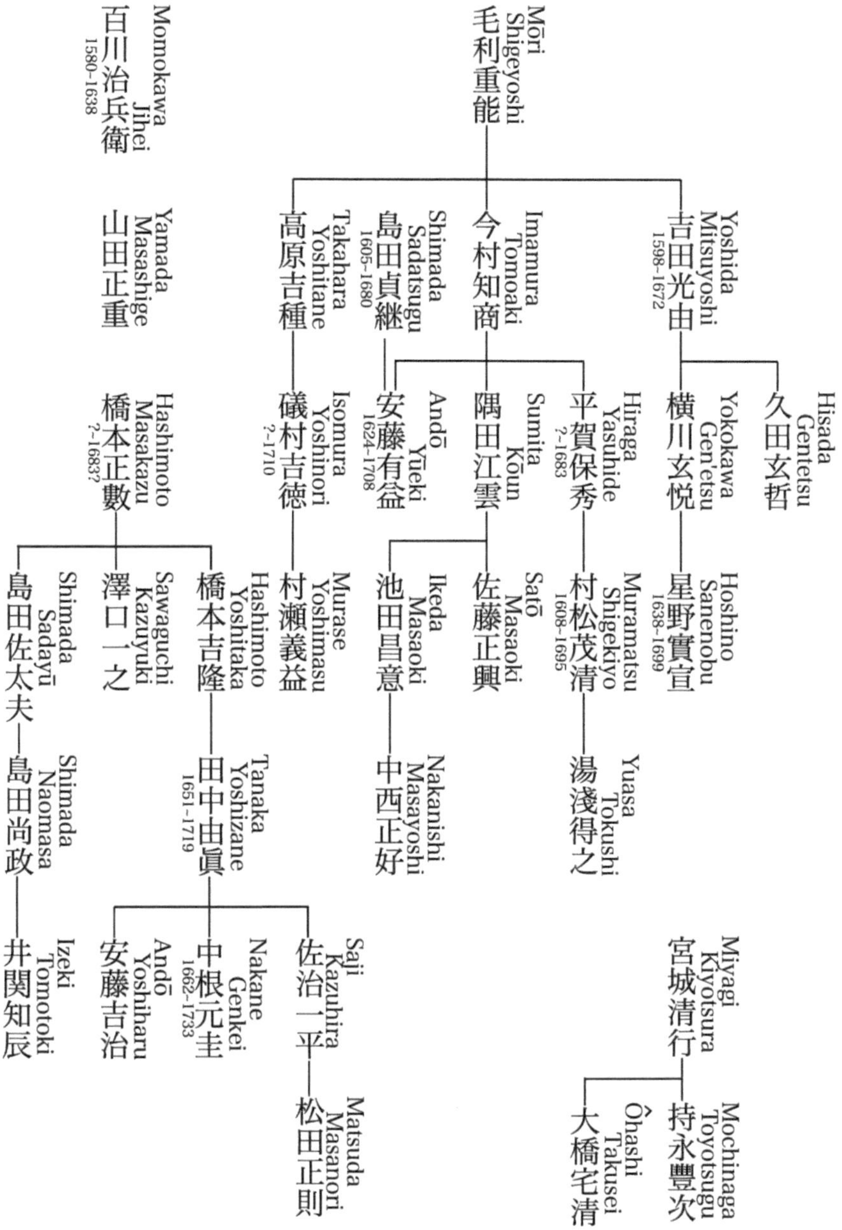

Lineage of Seki's School

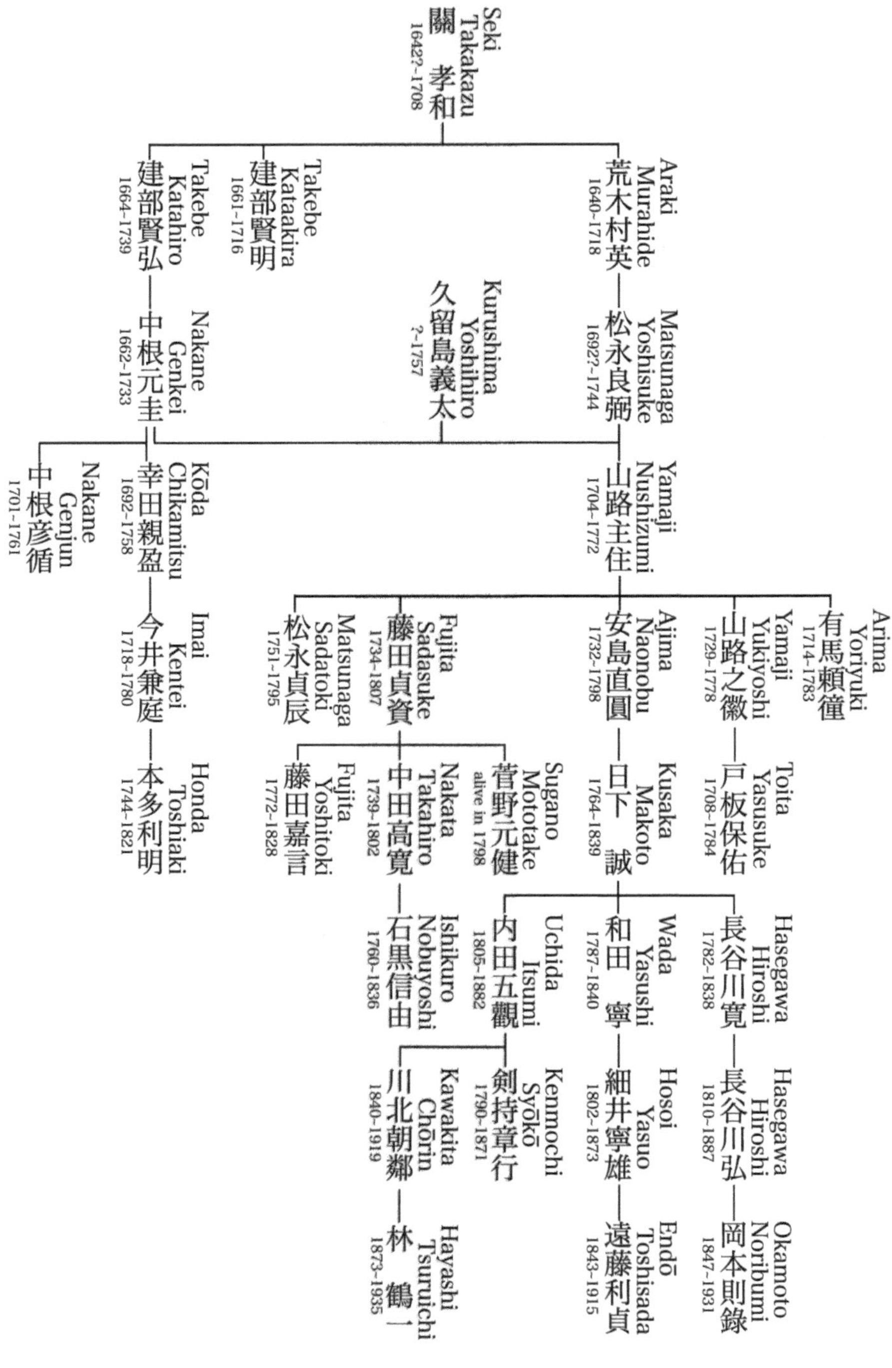

1.2 Mathematicians in Kyoto and Osaka

At the beginning of the Edo period Edo (present Tokyo) was a new city under construction and the old capital Kyoto and the trade center Osaka still kept the position of the center of culture in Japan. Most of famous mathematicians lived there.

The first successful mathematician was Yoshida Mitsuyoshi [吉田光由] (1598–1672) from a very rich family in Kyoto. His book From Trifling Dust to Eternal Time [塵劫記 Jingōki] (1627) [7] is the first million seller in Japan. The new edition (1641) contained 12 open problems, one of which asked a circular land cut by parallel lines with given areas as the three parts. This gave birth to the theory of circles [圓理 enri]. Those open problems are called bequeathed problems [遺題 idai]. When one solved all bequeathed problems in a book, he usually published a book of solutions and left his own open problems in his book. This is called a succession of bequeathed problems [遺題継承 idai keishō]. This habit certainly stimulated mathematicians in the Edo period.

Muramatsu Shigekiyo [村松茂清] (1608–1695) was born in Ibaragi [茨城] and moved to Akō [赤穂] following his Load Asano. His adopted son and grandson were among 47 Samurais of the famous revenge in 1702. His book Mathematical Cooking Boards [算俎 Sanso] (1663) [9] contains many original results. He computed the circumference of the regular 2^n-gon inscribed in a circle of diameter 1 for $n = 3, 4, \cdots, 15$ and obtained the approximate value of π as accurate as Zu Chong-Zhi though he did not claim so. He also computed four digits of the volume of a sphere with diameter 1 by slicing it into 100 equally thin plates. Seki refined these works and got much more accurate values as shown in his posthumous book Concise Collection of Mathematical Methods [括要算法][23].

Shimada Sadatsugu [島田貞継] and Andō Yūeki [安藤有益] were mathematicians invited by the Load of Aizu [会津] Hoshina Masayuki [保科正之] who intended to make Calendar Reform under the authority of Shōgunate. Shibukawa Harumi, who was born in Kyoto, joined them later and succeeded in it. Seki was also interested in the calendar reform but the chance was not left.

Introduction to Mathematics [算学啓蒙] was published in Japan with some commentaries by Hisada Gentetsu [久田玄哲] in 1658 and by Hoshino Sanenobu [星野實宣] in 1672. There remains, however, a doubt whether or not they really understood the heavenly element method [天元術]. Except for Seki Takakazu and his disciples, the first mathematicians in Japan who digested Introduction to Mathematics were Hashimoto Masakazu [橋本正数] and his group, in particular, Sawaguchi Kazuyuki [澤口一之] and Tanaka Yoshizane [田中由真] (1661–1729), and a little independently by Miyagi Kiyotsura [宮城清行] and their disciples.

1.3 Methods of Mathematics, Old and New by Sawaguchi

The Methods of Mathematics, Old and New [古今算法記 Kokon Sanpōki] [10] is a book of 7 volumes written by Sawaguchi in 1670. The first three volumes are written in Japanese. This part is a little advanced but standard textbook of elementary mathematics. Calculating tools are abacuses in Japanese style, i.e. with 4 beads under the beam and 1 above. The following four volumes are written in classical Chinese. The first three of them are allocated to the answers to the 150 bequeathed problems in Fundamental Mathematics [算法根源記 Sanpō Kongenki] (1669) by Satō Masaoki [佐藤正興]. He answered all but 16 problems, which he says are concerning circles and no known formulas in the theory of circles [圓理] are exact. Then, he closes his book by the seventh volume consisting only of his 15 bequeathed problems.

We consider here only the 85th problem of Fundamental Mathematics:

> Suppose that there is a right-angled triangle containing a rectangle as in figure. The length of the rectangle is 3*shaku* 2*sun*[1] and the width 6*sun*. Moreover, the sum of the hook, the leg and the chord of the right-angled triangle is 1*jō* 2*shaku*. Find the hook, the leg and the chord of the right-angled triangle.
> [今有鉤股弦内如圖縦横平。只云. 縦三尺二寸。横六寸。別. 鉤股弦寸各三和一丈二尺。問鉤股弦幾何]

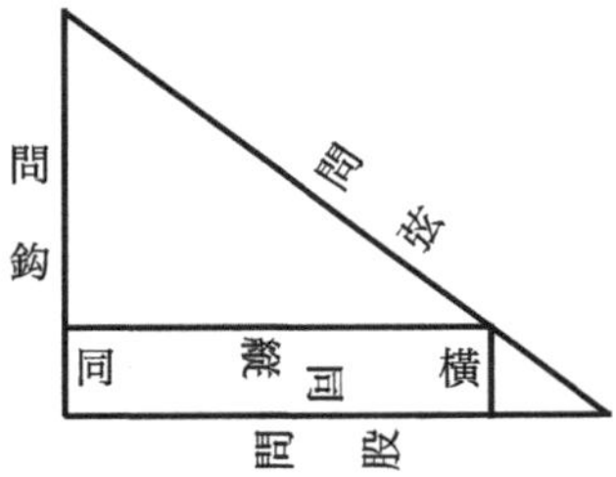

Fig. 1 Problem 85

> Answer says: The hook is 3*shaku*, the leg 4*shaku* and the chord 5*shaku*.
> [○答曰○鉤三尺○股四尺○弦五尺].

Explicit assumptions are the following three equations:

$$\text{Assume} \qquad \text{Length} = 32,\ \text{Width} = 6, \tag{1}$$

$$\text{Also} \qquad \text{Hook} + \text{Leg} + \text{Chord} = 120, \tag{2}$$

but we have implicitly

$$\text{Hook}^2 + \text{Leg}^2 = \text{Chord}^2, \tag{3}$$

$$(\text{Hook} - \text{Width}) \times \text{Leg} = \text{Length} \times \text{Hook}, \tag{4}$$

by the Pythagorean theorem and the proportionality.

If we admit (1), the problem is reduced to that of finding three unknowns from the remaining three equations. The procedure given by Sawaguchi is the following and does not depend on the data on the right-hand sides of (1).

術曰. 立天元一. 為鉤。内減横. 余為短鉤。寄甲位○列別云数. 内減鉤. 余為股弦和。自之. 得内減鉤冪. 余為因股弦和二段股。以甲位乗之. 為因股弦和因鉤二段縦。寄左○列縦. 以股弦和与鉤相乗之. 得数倍之. 与寄左相消. 得開方式。平方翻方開之. 得鉤。依前術得股弦。各合問。

As far as procedures are concerned, this is all what Sawaguchi wrote for this problem. It looks very different from the procedures [術 shù] stated in Introduction

[1] *sun*[寸] is a unit of length of about 3cm, 10*sun* is 1*shaku* [尺], and 10*shaku* is 1 *jō* [丈].

to Mathematics by Zhū Shì-Jié. The heavenly element method [天元術] is a way to derive an equation from which we compute the answer by the root-extraction method [開方術]. Therefore, Zhū Shì-Jié always wrote down a polynomial or an equation with numerical coefficients he obtained after each procedure he explained. But here we cannot find any of them.

In the following translation we supplement them but we allow polynomials as coefficients if necessary.

Set the heavenly element, and let it be Hook ; $\boxed{0}\boxed{1}$.

Subtract Width, to get Short Hook ; $\boxed{\text{- Width}}\boxed{1}$. Transfer it to Place A.

Set AlsoNo. ; $\boxed{\text{AlsoNo.}}$.

Subtract Hook, to get Leg + Chord ; $\boxed{\text{AlsoNo.}}\boxed{-1}$.

Square it; $\boxed{\text{AlsoNo.}^2}\boxed{\text{-2AlsoNo.}}\boxed{1}$.

Subtract Hook squared ; $\boxed{\text{AlsoNo.}^2}\boxed{\text{-2AlsoNo.}}$, to get twice the sum of Leg and Chord multiplied by Leg.

Multiply A; $\boxed{-\text{AlsoNo.}^2 \times \text{Width}}\boxed{\text{AlsoNo.}^2 + 2\text{AlsoNo.} \times \text{Width}}\boxed{-2\text{AlsoNo.}}$ to get $(\text{Leg} + \text{Chord}) \times 2\text{Hook} \times \text{Length}$. And transfer it to the left-hand side.

Set Length ; $\boxed{\text{Length}}$.

Multiply it by the sum of Leg and Chord, and by Hook, and then double the result ; $\boxed{0}\boxed{2\text{Length} \times \text{AlsoNo.}}\boxed{-2\text{Length}}$.

Cancel with the left-hand side, and get the equation;

$\boxed{-\text{AlsoNo.}^2 \times \text{Width}}\boxed{2\text{Length} \times \text{AlsoNo.} - \text{AlsoNo.}^2 - 2\text{AlsoNo.} \times \text{Width}}$

$\boxed{-2\text{Length} + 2\text{AlsoNo.}}$.

Extract the roots of the quadratic equation and obtain the Hook. By the former equations we obtain Leg and Chord.

Sawaguchi showed in this way that the expressions in classical Chinese can be as concise as or even more than modern notations invented by Descartes [8]. The same was claimed by Jock Hoe [33, 5] concerning Jade Mirror of Four Elements by Zhu Shi-Jie. More interesting is Sawaguchi's usage of the word 因 in 因股弦和二段股 and 因股弦和因鈎二段縦. It is here employed to mean the operator Product of · and · with two arguments in the Polish notation of logic as introduced by Łukasiewicz [30]. Thus we have

$$\begin{aligned} &\text{因股弦和因鈎二段縦} = \text{股弦和} \times \text{因鈎二段縦} \\ &= \text{股弦和} \times (\text{鈎二段} \times \text{縦}) = (\text{股} + \text{弦}) \times \{ (2\ \text{鈎}) \times \text{縦} \}. \end{aligned}$$

The advantage of this notation is that one can express any complicated combinations of operators uniquely without parentheses, which are necessary in Cartesian notations and difficult to translate into classical Chinese having few relative words. Seki Takakazu might have tried to introduce the operator 併 for addition. In oldest extant copies of his Method of Solving concealed Problems we find the passage 子三箇内減併丑二箇寅一箇.[2]

[2] See these proceedings p. 478.

2 Mathematical Methods without Secrets by Seki Takakazu

Mathematical Methods without Secrets [發微算法 Hatsubi Sanpō] (1674) [12] is the only printed book Seki Takakazu [関孝和] published in his lifetime. There he solved all the 15 open problems Sawaguchi left in his book Mathematical Methods, Old and New (1670). We consider here only the 14th problem.

The question says: Suppose that there are two flat cones. Set x, y, z, u, v and w *sun*, and cube them individually. The differences [of the cubes] are as follows. Compared with x number, y number is smaller by 271 cubic *sun*. Compared with the y number, the z number is smaller by 217 cubic *sun*. Compared with the z number, the u number is smaller by 60.08 cubic *sun*. Compared with the u number, the v number is smaller by 326.2 cubic *sun*. And compared with the v number, the w number is smaller by 61 cubic *sun*. Find x, y, z, u, v and w.

[今有両平錐只云列甲乙丙丁戊己寸各別別再自乘之各其差云則者從甲数而乙数者少寸立積二百七十一坪從乙数而丙数者少寸立積二百十七坪從丙数而丁数者少寸立積六十坪令八分從丁数而戊数者少寸立積三百二十六坪二分從戊数而己数者少寸立積六十一坪問甲乙丙丁戊己幾何]

The solution Seki gave in the book is the following curt one:

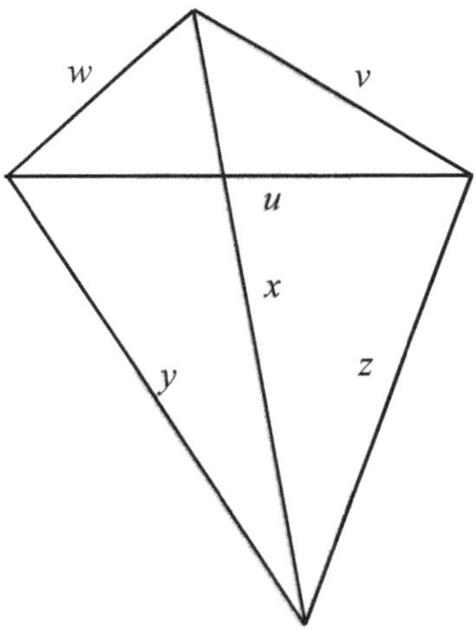

Fig. 2 Problem 14 of Sawaguchi

Answer says: The procedure of obtaining x is an equation of degree 1457. The process of elimination is complicated and needs a long writing. Therefore we omit it, whereas the outline of its starting procedure and its [elimination] process is the following. Let the heavenly element be x. Then, the cubes of y, z, u, v and w are represented thereby. Hence the process of eliminating w cubed leads to [an equation of] degree 17. The process of eliminating v cubed leads to [an equation of] degree 53. The process of eliminating u cubed leads to [an equation of] degree 161. And the process of eliminating z cubed leads to [an equation of] degree 485. Then, we start the process of eliminating y cubed to obtain two places which are identically equal. By canceling them, we obtain an equation of degree 1457. Solving it, we obtain x. This is a recursive method, which is a profound secret for solving difficult problems and is an essence that the students should endeavor to try.

[答曰得甲術千四百五十七乗方飜法也演脱多端而文繁故略之乃其起術演段大概日立天元一為甲依之得乙丙丁戊己各再自乗数從是脱己再自乗数演段至一十七乗方次脱戊再自乗数演段至五十三乗方次脱丁再自乗数演段至一百六十一乗方次脱丙再自乗数演段至四百八十五乗方於是起術而脱乙再自乗数求両位適等相消得開方式一千四百五十七乗方飜法開之得甲也是則循々誘入之意蓋解難問之奥妙也尤為学者当務之要也]

The degree of a power in Wasan is smaller than ours by one because they counted the number of multiplications.

This may look too concise an explanation to understand Seki's method. But we can guess it from what he wrote for the preceding problems. Denote by a, b, c, d and e the known numbers obtained as the sums of differences of cubes. Then we have to solve the system of equations:

$$y^3 = x^3 + a, \tag{5}$$

$$z^3 = x^3 + b, \tag{6}$$

$$u^3 = x^3 + c, \tag{7}$$

$$v^3 = x^3 + d, \tag{8}$$

$$w^3 = x^3 + e, \tag{9}$$

$$\begin{aligned} &x^2u^2(-x^2+y^2+z^2-u^2+v^2+w^2) \\ &+y^2v^2(x^2-y^2+z^2+u^2-v^2+w^2) \\ &+z^2w^2(x^2+y^2-z^2+u^2+v^2-w^2) \\ &-x^2y^2w^2-y^2z^2u^2-z^2x^2v^2-u^2v^2w^2=0. \end{aligned} \tag{10}$$

The last one is the condition under which 6 segments in the figure are on a plane and called the quadrilateral method [四斜法 shishahō] or the six obliques procedure [六斜術 rokusha jutsu], which was given as Problem 58 in Muramatsu [9, p. 137 in the Sato edition].

We rearrange it in the order of ascending powers in w and substitute (9) for w^3 in w^4. Then, we have

$$A + Bw + Cw^2 = 0, \tag{11}$$

where

$$\begin{aligned} A &= x^2u^2(-x^2+y^2+z^2-u^2+v^2) \\ &+y^2v^2(x^2-y^2+z^2+u^2-v^2)-y^2z^2u^2-z^2x^2v^2, \end{aligned}$$

$$B = -z^2(x^3+e),$$
$$C = x^2u^2+y^2v^2+z^2(x^2+y^2-z^2+u^2+v^2)-x^2y^2-u^2v^2.$$

Now, employing

$$P+Q+R=0 \Longrightarrow P^3+Q^3+R^3-3PQR=0^3 \tag{12}$$

for $P=A, Q=Bw$, and $R=Cw^2$. we have

$$A^3+(B^3-3ABC)w^3+C^3w^6=0 \tag{13}$$

as a consequence of (6) and (5). Since A, B and C have no w,

$$A^3+(B^3-3ABC)(x^3+e)+C^3(x^3+e)^2=0 \tag{14}$$

is an equation obtained from (10) and (9) by eliminating w. Since A, B and C are of degree 6, 5, and 4, respectively, this is an equation of degree 18.

Then, we rearrange it in the order of ascending powers in v, substitute (8) whenever we have the factor v^3 in the terms and repeat it until we have an equation of the form (11) with w replaced by v. Thus, we obtain an equation of the form (13) with w replaced by u which does not contain neither w nor v. Since the coefficients A, B and C are cubed in the process of converting (11) into (13), this is an equation in x, y, z and u of degree $18\times 3=54$.

Similarly we eliminate u, z and y in turn and obtain an equation only in x of degree $54\times 3^3=1458$.

Presumably the principle is understood but it is beyond human ability to carry it out. Neither Seki nor any other Wasan mathematicians did. Until recently nobody could. Only the recent development of symbolic calculus made it possible by the use of computers. Dr. Kinji Kimura's calculation [37] shows that the equation of degree 1458 is of the lowest degree, and that

$$\begin{array}{llll} x = 10.0000056403 & & y = 9.0000069815 \\ z = 8.0000083910 & & u = 7.6699093899 \\ v = 5.0000228360 & & w = 4.0000357240 \end{array} \tag{15}$$

is its only one solution with positive entries, and moreover, there are seven sets of real solutions and 1450 sets of complex solutions.[4]

Eleven years later, in 1685, Seki's pupil Takebe Katahiro [建部賢弘] published a book titled Commentaries in Japanese of Procedures in Mathematical Methods without Secrets [發微算法演段諺解] [16] and gave detailed commentaries on Seki's solutions written in the style of Sawaguchi. Takebe introduced here the byscript method [傍書法 bōsho hō] to express polynomials in several variables which is

[3] $= (P+Q+R)(P^2+Q^2+R^2-PQ-QR-RP)$

[4] Kimura solved the problem assuming that $z^3-u^3=60.8$ but not 60.08 according to Hirayama [32, p. 76]. Moreover the author erroneously gave 4.0000359240 as w in his talk.

suitable, in particular, for denoting the intermediate and final results in the heavenly element method, which Sawaguchi had to abandon. Therefore, it is generally believed that Algebra was established in Japan by this book.

3 Quadrilateral Method or Six Obliques Procedure

The quadrilateral relation which holds for the sides b, c, q, r and the diagonals a and p of a quadrilateral on a plane is a consequence of

$$a^2 + b^2 - r^2 = 2as, \tag{16}$$

$$a^2 + c^2 - q^2 = 2at, \tag{17}$$

$$b^2 = s^2 + u^2, \tag{18}$$

$$c^2 = t^2 + v^2, \tag{19}$$

$$p^2 = (s-t)^2 + (u+v)^2. \tag{20}$$

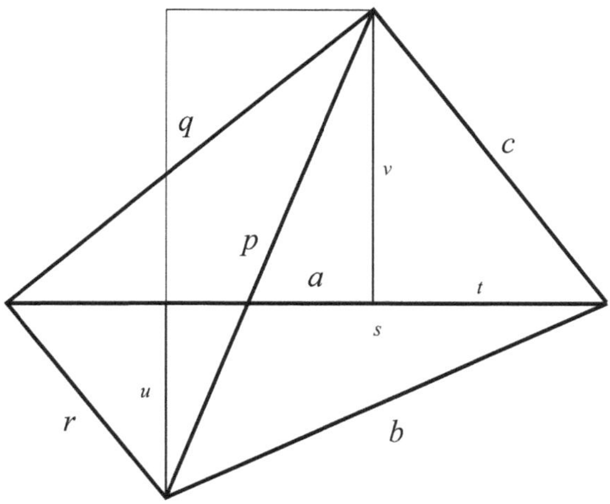

Fig. 3 Quadrilateral Method

The first two are the cosine law and the rest are the Pythagorean theorem. The result

$$\begin{aligned} &a^2p^2(-a^2+b^2+c^2-p^2+q^2+r^2) \\ &+b^2q^2(a^2-b^2+c^2+p^2-q^2+r^2) \\ &+c^2r^2(a^2+b^2-c^2+p^2+q^2-r^2) \\ &-a^2b^2r^2 - b^2c^2p^2 - c^2a^2q^2 - p^2q^2r^2 = 0. \end{aligned} \tag{21}$$

became famous among Japanese mathematicians after Seki Takakazu employed it in the solutions of Problems 12 as well as 14 of Sawaguchi under a little different configurations. Here we reproduce the proof of Seki and Takebe according to the sketch in Volume 10 of the Complete Books of Mathematics [大成算経][5].

Let $S_+ = av/2$ and $S_- = au/2$ be the areas of the upper and lower triangles, respectively. Then, it follows from (19) and (17) that

$$\begin{aligned} 16S_+^2 &= 4a^2c^2 - 4a^2t^2 \\ &= 4a^2c^2 - (a^2+c^2-q^2)^2 \\ &= 2a^2c^2 + 2c^2q^2 + 2a^2q^2 - a^4 - c^4 - q^4. \end{aligned} \tag{22}$$

Similarly (18) and (16) imply

$$16S_-^2 = 2a^2b^2 + 2b^2r^2 + 2a^2r^2 - a^4 - b^4 - r^4. \tag{23}$$

They are Heron's formula of the area of a triangle.

On the other hand, the square of the difference of (16) and (17)

$$b^2 - c^2 + q^2 - r^2 = 2a(s-t)$$

gives

$$(b^2 - c^2 + q^2 - r^2)^2 = 4a^2(s-t)^2. \tag{24}$$

Hence we have by (20)

$$\begin{aligned} 16(S_+ + S_-)^2 &= 4a^2(u+v)^2 = 4a^2p^2 - 4a^2(s-t)^2 \\ &= 4a^2p^2 - (b^2-c^2+q^2-r^2)^2. \end{aligned}$$

Subtracting (22) and (23) from both sides, we obtain

$$32S_+S_- = 2a^2(a^2-b^2-c^2+2p^2-q^2-r^2) + 2(b^2-r^2)(c^2-q^2).$$

Dividing both sides by 2 and then squaring them, we have

$$\begin{aligned} 16S_+^2 \cdot 16S_-^2 &= \{(a^2(a^2-b^2-c^2+2p^2-q^2-r^2) + (b^2-r^2)(c^2-q^2)\}^2 \\ &= a^8 + a^6\{-2b^2-2c^2+4p^2-2q^2-r^2\} \\ &+ a^4\{(b^4+4b^2c^2-4b^2p^2+2b^2r^2) + (c^4-4c^2p^2+2c^2q^2) \\ &\quad + (4p^4-4p^2q^2-4p^2r^2) + (q^4+4q^2r^2) + r^4\} \\ &+ a^2\{-2b^4c^2+2b^4q^2-2b^2c^4+4b^2c^2p^2-4b^2p^2q^2+2b^2q^4 \\ &\quad + 2c^4r^2+4p^2q^2r^2-2q^4r^2-2q^2r^4-4c^2p^2r^2+2c^2r^4\} \\ &+ \{b^4c^4-2b^4c^2q^2+b^4q^4-2b^2c^4r^4+4b^2q^2r^2-2b^2q^4r^2 \\ &\quad + c^4r^4-2c^2q^2r^4+q^4r^4\}. \end{aligned} \tag{25}$$

The same quantity is computed as the product of (22) and (23). Thus we have

[5] See these Proceedings pp. 527–529.

$$\begin{aligned}16S_+^2\cdot 16S_-^2 &= \{2a^2c^2+2c^2q^2+2a^2q^2-a^4-c^4-q^4\}\\&\times\{2a^2b^2+2b^2r^2+2a^2r^2-a^4-b^4-r^4\}\\&= a^8+a^6\{-2b^2-2c^2-2q^2-2r^2\}\\&+a^4\{(b^4+4b^2c^2+4b^2q^2-2b^2r^2)+(c^4-2c^2q^2+4c^2r^2)\\&\quad+(q^4+4q^2r^2+r^4)\}\\&+a^2\{-2b^4c^2-2b^4q^2-2b^2c^4+4b^2c^2q^2+4b^2c^2r^2-2b^2q^4\\&\quad+4b^2q^2r^2-2c^4r^2+4c^2q^2r^2-2c^2r^4-2q^4r^2-2q^2r^4\}\\&+\{b^4c^4-2b^4c^2q^2+b^4q^4-2b^2c^4r^2+4b^2c^2q^2r^2-2b^2q^4r^2\\&\quad+c^4r^4-2c^2q^2r^4+q^4r^4\}.\end{aligned}\tag{26}$$

The right hand side of (25) has 22 positive terms and 18 negative terms and that of (26) 18 and 18. We transfer (25) to the left place [寄左] and equate it with (26) [相消]. Then, we obtain the eliminated equation (21) times the non-zero factor $-4a^2$.

When a,b,c,p,q and r are the lengths of edges of a tetrahedron, the left hand side of (21) represents $144V^2$, where V is its volume. Seki and Takebe brothers used the negative of (21). Probably this means that they didn't notice the volume formula.

The first paper of Arthur Cayley (1821–1895) [27] was a proof of the quadrilateral method and its generalizations, which is reproduced in these proceedings on p.569.

4 Power Procedures or Diminishing and Stretching Method

In the above solution of Probelem 14 the most essential was the implication

$$A+Bw+Cw^2=0 \implies A^3+(B^3-3ABC)w^3+C^3w^6=0. \tag{27}$$

Most of the other problems of Sawaguchi can also be solved in a similar way. Sawaguchi and other mathematicians in Kyoto and Osaka area were developing the power procedures [冪乗演式 bekijō enshiki], that is, for a given number $n=2,3,4,\cdots$, they knew how to compute a new polynomial $F(X^n)$ in X^n of the lowest degree from a given equation

$$f(X)=a_0X^0+a_1X^1+\cdots+a_mX^m=0 \tag{28}$$

such that all roots of the above are also roots of

$$F(X^n)=A_0X^0+A_1X^n+\cdots+A_MX^{Mn}=0. \tag{29}$$

Miyagi Kiyotsura [宮城清行] published his result for $n=4,5$ and 6 in Lucid Methods of Mathematics [明元算法 Meigen Sanpō] (1689) [17], Andō Yoshiharu [安藤吉治] for $n=7$ in Monopolized Mathematical Method [一極算法 Ikkyoku Sanpō]

(1689) [18], and Nakane Genkei [中根元圭] for $n = 8$ in Procedure of the Seventh Powers [七乘冪演式 Shichijōbeki enshiki] (1691) [21] without any errors. In the last book 810 terms were published in two volumes.

Since these works are almost forgotten, we reproduce here their results. We may assume that $m < n$. The coefficients $a_0, a_1, a_2, \cdots$ are written $a, b, c, \cdots$, and $= 0$ at the end are omotted.

$n = 2$

$$a^2X^0 - b^2X^2. \tag{30}$$

$n = 3$

$$a^3X^0 + (b^3 - 3abc)X^3 + c^3X^6. \tag{31}$$

$n = 4$

$$\begin{aligned} &a^4X^0 + (-b^4 + 4ab^2c - 2a^2c^2 - 4a^2bd)X^4 \\ &+(c^4 - 4bc^2d + 2b^2d^2 + 4acd^2)X^8 - d^4X^{12}. \end{aligned} \tag{32}$$

$n = 5$

$$\begin{aligned} &a^5X^0 + (b^5 - 5ab^3c + 5a^2bc^2 + 5a^2b^2d - 5a^3cd - 5a^3be)X^5 \\ &+(c^5 - 5bc^3d + 5b^2cd^2 + 5ac^2d^2 - 5abd^3 + 5b^2c^2e \\ &-5ac^3e - 5b^3de - 5abcde + 5a^2d^2e + 5ab^2e^2 + 5a^2ce^2)X^{10} \\ &+(d^5 - 5cd^3e + 5c^2de^2 + 5bd^2e^2 - 5bce^3 - 5ade^3)X^{15} + e^5X^{20}. \end{aligned} \tag{33}$$

$n = 6$

$$\begin{aligned} &a^6X^0 + (-b^6 + 6ab^4c - 9a^2b^2c^2 + 2a^3c^3 - 6a^2b^3d \\ &+12a^3bcd - 3a^4d^2 + 6a^3b^2e - 6a^4ce - 6a^4bf)X^6 \\ &+(c^6 - 6bc^4d + 9b^2c^2d^2 + 6ac^3d^2 - 2b^3d^3 - 12abcd^3 + 3a^2d^4 + 6b^2c^3e - 6ac^4e \\ &-12b^3cde + 18ab^2d^2 + 3b^4e^2 + 9a^2c^2e^2 - 18a^2bde^2 + 2a^3e^3 - 6b^3c^2f + 12abc^3f \\ &+6b^4df - 18a^2c^2df - 12ab^3ef + 12a^3def + 9a^2b^2f^2 + 6a^3cf^2)X^{12} \\ &+(-d^6 + 6cd^4e - 9c^2d^2e - 6bd^3e^2 + 2c^3e^3 + 12bcde^3 + 6ad^2e^3 - 3b^2e^4 - 6ace^4 \\ &-6c^2d^3f + 6bd^4f + 12c^3def - 12ad^3ef - 18bc^2e^2f + 12abe^3f - 3c^4f^2 \\ &-9b^2d^2f^2 + 18acd^2f^2 + 18b^2cef^2 - 9a^2e^2f^2 - 2b^3f^3 - 12abcf^3 - 6a^2df^3)X^{18} \\ &+(e^6 - 6de^4f + 9d^2e^2f^2 + 6ce^3f^2 - 2d^3f^3 - 12cdef^3 \\ &-6be^2f^3 + 3c^2f^4 + 6bdf^4 + 6aef^4)X^{24} - f^6X^{30}. \end{aligned} \tag{34}$$

$n=7$

$$a^7X^0+(b^7-7ab^5c+14a^2b^3c^2-7a^3bc^3+7a^2b^4d-21a^3b^2cd+7a^4c^2d$$
$$+7a^4bd^2-7a^3b^3e+14a^4bce-7a^5de+7a^4b^2f-7a^5cf-7a^5bg)X^7$$

$$+(c^7-7bc^5d+14b^2c^3d^2+7ac^4d^2-7b^3cd^3-21abc^2d^3+7ab^2d^4+7a^2cd^4$$
$$+7b^2c^4e-7ac^5e-21b^3c^2de+7abc^3de+7b^4d^2e+35ab^2cd^2e-7a^2c^2d^2$$
$$-21a^2bd^3e+7b^4ce^2-7ab^2c^2e^2+14a^2c^3e^2-21ab^3de^2-14a^2bcde^2$$
$$+14a^3d^2e^2+14a^2b^2e^3-7a^3ce^3-7b^3c^3f+14abc^4f+14b^4cdf-14ab^2c^2df$$
$$-21a^2c^3df-21ab^3d^2f+35a^2bcd^2f-7a^3d^3f-7b^5ef+7ab^3cef-14a^2bc^2ef$$
$$+35a^2b^2def+7a^3cdef-21a^3be^2f+7ab^4f^2-7a^2b^2cf^2+14a^3c^2f^2$$
$$-21a^3bdf^2+7a^4ef^2+7b^4c^2g-21ab^2c^3g+7a^2c^4g-7b^5dg+7ab^3cdg$$
$$+35a^2bc^2dg-7a^2b^2d^2g-21a^3cd^2g+14ab^4eg-14a^2b^2ceg-21a^3c^2eg$$
$$+7a^3bdeg+7a^4e^2g-21a^2b^3fg+7a^3bcfg+14a^4dfg+14a^3b^2g^2+7a^4cg^2)X^{14}$$

$$+(d^7-7cd^5e+14c^2d^3e^2+7bd^4e^2-7c^3de^3-21bcd^2e^3-7ad^3e^3+7bc^2e^4$$
$$+7b^2de^4+14acde^4-7abe^5+7c^2d^4f-7bd^5f-21c^3d^2ef+7bcd^3ef+14ad^4ef$$
$$+7c^4e^2f+35bc^2de^2f-7b^2d^2e^2f-14acd^2e^2f-21b^2ce^3f-21ac^2e^3f+7abde^3f$$
$$+7a^2e^4f+7c^4df^2-7bc^2d^2f^2+14b^2d^3f^2-21acd^3f^2-21bc^3ef^2-14b^2cdef^2$$
$$+35ac^2def^2-14abd^2ef^2+14b^3e^2f^2+35abce^2f^2-7a^2de^2f^2+14b^2c^2f^3$$
$$-7ac^3f^3-7b^3df^3+7abcdf^3+14a^2d^2f^3-21ab^2ef^3-21a^2cef^3+7a^2bf^4$$
$$-7c^3d^3g+14bcd^4g-7ad^5g+14c^4deg-14bc^2d^2eg-21b^2d^3eg+7acd^3eg$$
$$-21bc^3e^2g+35b^2cd^2g-14ac^2de^2g+35abd^2e^2g-7b^3e^3g+7abce^3g-21a^2de^3g$$
$$-7c^5fg+7bc^3dfg-14b^2cd^2fg+35ac^2d^2fg+7abd^3fg+35b^2c^2efg$$
$$+7ac^3efg+7b^3defg-105abcdefg-14a^2d^2efg-14ab^2e^2fg+35a^2ce^2fg$$
$$-21b^3cf^2g-14abc^2f^2g+35ab^2df^2g-14a^2cdf^2g+35a^2bef^2g-7a^3f^3g$$
$$+7bc^4g^2-7b^2c^2dg^2-21ac^3dg^2+14b^3d^2g^2-14abc^2g^2+14a^2d^3g^2-21b^3ceg^2$$
$$+35abc^2eg^2-14ab^2deg^2+35a^2cdeg^2-7a^2be^2g^2+7b^4fg^2+35ab^2cfg^2$$
$$-7a^2c^2fg^2-14a^2bdfg^2-21a^3efg^2-7ab^3g^3-21a^2bcg^3-7a^3dg^3)X^{21}$$

$$+(e^7-7de^5f+14d^2e^3f^2+7ce^4f^2-7d^3ef^3-21cde^2f^3-7be^3f^3+7cd^2f^4$$
$$+7c^2ef^4+14bdef^4+7ae^2f^4-7bcf^5-7adf^5+7d^2e^4g-7ce^5g-21d^3e^2fg$$
$$+7cde^3fg+14be^4fg+7d^4f^2g+35cd^2ef^2g-7c^2e^2f^2g-14bde^2f^2g-21ae^3f^2g$$
$$-21c^2df^3g-21bd^2f^3g+7bcef^3g+7adef^3g+7b^2f^4g+14acf^4g$$

$$+7d^4eg^2-7cd^2e^2g^2+14c^2e^3g^2-21bde^3g^2+7ae^4g^2-21cd^3fg^2-14c^2defg^2$$
$$+35bd^2efg^2-14bce^2fg^2+35ade^2fg^2+14c^3f^2g^2+35bcdf^2g^2-7ad^2f^2g^2$$
$$-7b^2ef^2g^2-14acef^2g^2-21abf^3g^2+14c^2d^2g^3-7bd^3g^3-7c^3eg^3+7bcdeg^3$$
$$-21ad^2eg^3+14b^2e^2g^3-21ace^2g^3-21bc^2fg^3-21b^2dfg^3+7acdfg^3$$
$$+7abefg^3+14a^2f^2g^3+7b^2cg^4+7ac^2g^4+14abdg^4+7a^2eg^4)X^{28}$$

$$+(f^7-7ef^5g+14e^2f^3g^2+7df^4g^2-7e^3fg^3-21def^2g^3-7cf^3g^3+7de^2g^4$$
$$+7d^2fg^4+14cefg^4+7bf^2g^4-7cdg^5-7beg^5-7afg^5)X^{35}+g^7X^{42}. \tag{35}$$

$n=8$

$$a^8X^0+(-b^8+8ab^6c-20a^2b^4c^2+16a^3b^2c^3-2a^4c^4-8a^2b^5d+32a^3b^3cd$$
$$-24a^4bc^2d-12a^4b^2d^2+8a^5cd^2+8a^3b^4e-24a^4b^2ce+8a^5c^2e+16a^5bde$$
$$-4a^6e^2-8a^4b^3f+16a^5bcf-8a^6df+8a^5b^2g-8a^6cg-8a^6bh)X^8$$

$$+(c^8-8bc^6d+20b^2c^4d^2+8ac^5d^2-16b^3c^2d^3-32abc^3d^3+2b^4d^4+24ab^2cd^4$$
$$+12a^2c^2d^4-8a^2bd^5+8b^2c^5e-8ac^6e-32b^3c^3de+16abc^4de+24b^4cd^2e$$
$$+48ab^2c^2d^2e-16a^2c^3d^2e-32ab^3d^3e-32a^2bcd^3e+8a^3d^4e+12b^4c^2e^2$$
$$-16ab^2c^3e^2+20a^2c^4e^2-8b^5de^2-32ab^3cde^2-16a^2bc^2de^2+56a^2b^2d^2e^2$$
$$+16a^3cd^2e^2+8ab^4e^3+16a^2b^2ce^3-16a^3c^2e^3-32a^3bde^3+6a^4e^4-8b^3c^4f$$
$$+16abc^5f+24b^4c^2df-32ab^2c^3df-24a^2c^4df-8b^5d^2f-32ab^3cd^2f$$
$$+80a^2bc^2d^2f+16a^2b^2d^3f-32a^3cd^3f-16b^5cef+32ab^3c^2ef-32a^2bc^3ef$$
$$+48ab^4def-32a^2b^2cdef+32a^3c^2def-32a^3bd^2ef-48a^2b^3e^2f+32a^3bce^2f$$
$$+8a^4de^2f+4b^6f^2-8ab^4cf^2+8a^2b^2c^2f^2+16a^3c^3f^2-16a^2b^3df^2-32a^3bcdf^2$$
$$+20a^4d^2f^2+48a^3b^2ef^2-24a^4cef^2-8a^4bf^3+8b^4c^3g-24ab^2c^4g+8a^2c^5g$$
$$-16b^5cdg+32ab^3c^2dg+32a^2bc^3dg+24ab^4d^2g-80a^2b^2cd^2g-16a^3c^2d^2g$$
$$+32a^3bd^3g+8b^6eg-16ab^4ceg+16a^2b^2c^2eg-32a^3c^3eg-32a^2b^3deg$$
$$+64a^3bcdeg-24a^4d^2eg+16a^3b^2e^2g+8a^4ce^2g-16ab^5fg+32a^2b^3cfg$$
$$-32a^3bc^2fg+32a^3b^2dfg+16a^4cdfg-48a^4befg+8a^5f^2g+12a^2b^4g^2$$
$$+16a^4bcgh+16a^5dgh+20a^4b^2h^2+8a^5ch^2)X^{16}$$

$$+(-d^8+8cd^6e-20c^2d^4e^2-8bd^5e^2+16c^3d^2e^3+32bcd^3e^3+8ad^4e^3-2c^4e^4$$
$$-24bc^2de^4-12b^2d^2e^4-24acd^2e^4+8b^2ce^5+8ac^2e^5+16abde^5-4a^2e^6-8c^2d^5f$$
$$+8bd^6f+32c^3d^3ef-16bcd^4ef-16ad^5ef-24c^4de^2f-48bc^2d^2e^2f+16b^2d^3e^2f$$
$$+32acd^3e^2f+32bc^3e^3f+32b^2cde^3f+32ac^2de^3f-32abd^2e^3f-8b^3e^4f$$
$$-48abce^4f+8a^2de^4f-12c^4d^2f^2+16bc^2d^3f^2-20b^2d^4f^2+24acd^4f^2+8c^5ef^2$$
$$+32bc^3def^2+16b^2cd^2ef^2-80ac^2d^2ef^2+32abd^3ef^2-56b^2c^2e^2f^2-16ac^3e^2f^2$$

$$-16b^3de^2f^2+32abcde^2f^2-8a^2d^2e^2f^2+48ab^2e^3f^2+16a^2ce^3f^2-8bc^4f^3$$
$$-16b^2c^2df^3+32ac^3df^3+16b^3d^2f^3-32abcd^2f^3-16a^2d^3f^3+32b^3cef^3$$
$$+32abc^2ef^3-32ab^2def^3+32a^2cdef^3-48a^2be^2f^3-6b^4f^4-8ab^2cf^4$$
$$-20a^2c^2f^4+24a^2bdf^4+8a^3ef^4+8c^3d^4g-16bcd^5g+8ad^6g-24c^4d^2eg$$
$$+32bc^2d^3eg+24b^2d^4eg-16acd^4eg+8c^5e^2g+32bc^3de^2g-80b^2cd^2e^2g$$
$$+16ac^2d^2e^2g-32abd^3e^2g-16b^2c^2e^3g-32ac^3e^3g+32b^3de^3g+64abcde^3g$$
$$+16a^2d^2e^3g-24ab^2e^4g+8a^2ce^4g+16c^5dfg-32bc^3d^2fg+32b^2cd^3fg$$
$$-32ac^2d^3fg-16abd^4fg-48bc^4efg+32b^2c^2defg+64ac^3defg-32b^3d^2efg$$
$$+64abcd^2efg+32a^2d^3efg+32b^3ce^2fg+32abc^2e^2fg-32ab^2de^2fg$$
$$-160a^2cde^2fg+32a^2be^3fg+48b^2c^3f^2g-24ac^4f^2g-32b^3cdf^2g-32abc^2df^2g$$
$$+16ab^2d^2f^2g+48a^2cd^2f^2g-8b^4ef^2g-96ab^2cef^2g+16a^2c^2ef^2g+32a^2bdef^2g$$
$$+16a^3e^2f^2g+32ab^3f^3g+32a^2bcf^3g-32a^3df^3g-4c^6g^2+8bc^4dg^2-8b^2c^2d^2g^2$$
$$+16ac^3d^2g^2-16b^3d^3g+32abcd^3g^2-20a^2d^4g^2+16b^2c^3eg^2+8ac^4eg^2$$
$$+32b^3cdeg^2-160abc^2deg^2+48ab^2d^2eg^2+16a^2cd^2eg^2-20b^4e^2g^2+16ab^2ce^2g^2$$
$$+56a^2c^2e^2g^2-16a^2bde^2g^2-16a^3e^3g^2-48b^3c^2fg^2+32abc^3fg^2+24b^4dfg^2$$
$$+32ab^2cdfg^2-16a^2c^2dfg^2-48a^2bd^2fg^2+32ab^3efg^2+32a^2bcefg^2$$
$$+32a^3defg^2-56a^2b^2f^2g^2-16a^3cf^2g^2+8b^4cg^3+16ab^2c^2g^3-16a^2c^3g^3$$
$$-32ab^3dg^3+32a^2bcdg^3+16a^3d^2g^3-16a^2b^2eg^3-32a^3ceg^3+32a^3bfg^3$$
$$-2a^4g^4-8c^4d^3h+24bc^2d^4h-8b^2d^5h-16acd^5h+16c^5deh-32bc^3d^2eh$$
$$-32b^2cd^3eh+32ac^2d^3eh+48abd^4eh-24bc^4e^2h+80b^2c^2de^2h-32ac^3de^2h$$
$$+16b^3d^2e^2h-32abcd^2e^2h-48a^2d^3e^2h-32b^3ce^3h+32abc^2e^3h-32ab^2de^3h$$
$$+32a^2cde^3h+8a^2be^4h-8c^6fh+16bc^4dfh-16b^2c^2d^2fh+32ac^3d^2fh$$
$$+32b^3d^3fh-64abcd^3fh+24a^2d^4fh+32b^2c^3efh+16ac^4efh-64b^3cdefh$$
$$-64abc^2defh-32ab^2d^2efh+32a^2cd^2efh+24b^4e^2fh+32ab^2ce^2fh$$
$$-16a^2c^2e^2fh+96a^2bde^2fh-32a^3e^3fh-16b^3c^2f^2h-32abc^3f^2h-8b^4df^2h$$
$$+160ab^2cdf^2h-48a^2c^2df^2h-16a^2bd^2f^2h-32ab^3ef^2h-32a^2bcef^2h$$
$$-32a^3def^2h-16a^2b^2f^3h+32a^3cf^3h+16bc^5gh-32b^2c^3dgh-48ac^4dgh$$
$$+32b^3cd^2gh+32abc^2d^2gh-32ab^2d^3gh+32a^2cd^3gh-32b^3c^2egh+64abc^3egh$$
$$-16b^4degh+64ab^2cdegh+32a^2c^2degh-32a^2bd^2egh+32ab^3e^2gh-160a^2bce^2gh$$
$$+32a^3de^2gh+48b^4cfgh-32ab^2c^2fgh+32a^2c^3fgh-64ab^3dfgh-64a^2bcdfgh$$
$$-32a^3d^2fgh+32a^2b^2efgh+64a^3cefgh+32a^3bf^2gh-8b^5g^2h-32ab^3cg^2h$$
$$-16a^2bc^2g^2h+80a^2b^2dg^2h-32a^3cdg^2h+32a^3beg^2h-24a^4fg^2h-12b^2c^4h^2$$
$$+8ac^5h^2+16b^3c^2dh^2+32abc^3dh^2-20b^4d^2h^2+16ab^2cd^2h^2-56a^2c^2d^2h^2$$
$$-16a^2bd^3h^2+24b^4ceh^2-80ab^2c^2eh^2-16a^2c^3eh^2+32ab^3deh^2+32a^2bcdeh^2$$
$$+48a^3d^2eh^2-8a^2b^2e^2h^2+16a^3ce^2h^2-8b^5fh^2-32ab^3cfh^2+80a^2bc^2fh^2$$
$$-16a^2b^2dfh^2+32a^3cdfh^2-32a^3befh^2-12a^4f^2h^2+24ab^4gh^2+48a^2b^2cgh^2$$

$$-16a^3c^2gh^2-32a^3bdgh^2-24a^4egh^2-16a^2b^3h^3-32a^3bch^3-8a^4dh^3)X^{24}$$

$$+(e^8-8de^6f+20d^2e^4f^2+8ce^5f^2-16d^3e^2f^3-32cde^3f^3-8be^4f^3+2d^4f^4$$
$$+24cd^2ef^4+12c^2e^2f^4+24bde^2f^4+8ae^3f^4-8c^2df^5-8bd^2f^5-16bcef^5$$
$$-16adef^5+4b^2f^6+8acf^6+8d^2e^5g-8ce^6g-32d^3e^3fg+16cde^4fg+16be^5fg$$
$$+24d^4ef^2g+48cd^2e^2f^2g-16c^2e^3f^2g-32bde^3f^2g-24ae^4f^2g-32cd^3f^3g$$
$$-32c^2def^3g-32bd^2ef^3g+32bce^2f^3g+32ade^2f^3g+8c^3f^4g+48bcdf^4g$$
$$+24ad^2f^4g-8b^2ef^4g-16acef^4g-16abf^5g+12d^4e^2g^2-16cd^2e^3g^2+20c^2e^4g^2$$
$$-24bde^4g^2+8ae^5g^2-8d^5fg^2-32cd^3efg^2-16c^2de^2fg^2+80bd^2e^2fg^2$$
$$-32bce^3fg^2+32ade^3fg^2+56c^2d^2f^2g^2+16bd^3f^2g^2+16c^3ef^2g^2-32bcdef^2g^2$$
$$-80ad^2ef^2g^2+8b^2e^2f^2g^2+16ace^2f^2g^2-48bc^2f^3g^2-16b^2df^3g^2-32acdf^3g^2$$
$$+32abef^3g^2+12a^2f^4g^2+8cd^4g^3+16c^2d^2eg^3-32bd^3eg^3-16c^3e^2g^3+32bcde^2g^3$$
$$-16ad^2e^2g^3+16b^2e^3g^3-32ace^3g^3-32c^3dfg^3-32bcd^2fg^3+32ad^3fg^3$$
$$+32bc^2efg^3-32b^2defg^3+64acdefg^3-32abe^2fg^3+48b^2cf^2g^3+16ac^2f^2g^3$$
$$+32abdf^2g^3-16a^2ef^2g^3+6c^4g^4+8bc^2dg^4+20b^2d^2g^4-24acd^2g^4-24b^2ceg^4$$
$$+8ac^2eg^4+16abdeg^4+20a^2e^2g^4-8b^3fg^4-48abcfg^4-24a^2dfg^4+8ab^2g^5$$
$$+8a^2cg^5-8d^3e^4h+16cde^5h-8be^6h+24d^4e^2fh-32cd^2e^3fh-24c^2e^4fh$$
$$+16bde^4fh+16ae^5fh-8d^5f^2h-32cd^3ef^2h+80c^2de^2f^2h-16bd^2e^2f^2h$$
$$+32bce^3f^2h-32ade^3f^2h+16c^2d^2f^3h+32bd^3f^3h-32c^3ef^3h-64bcdef^3h$$
$$+32ad^2ef^3h-16b^2e^2f^3h-32ace^2f^3h+24bc^2f^4h-8b^2df^4h-16acdf^4h$$
$$+48abef^4h-8a^2f^5h-16d^5egh+32cd^3e^2gh-32c^2de^3gh+32bd^2e^3gh$$
$$+16bce^4gh-48ade^4gh+48cd^4fgh-32c^2d^2efgh-64bd^3efgh+32c^3e^2fgh$$
$$-64bcde^2fgh+32ad^2e^2fgh-32b^2e^3fgh+64ace^3fgh-32c^3df^2gh$$
$$-32bcd^2f^2gh-32ad^3f^2gh+32bc^2ef^2gh+160b^2def^2gh+64acdef^2gh$$
$$-32abe^2f^2gh-32b^2cf^3gh+32ac^2f^3gh-64abdf^3gh-32a^2ef^3gh-48c^2d^3g^2h$$
$$+24bd^4g^2h+32c^3deg^2h+32bcd^2eg^2h+32ad^3eg^2h-16bc^2e^2g^2h$$
$$-48b^2de^2g^2h+32acde^2g^2h+32abe^3g^2h+8c^4fg^2h+96bc^2dfg^2h-16b^2d^2fg^2h$$
$$-32acd^2fg^2h-32b^2cefg^2h-160ac^2efg^2h-64abdefg^2h-16a^2e^2fg^2h$$
$$-16b^3f^2g^2h+32abcf^2g^2h+80a^2df^2g^2h-32bc^3g^3h-32b^2cdg^3h+32ac^2dg^3h$$
$$-32abd^2g^3h+32b^3eg^3h+64abceg^3h-32a^2deg^3h+32ab^2fg^3h+32a^2cfg^3h$$
$$-24a^2bg^4h+4d^6h^2-8cd^4eh^2+8c^2d^2e^2h^2-16bd^3e^2h^2+16c^3e^3h^2$$
$$-32bcde^3h^2+48ad^2e^3h^2+20b^2e^4h^2-24ace^4h^2-16c^2d^3fh^2-8bd^4fh^2$$
$$-32c^3defh^2+160bcd^2efh^2-32ad^3efh^2-48bc^2e^2fh^2-16b^2de^2fh^2$$
$$-32acde^2fh^2-32abe^3fh^2+20c^4f^2h^2-16bc^2df^2h^2-56b^2d^2f^2h^2$$
$$+16acd^2f^2h^2+16b^2cef^2h^2+48ac^2ef^2h^2-32abdef^2h^2+56a^2e^2f^2h^2$$
$$+16b^3f^3h^2-32abcf^3h^2+16a^2df^3h^2+48c^3d^2gh^2-32bcd^3gh^2$$

$$
\begin{aligned}
&-8ad^4gh^2-24c^4egh^2-32bc^2degh^2+16b^2d^2egh^2-96acd^2egh^2\\
&+48b^2ce^2gh^2++16ac^2e^2gh^2+32abde^2gh^2+16a^2e^3gh^2-32bc^3fgh^2\\
&-32b^2cdfgh^2+32ac^2dfgh^2+160abd^2fgh^2-32b^3efgh^2+64abcegh^2\\
&-32a^2defgh^2+16ab^2f^2gh^2-80a^2cf^2gh^2+56b^2c^2g^2h^2+16ac^3g^2h^2\\
&+16b^3dg^2h^2-32abcdg^2h^2+8a^2d^2g^2h^2-80ab^2eg^2h^2+16a^2ceg^2h^2\\
&-48a^2bfg^2h^2+16a^3g^3h^2-8c^4dh^3-16bc^2d^2h^3+16b^2d^3h^3+32acd^3h^3\\
&+32bc^3eh^3-32b^2cdeh^3+32ac^2deh^3-32abd^2eh^3-16b^3e^2h^3\\
&+32abce^2h^3-48a^2de^2h^3+16b^2c^2fh^3-32ac^3fh^3+32b^3dfh^3\\
&-64abcdfh^3-16a^2d^2fh^3+32ab^2efh^3-32a^2cefh^3+16a^2bf^2h^3\\
&-32b^3cgh^3-32abc^2gh^3-32ab^2dgh^3+32a^2cdgh^3+32a^2begh^3\\
&+32a^3fgh^3+2b^4h^4+24ab^2ch^4+12a^2c^2h^4+24a^2bdh^4+8a^3eh^4)X^{32}
\end{aligned}
$$

$$
\begin{aligned}
&+(-f^8+8ef^6g-20e^2f^4g^2-8df^5g^2+16e^3f^2g^3+32def^3g^3+8cf^4g^3\\
&-2e^4g^4-24de^2fg^4-12d^2f^2g^4-24cef^2g^4-8bf^3g^4+8d^2eg^5+8ce^2g^5\\
&+16cdfg^5+16befg^5+8af^2g^5-4c^2g^6-8bdg^6-8aeg^6-8e^2f^5h\\
&+8df^6h+32e^3f^3gh-16def^4gh-16cf^5gh-24e^4fg^2h-48de^2f^2g^2h\\
&+16d^2f^3g^2h+32cef^3g^2h+24bf^4g^2h+32de^3g^3h+32d^2efg^3h\\
&+32ce^2fg^3h-32cdf^2g^3h-32bef^2g^3h-32af^3g^3h-8d^3g^4h-48cdeg^4h\\
&-24be^2g^4h+8c^2fg^4h+16bdfg^4h+16aefg^4h+16bcg^5h+16adg^5h\\
&-12e^4f^2h^2+16de^2f^3h^2-20d^2f^4h^2+24cef^4h^2-8bf^5h^2+8e^5gh^2\\
&+32de^3fgh^2+16d^2ef^2gh^2-80ce^2f^2gh^2+32cdf^3gh^2-32bef^3gh^2\\
&+24af^4gh^2-56d^2e^2g^2h^2-16ce^3g^2h^2-16d^3fg^2h^2+32cdefg^2h^2\\
&+80be^2fg^2h^2-8c^2f^2g^2h^2-16bdf^2g^2h^2+48aef^2g^2h^2+48cd^2g^3h^2\\
&+16c^2eg^3h^2+32bdeg^3h^2-16ae^2g^3h^2-32bcfg^3h^2-32adfg^3h^2\\
&-12b^2g^4h^2-24acg^4h^2-8de^4h^3-16d^2e^2fh^3+32ce^3fh^3+16d^3f^2h^3\\
&-32cdef^2h^3+16be^2f^2h^3-16c^2f^3h^3+32bdf^3h^3-32aef^3h^3\\
&+32d^3egh^3+32cde^2gh^3-32be^3gh^3-32cd^2fgh^3+32c^2efgh^3\\
&-64bdefgh^3-32ae^2fgh^3+32bcf^2gh^3-32adf^2gh^3-48c^2dg^2h^3\\
&-16bd^2g^2h^3-32bceg^2h^3+32adeg^2h^3+16b^2fg^2h^3+32acfg^2h^3\\
&+32abg^3h^3-6d^4h^4-8cd^2eh^4-20c^2e^2h^4+24bde^2h^4+8ae^3h^4\\
&+24c^2dfh^4-8bd^2fh^4-16bcefh^4+48adefh^4-20b^2f^2h^4+24acf^2h^4\\
&+8c^3gh^4+48bcdgh^4-8ad^2gh^4+24b^2egh^4-16acegh^4-16abfgh^4\\
&-20a^2g^2h^4-8bc^2h^5-8b^2dh^5-16acdh^5-16abeh^5-8a^2fh^5)X^{40}
\end{aligned}
$$

$$
\begin{aligned}
&+(g^8-8fg^6h+20f^2g^4h^2+8eg^5h^2-16f^3g^2h^3-32efg^3h^3-8dg^4h^3+2f^4h^4\\
&+24ef^2gh^4+12e^2g^2h^4+24dfg^2h^4+8cg^3h^4-8e^2fh^5-8df^2h^5-16degh^5\\
&-16cfgh^5-8bg^2h^5+4d^2h^6+8ceh^6+8bfh^6+8agh^6)X^{48}-h^8X^{56}. \qquad (36)
\end{aligned}
$$

Seki and Takebe called the same procedure the Diminishing and Stretching Method [消長法 shōchō hō]. This word appears in Takebe's Commentaries in Japanese [16] as a powerful means of elimination but with no explanations. In Complete Book of Mathematics [22] it appears in Volume 3 Modification Techniques [変技] [of algebraic equations] and Volume 17 Solutions of Well-posed Problems [全題解]. In Subsection Powered Roots [冪商] of Volume 3, Section 2, it is used to construct a new equation $F(Y) = 0$ from a given equation $f(X) = 0$ such that the power ξ^n of any root ξ of $f(X) = 0$ is a root of $F(Y) = 0$. Section 5 Alternating Multiplication [交乗] of Volume 17 deals with determinants. Here they computed the above equations for $n = 2,3,4$ and 5. Thus they understood that Equation (29) was the resultant of the system of original Equation (28) and $Y - X^n = 0$, then with Y replaced by X^n, but what they wrote was more primitive. In order to prove (31), they cubed the both sides of $-a = bx + cx^2$, and then employed $3b^2cX^4 + 3bc^2X^5 = -3abcX^3$. They made a miscalculation for $n = 4$ and their result for $n = 5$ contained 3 errors in numerical coefficients and 1 sign.

Compared with these, the above error-free computations by Miyagi [17], Ando [18] and Nakane [21] are miraculous. However, they left no descriptions how they computed these resuts in their books except that Andō writes "There are two procedures, one is direct and the other reduced. If I had taken the direct procedure, I would have had two equations each of which had 5538 terms, which would be too long, so that I took the reduced one to get this result." Nakane writes that his computation took 30 days.

5 Resolution of Entanglements in Mathematics by Tanaka

Their master Tanaka Yoshizane [田中由眞] (1651–1719) is a mathematician who lived in Kyoto. He was a strong opponent of Seki Takakazu. In 1679 he published Clearly Explained Methods of Mathematics [算法明解 Sanpō Meikai] [13] in two volumes and solved all problems of Sawaguchi by different methods from Seki's Mathematical Methods without Secrets [12].

It seems, however, that neither Seki nor Tanaka was satisfied with their methods. They started to seek for more general procedures to eliminate auxiliary variables from arbitrary systems of algebraic equations. By 1683 Seki finished his Methods of Solving Concealed Problems [解伏題之法 Kaihukudai no Hō] [14] in which he introduced resultants with the use of determinants. (See Goto-Komatsu [35] reproduced in these Proceedings with a supplement of mathematical notes.) He should have thought that mathematics was completed and in the same year started to write a treatise of all mathematics of his realm with the cooperation of Takebe Katahiro [建部賢弘] and Kataakira [賢明] brothers. Twenty eight years later they wrote up the twenty volumes of Complete Book of Mathematics [大成算經 Taisei Sankei][22].

Tanaka's book Resolution of Entanglements in Mathematics [算学紛解 Sangaku Funkai] [19] is his counterpart. The first four volumes out of eight are devoted to the elimination theory. Except for the date 1683 in the last Volume 8 no dates are written. Volume 2 mentions the above books by Miyagi [17], Andō [18] and Nakane [21]. Thus this part was certainly written later than 1691. On the other hand, the disciples Ando and Nakane of Tanaka might have written their books by Tanaka's theory described here.

Tanaka could have reached the concept of the resultant of two polynomial equations, earlier than Seki, independent of the determinant theory. On the other hand, his theory of determinants looks like a modification of what he learned from someone else. Following Fujiwara [29] we will explain his main results on elimination. The general theories are developed in the first volume and various methods for power procedures in Volumes 2–4.

At the beginning of Voume 2 Tanaka describes a very primitive method to obtain $F(X^n)$ in (29) as a polynomial of $f(X)$. This part should have been written before 1689. He still uses Seki's byscript method, but in the rest he switches to his own method.

The first section of Volume 1 is titled Unilateral Procedure for two equations [双式一貫之術 Sōshiki Ikkan no jutu]. Here he assures that the elimination is always possible. In fact, given two equations

$$f(x) = a_0 + a_1 X + \cdots + a_m X^m = 0, \tag{37}$$

$$g(x) = b_0 + b_1 X + \cdots + b_n X^n = 0, \tag{38}$$

in a common unknown X with $m \leq n$, multiply (37) and (38) by $b_n X^{n-m}$ and by a_m respectively. Then, the difference is an equation of degree $< n$. Replacing (38) by the difference, we continue the same procedure until we get an equation with the constant term only

$$h(x) = c_0 = 0, \tag{39}$$

which is an eliminated equation. Equation (39) is certainly a necessary condition in order that equations (37) and (38) have a common root but it is not sufficient in general, however.

In the next section Standard Procedure for two equations [双式定格術] Tanaka gives a distorted version of Seki's Methods of Solving Concealed Problems [14] restricting himself to the case $m = n$ for two equations (37) and (38).

We concider here only a system of two cubic equations

$$f(X) = a + bX + cX^2 + dX^3 = 0, \tag{40}$$

$$g(X) = p + qX + rX^2 + sX^3 = 0. \tag{41}$$

Similarly to Seki [14] Tanaka constructs the following three equations :

$$\begin{cases} h_1^T(X) = (ag(X) - f(X)p)/X = 0, \\ h_2^T(X) = (bg(X) - f(X)q + h_1^T(X))/X = 0, \\ h_3^T(X) = f(X)s - dg(X) = 0, \end{cases} \tag{42}$$

or

$$\begin{cases} h_1^T(X) = (aq-bp)+(ar-cp)X+(as-dp)X^2 = 0, \\ h_2^T(X) = (ar-cp)+\{(br-cq)+(as-dp)\}X+(bs-dq)X^2 = 0, \\ h_3^T(X) = (as-dp)+(bs-dq)X+(cs-dr)X^2 = 0, \end{cases} \tag{43}$$

and claims that the determinant of the coefficients of those equations is the resultant of the given equations.

In Seki [14, Sheets 9–11] the first transformed equation [換式] $h_1(X) = dg(X) - f(X)s$ was the result of eliminating the terms of the highest degree, whereas Tanaka's $h_1^T(X)$ is the result of eliminating the constant terms, which he never tried elsewhere. The definition of determinants is also different. Seki defines them by the cofactor expansions with respect to the constant terms of the transformed equations, which naturally gives the equation that a constant, independent of X, $= 0$ as a consequence of the original equations. On the other hand, Tanaka defines here the same by the cofactor expansion with respect to the coefficients of the first equation $h_1^T(X)$ for which we don't think it easy to give a meaning that the determinant $= 0$ is a condition for the existence of a common root of $f(X) = g(X) = 0$.

However, determinants are invariant under the transposition of the rows and the columns, though never stated explicitly so in Japan at that time, and moreover, Seki's system of transformed equations $\{h_j(X) = 0\}$ and Tanaka's $\{h_j^T(X) = 0\}$ are the same up to the ordering of equations and signs. Hence, two theories are equivalent in practice.

An Exhibition of Mathematical Methods [算法発揮 Sanpō Hakki] (1690) [20] by Iseki Tomotoki [井関知辰] is the first printed book in the world on the elimination theory. Iseki adopted there the same procedures as Tanaka. Moreover, even in Complete Book of Mathematics Takebe brothers abandoned Seki's definition of determinants and adopted the cofactor expansion with respect to the coefficients of the first equation. We do not know who invented these methods. The way Tanaka writes this part looks different from his other writings. For example, his alphabetical naming of coefficients are usually starts from the coefficient of the highest degree but here it starts from the constant term, etc.

In the rest of Volume 1 Tanaka [19] gives the resultants of two general equations of degree ≤ 3 and his original method to compute them under the title Fundamental Method to find resultants of two equations of different degrees [双式異乗之陰陽率并求根源術].

Let (37) and (38) be two equations. Then, the resultant $R(f,g)$ of Bézout–Sylvester[6] is expressed as

$$R(f,g) = {a_m}^n {b_n}^m \prod_{i=1}^{m} \prod_{j=1}^{n} (\xi_i - \eta_j), \tag{44}$$

where ξ_i and η_j are the roots of $f(X) = 0$ and $g(X) = 0$, respectively, in an algebraically closed field containing all coefficients of the equations. Hence it follows that when developed as a polynomial of the coefficients a_i and b_j, each monomial

[6] Seki's resultant $\mathscr{R}(f,g)$ is its multiple by $(-1)^{mn}$. See these Proceedings p. 557 in [35].

in the development is homogeneous of degree n in a_i, homogeneous of degree m in b_j and isobaric of weight mn in a_i and b_j (See Takagi[28, §28] or van der Waerden [31, §5.9].). Tanaka and probably Seki knew the following practical method to find all such monomials by experience.

For example, concider the system of a quadratic and a cubic equations

$$f(X) = c + bX + aX^2 = 0, \tag{45}$$

$$g(X) = s + rX + qX^2 + pX^3 = 0. \tag{46}$$

We raise the former equation to the power equal to the degree of the latter without regarding the numerical coefficients and signs. Thus we have

$$c^3X^0 + bc^2X^1 + (ac^2 + b^2c)X^2 + (abc + b^3)X^3 + (a^2c + ab^2)X^4 + a^2bX^5 + a^3X^6.$$

Then, we raise the latter equation to the power equal to the degree of the former without regarding the numerical coefficients and signs, and arrange it in the reversed order:

$$p^2X^6 + pqX^5 + (pr + q^2)X^4 + (ps + qr)X^3 + (qs + r^2)X^2 + rsX^1 + s^2X^0. \tag{47}$$

The monomials constituting the resultant are obtained as the products of the coefficients of the upper and the lower formulas of total degree equal to mn. Thus we have

$$\begin{aligned} R(f,g) = & C_1c^3p^2 + C_2bc^2pq + C_3ac^2pr + C_4ac^2q^2 + C_5b^2cpr \\ & + C_6b^2cq^2 + C_7abcps + C_8abcqr + C_9b^3ps + C_{10}b^3qr \\ & + C_{11}a^2cqs + C_{12}a^2r^2 + C_{13}ab^2qs + C_{14}ab^2r^2 + C_{15}a^2brs + C_{16}a^3s^2 \end{aligned} \tag{48}$$

with numerical coefficients C_k, which are determined by equating the right-hand side identically to 0 after we substitute $c = -bX - aX^2$ and $s = -rX - qX^2 - pX^3$.

This is a feasible but clumsy way. The resultant of two general cubic equations given in [19] contains many numerical errors, which may be caused by this method.

Volumes 2–4 of [19] are mainly devoted to various methods of computing power procedures, that is, the resultants of the system

$$f(X) = a_0X^0 + a_1X^1 + \cdots + a_{n-1}X^{n-1} = 0, \tag{49}$$

$$Y - X^n = 0, \tag{50}$$

In this case the counterpart of (47) and hence that of (48) are very sparse polynomials. He should have found this method here and later extended it to the case of general resultants.

In Methods of Solving Concealed Problems [14] Seki gives an estimate of the degree of the eliminated equation, that is, the resultant $= 0$ by this principle. (See these Proceedings pp. 475–476.)

At the end of Volume 2 Tanaka gives the following expression of the power procedure by the determinant:

$$\begin{vmatrix} a_0 & a_1 & a_2 & \cdots & \cdots & a_{n-2} & a_{n-1} \\ a_{n-1}X^n & a_0 & a_1 & a_2 & \cdots & a_{n-3} & a_{n-2} \\ a_{n-2}X^n & a_{n-1}X^n & a_0 & a_1 & \cdots & \cdots & a_{n-3} \\ \vdots & \ddots & \ddots & & & & \\ a_iX^n & \cdots & a_{n-1}X^n & a_0 & \cdots & \cdots & a_{i-1} \\ & \ddots & \ddots & \ddots & \ddots & \ddots & \\ a_2X^n & & & a_{n-1}X^n & a_0 & a_1 & \\ a_1X^n & a_2X^n & \cdots & \cdots & \cdots & a_{n-1}X^n & a_0 \end{vmatrix}, \qquad (51)$$

To be exact, this is different from Seki's resultant defined in [14] for the system (49) and (50), but coincides with Bézout's. This may mean that Tanaka understood Bézout's Lemme 1 saying that a determinant vanishes if and only if its columns are linearly dependent, which is given just after the definition of determinants in Bézout [25].

Tanaka was very much proud of this expression. In case n is prime, this may serve as the quickest way to compute the power procedure.

In Volume 4 Tanaka treats the case in which the exponent n is a compound number pq and shows that the n-th power procedure is computed by the repetition of the p-th power procedure and the q-th power procedure.

For example, if $n = 4$, we start with

$$f(X) = a + bX + cX^2 + dX^3 = (a + cX^2) + (b + dX^2)X = 0. \qquad (52)$$

Applying the square procedure, we have

$$(a + cX^2)^2 - (b + dX^2)^2X^2 = a^2 - b^2X^2 + 2acX^2 + c^2X^4 - 2bdX^4 - d^2X^6 = 0.$$

Hence we obtain the fourth power procedure (32) by multiplying the last equation by

$$a^2 + b^2X^2 - 2acX^2 + c^2X^4 - 2bdX^4 + d^2X^6.$$

We are easily marveled at Nakane's computation of the 8-th power procedure with 810 terms (36), but what he actually did may be only the computation of two squares of polynomials each of which has 43 terms and of their difference.

6 Descartes' too Optimistic View

Abstract The following is an English translation of the first five pages of the famous "La Géométrie" [8] by Lené Descartes published in 1637. He gives a very optimistic view of how to reduce any problem in Geometry to systems of algebraic equations in the lengths of line segments in the problem and how to solve it by reducing the system to algebraic equations in one unknown by elimination.

6.1 Problems that can be constructed only by means of circles and straight lines

All problems of Geometry can easily be reduced to such expressions that after these [reformulations] we need only to know the length of some straight lines [= segments] in order to construct the solutions.

How the calculi of Arithmetic are related to the operations in Geometry. As the whole Arithmetic[= algebra] is composed of only four or five operations, that is, addition, subtraction, multiplication, division and the extraction of roots [of single algebraic equations], which can be considered a kind of division, so we do, in Geometry in the process of preparing the segments in question to be known, nothing other than to add them with others, or to remove them from others, or, having a segment fixed which we will name the unit in order to relate it as much as possible with numbers, and which we can choose at our disposal in the ordinary case, and then given two other segments, to find a fourth segment which is to one of these two as the other is to the unit, that is the same as multiplication; or to find a fourth segment which is to one of these two as the unit is to the other, that is the same as division; or finally to find one or two or several proportional means between the unit and some other segment; that is the same as to take the square, or cube roots, etc. So I will not hesitate to introduce these expressions of Arithmetic into Geometry, to make me more intelligible.

The Multiplication.

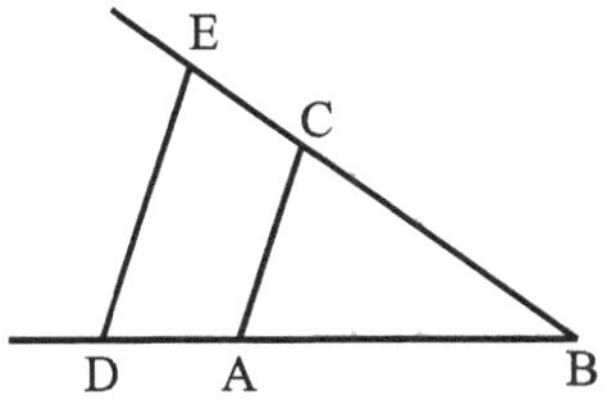

Let, for example, AB be the unit. In order to multiply BD by BC, I have only to join points A and C, and draw DE parallel to CA. Then BE is the result of this multiplication.

The Division.
Or else, if I have to divide BE by BD, after joining points E and D, I draw AC parallel to DE. Then BC is the result of this division.

The Extraction of the square root.

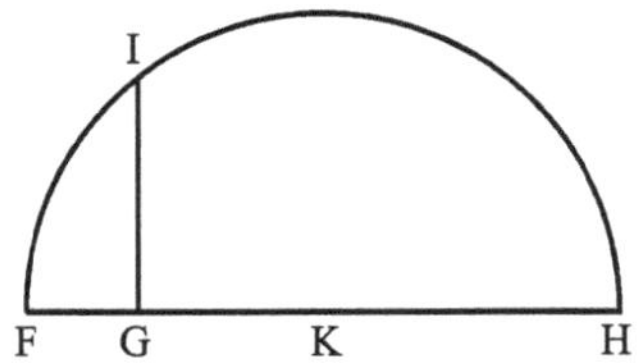

Or, if it is necessary to draw the square root of GH, I add to it a segment FG which is the unit, and dividing FH into two equal parts at the point K, I draw the circle FIH with the center at K, and then raise from the point G a straight line up to I with the right angle over FH.

This GI is the required root. I do not say.here nothing of the cube roots, nor of the other roots because I will tell of them more conveniently later.

How we can use letters in Geometry. However, it is often not necessary to trace these segments on paper, but sufficient to indicate by some letters each by a different one. As I add the segment BD to the segment GH, I name one by a and the other by b and write $a+b$; And $a-b$ to subtract b from a; and ab to multiply one by the other; And $\frac{a}{b}$ to divide a by b; And aa, or a^2 to multiply a by itself; And a^3 to multiply one more time, and so on to infinity; And $\sqrt{a^2+b^2}$ to draw the square root of a^2+b^2; And $\sqrt{C.a^3-b^3+abb}$ to draw the cube root of a^3-b^3+abb, and so on. Here it should be remarked that by a^2 or b^3 or the like I usually conceive only the simple segments, although I name them squares or cubes etc. as employed in Algebra.

It is also remarked that all parts of the same segment should usually be expressed by one and the same dimension as far as the unit is not determined in the question, as a^3 has the same dimension as abb or b^3 which compose the segment I named $\sqrt{C.\, a^3-b^3+abb}$; this is not, however, the case when the unit is determined since the unit can always be understood tacitly whenever there is a too many or too few dimensions: as in the case where we take the cube root of $aabb-b$, we must consider that the quantity $aabb$ is divided once by the unit, and that the other quantity b is multiplied twice by the same [unit].

Finally in order not to fail to remember the names of those segments, it is always necessary to make a separate list according as we set them or as we change, and we write, for example,

$$\text{AB} \propto 1, \quad \text{that is, AB is equal to 1.}$$
$$\text{GH} \propto a$$
$$\text{BD} \propto b, \text{ etc.}$$

How to arrive at Equations which serve to solve the problems. If we wish to solve some problem in this way, we have first to consider it as already done, and give names to all segments which seem necessary to construct the solution as well to those which are unknown as to the knowns. Then without considering any distinction between those segments which are known and unknown, we must survey the discrepancy, according to the order which exhibits the most naturally of all in which manner the segments depend mutually one on the others, until we find means to express one and the same quantity in two ways: That is called an Equation; because the expression of one of these two ways is equal to that of the other. And we have to find as many of such Equations as [the number] we suppose of the segments which should be unknown. Or else, if we cannot find so many [equations] notwithstanding we omit nothing which are required in the question, then it testifies that the question is not entirely determined. Then, we can take those segments as known by our choice, for all the unknowns to which no equation corresponds.

If there remain several [equations] after that, it is necessary to employ by order each of the remaining Equations either by considering it alone or else by comparing it with the others so as to explain each of those unknown segments; And make, by disentangling them, so that it stays to the only one, equal to some which is known, or else its square, or the cube, or the square of square, or the sursolid, or the square of cube etc. is equal to what is produced by the addition, or the subtraction of two or several other quantities, one of which is known, and the others are composed by some proportional means between the unit, and this square, or cube, or, square of square, etc. multiplied by other knowns. I write this in the manner.

$$
\begin{aligned}
z &\propto b, \quad \text{or} \\
z^2 &\propto -az+bb, \text{ or} \\
z^3 &\propto az^2+bbz-c^3, \text{ or} \\
z^4 &\propto az^3-c^3z+d^4, \text{ etc.}
\end{aligned}
$$

Namely, z, which I take for the unknown quantity, is equal to b, or the square of z is equal to the square of b minus a multiplied by z. Or the cube of z is equal to a multiplied by the square of z plus the square of b multiplied by z minus the cube of c . And similar for the other.

And *we can always reduce the unknown quantities in this way to only one,* when the problem can be constructed by circles and straight lines, or also by conic sections, or even by other lines which is by one or two degrees more complicated. But I do not stop here to explain this in more detail, because [then] I would deprive you of the pleasure of learning by yourself, and of the usefulness of cultivating your spirit while you are exercising there, which is, in my opinion, the principal utility that we can draw from this science. Moreover, *I notice here nothing so difficult that anyone who is a little versed in the common Geometry and Algebra, and who takes care of all that is in this treatise, can not discover.*

7 Applications to Geometry

At the beginning quarter of the seventeenth century Japan was an open country. The family of Yosida Mitsuyoshi had a license to engage in foreign trade with Annam (the present Vietnam) and the Jesuit Missionaries had a Collegio or a Seminaryo in Kyoto, Nagasaki and Bungo in Kyūshū, where they taught Mathematics, Astronomy. Music and Latin as an introduction to Christianity to the common people.

The year 1637 was the worst time, however. In the same year a large revolt took place in the Shimabara peninsula in Kyūshū and the government force of 100,000 had to spend three months to defeat the siege of 37,000 peasants led by a young Christian. The seclusion policy, which started a few years before, was strengthened and all books from Europe were strictly banned to import.

In spite of this unhappy incident Descartes' Program was first realized in Japan. In Volume 10, Geometry, of Complete Book of Mathematics. Seki and Takebe brothers applied the elimination theory to Geometry to obtain the algebraic relation of 5 sides and three diagonals of a general pentagon.

Pentagonal Method

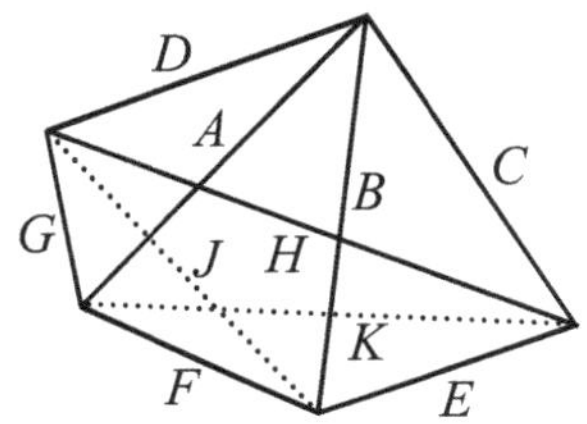

Problem at the end of Sheet 29 in Volume 10 (See these Proc. pp. 529–534) : Suppose in a pentagon that A is 21*sun*, B is 20*sun*, C is 18*sun*, D is 17*sun*. E is 14*sun*, F is 13*sun* and G is 10*sun*.
Find H.
Answer says : H is 28*sun*.281434 plus a remainder.

They apply the quadrilateral method (21) to the quadrilaterals $CDJE$ and $BDGF$ in the orders $HECBDJ$ and $ABDJGF$, respectively, and eliminate the common variable J in the reduced form, that is, taking the square of each variable as an independent variable. Then, the quadrilateral method is a quadratic equation in each variable, so that the eliminated equation is represented by a 2×2 symmetric determinant$= 0$, which may be called the pentagonal method [五斜術]. The structure of the quadrilateral method is simple but it is an equation with 22 terms. Therefore, the calculation which ended up as an equation with 843 terms was painstaking. In his book [24] Takebe Katahiro recalled "Kataakira once tried to deparenthesize on the pentagon problem and met great complexity. He said that even if ten thousand terms were necessary, it would not take more than one hundred days if he computed one hundred terms a day, and really completed the work in a little more than a month." [曽テ五斜ノ括術ヲ爲ント欲シテ甚繁雑セリ假ニ萬位ニ及フトモ一日ニ百位ヲ造サハ徐ク百日ニシテ畢テント言テ果シテ月餘ニシテ悉ク成セリ]

They also considered similar problems for hexagons and showed that the eliminated equation was written as a 4×4 symmetric determinant $= 0$. Kimura [38] shows that it has in the reduced form 273,123 terms. Elimination is a very powerful method but its practice easily goes beyond human capabilities.

Only the recent progress of computers and mathematical techniques to use them is making it possible. The attempt Professor Wu Wen-tsun and his group started during the Cultural Revolution has now grown up to the first stage so that Descartes' dream of mechanical acquisition of new knowledge is coming true (see Wu [34]).

References

1. Karine Chemla et Guo Shuchun [郭書春]: *Les Neuf Chapitres : le classique mathématique de la Chine ancienne et ses commentaires*, Dunod, Paris (2004).
2. Qín Jiǔsháo [秦九韶] : Book of Mathematics in Nine Chapters [数書九章 Shùshū Jiǔzhāng] (1247); YiJiaTang [宜稼堂] (1616).
3. Yáng Huī [楊輝]: Yang Hui's Methods of Mathematics [楊輝算法 Yáng Huī Suàn-Fǎ] (1275), transcribed by Seki Takakazu in 1661; A later copy by Ishikuro Nobuyoshi [石黒信由] is preserved in Shinminato Museum, Toyama.
4. Zhū Shì-Jié [朱世傑]: Introduction to Mathematics [算学啓蒙 Suàn-Xué Qǐ-Méng] (1299) ; with commentaries by Takebe Katahiro [建部賢弘], Ibaraki Hōdō, Kyoto (1690).
5. J. Hoe : The Jade Mirror of the Four Unknowns [四元玉鑑 Sìyuán Yùjiàn] (1303) *by Zhū Shijié*, Mingming Bookroom, P. O. Box 29-316, Christchurch, New Zealand (2007).
6. Chéng Dàwèi [程大位]: Summary of Mathematical Methods [算法統宗 Suànfǎ Tǒngzōng] (1592).
7. Yoshida Mitsuyoshi [吉田光由]: From Trifling Dust to Eternal Time [塵劫記 Jingōki], Kyōto (1627); New Edition (1641).
8. René Descartes: *La Géométrie*, Leiden (1637); in David Eugene Smith and Marcia L. Latham: *The Geometry of René Descartes*, Dover, New York (1954).
9. Muramatsu Shigekiyo [村松茂清]: Mathematical Cooking Boards [算俎 Sanso], Nishimura Mataemon, Edo (1663); Ken'ichi Sato [佐藤健一], translated into modern Japanese with commentaries, Kenseisha [研成社], Tokyo (1987).
10. Sawaguchi Kazuyuki [澤口一之]: Methods of Mathematics, Old and New [古今算法記 kokonsanpōki], Kyoto (1670); Collated by Nunoo Shimizu [清水布夫], Kenseisha [研成社], Tokyo (1993).
11. Seki Takakazu [関孝和]: *Takakazu Seki's Collected Works edited with Explanations* [関孝和全集], Osaka Kyōiku Tosho [大阪教育圖書], Osaka (1974).
12. Seki Takakazu [関孝和]: Mathematical Methods without Secrets [發微算法 Hatsubi Sanpō] (1674) ; reprint by Wasan Institute [和算研究所], Tokyo (2003); [11, pp. 103–120].
13. Tanaka Yoshizane [田中由眞] : Clearly Explained Methods of Mathematics [算法明解 Sanpō Meikai] in 2 vols. but only hand copies of vol. 1 (1679) are extant, Okamoto Manuscript [岡本写] C. 06, Tōhoku University.
14. Seki Takakazu [関孝和]: Methods of Solving Concealed Problems [解伏題之法 Kaihukudai no Hō] (1683) ; [11, pp. 142–158].
15. Seki Takakazu [関孝和]: Methods of Equation Modifications [開方翻変之法 Kaihō Honpen no Hō] (1685) ; [11, pp. 159–170].
16. Takebe Katahiro [建部賢弘]: Commentaries in Japanese of Procedures in Hatsubi Sanpō [發微算法演段諺解 Hatsubi Sanpō Endan Genkai] (1685) ; [11, pp. 509–572].
17. Miyagi Kiyotsura [宮城清行] : Lucid Methods of Mathematics [明元算法 Meigen Sanpō] in 2 vols., Nishimura Shichirobei, Kyoto (1689); Hayashi Collection 0409, Tōhoku University.
18. Andō Yoshiharu [安藤吉治] : Monopolized Mathematical Method [一極算法 Ikkyoku Sanpō], Nakamura Yahei (1689), Hayashi Collection 0042; Tōhoku University.
19. Tanaka Yoshizane [田中由真] : Resolution of Entanglements in Mathematics [算学紛解 Sangaku Funkai] in 8 vols. (1683?); Manuscript 618/204, Nakanoshima Library, Osaka.
20. Iseki Tomotoki [井関知辰]: An Exhibition of Mathematical Methods [算法発揮 Sanpō Hakki], Nagano Hikouemon, Osaka (1690); Okamoto Prints 0063, Tōhoku University.
21. Nakane Genkei [中根元圭] : Procedure of the Seventh Power [七乘冪演式], Umemura Yaemon, Kyoto (1691); Okamoto Prints 065, Tōhoku University.
22. Seki Takakazu, Takebe Kataakira and Takebe Katahiro [関孝和, 建部賢明, 建部賢弘]: Complete Book of Mathematics [大成算経 Taisei Sankei], 20 Vols. (1683–1711); Manuscripts 物-423∼425 in the Modern Science Museum, Tokyo University of Science.
23. Seki Takakazu [関孝和]: Concise Collection of Mathematical Methods [括要算法], 4 vols., Kawakatsu et al., Kyoto (1712) ; [11, pp. 345–370].

24. Takebe Katahiro [建部賢弘]: Master Takebe's Composing Methods [建部先生綴術] (1722) ; T20/74, Main Library, University of Tokyo.
25. E. Bézout : Recherches sur le degré des équations résultantes de l'évanouissement des inconnues, et sur les moyens qu'il convient d'employer pour trouver ces équations, *Mémoires Acad. Royale Sci. Paris*, pp. 288–338 (1764).
26. J. J. Sylvester : A method of determining by mere inspection the derivatives from two equations of any degree, *Phil. Magazine*, vol. 16, pp. 132–135 (1840).
27. A. Cayley : On a theorem in the geometry of position, *Cambridge Math. J.*, vol. 2, pp. 267–271 (1841).
28. Teiji Takagi [高木貞治] : *A Course of Algebra* [代数学講義], Kyōritsu Shuppan [共立出版], Tokyo (1948).
29. Matsusaburo Fujiwara [藤原松三郎], ed. Japan Academy : *History of Mathematics in Japan before Meiji* [明治前日本数学史], vol. III, Iwanami [岩波書店], Tokyo (1957).
30. Jan Łukasiewicz : *Elements of Mathematical Logic*, Pergamon Press, Oxford (1963).
31. B.L, van der Waerden : *Algebra*, Volume I, Springer-Verlag, Berlin (1966).
32. Akira Hirayama [平山諦] : *Seki Takakazu* [関孝和], Enlarged Ed. [増補訂正版], Kōseisha [恒星社], Tokyo (1974).
33. J. Hoe : *Les Systèmes d'Équations Polynômes dans le Sìyuán Yùjiàn*, Collège de France (1977).
34. Wu Wen-tsun [呉文俊]: *Mathematics Mechanization, Mechanical Geometry Theorem-proving, Mechanical Geometry Problem-Solving and Polynomial Equation-Solving*, Science Press, Beijing, and Kluwer Academic Publishers, Dordrecht-Boston-London (2000).
35. Takefumi Goto [後藤武史] and Hikosaburo Komatsu [小松彦三郎] : Determinants, resultants and discriminants in Japan in the seventeenth century and in Europe in the eighteenth and nineteenth centuries, *J. Northwest University* (Natural Science Edition), **33**, pp. 363–367 (2003) ; *These Proceedings*, pp. 553–559.
36. Takefumi Goto and Hikosaburo Komatsu : A Correction of Seki's error in the expansion of determinants, *J. Northwest University* (Natural Science Edition), **33**, pp. 376–380 (2003) ; *These Proceedings*, pp. 559–565.
37. Kinji Kimura [木村欣司], Teruhiko Hirano [平野照比古] and Kazuhiro Yokoyama [横山和弘]: Solutions of a problem of Seki Takakazu [関孝和の問題を解く], *J. Japan Society for Symbolic and Algebraic Computation* [数式処理], vol. 11 (no. 3, 4), pp. 33–42 (2005).
38. Kinji Kimura : Computer algebra at the critical performance vs. Wasan [限界性能を追求した計算機代数 vs 和算], *17th Symposium on the History of Mathematics (2006), Reports of Institute for Mathematics and Computer Science*, vol. 28, Tsuda College, Tokyo, pp. 48–63 (2007).

Some Questions and Observations Around the Mathematics of Seki Takakazu

Silke Wimmer-Zagier and Don Zagier

Abstract This informal paper, an expanded version of the short talk given by the second author at the conference on the occasion of the 300th anniversary of Seki's death, is slightly non-standard in nature and perhaps requires a short explanatory preamble. The authors are not professional historians of mathematics, and no attempt has been made to interpret the material discussed from a historical viewpoint. Instead, the first section contains several specific mathematical comments, from the point of view of a contemporary professional mathematician (D.Z.), on a few of the problems and solutions of Seki and of his predecessors in China and Japan, pointing out places where the mathematical content is unexpectedly naive or unexpectedly sophisticated, or where particular mathematical features of the problems permit deductions about their authors' methods or views.

The second section concerns the thorny question of possible contacts that Seki or his disciple Takebe Katahiro may have had with European mathematics as a result of the Dutch presence in Dejima and their yearly visit to the Edo court. In particular, we describe the results of a search (by S.W.-Z.) through the archives of the Dutch East India Company for the years in question that yielded details of a meeting between Takebe and the Dutch but show clearly that there was, at least on this occasion, no serious discussion of any scientific or mathematical questions. We also mention a few other arguments militating against the thesis that there was any direct impact of European mathematics (prior to the partial lifting of the ban on Western books in 1720) on the work of these two scientists.

S. Wimmer-Zagier and D. Zagier
Max-Planck-Institut für Mathematik, Vivatsgasse 7, 53111 Bonn, Germany
e-mail: don.zagier@mpim-bonn.mpg.de

1 The mathematics of Seki and his predecessors seen from a modern standpoint

In this section we discuss some of the aspects of the Japanese mathematics of the 17th century, and of the Chinese mathematics on which it is based, which are surprising from a contemporary point of view.

It is of course a commonplace that the nature of scientific progress, like that of any other human activity, is heavily dependent on the culture and social context in which it takes place, and it is in no way surprising that the Japanese mathematics of the Edo period, or the earlier Japanese and Chinese mathematics on which it is based, should often be very different from ours. But there are specific aspects which are startling to a contemporary mathematical eye, either because seemingly simple points are often overlooked or ignored, or, conversely, because the level of sophistication sometimes appears unexpectedly high. Of course any such reaction from a modern mathematician is anachronistic and based entirely on hindsight. Yet mathematics has an intrinsic rhythm and internal "necessariness" which all its practitioners feel, and particular interest therefore attaches to instances where the background or traditions of earlier mathematicians led them to see things in a way strange to us. We will list a number of such cases occurring in the works used or written by Seki, and will formulate a number of specific questions whose answers will doubtless be known to specialists in some cases and unknowable in others, but some of which may perhaps suggest interesting topics for further investigation.

Essentially all of our examples are taken from the book [7] by Annick Horiuchi, where they are discussed in their historical context. The English translations of the problems and of passages from this book are taken from [8].

1.1 A problem on volumes

Our first example comes from the Introduction to Mathematics [算学啓蒙 *Suanxue Qimeng*] by Zhu Shijie [朱世傑], published in 1299, which was one of the most influential Chinese mathematical texts in the early Edo period and of which a Japanese version with detailed commentary would be published by Takebe in 1690.

Problem 34 of the last chapter Unlocking of Roots [開方釋鎖 *kaifang shisuo*], discussed in [8, pp. 74–75], reads as follows in the original Chinese and in English translation:

今有立方. 立圓. 平方. 古圓田. 徽圓田. 各一。共積三万三千六百二十二尺二百分尺之三十七。只云 立方面不及立圓径四尺. 多如徽圓径三尺。立圓径如平方面三分之一。古圓周与立方面適等。問五事各幾何。
答曰 立方面二十四尺。立圓径二十八尺。平方面八十四尺。古圓周二十四尺。徽圓径二十一尺。

> One now has a cube, a sphere, a square, an old circular field and a Hui circular field. The total accumulation is 33622 *chi*[1] and 37 200ths of a *chi*. It is said only that the side of the cube is 4 *chi* less than the diameter of the sphere and 3 *chi* more than the diameter of the Hui circle, that the diameter of the sphere is equal to a third of the side of the square, and that the circumference of the old circle and the side of the cube are equal. One asks for the values of the five quantities.
> Answer : Side of cube is 24 *chi*, diameter of sphere 28 *chi*, side of square 84 *chi*, circumference of old circle 24 *chi*, diameter of Hui circle 21 *chi*.

Here the word "accumulation" means the sum of all the areas and volumes concerned. The phrase "old circular field" means a field in the shape of a circle with the *lü* [率] (fixed ratio, the Chinese word for π) being taken to be the ancient value of 3, while the "Hui circular field" is one where π is taken to be the value 3.14 ascribed to the third century mathematician Liu Hui [劉徽]. In modern terminology, the problem can therefore be stated as $A+B+C+D+E = 33622\,\frac{37}{200}$ with

$$A = a^3\,, \quad B = t\,b^3\,, \quad C = c^2\,, \quad D = \frac{1}{4\cdot\pi_{\text{old}}}\,d^2\,, \quad E = \frac{\pi_{\text{Hui}}}{4}\,e^2$$

together with the supplementary conditions $a = b-4 = e+3$, $b = c/3$, $d = a$, where $\pi_{\text{old}} = 3$ and $\pi_{\text{Hui}} = 3.14$ are the "old *lü*" and the "*lü* of Liu Hui" and t is the fixed ratio (=ratio of the volume to the cube of the diameter) for the sphere.

Seen from a modern perspective, the problem and its solution have a number of peculiarities. The most striking, of course, is the use of two different values for π in the same problem, which is quite incomprehensible for us. It is already strange to find Zhu using the "ancient" value of 3 when far more accurate values were already given in the Nine Chapters [九章算術 *Jiuzhang Suanshu*] and commentaries to them (e.g., Liu Hui himself gave the value 3.1416 and not just 3.14 as Zhu assumes; cf. [5, pp. 145–148]. But the simultaneous occurrence of two circles with differing values of π indicates a rather nebulous understanding of the relationship between the circumference and area of circles. Of course, this may have been merely an homage to the Ancients, or simply a more poetic way of writing the algebraic problem "$a^3+tb^3+c^2+\frac{1}{12}d^2+\frac{3.14}{4}e^2 = 33622\frac{37}{200}$" in words, in which case it is an entirely legitimate procedure. In any case, this passage suggests our first question:

Q1. How did the idea of π (or of other fixed geometric ratios [定法 *teihō* in Japanese]) evolve in China and Japan? At what stage was it clearly realized that π is a single well-defined number which in principle can be calculated to any desired accuracy?

The next point is the addition of areas and volumes, something that is not only not correct, but also rather unnatural, since figures of different dimensions cannot be juxtaposed, and was, for instance, not a possible operation in classical Greek mathematics, where figures were considered geometrically rather than being expressed in terms of units like the Chinese *chi*. (One can even speculate that this difference of approaches may have actually impeded the development of algebraic notions or notations in Greece, since it was not until Diophantus that polynomials were consid-

[1] *chi* [尺 *shaku* in Japanese] is a unit of length, which is an equivalent of a foot. (editor)

ered, and may perhaps have correspondingly abetted the rather early development of algebraic ideas in China.) Our second question is therefore:

Q2. At what point was it clearly realized that lengths, areas and volumes are incommensurable and must be given in different units (here, *chi*, chi^2 and chi^3), and that expressions involving mixed exponents must be expressed in terms of pure numbers rather than lengths?

The third point is the startling disparity between the very approximate values of π used (even the "lü of Liu Hui" is only a 3-digit approximation) and the very precise value $33622\frac{37}{200}$ given for the total volume. It is clear to us, but evidently not to the Chinese mathematicians of the 13th century, that it makes little sense to give the latter to 8 significant digits when the former is given to only 3- or even 1-digit accuracy, and we will see the same lack of understanding of the meaning of significant figures recurring in some of the other problems we consider below, even by much later mathematicians. This suggests our third question:

Q3. At what point were the rules for calculating with approximate values first understood, viz., that numbers that are being multiplied or divided ought to be specified to the same *relative* precision (number of significant digits), and numbers that are being added or subtracted to the same *absolute* precision (position of the first uncertain digit)? In any case, the author (ostensibly Seki, but probably Takebe Katahiro: cf. [8, p. 183]) of the Configurations for the Extraction of Roots [開方算式 *Kaihō Sanshiki*] [1, pp. 257–268] understood these rules well and used them absolutely correctly when describing the numerical determination of the root of $11+8x+x^2=0$ to high precision [8, p. 184].

The next point again concerns "fixed ratios," but this time for the sphere. Like other Chinese mathematicians of the period, Zhu gives values for the numbers solving his problem, but no indication of how one obtains them, nor any verification that they actually satisfy the terms of the problem. Doing the calculation, we find that they do so only if the ratio t is taken to be 9/16. If we assume the formula $t=\pi/6$ for the ratio of the volume of the sphere to the cube of its diameter, this would correspond to a value of $\pi=(3/2)^3=3.375$, a *lü* which has certainly never occurred in the literature and which is clearly much too big. From this we can see that the relation of the ratios associated to the circle and the sphere was not known at this time, and can deduce that the value for the sphere which is being used is $9/16$ (as opposed to $1/2$ or 0.523, which is what one would get with $t=\pi/6$ if one used the "old" or the "Hui" value for π, respectively). And indeed, this deduction can be confirmed: Problem 53-1 of the famous Jinkōki [塵劫記][2] by Yoshida Mitsuyoshi [吉田光由] gives 48 *chi* as the diameter of a sphere of volume 62208 chi^3, again corresponding to a value of 9/16 of the ratio $t=V/d^3$ (and again illustrating the invalid mixing of approximate and precise numbers). In the commentary to this problem in the modern Japanese edition of the Jinkōki [3, p. 149] it is stated that this value is taken from the Unified Foundations of Mathematics [算法統宗 *Suanfa Tongzong*] of Cheng Dawei [程大位], which was published in 1592, 35 years before the Jinkōki but almost 300 years after the Introduction to Mathematics, while according to the

[2] Jinkō [塵劫] is a Buddhist term meaning an eternal time or an extremely large number. (editor)

English edition [4, p. 178] it can be found already in the Nine Chapters. Tables of the successive (Occidental or Oriental) values of π can, of course, be found in many places in the literature, but they are always based on the "π" of the circle (defined either as the circumference divided by the diameter or as the area divided by the square of the radius, the equality of these two definitions having been known for a very long time), not for the sphere. This suggests the following questions:

Q4. When was the value 9/16 for t first used, and where did it come from? (It is not a particularly good value, being considerably further from the correct value $\pi/6$ than the simpler fraction 1/2.) When was it first improved?

Q5. When was the relation $C = 4\pi/3$ between the constants for the sphere and for the circle first found (the relationship $V = A/3$ between the volume and the surface area of a sphere of radius 1, which follows by considering an inscribed polyhedron with small triangular faces, is much easier and was presumably known earlier), and was it discovered independently in both the East and the West? In any case the correct formulas were known to Takebe [8, p. 266].

The final—and perhaps most interesting—aspect of Zhu's problem which we wish to discuss is that a very complicated input, with values of 3 and 3.14 for π and a total "accumulation" of $33622\frac{37}{200}$, leads in the end to a very simple answer in which $a, \ldots, e$ are all two-digit whole numbers. This, too, is typical of the ancient Chinese and Japanese mathematical texts: the problems are almost always "fixed" so that, even when the statements are very complicated, the numerical solutions are simple. This is on the one hand nice for the solver, who knows when he has found the right answer, but on the other hand unfair since the author, who usually gives no method of derivation for the solution, starts out knowing the answer and therefore does not actually need to possess any method that would work in general. Moreover, problems "fixed" in this way can also be solved in an easy way, surely not intended: if one knows that the answer is going to come out in small integers, one can find it without really "solving" the problem at all by trial and error, e.g. here by taking one of the unknowns (say a, the side of the cube) as celestial element [天元 *tianyuan*] or chosen independent variable, expressing the other unknowns in terms of this one (here $b, c, d, e = a + 4, 3a + 12, a, a - 3$), and then trying each integer value $a = 4, 5, \ldots$ in turn until finding the one for which the total accumulation $a^3 + tb^3 + \ldots$ takes on the specified value. From our point of view the whole process seems questionable—the author has "cheated" by working from the answer to the problem rather than inversely, and has given his readers or students the possibility of "cheating" by guessing rather than calculating the answer required—but presumably it did not seem so at the time. In any case, we can ask:

Q6. At what point or by what stages did the shift between problems constructed from known solutions and problems solvable by generally applicable methods occur? Was this distinction ever recognized explicitly by mathematicians of the period?

1.2 The formula for arc length in the Jugairoku

In the Jugairoku [竪亥録][3] by Imamura Tomoaki [今村知商], published in 1639, formulas are given relating the diameter, chord, sagitta ("arrow"), and arc of a circle segment. These formulas, discussed in detail by Horiuchi [8, pp. 34–38] are as follows:

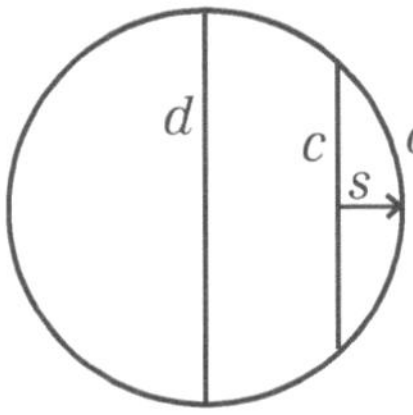

$d =$ diameter [径 *kei*]	$d = s + c^2/4s\,,$
$c =$ chord [弦 *gen*]	$c = \sqrt{4s(d-s)}\,,$
$s =$ sagitta [矢 *shi*]	$s = \big(d - \sqrt{d^2 - c^2}\big)/2\,,$
$a =$ arc [弧 *ko*]	$a^2 = 4s\,\big(d + s/2\big)\,.$

The first three of these are equivalent to the formula $(c/2)^2 + (d/2 - s)^2 = (d/2)^2$, which follows from Pythagoras's theorem, and are exact, whereas the last is an approximation. (The exact formula is of course transcendental and requires infinite series, as Takebe discovered in 1722.) No derivation for this formula is given. In [7], a partial explanation is suggested, namely, that this formula yields the values $a = 0$ for $s = 0$, which is obviously correct, and $a^2 = 5d^2/2$ if $s = d/2$, which is correct if one uses $\pi = \sqrt{10}$, as Imamura did. But this explanation is not sufficient: there are even simpler formulas which would give these two special cases (e.g. $a^2 = 10s^2$), but they would give very poor approximations for other values of s, whereas Imamura's formula is uniformly very good, with an error never exceeding one part in 75. Horiuchi also goes on to say that the Chinese tradition was to use quadratic interpolations, but even then one has a choice of formulas, since any formula of the form $a^2 = ksd + (10 - 2k)s^2$ would give the "right" special values for $s = 0$ and $s = d/2$. One can then ask on what mathematical basis, short of carrying out the much more complicated analysis later given by Takebe and others, one might obtain this simple and very good formula. Here we can offer two explanations, with the hope that experts will be able to say whether one of them might correspond to the procedure Imamura actually used. The first is to consider the asymptotic behavior when s tends to 0, rather than merely the value at $s = 0$.

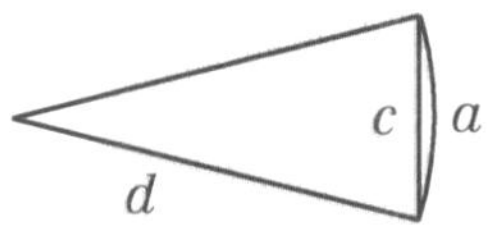

It is obvious from a picture that, when s is very small, a is approximately equal to c, so Imamura's second formula gives $a^2 \approx c^2 \approx 4sd$ and hence $k = 4$.

[3] Jugai [竪亥] is the name of a legendary person in China. (editor)

This explanation seems simple and palatable to us, but it is not at all clear that such an argument would have been natural to a Japanese mathematician of the period, and indeed from the talk given by Prof. Morimoto at the Seki memorial conference it seems that even the far more sophisticated Takebe did not employ such asymptotic arguments, but used only precise values based on polygons [12, p. 355]. In the case at hand, one can imagine that one simply looks at the pictures for a square and a hexagon, respectively, embedded in a circle. In both cases all the dimensions involved are obvious either *a priori* or from Pythagoras's theorem, and (using Imamura's value $\pi = \sqrt{10}$) one obtains the two further known values

$$\frac{d}{c} = \sqrt{2} \quad \Rightarrow \quad \frac{a}{c} = \frac{\pi\sqrt{2}}{4} = 1.118, \qquad \frac{d}{c} = 2 \quad \Rightarrow \quad \frac{a}{c} = \frac{2\pi}{6} = 1.054.$$

Imamura's formula gives 1.122 and 1.052, respectively, for these two special cases, in excellent agreement with these "exact" values.

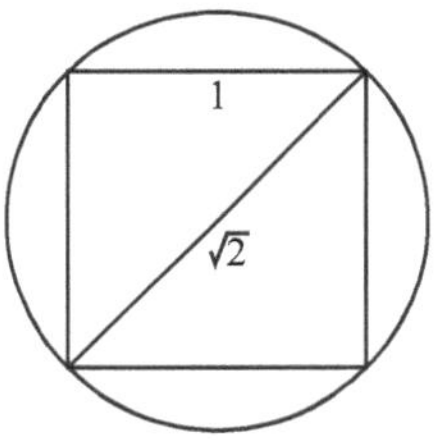

$$d = \sqrt{2},\quad c = 1,\quad s = \frac{\sqrt{2}-1}{2},\quad a = \frac{\pi\sqrt{2}}{4},$$

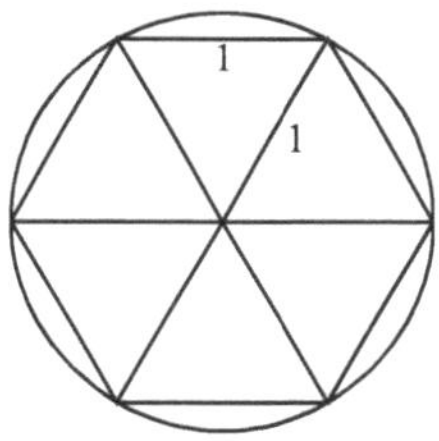

$$d = 2,\quad c = 1,\quad s = 1 - \frac{\sqrt{3}}{2},\quad a = \frac{2\pi}{6}.$$

Of course, it may be that neither of these explanations is correct. Horiuchi mentions that the formula $a^2 = c^2 + 6s^2$, equivalent to Imamura's, is given by Isomura Yoshinori [礒村吉徳] in the Mathematical Methods without Doubts [算法闕疑抄 *Sanpō Ketsugishō*] of 1659 as being based on an unspecified "rule of augmentation" [増術 *zōjutsu*], and also suggests an earlier interpolation method of the Chinese calendarists as a possible, but non-verifiable, source for Imamura's formula. In any case, we can formulate:

Q7. Is there any evidence that could help one decide where the formula for the arc length in the Jugairoku came from, and in particular whether it was based on the considerations of the special cases $c/d = 1/\sqrt{2}$ and $c/d = 1/2$ or on the limiting behavior as c/d tends to 0?

1.3 The value of π in the Sanso

In the Sanso [算組] (or Mathematical Chopping-Board), published in 1663, Muramatsu Shigekiyo [村松茂清] computes π by using polygons of 8, 16, 32, ..., $2^{15} = 32768$ sides to obtain the series of approximations:

$$
\begin{aligned}
p_3 &= \underline{3.}061467\\
p_4 &= \underline{3.}121445\\
p_5 &= \underline{3.1}36548\\
&\ \vdots\\
p_{13} &= \underline{3.141592}5765\\
p_{14} &= \underline{3.141592}63433\\
p_{15} &= \underline{3.1415926}487
\end{aligned}
$$

The method is fine, although not original (it goes back to the Nine Chapters, and of course in the West to Archimedes), but there are several odd things about Muramatsu's use of it that seem worth commenting on:

(a) Muramatsu calculates to a very high precision, ending up with 8 correct digits of π, but, at least according to the discussion in [7], only seems to care about the first three digits, writing that we can neglect the digits 1 and 6 (of 3.1416) but that we do not find any deviation from 3.14 (the value that he ascribes to "Master Meng [孟] from Jin [晋] and Liu Hui from Wei [魏]," although, as already mentioned, Liu Hui actually also gave 3.1416 and Zu Chongzhi [祖沖之] had given the much more accurate approximation 355/113 in the 5th century). To a modern mind it seems incredible that one would go to such lengths just to decide between two much older and primitive values, and discard most of what one had calculated with such efforts, bringing us back to the question **Q1** formulated in connection with Zhu Shijie's use of π.

(b) The calculations are carried out to unnecessarily high precision: Muramatsu calculates the sides of each of the successive polygons to 20 decimal digits, even though he could surely have seen after the first few steps that he was gaining only about half a digit of precision with each iteration and therefore, if he was planning to stop with the 2^{15}-gon, was not going to get more than about 8 digits anyway. This is a further instance of the same lack of understanding of the meaning of precision that we already saw in the problem by Zhu Shijie (question **Q3**).

(c) On top of this he was very lucky, because he made three numerical mistakes in his calculation (each of which, of course, persists through all further iterations of the doubling procedure), but each of them occurred only after the 10th digit and therefore did not affect the correctness of the 8 digits which his calculation was in principle capable of producing.

(d) Most importantly, he failed to give any discussion or do any analysis, even a crude one, of the numbers he obtained, not even saying explicitly that they seemed to be converging to a well defined limit or that the persistence of certain digits (those shown underlined in the display above) suggested the correctness of the first 8 digits he had obtained. Had he been the genius that Takebe was, he would have thought of looking at the successive differences (and then perhaps even at the differences of these differences, and then again at the differences of these) and recognized the rules governing these differences, thus obtaining, as Takebe did, many more digits of π without needing to calculate polygons of any larger size than he already had calculated.

1.4 Seki's solution of a problem of the Kokon Sanpōki

In the Hatsubi Sanpō [發微算法] (translatable as something like "Mathematics with Humble Determination"), the only work which Seki Takakazu [関孝和] published during his lifetime, he gave solutions for all of the fifteen bequeathed problems [遺題 *idai*] that had been posed by Sawaguchi Kazuyuki [澤口一之] in the Treatise of Ancient and Modern Mathematics [古今算法記 *Kokon Sanpōki*] of 1671 as a challenge to future mathematicians. We make a few observations here on his solution to the 4th problem, and then in Subsection 1.5 discuss some peculiar features of the problem itself as well as of Problem No. 14, the most complicated of the fifteen.

Problem 4 is stated in the Treatise of Ancient and Modern Mathematics in the following terms:

> 今有 甲乙丙立方各一。只云 甲積与乙積相併共寸立積十三万七千三百四十坪。又乙積与丙積相併共寸立積十二万千七百五十坪。別甲方面寸為實開平方之見商寸与乙方面寸為實開立方之見商寸及丙方面寸為實開三乗方之見商寸各三和一尺二寸。問甲乙丙方面各幾何。

> We have now A, B and C which are each a cube. It is told only that the volume of A and the volume of B together are 137 340 *tsubo*[4], and also that the volume of B and the volume of C together are 121 750 *tsubo*. Furthermore it is told that the quotient in *sun*[5] obtained by placing the side of A as the dividend and opening the square, the quotient in *sun* obtained by placing the side of B as the dividend and opening the cube, and the quotient in *sun* obtained by placing the side of C as the dividend and opening the square multiplied three times, make together 1 *shaku* and 2 *sun*. One asks for the values of the sides of A, B and C.

Stated in modern notation, the problem asks us to find (numerically) the values of three numbers a, b and c satisfying the simultaneous equations

$$a^3 + b^3 = 137340, \qquad b^3 + c^3 = 121750, \qquad \sqrt[2]{a} + \sqrt[3]{b} + \sqrt[4]{c} = 12. \qquad (1)$$

Seki's solution, discussed in detail in [7, Chapter 6] (whose notations we follow), is truly amazing. He first replaces the three concrete numbers 137340, 121750 and 12 by letters, say N, N' and N''. He then chooses as *tianyuan* [天元] or basic independent variable the quantity $x = \sqrt[4]{c}$ and defines

$$m = N'' - x, \qquad n = N' - x^{12}, \qquad o = N - n$$

so that $m = \sqrt[2]{a} + \sqrt[3]{b}$, $n = b^3$, $o = a^3$ and the problem has been reduced to finding x such that $m = \sqrt[6]{o} + \sqrt[9]{n}$. So far, nothing very surprising. But now he introduces the six new quantities

---

[4] *tsubo* [坪] is a unit of area and volume. *A tsubo* means normally about $4m^2$ but it is used here to mean 1 sun^3. (editor)

[5] *sun* [寸 *cun* in Chinese] is a unit of length equal to 0.1 *shaku*≒ 3cm. (editor)

$$
\begin{aligned}
p &= 36m^{14} + 9m^2o^2\,, \\
q &= 252m^5n + 126m^8o\,, \\
r &= 126m^{10}o\,, \\
s &= 9m^{16} + 72m^7n + 18mno + 36m^4o^2\,, \\
t &= m^{18} + 84m^6o^2 + n^2\,, \\
u &= 2m^9n + 168m^3no + 84m^{12}o + o^3
\end{aligned}
$$

and then gives the answer to the problem in the form of the remarkable formula

$$
\begin{aligned}
&o^2p^3 + 3o^2pq^2 + 3opru + 3opst + 3oqrt + 3oqsu + 3ors^2 + or^3 + t^3 + 3tu^2 \\
&= 3o^2p^2q + o^2q^3 + 3oprt + 3opsu + 3oqru + 3oqst + 3or^2s + os^3 + 3t^2u + u^3
\end{aligned}
$$

(here some misprints in [7] have been corrected), which when multiplied out gives an equation of degree 108 for x. Of course he gives no justification for the correctness of this solution, let alone any indication of how one should go about finding it—this was done later by Takebe Katahiro [建部賢弘] in the *Hatsubi Sanpō Endan Genkai* [發微算法演段諺解] or "Commentaries in the Vernacular on the *Hatsubi Sanpō*," published in 1685—but it is indeed correct. The following aspects strike us particularly:

(a) The solution is incredibly complicated. Even verifying its correctness is tedious and would be seen by most mathematicians today as something that cannot reasonably be done without using an electronic computer, while to find this solution from scratch would appear to a modern mathematician to require modern algebraic tools like Gröbner bases.

(b) Seki doesn't actually write out the full equation of degree 108 for x, but merely says that one can do so if so desired and then solve it by the standard Chinese method. One can wonder—and the question is posed explicitly in [7]—whether he himself wrote out or solved this equation. The answer is surely negative, for at least three reasons: First of all, the calculation is so complicated that it would surely have defied even his extraordinary computational abilities, since the polynomial equation in question has the form

$$
\begin{aligned}
&x^{108} - 9x^{102} + 648x^{101} - 19440x^{100} + 312030x^{99} - 2835000x^{98} + \cdots \\
&\cdots + 812692048405755416419319591620935800302655616965774072340 48x \\
&- 1795038440510576073588274626088072882897678805742137464 3904 = 0\,.
\end{aligned}
$$

Secondly, no matter how modest he was, he would surely have given some indication of having done such a mammoth calculation if he had actually performed it. Thirdly, the solution as he wrote it out contains one or two minor inaccuracies (one coefficient and one exponent are written incorrectly), so that if he had actually multiplied everything out and found the value of x numerically "by the Chinese method" he would have discovered that it failed to satisfy the conditions of the original problem.

(c) Less trivial is the remark that it would in any case have been superfluous to do this numerical computation, and that for the same reason Seki's entire solution, brilliant though it is, is totally unnecessary if one merely wants to solve the given problem numerically, which is ostensibly his goal (and anyway is the best one can hope for, since, as we know today, polynomial equations of high degree cannot in general be solved exactly in closed form). Indeed, suppose that one were to make the effort and write out the polynomial $P(x)$ of degree 108 completely. The "Chinese method" consists of either binary interpolation or some form of Newton's method, so in its crudest (but numerically quite sufficient) form would consist in finding two values of x for which $P(x)$ takes on opposite signs and then repeatedly bisecting the interval they define and retaining only the half which includes a sign change. But this method works perfectly well for the original problem, without the necessity of any elimination theory at all! Indeed, taking the "*tianyuan*" to be (say) $x = \sqrt[2]{a}$ (here Seki's choice $x = \sqrt[4]{c}$ would serve equally well), we can rewrite the original problem as $f(x) = 12$, where $f(x) := x + \sqrt[9]{137340 - x^6} + \sqrt[12]{x^6 - 15590}$. The algebraic function $f(x)$ could be calculated numerically just as easily as the polynomial function $P(x)$, since the four arithmetic operations and the numerical extraction of square and cube roots with the use of counting rods [算木 *sangi*] were familiar procedures at Seki's time, and by calculating successively the values

$$\begin{array}{lll} f(5) \approx 10.02 & f(6.0) \approx 11.9228 & f(6.06) \approx 11.9877 \\ f(6) \approx 11.92 & f(6.1) \approx 12.0300 & f(6.07) \approx 11.9983 \\ f(7) \approx 12.61 & f(6.2) \approx 12.1327 & f(6.08) \approx 12.0089 \\ \vdots & \vdots & \vdots \end{array}$$

one would obtain after just a few steps an accurate numerical value for x and hence also for a, b and c:

$$\begin{aligned} x &= \ \ 6.0715851750416302735 7\cdots \\ a &= 36.8641465377853041287 2\cdots \\ b &= 44.3516761876674573034 9\cdots \\ c &= 32.5563821163811031295 8\cdots \end{aligned}$$

In other words, it is just as easy to solve the original problem as it is to solve the "simpler" one to which Seki reduces it, and by exactly the same method. Whether Seki was aware of this, of course, must remain moot. We can nevertheless ask:

Q8. Were interpolation methods or "Newton's method" ever used by *wasan* mathematicians for the numerical solution of (non-polynomial) algebraic equations?

(d) Finally, we pose a question concerning an algebraic aspect of Seki's solution. To perform his elimination, he needs to repeatedly transform algebraic equations of the form $Q(x, \sqrt{y}) = 0$ or $Q(x, \sqrt[3]{y}) = 0$, where y is a polynomial in x, into purely polynomial equations for x. The method to do this (stated by Takebe in the *Endan* to the *Hatsubi Sanpō* and mentioned by Prof. Komatsu in his talk at this conference [11]) is based on the two algebraic identities

$$A+B=0 \Rightarrow A^2-B^2=0, \tag{2}$$
$$A+B+C=0 \Rightarrow A^3+B^3+C^3-3ABC=0. \tag{3}$$

Then in the first case we can split up the polynomial $Q(x,\eta)$ into even and odd powers of $\eta=\sqrt{y}$ to rewrite the given equation as $Q_0(x,\eta^2)+\eta Q_1(x,\eta^2)=0$ and use (2) to replace this by the purely polynomial equation $Q_0(x,y)^2-yQ_1(x,y)^2=0$, and similarly in the second case split up $Q(x,\eta)$ according to the values modulo 3 of the exponents of $\eta=\sqrt[3]{y}$ to rewrite the equation as $Q_0(x,\eta^3)+\eta Q_1(x,\eta^3)+\eta^2 Q_2(x,\eta^3)=0$ and use (3) to replace this by the polynomial equation $Q_0(x,y)^3+yQ_1(x,y)^3+y^2Q_2(x,y)^3-3yQ_0(x,y)Q_1(x,y)Q_2(x,y)=0$. The identities (2) and (3) can be verified easily by using the factorizations $A^2-B^2=(A+B)(A-B)$ and $A^3+B^3+C^3-3ABC=(A+B+C)(A^2+B^2+C^2-AB-AC-BC)$, but the second of these is not obvious and in any case does not explain on what basis the expression $A^3+B^3+C^3-3ABC$ was found originally. More synthetic proofs of (2) and (3), not requiring one to know the answers in advance, can be given using substitution rather than factorization:

$$\begin{aligned} B=-A \quad &\Rightarrow \quad B^2=(-A)^2=A^2, \\ C=-A-B \quad &\Rightarrow \quad C^3=(-A-B)^3=-A^3-3A^2B-3AB^2-B^3 \\ &\qquad\quad =-A^3-B^3+3ABC. \end{aligned}$$

Q9. Can one determine whether the equations (2) and (3) for elimination of square and cube roots were first obtained by *wasan* mathematicians by factorization, by substitution, or by some other method? Cf. [8, p. 150].

1.5 On Problems 4 and 14 of the Kokon Sanpōki

In the last subsection we discussed Seki's solution of Problem 4 of the *Kokon Sanpō ki*, but not where the problem itself comes from. We observed above that the Chinese and Japanese tradition usually involved "fixing" problems in advance so that, even if the data in the problem was complicated (like the number $33622\frac{37}{200}$ in Zhu's problem), the answer came out in simple integers (like 24, 28, … in that problem). But here the situation is different: the two numbers 137340 and 121750 in equation (1) are just as complicated as the number occurring in Zhu's problem, but the solution, as we saw above, is very far from being integral. Why, then, did Sawaguchi choose the specific numbers 137340 and 121750? The following simple numerical experiment suggests that they were not random choices. We look at the decompositions of 12 as a sum of three positive integers (there are only 55 of them, so this is easily done), and in each case define a, b and c as the square, cube, and fourth power, respectively, of the three summands. Then in one case, viz., $a=49$, $b=27$, $c=16$, we find that two of the sums of their cubes differ by only a few units from the numbers in Sawaguchi's problem. True, the sums of cubes in question are the wrong ones, since for these values of a, b and c one has

$$a^3 + b^3 = 137332\,, \qquad a^3 + c^3 = 121745\,, \qquad \sqrt[2]{a} + \sqrt[3]{b} + \sqrt[4]{c} = 12 \qquad (4)$$

instead of (1), with a rather than b in the second equation. But the probability of this near-coincidence being accidental is astronomically small, so we can ask:

Q10. Was Problem 4 in the *Kokon Sanpōki* in some way based on equation (4)?

Obviously, there is no way to decide this for sure (Sawaguchi, like everybody else in this business, kept his secrets well), but if we take as a working hypothesis that the two equations are in fact connected and try to imagine how the connection might look, we may be led to relationships that can be tested in others of Sawaguchi's problems. We first observe that, if Sawaguchi had simply started out (*à la* Zhu) with the integral values $\sqrt[2]{a} = 7$, $\sqrt[3]{b} = 3$, $\sqrt[4]{c} = 2$ to obtain the three relations (4), the problem obtained would have been much too simple for him to include in his collection of 15 *idai*: a potential solver would only need to calculate $\sqrt[3]{121745} = 49.5\cdots$ and then try subtracting 49^3 from 137332 and 121745 to discover that the differences were a 9th and a 12th power, respectively, making the solution immediate. So we can imagine that Sawaguchi might have changed the problem a little to make it non-trivial, in one of two ways:

(a) replace the numbers "137332" and "121745" in (4) by very similar numbers, say 137340 and 121750;

(b) and/or replace the left-hand side of the second equation in (4) by $b^3 + c^3$.

If he had done merely (a), this would have had the advantage that he could be fairly sure that the new problem, even if he did not know how to solve it analytically, was non-defective (i.e., had a unique real solution). Doing (b) as well, to give the problem as actually stated in the *Kokon Sanpōki*, would not have this property, and it is quite possible that Sawaguchi simply made a mistake in transcription. In that case, the problem he meant to give (namely, (1) with $a^3 + c^3$ in the middle equation) would have had a solution in numbers very close to integers:

$$\sqrt[2]{a} = 7.00007044\cdots\,, \qquad \sqrt[3]{b} = 3.00001517\cdots\,, \qquad \sqrt[4]{c} = 1.99991437\cdots\,.$$

If this hypothesis is true, then we would expect some others of his problems to have the same property, and indeed, we will now see that this is the case.

Problem 14, discussed in detail at the lecture by Professor Komatsu [11] at this conference, is the most complicated of the 15 *idai* in the *Sanpōki*, since it leads (as Seki showed) to a polynomial equation of degree $2 \times 3^6 = 1458$. The problem requires finding four points in the plane such that the differences of the cubes of the six distances x, y, z, u, v, w between them have given values:

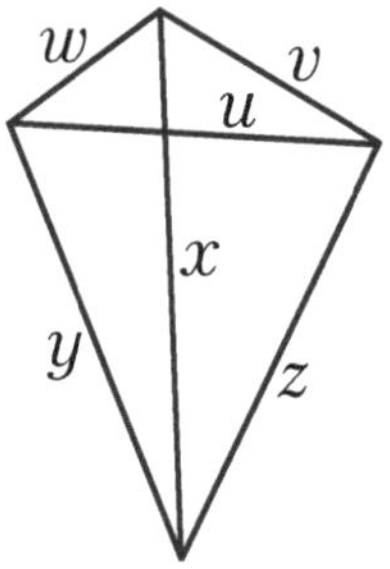

$$\begin{aligned} x^3 - y^3 &= 271\,, \\ y^3 - z^3 &= 217\,, \\ z^3 - u^3 &= 60.8\,, \\ u^3 - v^3 &= 326.2\,, \\ v^3 - w^3 &= 61\,. \end{aligned} \tag{5}$$

These five equations alone would lead to an indeterminate problem in the six unknowns and must be supplemented by a sixth equation

$$\begin{aligned} P(x,\ldots,w) := (u^2+v^2-w^2)(x^2y^2+w^2z^2) + (v^2+w^2-u^2)(y^2z^2+u^2x^2) \\ +(w^2+u^2-v^2)(z^2x^2+v^2y^2) - u^2x^4 - v^2y^4 - w^2z^4 - u^2v^2w^2 = 0 \end{aligned}$$

expressing the fact that the four vertices are coplanar. (The polynomial P, divided by 144, gives the square of the volume of a tetrahedron with edges of length $x,\ldots,w$ and hence vanishes if the vertices lie in a plane.) This last equation, as discussed in [11], is contained in the *Sanso* and hence was known to both Sawaguchi and Seki.

In his talk, Professor Komatsu cited the numerical solution

$$\begin{aligned} &x = 10.0000056403\,, && y = 9.0000069815\,, && z = 8.0000083910\,, \\ &u = \;7.6699093899\,, && v = 5.0000228360\,, && w = 4.0000359240\,. \end{aligned}$$

calculated by Dr. Kinji Kimura (cf. [9]) using Gröbner bases and a numerical computation library. These approximate values certainly fit in with the prediction of being, in five cases out of six, very close to integers. Let us suppose that Sawaguchi started with the arbitrarily chosen simple integral values

$$x = 10, \quad y = 9, \quad z = 8, \quad v = 5, \quad w = 4 \tag{6}$$

for five of the unknowns. Then substituting them into the equation $P(x,\ldots,w) = 0$ he would have ended up with the quadratic equation

$$100\,u^4 - 6065\,u^2 + 10726 = P(10,9,8,u,5,4) = 0$$

for u^2, giving the numerical value

$$u = \sqrt{\frac{1213+69\sqrt{273}}{40}} = 7.6698551213\cdots \tag{7}$$

for the sixth unknown u. These numbers would have led in turn to the values

$$\begin{gathered} x^3 - y^3 = 271\,, \quad y^3 - z^3 = 217\,, \quad z^3 - u^3 = 60.80790567\cdots, \\ u^3 - v^3 = 326.19209432\cdots, \quad v^3 - w^3 = 61 \end{gathered}$$

for the data of the problem. These are strikingly similar to the numbers in (5), the only difference, apart from the degree of precision chosen for the non-integral numbers, being that one has 60.8 instead of 60.08. This suggests:

Q11. Was the number 60.08 in Sawaguchi's problem originally 60.8 ? Was the intended solution the one given by (6) and (7)?

Here we have a definite prediction that can be checked against the documentary evidence. In the facsimile copy of the *Hatsubi Sanpō* [2], the number in question is given as 60.08, but of course Seki might easily have made an error of transcription in copying from the *Kokon Sanpōki*. To check this, Prof. Komatsu kindly asked the director of a *wasan* exhibition at Tokyo University of Science that was being held concurrently with the conference to show us an original edition of the *Sanpōki* that was on display there. Rather disappointingly, the number there also had a clear "0" [零 *rei*] before the "8". However, later we noticed that the problem as cited in 1914 by Smith and Mikami [15, p. 101] indeed gave the relevant number as "60.8," which suggests that there must be different original editions of the *Sanpōki* and that there was a copying error at some stage. Furthermore, the numerical solution of the problem as given in (5) is

$$x = 9.9977624076\,,\qquad y = 8.9972373104\,,\qquad z = 7.9965030158\,,$$
$$u = 7.6660952225\,,\qquad v = 4.9910355706\,,\qquad w = 3.9859690172\,,$$

while the solution for the problem with 60.08 replaced by 60.8 is

$$x\ = 10.0000057162\,,\quad y = 9.0000070570\,,\qquad z = 8.0000089315\,,$$
$$u = 7.6699096344\,,\qquad v = 5.0000228647\,,\qquad w = 4.0000357259\,.$$

The latter numbers are very much closer to the ones given by Kimura, suggesting that he, too, must have used a source in which the third number in Sawaguchi's problem appeared as 60.8 rather than 60.08.

We have given this numerical analysis at some length because the details of the various numerical coincidences give concrete and rather convincing support to hypotheses which would otherwise by their nature be somewhat speculative. To go further, of course, one should search for more cases of the same phenomenon. Our next question is therefore:

Q12. Are there other problems, by Sawaguchi or other authors of the time, whose solutions involve numbers very close to integers and which thus may have been constructed starting from integral solutions?

We mention one final point. If the above reconstruction is correct and Sawaguchi really intended his problem to have the solution given by equations (5) and (6), then one can wonder whether he might not have attempted to give a version having a solution consisting *entirely* of integers. To do this, he would have had to find a solution in integers of the indeterminate problem $P(x,\ldots,w) = 0$, and could then simply have given the values of the differences $x^3 - y^3, \ldots, v^3 - w^3$ as the data in the other five equations of his problem. In fact the equation $P(x,\ldots,w) = 0$ does have various integral solutions, examples being $(x,y,z,u,v,w) = (17,13,8,9,11,5)$,

$(22,17,19,4,8,6)$, or, with $x > y > z > u > v > w$, $(27,24,17,13,11,9)$. Of course, these are not particularly easy to find without a computer, although they are easy enough to verify numerically, but there are many other problems (in particular, those concerning Pythagorean triples, i.e., right triangles with sides of integral length) whose formulations in terms of geometry that was familiar at this epoch would have been quite natural and which can be solved by hand, either by systematic analysis or by trial and error. This suggests our final question:

Q13. Were any problems of what we now call Diophantine analysis ever considered by Japanese mathematicians during the Edo period?

2 Did Seki or Takebe learn any mathematics from the Dutch?

The mathematics of Seki and Takebe, seen from the point of view of the twenty-first century, is a "quantum jump" beyond of that of their Chinese or Japanese predecessors, and it is of course a very natural question whether this progress was due entirely to the natural genius of these two mathematicians or whether they were directly or indirectly inspired by the mathematics done in Europe during the preceding decades. Western authors in the early twentieth century tended to say that there of course must have been such an inspiration. But they give little evidence, and a modern reader cannot help feeling that this is often simply a reflection of their *a priori* belief that the miracles of Western science could not have been discovered by the members of such a different culture.

For instance, George Sansom, in his well-known book *Japan, A Short Cultural History* [13], quotes Engelbert Kaempfer, who came to Nagasaki in 1690 as a physician for the Dutch, as having said that the Japanese "had little taste for speculative philosophy, which ... they thought an amusement proper for lazy monks" and then goes on to say

> The crowning triumph of the Western intellect, the great gift which at that time Europe might have made to them, they were either unprepared or unwilling to receive, for, to quote the same authority, they knew "nothing of mathematics, more especially of its deeper and speculative parts" [13, p. 421].

In a footnote [13, p. 428], he adds that—whereas the Chinese seemed to have made few advances after their discoveries of the 12th and 13th centuries—

> The Japanese worked out an original method of the differential calculus from hints coming through the Dutch, and in general they appear to have displayed remarkable ingenuity in application of a limited knowledge; but Kaempfer's judgement as to their backwardness in theory seems to have been correct.

But he gives no evidence at all for his assertion that the progress made by the Japanese was based on "hints coming from the Dutch."

Similarly, in the well-known history of Japanese mathematics by Smith and Mikami, we find a somewhat vague reference to "others whose names are not now remembered" who "might have formed a possible medium of communication with the West in the time of Seki," followed by the more specific passage

> . . . we have the record of two men who were in touch with Western mathematics. These men were Hayashi Kichizaemon and his disciple Kobayashi Yoshinobu, both of them interpreters in the open port of Nagasaki. Each of these men knew the Dutch language, and each was interested in the sciences, the latter being well versed in the astronomy of the West. ... While it is probable that these men did not know much of the European mathematics of the time, it is inconceivable that they were unaware of the general trend of the science, and that they should fail to give to inquirers some hint as to the nature of this work. [15, pp. 140–141]

and then, in summary, the following:

> The conclusion appears from present evidence to be that some knowledge of European mathematics began to find its way into Japan in the seventeenth century; that we have no definite information as to the nature of this work beyond the fact that mathematical astronomy was part of it; that there is no evidence that Seki or his school borrowed their methods from the West; but that Japanese mathematicians of that time might very well have known the general trend of the science and the general nature of the results attained in European countries [15, pp. 141–142].

But the authors give little evidence, either here or later, to substantiate the last of these assertions, and in general it has to be said that this book, despite the eminence of the first author as a historian of Western mathematics and the preeminence of the second author as a historian of Japanese mathematics, is marred in many places by the Western prejudices of its first author (cf. [8, p. XXVI, footnote 3]).

The question about Western influence obviously cannot be resolved easily, at least without further evidence coming to light, and certainly not by the present authors. But we would like to add a few intrinsic arguments concerning the various "hints from the Dutch" theses, and then give some concrete documentary evidence supporting the opposing viewpoint, that Seki and Takebe had not had any direct contact with the mathematics which had been done in Europe between the closing of Japan in 1639 and the lifting of the partial ban on foreign books in 1720.

The first and most obvious argument is that at least some of the discoveries of Seki and Takebe cannot have come from Western sources simply because they predated them. The most notable of these is of course Seki's independent discovery of determinants, which were found also by Leibniz in 1693 but published only after his (and *a fortiori* after Seki's) death, and which were treated in much greater generality by Seki; here even Smith and Mikami speak of a "marked proof of Seki's genius" and concede that "Seki was not only the discoverer but that he had a much broader idea than that of his great German contemporary" [15, pp. 124–125]. A further example are the Bernoulli numbers, which Seki discovered independently of Jacob Bernoulli and actually published earlier (the publications by Seki and Bernoulli, both posthumous, are from 1712 and 1713, respectively). This example is in some ways even more striking than the first, since determinants are an essential mathematical tool which must inevitably be discovered when one studies systems of linear equations (as had been done in the East for centuries), while Bernoulli numbers constitute a far more specific discovery that is not an inevitable consequence of any particular "general trend in the science." Seki's theory of elimination, too, although undoubtedly based in part on Chinese precedents, goes much further than anything that European mathematicians could have done at the time, or indeed for the next hundred years. Knowing that in these cases the discoveries were his, why should we doubt that other discoveries that Seki made, even when these occurred later than in the West, were wholly his own?

Secondly, and no less importantly, the level of the mathematical advances we are speaking of is so high that it is hard to imagine how they could have been transmitted by general osmosis or "hints." Before the lifting of the ban, these hints would have had to come via the Dutch. But, as Goodman, who studied the Dutch-Japanese interaction more intensively than any other historian, says,

> Further, the Dutch who came to Japan were hardly trained scientists and were no more able to respond to scientific inquiries abroad than they would have been at home ... the overwhelming proportion of Japanese contact with Hollanders was with the average employees of the Dutch East India Company ... [6, pp. 64–65].

And even if one imagines, despite the inherent unlikeliness and the lack of any evidence to this effect, that at some point there was a member of the Dutch contingent who was versed in the state-of-the-art mathematics of the day and had Japanese interlocutors capable of and interested in learning this material, it would still have been impossible for the former to transmit his knowledge to the latter because of the linguistic barriers involved. One must remember that, while in the 16th century there were a number of Europeans (principally Jesuit missionaries) who had spent years in Japan and had mastered the language, and of Japanese who had become fluent in various European languages, the situation was completely different during the first half of the Edo period, when the *bakufu* followed a deliberate policy of preventing linguistic competence on either side, demanding that members of the Dutch contingent (in particular, the *opperhoofd* or ship's captain) be replaced every year so that they could not learn too much about Japan and themselves providing Japanese interpreters with a very inadequate knowledge of the Dutch language. There was no serious training of interpreters before the lifting of the ban, and no Dutch-Japanese dictionary until 1745. That sophisticated abstract mathematical ideas like interpolation or the series development of functions could have been discussed in such a context seems very unlikely.

In a related vein, one must remember that not only the language, but also the backgrounds, the styles of presentation and above all the aims of the research itself were so different in the two cultures that it is by no means clear that a Japanese mathematician, exposed in an unsystematic way to a piece of Western mathematics, would have been able to appreciate it (just as, of course, a Western mathematician would not have been able to appreciate Japanese mathematics without long exposure to it). And indeed, even as late as 1811 a Japanese mathematician who *had* seen Western mathematics was able to write with obvious conviction that "foreign mathematics is not on so high a plane as the mathematics of our own country," (quoted in [14, p. 172]).

Of course, European mathematical knowledge transmitted in *written* form might have been understood by a qualified mathematician working imaginatively against the language barrier. But despite the vague comments of Sansom or of Smith–Mikami, there seems to be no evidence that such texts were available before 1726, the date of the importation of the Encyclopedia of Calendrical Mathematics [暦算全書 *Lisuan Quanshu*] by Mei Wending [梅文鼎], into which Takebe had the marks necessary for Japanese to read the Chinese text inserted. Thus, when Smith (we assume that it was he) challenges the credit given to Takebe for his infinite series for the square of the arc, writing

> The series seems, however, to have been given by Pierre Jartoux, a Jesuit missionary, resident in Peking ... There is a tradition that Jartoux gave nine series, of which three were transmitted to Japan, and it seems a reasonable conjecture that Western learning was responsible for his work, that he was responsible for Takebe's series, and that Takebe explained the series as best he could [15, pp. 154–155],

he gives a few internal arguments for this thesis, such as the somewhat awkward and unconvincing presentation of the series in Takebe's text, but again no documentation that Takebe had seen the series that were transmitted to Japan. (In fact, a footnote to the passage says that the three series in question appeared in a book by Mei Kucheng [梅瑴成] with unknown date and *without* evidence that they reached Japan in this period, i.e., before Takebe's *Fukyū Tetsujutsu* [不休綴術] containing his infinite series was published.) In the book by Horiuchi, as well as a much more careful analysis of the Jartoux issue [8, pp. 296–298], we find a different possibility for the path that might have led Takebe to infinite series:

> ... Seki's treatise on faulty problems bristles with ideas, more or less well-mastered, on the relations between the coefficients and the roots. It is also the place where Seki proceeds for the first time to the extraction of a literal magnitude in an equation with literal coefficients. So we are in the presence here of a new extension of the use of literal notation, an extension that will later inspire Takebe to express the square root of a quadratic equation in the form of an infinite series [8, p. 182].

Later, there is a careful analysis of Seki's procedure of the arc, giving a coherent background on which Takebe's series could have arisen without any Western input, and ending with the remarks

> Here again, the methods brought into play by Seki to treat the problem are incomparable with those of his predecessors ... We thus see that Seki extended the use of this algebraic tool to contexts where it was assumed a priori that there existed a functional relation between two geometric quantities even though the expression for this relation was not yet known. Here we touch an essential feature of Seki's works, that of having considerably enlarged the domain of use of algebraic techniques. Mikami very early stressed the historical importance of the solution proposed by Seki, which he considered more important than the exact solution later obtained by Takebe ... Mikami here forcefully asserts that, if Seki had not introduced algebraic techniques into this domain, Takebe's solution would never have seen the light of day [8, p. 251].

One final argument—admittedly of a more subjective nature—is on the level of mathematical style and of the personalities of the protagonists. Written mathematics is characterized not only by its contents, but by the way in which the author sees and presents the discoveries that he is expounding. Seki's approach to algebraic problems, as indicated in the passage just quoted, was extraordinarily original and innovative, but Seki himself, who was steeped in and deeply respectful of the Chinese tradition, typically couched his exposition in the context of this tradition, and indeed may well have believed that he was working within it even when in fact he was doing something very new. We saw one example of this in the first part of this paper, where he formulated his solution of a problem of Sawaguchi as if he were presenting an algorithm for the numerical determination of the sought-for root, but in fact showed no interest in actually finding this root but instead describes a purely algebraic procedure for expressing the solution of the problem as the root of a polynomial—a modern mathematician *malgré lui*. He presents the elimination theory in the traditional language of *sangi* [算木] but uses these in a new way, very different from his predecessors in Japan or China, and of course even more different from the way in which any European mathematician, had he achieved the same results, would have formulated them. If Seki had been exposed, even tangentially, to Western mathematical thinking, then surely one can speculate that this might have affected the presentation, as well as the contents, of what he wrote. As to Takebe, it can be mentioned that he was particularly noted for his honesty, that he was far more inclined to speak of his own weaknesses in mathematics than to take unearned credit for things he had not done, and that he explicitly expressed his enthusiasm about ideas in astronomy coming from Europe when he learned about them after the lifting of the ban [8, pp. 225–226]. There seems to be no reason to think that he would have kept silent about recent beautiful mathematics done by Europeans if he had known about or made use of it.

We now turn to the documentary evidence. We have already mentioned that the *bakufu* [幕府], fearful of the Hollanders' acquiring too much knowledge about Japan, required that the ship's captain be replaced every year. The East India Company, which was just as interested in ensuring that precisely this knowledge should be available, therefore had each *opperhoofd* keep a very detailed diary and made sure that several transcriptions of it should be made, so that at least one copy would survive the perilous return trip to Holland. These diaries, preserved in their entirety in the Dutch National Archives in The Hague, constitute a miraculous record, of a dimension perhaps unparalleled by any other historical document, of the relations between the Dutch contingent and the Japanese during the entire Edo period. A search through the diaries of the relevant years failed to turn up any evidence of an actual exchange of information on any kind of mathematical question whatsoever. There was one fully documented personal encounter between Takebe and the Hollanders in 1727 in which Takebe put detailed questions to the Dutch. These questions concerned such issues as the way the Dutch put out fires, how they named their sons, whether they used matches, fans, or ear-picks, and whether they were acquainted with black magic (the complete list of questions is given in the Appendix), but absolutely nothing specifically scientific, let alone mathematical. Indeed, there is no indication anywhere in the diaries that the Dutch were even aware that Takebe (whom they knew only as an ambassador of the Shogun and consistently referred to as "the Imperial Minion") was a mathematician at all. The only thing indicating that they had recognized Takebe's ability is a diary entry a few pages later in which the captain tells how a defective Dutch watch was said to have been repaired by Takebe and that he "gladly believed this" because, on the occasion when Takebe had asked him the thirty questions, he had observed "that nature had not spared in dispensing knowledge to him, but that he is versed in different sciences and is a fine erudite man" ("in verschejde wetenschappen ervaaren en een fijn doorsleepen man is"). But this is still very far from a technical discussion of mathematics! And indeed, in the entry for the very day on which Takebe put his thirty questions, the captain describes how at the end of their meeting the Japanese continued with various "mathematical, astronomical and geometric propositions, ... to which ⟨I⟩ responded never to have learned the said Sciences, with which the interrogation came to an end." This agrees well with the comment of Goodman cited above.

In summary, the documentary evidence, though admittedly inconclusive, seems to us to give more support to the view that the Japanese did *not* learn any mathematically interesting facts from the Dutch than to the view that they did do so.

Appendix: Takebe's questions to the Dutch in Dejima

We give here a translation of the part of the diary entry for March 25, 1727, that enumerates the questions put by Takebe. The rest of the text, describing the sea-bass that the Japanese presented to the Dutch and wished to watch them to eat, a meeting with the Imperial watchmaker, and the captain's inability to respond to the "mathematical, astronomical and geometric propositions" put forward by the Japanese delegation, is omitted.

Tuesday 25th

In the morning around ten o'clock came the Senior Interpreter Kizits with Lord Takebe Fiko Sira Sama accompanied by three of the First Servants of his Imperial Majesty named [Master] Faomi Foukan, Maayeda Kioriso-o and Ito ga au, arrived, mutually having exchanged some compliments, ⟨they⟩ sat down, ⟨and⟩ the following questions were asked, such as

Firstly — the four elements, whether they are known to us, and how they were named;
Secondly — the Elements, whether we don't apply them to the human body;
Thirdly — the twelve Zodiac signs, whether we know them, how they are named, and whether these are applied to anything;
Fourthly — compasses, what about the compass which points wrongly, whether this by some means can be discovered and demonstrated;
Fifthly — whether without compass, East can be shown and by what means;
Sixthly — whether East and West are known;
Seventhly — whether we have lanterns in use like the Japanese: if so, how and from which kind of material they are made;
Eighthly — whether witchcraft, or black magic, is known to us and is in use;
Ninthly — on which day we rest, and on which ones we work, ⟨and⟩how these are called;
Tenthly — Sunday, how long it has been in use, and why so named;
Eleventhly — the measure of an *ikken* [1.92 m.], and whether others are in use and how they are named;
Twelfthly — ⟨whether⟩ the *gantang* [measure used for rice and pepper, ca. 8 1/2 liters] is used by us, or whether ⟨there are⟩ others and how these are then named;

Thirteenthly — the *dai ching* [steelyard balance], whether it is used for the money that is paid out daily, or for what;
Fourteenthly — Dutch houses, whether they are like the Japanese, then how they are built;
Fifteenthly — warehouses, how they are built by the Hollanders;
Sixteenthly — the Lords' or big manor houses, whether they aren't distinguishable from ordinary men's houses;
Seventeenthly — whether when there is a fire in Holland, whole streets burn down at once like in Japan;
Eighteenthly — how and by what a fire is extinguished;
Nineteenthly — rice, wheat, barley, buckwheat and other grains, how they are presented;
Twentiethly — rice as well as other grains, whether they can be stored in Holland and on Jakarta longer than a year;
Twenty-firstly — the names which are born by somebody for a thousand years or less, whether the descendants may continue to bear them;
Twenty-secondly — whether the Daimyo or other great ⟨persons⟩ don't change their names when they get another function;
Twenty-thirdly — whether the Hollanders like the Japanese have two names in use, i.e. how one names each other daily and the other in writing, that is when one signs one's name;
Twenty-fourthly — when to the father or to the master of a house a son is born, whether he himself gives the name or whether the name is given by Dutch priests, and whether like with the Japanese the name then is used for a signature seal;
Twenty-fifthly — the sulfur-match, whether it is used by us Hollanders, and from what it is made;
Twenty-sixthly — fire, when one wants to keep it, how this is done;
Twenty-seventhly — ink, from what substance it is made;
Twenty-eighthly — quills, from what they are made;
Twenty-ninthly — ear-picks, whether they are used by us;
Thirtiethly — fans, whether they are in use by the Hollanders.
Which above-mentioned questions ⟨I⟩ answered according to my knowledge of science shortly and concisely, . . .

References

1. Seki Takakazu [関孝和]: *Takakazu Seki's Collected Works edited with Explanations* [関孝和全集], Osaka Kyōiku Tosho [大阪教育図書], Osaka (1974).
2. Seki Takakazu [関孝和]: *Hatsubi Sanpō* [發微算法] (translatable as something like "Mathematics with Humble Determination") (1674); [1, pp. 103–120]; Facsimile edition, Wasan Institute (2003).
3. Yoshida Mitsuyoshi [吉田光由] : *Jinkōki* [塵劫記] (Inalterable Treatise) in four volumes (1627) ; Modern Japanese edition with commentaries based on the fifth edition in three volumes (1641) ; [現代語塵劫記 *Gendaigo Jinkōki*], Wasan Institute (2000).
4. Yoshida Mitsuyoshi : *Jinkōki* [塵劫記], English translation of [3] by the Committee headed by O. Takenouchi, with the facsimile of the original print, Wasan Institute (2000).
5. K. Chemla and S. Guo : *Les Neuf Chapitres, Le Classique mathématique de la Chine ancienne et ses commentaires*, Dunod, Paris (2004).
6. G. K. Goodman: *Japan and the Dutch (1600–1853)*, Curzon Press, Richmond (2000).
7. A. Horiuchi: *Les mathématiques japonaises à l'époque d'Edo (1600–1868) : une étude des travaux de Seki Takakazu (? –1708) et de Takebe Katahiro (1664–1739)*, J. Vrin, Paris (1994).

8. A. Horiuchi: *Japanese Mathematics in the Edo Period (1600–1868). A study of the works of Seki Takakazu (?–1708) and Takebe Katahiro (1664–1739)*, English translation of [7] by S. Wimmer-Zagier, Birkhäuser, Basel (2010).
9. K. Kimura et al.: Solutions of a problem of Seki Takakazu [関孝和の問題を解く], *J. Japan Society for Symbolic and Algebraic Computation* [数式処理], vol. 11 (no. 3,4), pp. 33–42 (2005); (http://www.jssac.org/Editor/Suushiki/V11/No3/V11N3,4_105.pdf).
10. T. Kobayashi: Influence of European mathematics on pre-Meiji Japan, Lecture at the International Conference on History of Mathematics in Memory of Seki Takakazu, 2008, *These Proceedings*, pp. 356–374.
11. H. Komatsu: Algebra, elimination and Complete Book of Mathematics, Lecture at the International Conference on History of Mathematics in Memory of Seki Takakazu, 2008, *These Proceedings*, pp. 245–274.
12. M. Morimoto: Takebe Katahiro's algorithms for finding the circular arc length, Lecture at the International Conference on History of Mathematics in Memory of Seki Takakazu, 2008, *These Proceedings*, pp. 331–342.
13. G. B. Sansom: *Japan, A Short Cultural History* (Revised edition of 1952), Charles E. Tuttle, Tokyo (1997).
14. C. Sasaki : The adoption of Western mathematics in Meiji Japan, 1853–1903, *The Intersection of History and Mathematics* (eds. C. Sasaki, M. Sugiura and J.W. Dauben), Birkhäuser, Basel/Boston/Berlin, pp. 165–186 (1994).
15. D. E. Smith and Y. Mikami: *A History of Japanese Mathematics* (Republication of the 1914 edition), Dover, Mineola (2004).
16. Verenigde Oost-Indische Compagnie : *Dagregister, 1727. Japan in de Hoffstad Jedo* [Diaries], preserved in the National Archive, The Hague.

Ming Antu and His Power Series Expansions

Luo Jianjin

Abstract Ming Antu (1692?–1763?), of Mongolian nationality, is a famous Chinese astronomer, mathematician and topographist. He began to work in the Imperial Observatory in 1713. He participated in the work of compiling and editing three very important books in astronomy and joined the team of China's area measurement.

Ming wrote his book Quick Methods for Accurate Values of Circle Segments in 1730–1763, but it was published only in 1839. He was the first person in China who calculated infinite series and obtained more than 10 formulae. He built a system of conceptions and symbols for infinite series, and created a new algorithm for infinite series, including addition, subtraction, multiplication and inversion. He created a method of powers of an infinite series. Famous Japanese scholar Yoshio Mikami wrote in 1910: Ming Antu was the first Chinese who had ever entered into the analytical study of circle measurement.

In 1730's, he first introduced and used Catalan numbers C_n: Computer Scientist Prof. D. E. Knuth of Stanford University holds in his famous book that Ming is the first inventor of Catalan numbers in the world. Mathematician Dr. P. J. Larcombe of Derby University published 7 papers on Ming and C_n. We hold that infinite series with Catalan numbers should arouse more attention.

Ming's achievements in astronomy were accepted and appreciated in the international academic circles of astronomy. On May 26, 2002, the Nomination Committee of Minor Planet Center under the International Astronomical Union announced that Minor planet No.28242 was named "Ming Antu Star."

Li Yan [李儼] and Qian Baocong [錢宝琮] had done some foundational research work on Ming. J. Needham, Li Di [李迪] and C. Jami gave very high comments to

Luo Jianjin
Institute for the History of Science, Inner Mongolia Normal University, Huhhot, 010022, and
Institute of Science, Technology and Culture, Zhejiang University, Hangzhou, 310028, China
e-mail: lluojj@163.com

Ming's works.

The year 2002 was the 310th anniversary of Ming's birth. Ming Antu Star's Nomination Ceremony and traditional meeting were held in Ming's hometown in August 2002. More than 500 delegates and 20,000 local residents gathered together to celebrate, and a conference on "the Science Contribution of Ming Antu" was held. The government named Ming's hometown as "Ming Antu Town."

1 Ming Antu's Life and Works

Ming Antu [明安圖] (1692?–1763?) is of Mongolian nationality. "He was born in Zheng Bai Qi [正白旗] of Mongolia (in today's Inner Mongolia)" [23]. Ming Antu was a famous Chinese astronomer, mathematician, and topographer.

According to the decree of Emperor Kangxi [康熙] in 1670, a royal academic college was founded; Before 1710 Ming was selected to enter it and began to study science.

Emperor Kangxi loved natural science and mathematics. He often taught students by himself. He was Ming Antu's mathematics teacher. Ming worked hard and won Emperor's favor.

Fig. 1 The Statue of Ming Antu

1.1 Ming Antu was a top astronomer in China

Ming finished his schooling in 1713 and worked in the Imperial Observatory. He later became a top astronomer in China. He participated in the work of compiling and editing three very important books in astronomy:

1. The Source of Tuning and Calendars [律暦淵源, *Lu Li Yuan Yuan*] 100 Vols. (1713–1723)
2. The Calendars and Astronomical Phenomena [暦象考成后編, *Li Xiang Kao Cheng Hou Bian*] 10 Vols. (1737–1742)
3. The Astronomical Instruments and Star Maps [儀象考成, *Yi Xiang Kao Cheng*] 32 Vols. (1744–1752)

Ming Antu became the top astronomer in China in the middle of the 18th century and was promoted to the Leader of the Imperial Observatory [欽天監監正] His academic idea was influenced by Tycho Brahe [第谷] (1546–1601). The Mongolian Planisphere (sky map) carved on a stone in the Five Pagodas Temple in Huhhot City might have been drawn by Ming Antu.

1.2 Ming's work on the area measurement in China

The government of the Qing [清] Dynasty tried to make a survey and calculate China's area in 1708–1716, but the work was only half done at that time. Ming Antu joined the team of area measurement and went to Xinjiang [新疆] Province to measure the longitude and latitude in 1755–1756. In 1759, he visited Xinjiang and Tashkent again as the team leader. Based on the results of several measurements, a map of the whole China was drawn, and it was a foundation of later maps of China.

1.3 Ming Antu's main contributions to mathematics

Traditional mathematics in China had been declining since Ming [明] Dynasty (1368–1644). Many mathematics books from ancient times were lost. In the West, the Renaissance occurred, and many famous mathematicians were active. The difference between China and the West was obvious. At that time, Ming Antu could not receive any information from the West. Under such circumstance, Ming made great efforts to push forward traditional mathematics in China.

Ming Antu wrote his book Quick Methods for Accurate Values of Circle Segments [割圜密率捷法, *Ge Yuan Mi Lu Jie Fa*] [21] in 1730–1763, but he did not finish this book in his life. Before his death in 1763, he requested his son and students to complete the book, which was done in 1774. Its hand-copies were spreading but the book was published much later in 1839.

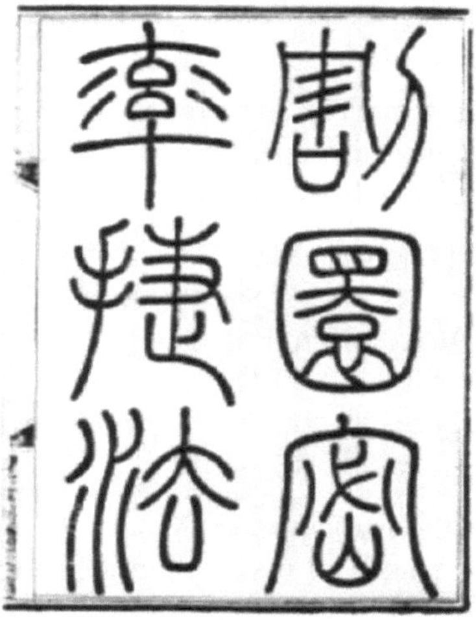

Fig. 2 Quick Methods for Accurate Values of Circle Segments

如圖甲爲圜心甲乙類爲半徑乙丙
丁爲圜周之一弧乙丙弧丙丁弧俱
爲二分弧之一爲一分弧乙丁爲全
弧通弦乙丙丙丁俱爲一分弧通弦
今命甲乙類爲連比例第一率乙丙
類爲連比例第二率求乙丁當連比
例之率數是三線本非一連比例欲
設法以取之也試任平分一分弧乙

Fig. 3 Quick Methods Vol. 3, p. 3

Ming Antu's influence on mathematics in Qing Dynasty lasted more than one hundred years and "Ming's School" was formed under this influence.

Ming Antu's main mathematics works include:

1. He was the first person in China who calculated infinite series and obtained several formulae independently;
2. He was the first person who established and used Catalan numbers in the history of world mathematics;
3. He founded the theory and algorithms of inverse functions and got 4 groups of inversion formulae;
4. Ming pushed forward the theory that geometric figures and numbers can be transformed into each other[1]. Li Yan [李儼] gives the comment that he rivals Descartes in starting analytic geometry;
5. Ming pushed forward the limit theory that curves and lines can finally reach the same in case of infinite division;
6. Ming was the first person who applied the infinite series of trigonometric functions to astronomical calculations in China.

In this paper we can only discuss his first and second achievements listed above.

2 The Power Series Expansions by Ming Antu

In the early stage of the 18th century, a French Jesuit missionary P. Jartoux (1668–1720) introduced the following three formulae of infinite series to China [22, p. 301]:

$$r\sin\frac{a}{r} = a - \frac{a^3}{3!r^2} + \frac{a^5}{5!r^4} - \frac{a^7}{7!r^6} + \cdots, \qquad \text{J. Gregory (1667)}, \quad (1)$$

$$r\text{Vers}\,\frac{a}{r} = \frac{a^2}{2!r} - \frac{a^4}{4!r^3} + \frac{a^6}{6!r^5} - \cdots. \qquad \text{J. Gregory (1667)}, \quad (2)$$

Newton's expansion in 1676:

$$\pi = 3\left(1 + \frac{1^2}{4\cdot 3!} + \frac{1^2\cdot 3^2}{4^2\cdot 5!} + \frac{1^2\cdot 3^2\cdot 5^2}{4^3\cdot 7!} + \cdots\right), \qquad \text{I. Newton (1676)}. \quad (3)$$

Using these three formulae, Chinese astronomers got accurate results. In traditional mathematics in China, there were no such formulae. However, P. Jartoux did not bring their proofs into China, and this aroused Ming's interest. He worked to prove three formulae and obtained other infinite series formulae of trigonometric functions.

[1] See Fig. 3, Mikami [20, p. 145ff.] or Jami [6], [7].

2.1 Ming Antu obtained six formulae of infinite series

Let the radius be r and the arc be a with the central angle $\alpha = a/r$; let the chord of $2a$ be c and the arrow of $2a$ and c be b. Ming Antu had got (in modern form) the following [15]:

$$c = 2r\sin\alpha = 2a - \frac{(2a)^3}{4\cdot 3!r^2} + \frac{(2a)^5}{4^2\cdot 5!r^4} - \frac{(2a)^7}{4^3\cdot 7!r^6} + \cdots, \tag{4}$$

or

$$c = 2r\sin\alpha = \sum_{n=0}^{\infty} \frac{(-1)^n(2a)^{2n+1}}{4^n(2n+1)!r^{2n}}.$$

$$b = r\text{Vers}\,\alpha = \frac{(2a)^2}{4\cdot 2!r} - \frac{(2a)^4}{4^2\cdot 4!r^3} + \frac{(2a)^6}{4^3\cdot 6!r^5} - \frac{(2a)^8}{4^4\cdot 8!r^7} + \cdots, \tag{5}$$

or

$$b = r\text{Vers}\,\alpha = \sum_{n=1}^{\infty} \frac{(-1)^{n+1}(2a)^{2n}}{4^n(2n)!r^{2n-1}}.$$

$$2a = c + \frac{1^2\cdot c^3}{4\cdot 3!r^2} + \frac{1^2\cdot 3^2\cdot c^5}{4^2\cdot 5!r^4} + \frac{1^2\cdot 3^2\cdot 5^2\cdot c^7}{4^3\cdot 7!r^6} + \cdots, \tag{6}$$

or

$$2a = \sum_{n=0}^{\infty} \frac{[(2n-1)!!]^2 c^{2n+1}}{4^n(2n+1)!r^{2n}}.$$

$$a = \sin\alpha = \frac{1^2(r\sin\alpha)^3}{3!r^2} + \frac{1^2\cdot 3^2(r\sin\alpha)^5}{5!r^4} + \cdots, \tag{7}$$

or

$$a = \sum_{n=1}^{\infty} \frac{[(2n-1)!!]^2(r\sin\alpha)^{2n+1}}{(2n+1)!r^{2n}}.$$

$$a^2 = r^2\text{Vers}\,\alpha + \frac{1^2(2r\text{Vers}\,\alpha)^2}{4!} + \frac{1^2\cdot 2^2(2r\text{Vers}\,\alpha)^3}{6!r} + \cdots, \tag{8}$$

or

$$a^2 = \sum_{n=0}^{\infty} \frac{(n!)^2(2r\text{Vers}\,\alpha)^{n+1}}{(2n+2)!r^{n-1}} = \sum_{n=0}^{\infty} \frac{(n!)^2(2b)^{n+1}}{(2n+2)!r^{n-1}}.$$

$$(2a)^2 = (8b)r + \frac{1^2(8b)^2}{4\cdot 4!} + \frac{1^2\cdot 2^2(8b)^3}{4^2\cdot 6!r} + \cdots, \tag{9}$$

or

$$(2a)^2 = \sum_{n=0}^{\infty} \frac{(n!)^2(8r\text{Vers}\,\alpha)^{n+1}}{4^n(2n+2)!r^{n-1}} = \sum_{n=0}^{\infty} \frac{(n!)^2(8b)^{n+1}}{4^n(2n+2)!r^{n-1}}.$$

These formulae are very famous in the history of mathematics in China. For quite a long time, these nine formulae above mentioned were named as "P. Jartoux's formulae," which is a historical mistake.

2.2 Catalan numbers in the infinite power series expansions

Ming Antu had got other important power series expansions as follows [16]:

$$(\sin\frac{\alpha}{2})^2 = \sum_{n=1}^{\infty} C_n(\frac{\sin\alpha}{2})^{2n}. \tag{10}$$

$$\sin 2\alpha = 2\sin\alpha - \sum_{n=1}^{\infty} \frac{C_n(\sin\alpha)^{2n+1}}{4^{n-1}}. \tag{11}$$

$$\sin 4\alpha = 4\sin\alpha - 10(\sin\alpha)^3 + \sum_{n=1}^{\infty} \frac{(16C_n - 2C_{n+1})(\sin\alpha)^{2n+3}}{4^n}. \tag{12}$$

He also had obtained infinite series of $\sin 10\alpha, \sin 100\alpha, \sin 1000\alpha$ and $\sin 10000\alpha$. British mathematician P. J. Larcombe [9] extended it to the case of $\sin 2^k\alpha$ in the year 2000.

In these formulae C_n are the Catalan numbers $1, 1, 2, 5, 14, 42, 132, 429, \cdots$. Catalan numbers are famous counting function in today's discrete mathematics (combinatorics and graph theory).

E. Catalan (1814–1894, Belgian) published a paper [4] involving such numbers in 1838. Catalan numbers then were named after him. Now we know that L. Euler (1707–1783) studied such numbers in 1758 [5].

The formulae of Catalan numbers are:

$$C_0 = 1, \quad C_1 = 1, \quad C_n = \frac{1}{n+1}\binom{2n}{n} \qquad (n \geq 0).$$

The convolutive recurrence formula of Catalan numbers is:

$$C_1 = 1, \quad C_2 = 1, \quad C_n = \sum_{k=1}^{n-1} C_{n-k}C_k \qquad (n \geq 2).$$

I. Newton's binomial theorem gives, when the exponent is equal to $1/2$,

$$(1+z)^{\frac{1}{2}} = 1 + \sum_{n=1}^{\infty}(-1)^{n-1}\frac{C_n z^n}{2^{2n-1}} \qquad (|z| < 1).$$

J. Binet [2] gave in 1839 the generating function of Catalan number:

$$\frac{1}{2} - \frac{1}{2}\sqrt{1-4z} = \sum_{n=1}^{\infty} C_n z^n \qquad (|z| < 1/4).$$

Brown [3] introduced the history of Catalan numbers and Alter [1] its developments. However, until recently the western researchers did not know that Ming's work held a safe lead in the world.

Now we know that there are 6∼7 definition formulae of Catalan numbers, and more than 50 combinatorial background meanings. Up to now, more than 600 papers have been published about them.

In 1730's, Ming Antu already encountered and used such numbers. In his book Quick Methods for Accurate Values of Circle Segments, Vol. 3, Ming used three methods and obtained the same Catalan numbers as well as two new formulae.

Formula A *The recurrence formula of Catalan numbers:*

$$C_1 = 1, \quad C_2 = 1, \quad C_{n+1} = \sum_{k \geq 0} (-1)^k \binom{n-k}{k+1} C_{n-k} \,. \tag{13}$$

This was a new formula, and nobody had ever known it before 1988 in the West [17].

Formula B *The finite polynomial generating function of Catalan numbers:*

$$\begin{aligned} M_1 &= (1), M_2 = (0,1), \\ M_3 &= (2M_1 + M_2)M_2 = [2(1) + (0,1)](0,1) \\ &= [(2) + (0,1)](0,1) = (2,1)(0,1) = (0,0,2,1)\,, \\ &\cdots\cdots\cdots\cdots\cdots \\ M_{n+1} &= \Big(2\sum_{k=1}^{n-1} M_k + M_n\Big)M_n, \\ MC_n &= \sum_{k=1}^{n} M_k = (1,1,2,5,14,42,132,429,\cdots,C_n)\,. \end{aligned} \tag{14}$$

This is a new formula. P. J. Larcombe gave a proof [10] in the year 1999.

In Ming Antu's infinite series expansions, the coefficients were expressed as Catalan numbers C_n. This is very important. In the past researches, this point has been neglected. This paper intends to stress that the infinite series with Catalan numbers should arouse more attention.

2.3 Ming Antu's method of calculating infinite series

In traditional mathematics in China, there were no infinite series. Ming Antu had to create a new method to calculate infinite series. His work is listed as follows [18]:

1. He built a system of conceptions and symbols for infinite series;
2. He created a new algorithm for infinite series, including addition, subtraction and multiplication;
3. He made calculation by himself and obtained right results of some infinite series.

Taking infinite series y_{10} as an example, this paper introduces his method of squaring an infinite series. For this purpose, we translates Ming's algorithm into current mathematical language based on the content of pp. 34–35 in Vol.3 of his book Quick Methods for Accurate Values of Circle Segments. Let

$$x = 2\sin\alpha \quad (0<\alpha<\pi/4), \quad y_m^n = (2\sin m\alpha)^n \quad (m \geq 1)\,.$$

Ming had got (processes are omitted):

$$\begin{aligned}y_{10} &= 5y_2 - 5y_2^3 + y_2^5 \\ &= 10x - 165x^3/4 - 3003x^5/4^3 - 21450x^7/4^5 - 60775x^9/4^7 \\ &\quad - \sum_{n=1}^{\infty}(16^4C_n - 8\cdot 16^3C_{n+1} + 21\cdot 16^2C_{n+2} - 20\cdot 16C_{n+3} + 5C_{n+4})x^{2n+9}/4^{2n+7}.\end{aligned}$$

Here C_n are Catalan numbers. When $n = 1,2,3$, the coefficients in the parentheses are 41990, 22610, 29716.

Now we rewrite this infinite series as follows:

$$y_{10} = a_0x + \sum_{n=1}^{\infty}\frac{a_nx^{2n+1}}{4^{2n-1}}\,. \tag{15}$$

When $n = 0,1,2,\cdots,7$, $a_n = 10$, -165, -3003, -21450, -60775, -41990, -22610, -29716, respectively.

Ming Antu had to calculate the self multiplication of y_{10}, namely $y_{10}\times y_{10}$. His method is as follows:

$$\begin{aligned}y_{10}^2 &= \left(a_0x + \sum_{n=1}^{\infty}a_nx^{2n+1}/4^{2n-1}\right)^2 \\ &= (a_0x)^2 + 2\sum_{n-1}^{\infty}a_0a_nx^{2n+2}/4^{2n-1} + \left(\sum_{n=1}^{\infty}a_nx^{2n+1}/4^{2n-1}\right)^2.\end{aligned} \tag{16}$$

The key of this operation is the last term in Formula (16). He created a method as follows:

$$\left(\sum_{n=1}^{\infty}a_nx^{2n+1}/4^{2n-1}\right)^2 = \sum_{n=2}^{\infty}\left(\sum_{k=1}^{n-1}4a_{n-k}a_k\right)x^{2n+2}/4^{2n-1}. \tag{17}$$

And he got

$$y_{10}^2 = b_0x^2 + \sum_{n=1}^{\infty}b_nx^{2n+2}/4^{2n-1}, \tag{18}$$

where

$$b_0 = a_0^2,\ b_1 = 2a_0a_1;\ \text{when } n \geq 2\text{: } b_n = 2a_0a_n + \sum_{k=1}^{n-1}4a_{n-k}a_k\,.$$

In this way, Ming Antu resolved this difficult problem of self multiplication of infinite series.

3 Our Commemoration

3.1 The researches on Ming Antu by historians and mathematicians

Famous Japanese scholar Yoshio Mikami [三上義夫] (1863–1950) wrote in 1910 [20, p. 149]: Antu Ming was the first Chinese who had ever entered into the analytical study [of the circle measurement].

Li Yan [李儼] (1892–1963) and Qian Baocong [銭宝琮] (1892–1974) had done some foundational research work on Ming Antu's achievements. Dr. J. Needham (1900–1995) gave very high comments to Ming Antu's work.

Li Di [李迪] (1927–2006) published Ming's biography [14] and more than 10 papers on him. He Shaogeng [何紹庚], Luo Jianjin [羅見今] and Tegus [特古斯] had done other works further in depth. In China, more than 80 papers or books were published on Ming Antu.

French scholar Catherine Jami writes: the virtue of Ming Antu's work is that he integrated two traditions of mathematics, Chinese and Western into one frame [6].

In his famous book The Art of Computer Programming [8, p. 407], the leading figure of Algorithms and Programming, Prof. D. E. Knuth of Stanford University introduces the history of Catalan numbers. He said: "A Mongolian Chinese mathematician, An-Tu Ming, had encountered the Catalan numbers before 1750 in his study of infinite series, ..." [17, 19].

P. J. Larcombe of Derby University published 7 papers on Ming and Catalan numbers, such as: *"On the history of the Catalan numbers: a first record in China"* [11]; *"The 18th century Chinese discovery of the Catalan numbers"* [12] and *"On expanding the sine function with Catalan numbers: A note on a role for hypergeometric functions"* [13]; etc.

3.2 The nomination of Ming Antu Star

In 1992, on the occasion of the 300th anniversary of the birth of Ming Antu, an academic conference was held in Huhhot, Inner Mongolia. Wang Shouwan, the academician of National Astronomical Observatory and more than 30 international scholars took part in this conference. Wang wrote a poem:

科学巨星明安圖 冉冉昇起乾康初 精修曆象制皇輿 割圓密率冠前騙
碧野連空故上都 循公往迹纂公書 欲上青天攬公裾 今朝興国并興区

Ming Antu's achievements in astronomy were accepted and appreciated in the international academic field of astronomy. In 1999, Chinese astronomers discovered a new minor planet No. 28242. On May 26, 2002, the Nomination Committee of Minor Planet Center under the International Astronomical Union announced that minor planet No. 28242 was named "Ming Antu Star."[2]

The year 2002 was the 310th anniversary of Ming Antu's birth. Ming Antu Star's Nomination Ceremony and traditional Nadamu[3] were held in Ming's hometown in August, 2002. The academician Wang of National Astronomical Observatory and academician Teng Jiwen of Geography and Geologic Research Institute of China Academy of Sciences took part in this celebration. More than 500 delegates and 20,000 local residents gathered together to celebrate and a conference on "the Science Contribution of Ming Antu" was held. The government named Ming's hometown "Ming Antu Town," and also announced the decision to build "Ming Antu Museum of Science and technology."

References

1. R. Alter: Some remarks and results on Catalan numbers, *Proc. 2nd Louisiana Conf. on Combinatorics, Graph Theory and Computing*, pp. 109–132 (1971).
2. J. Binet: Réflexions sur le problème de déterminer le nombre de manières dont une figure rectiligne peut être partagée en triangles au moyen de ses diagonals, *J. Math. Pures Appl.*, vol. 4, pp. 79–91 (1839).
3. W. G. Brown: Historical note on a recurrent combinatorial problem, *Amer. Math. Monthly*, vol. 72, pp. 973–977 (1965).
4. E. Catalan: Note sur une équation aux différences, *J. Math. Pures Appl.*, vol. 3, no. 1, pp. 508–516 (1838).
5. L. Euler: *Novi Commentarii Acade Scientiarum Imperialis Petropolitanae*, vol. 7, pp. 13–14 (1758–59).
6. Catherine Jami: Western influence and Chinese tradition in an eighteenth century Chinese mathematical work, *Historia Mathematica*, vol. 15, pp. 311–331 (1988).
7. Catherine Jami: *Les Méthodes rapides pour la trigonométrie et le rapports du cercle* (1774), *Tradition chinoise et apport occidental en mathématiques*, Collège de France, Institut des Hautes Études Chinoises, Paris (1990).
8. Donald E. Knuth: *The Art of Computer Programming*, vol. 1, Fundamental Algorithms, Third Edition, Addison-Wesley (1997); Tsinghua University Press (2002).
9. P. J. Larcombe: On Catalan numbers and expanding the sine function, *Bull. Inst. Combin. Appl.*, vol. 28, pp. 39–47 (2000).
10. P. J. Larcombe: On a finite polynomial generating function for Catalan subsequences: An 18th century observation proved, *Congr. Numer.*, vol. 141, pp. 49–60 (1999).
11. P. J. Larcombe: On the history of the Catalan numbers: a first record in China, *Math. Today (Southend-on-Sea)*, vol. 35, no. 3, p. 89 (1999).
12. P. J. Larcombe: The 18th century Chinese discovery of the Catalan numbers, *Mathematical Spectrum*, vol. 32, no. 1, pp. 5–7, (1999/2000).

[2] The document of "2002 May 26, M. P. C. 45750 New Name of Minor Planet" said: "Ming Antu(1692–1765?) was a Chinese astronomer and mathematician of Qing Dynasty. During the decades of his service in the Imperial Observatory, he participated in compiling and editing three very important astronomical works."

[3] 'Nadamu' is a traditional meeting of Mongolian nationality on grassland.

13. P. J. Larcombe and David R. French: On expanding the sine function with Catalan numbers: A note on a role for hypergeometric functions, *J. Combin. Math. Combin. Comput.*, vol. 37, pp. 65–74 (2001).
14. Li Di [李迪]: *Mongolian Scientist Ming Antu* [蒙古族科学家明安圖], Inner Mongolia People Press [内蒙古人民出版社] (1978).
15. Luo Jianjin [羅見今]: On Ming Antu's formulae [明安圖公式辨正], *J. Inn. Mong. Norm. Univ. Nat. Sci.*, [内蒙古師範大学報 (自然科学版)], vol. 1, pp. 42–48 (1988).
16. Luo Jianjin (translator and glossographer): *The Quick Methods for Accurate Values of Circle Segments* [明安圖原著, 羅見今訳注, 割圜密率捷法訳注], Inner Mongolia Education Press [内蒙古教育出版社] (1998).
17. Luo Jianjin [羅見今]: Ming Antu: the first inventor of Catalan Numbers in the world [明安圖是卡塔蘭数的首創者], *Acta Scientiarum Naturalium Universitatis Intramonglicæ* [内蒙古大学学報 Neimenggu Daxue Xuebao], vol. 19, no. 2, pp. 239–245 (1988).
18. Luo Jianjin: A study on Ming Antu's method of operation of infinite series [明安圖計算無窮級数的方法分析], *Stud. Hist. Nat. Sci.* [自然科学史研究], vol. 9, no. 3, pp. 197–207 (1990).
19. J. Luo: Catalan numbers in the history of mathematics in China, *Combinatorics and Graph Theory*, World Sci. Publ., pp. 68–70 (1993).
20. Yoshio Mikami [三上義夫]: *The Development of Mathematics in China and Japan*, Teubner, Leibzig (1913); Second Edition, Chelsea (1974).
21. Ming Antu [明安圖]: The Quick Methods for Accurate Values of Circle Segments [割圜密率捷法], 石梁岑氏校刊 (1839).
22. Qian Baocong (editor in chief) [銭宝琮主編]: *History of Mathematics in China* [中国数学史], Science Press [科学出版社], p. 301 (1964).
23. Zhao Erxun [趙爾巽] et al.: *The Draft of History of the Qing Dynasty*, book 46, vol. 506, Ming Antu [清史稿 46 冊 506 卷, 明安圖], Zhonghua Press [中華書局], pp. 13961–13964 (1977).

Standing on the Shoulders of the Giant Influence of Seki Takakazu on Takebe Katahiro's Mathematical Achievements

Xu Zelin and Zhou Chang

Abstract Seki Takakazu (ca. 1642–1708) and his pupils develop the Wasan [Japanese mathematics] into the most advanced mathematics in the world outside Europe in the Edo period (1603–1867), and mathematical achievements of Seki Takakazu and Takebe Katahiro (1664–1739) laid foundation for the Wasan. In order to commemorate this outstanding mathematician, this paper surveys Seki's influence on Takebe Katahiro, who was his excellent pupil.

1 Academic experiences of Takebe Katahiro as pupil and cooperator of Seki Takakazu

Takebe Katahiro [建部賢弘] was born in an intellectual family of a retainer of Tokugawa shōgunate [幕府 bakufu]. His great-grandfather Takebe Shōkō [建部昌興] was a clerk of the first Shōgun Tokugawa Ieyasu [徳川家康] (1543–1616), grandfather Takebe Naomasa [建部直昌] was a secretary of the third Shōgun Tokugawa Iemitsu [徳川家光] (1604–1651), and father Takebe Naotsune [建部直恒] was also a clerk of Tokugawa Iemitsu. Takebe Naotsune had four sons; the eldest son was Takebe Katao [建部賢雄] (1654–1723), the second son was Takebe Kataakira [建部賢明] (1661–1716), the third son was Takebe Katahiro, and the youngest son was Takebe Katamitsu [建部賢充]. Katahiro's elder brothers Katao and Kataakira held office in the shōgunate due to the Japanese feudal hereditary system, Katao as heir of his fa-

Xu Zelin
College of Humanities, Donghua University,
2999 Renminbei Road, Songjiang District, Shanghai 201620, P. R. China
e-mail: zelinxu@126.com

Zhou Chang
Department of Applied Mathematics & Applied Physics,
Xi'an institute of Post Telecommunications, Xi'an 710121, P. R. China
e-mail: maytheday@163.com

ther and Kataakira as adopted heir of his uncle, who was the head of a branch family of Takebe.

Tsukane Ogawa divided Takebe Katahiro's academic career into three stages [3]. Earlier stage (13-year-old—40-year-old): he learned and studied mathematics with Seki Takakazu [関孝和]; Middle stage (41-year-old—53-year-old): served as shōgunate retainer [幕臣 bakushin] of Tokugawa Ienobu [徳川家宣] (1662–1712) and Tokugawa Ietsugu [徳川家継] (1709–1716) and suspended mathematical study; Late stage (54-year-old—70-year-old): served for Tokugawa Yoshimune [徳川吉宗] (1684–1751) and engaged in mathematical and calendrical studies.

We propose to divide Takebe Katahiro's academic career into five stages in more detail:

Stage One: from 1664 (one-year-old) to 1675 (12-year-old). This is his childhood, when he lived with his father in Edo.

Stage Two: from 1676 (13-year-old) to 1695 (32-year-old). He learned and studied mathematics with Seki.

Stage Three: from 1696 (33-year-old) to 1715 (52-year-old). He worked for Ienobu and Ietsugu, and he seldom studies mathematics due to busy official business.

Stage Four: from 1716 (53-year-old) to 1733 (70-year-old). He worked for Yoshimune, engaged in astronomical-calendrical studies as calendrical consultant in the shōgunate, and sorted his own previous results of mathematical research.

Stage Five: from 1733 (70-year-old) to 1739 (76-year-old). He almost stopped the mathematical research during this retirement period.

His academic life was divided in this way because Katahiro seldom cooperated with Seki in mathematical research after the first twelve volumes of the Complete Book of Mathematics [大成算経 Taisei Sankei] were completed in 1695.

In 1676, as the hereditary *bushi*, Seki served for Tokugawa Tsunashige [徳川綱重] (1644–1678) of the Kōfu clan [甲府藩 Kōfu han]. In 1678, he served for Tokugawa Tsunatoyo [徳川綱豊], who was the heir of Tsunashige, as examiner of the division of accounts [堪定吟味 kanjō ginmi] engaging in work relating to land survey. Tsunashige and Tsunatoyo did not live in Kōfu at that time, but in Sakurada of Edo. Meanwhile, Naotsune, Katahiro's father, was serving for the third Shōgun Tokugawa Iemitsu; therefore, Katahiro also lived in Edo. In 1674 Seki Takakazu published his book Mathematical Methods for Exploring Subtle Points [発微算法 Hatsubi Sanpō], which made Seki famous, and Seki's prominence in mathematics became well-known [7].

In 1676, 13-year-old Katahiro and his elder brothers Katao and Kataakira formally became pupils of Seki, and studied mathematics until December 1704, when Tokugawa Tsunatoyo became the successor of the fifth Shōgun Tokugawa Tsunayoshi [徳川綱吉] (1646–1709) and moved to West Castle [西の丸 Nishinomaru] which is the residence of the Shōgun's successor. When both Seki and Takebe Katahiro became retainers of the shōgunate, they have studied mathematics together for 28 years. From then on and until November 1706 (Seki's resignation), they had chances to study together.

During Genroku [元禄] (1688–1704) and Hōei [宝永] (1704–1711) periods, celestial element method [天元術 tianyuan shu] became popular in Japan and many people started to study it eagerly. Seki transformed the celestial element method and thus created the byscript notation [傍書法 bōshohō] and method of calculating equations [演段術 endan-jutsu]. During Jōkyō [貞享] period (1684–1687), the Seki showed greatest mathematics creativity. Takebe brothers not only learned Seki's mathematics achievements directly, but also became his valuable assistants in mathematical study. In 1681, Matsuda Masanori [松田正則], a pupil of Saji Kazuhira [佐治一平], wrote the book Introduction to Mathematical Methods [算法入門 Sanpō Nyūmon] to answer 49 unsolved problems in Mathematical Textbook on Multiplication and Division [数学乗除往来 Sūgaku Jōjo Ōrai] (1674). Because of not understanding Seki's method of calculating equations, so they censured the Mathematical Methods for Exploring Subtle Points. Therefore, 20-year-old Takebe Katahiro wrote the Mathematical Methods for Clarifying Slight Signs [研幾算法 Kenki Sanpō] in 1683 to defend the mathematical methods of Seki. In order to explain the main idea of Seki's method of calculating equations further, Takebe Katahiro wrote Colloquial Commentary on the Operations in the Mathematical Methods for Exploring Subtle Points [発微算法演段諺解 Hatsubi Sanpō Endan Genkai] (1685) to explain operations in Seki's solution process in the Mathematical Methods for Exploring Subtle Points in detail.

In 1683, Seki, Takebe Kataakira and Katahiro planned to compile jointly the Complete Book of Mathematics. In first 12 years from 1683 to 1695 Takebe Katahiro devoted to the writing of mathematics; during this period, he took the main compilation work for the first 12 volumes of the Complete Book of Mathematics, and at the time in 1690 he published the Complete Colloquial Commentary on the Introduction to Mathematics [算学啓蒙諺解大成 Sangaku Keimō Genkai Taisei] which explained the Introduction to Mathematics [算学啓蒙 Suanxue Qimeng] of Zhu Shijie [朱世傑].

In the winter of this same year 1690, Katahiro was adopted by Hōjō Gengoemon [北条源五衛門] as heir and summoned to work for Tokugawa Tsunatoyo. In the autumn of 1703, Katahiro divorced himself from the Hōjō, returned to his original family. But he was summoned to the shōgunate to take a post of ceremonies in the household [納戸役 nando-yaku]; then he moved to West Castle in 1704 together with Seki and Tokugawa Tsunatoyo.

At that time, Seki had not much energy for the research of mathematics for he was weak because of age and illness, then he resigned in November, 1706, and died of disease 2 years later, that is 1708. Katahiro engaged in official duties as the retainer of Tsunatoyo in 1692, then he served Ienobu and Ietsugu, and promoted frequently, as a result, he had to stop the compilation of the Complete Book of Mathematics in 1695, his elder brother Kataakira continued the compilation independently, 20 volumes of the Complete Book of Mathematics was completed finally in 1711. Therefore, Katahiro and Seki had little time together for the research during the time of 13 years from 1695 to 1708.

In 1716, Ietsugu died and Yoshimune became the 8th Shōgun. Since then, Takebe Katahiro dedicated himself again to the mathematics and calendrical study as the astronomical-calendrical consultant of Yoshimune. Katahiro was instructed to make the map of Japan in 1719, and then received orders to engage in the geodetical measurement out of the need of drawing maps in 1720. The Mathematical Treatise on the Technique of Linkage [綴術算経 Tetsujutsu Sankei] (1722), which he compiled in 1722 and presented to Yoshimune, was the peak of his mathematical research.

In 1726, Mei Wending's [梅文鼎] (1633– 1721) Complete Treatise on Calendar and Computation [暦算全書 Lisuan Quanshu] (1723) was imported to Nagasaki, Yoshimune ordered Katahiro to translate and describe it, Katahiro entrusted his disciple Nakane Genkei [中根元圭] (1662–1733) who has a profound accomplishment in Sinology for the translation, after the translation was completed in 1728, it was named New Version of Complete Treatise on Calendar and Computation [新写訳本暦算全書 Shinsha Yakuhon Rekisan Zensho], 46 volumes in total; Katahiro wrote a preface for it, and presented it to Yoshimune in 1733. He lived in solitude after resigned in 1733, and almost stopped his activities in mathematical research; he died of disease in July 20, 1739.

A comprehensive survey of Katahiro's plain career as *samurai* and academic experience shows that his 20 years' academic association with Seki was of great importance, which laid a foundation for him to become one of the most outstanding mathematicians in the history of Wasan.

2 Takebe Katahiro's mathematical achievement and Seki's mathematical heritage

The age of Seki and Takebe Katahiro is the formation stage of Wasan research pattern, and they are the most important founders of this research pattern. Seki is the pioneer of Wasan algorithms, while Takebe Katahiro is the most important promoter based on Seki's work, his mathematics achievements are closely related with Seki's work.

2.1 Study and popularization of celestial element method, byscript notation and method of calculating equations

The most important mathematical method in Wasan is the procedure in operations of celestial element [天元演段術 tengen endan jutsu], which is a character algebraic method that is formed by reforming algebraic method in Song-Yuan dynasties of China; the mathematical research of Wasan mathematicians led to the road of algebra analysis by broadly applying this Algebraic method, and Seki is the most important founder of this method. The established time of Seki's byscript notation and method of calculating equations should be before or after the work Mathematical Methods for Exploring Subtle Points of 1674, and 2 years after, that book was published, Takebe Katahiro started to study with him. when Takebe wrote the Mathematical Methods for Clarifying Slight Signs and the Colloquial Commentary on the Operations in the Mathematical Methods for Exploring Subtle Points, he had acquired completely Seki's method of calculating equations, which laid a solid of mathematical foundation for his research in the future, and these two books played a very important role in the popularization of Seki's Algebraic method.

2.2 Improvement of infinitesimal analysis

2.2.1 Romberg algorithm for evaluating π: from Seki's method of one-time accelerated approximation to Takebe's method of successive accelerated approximation

The method of evaluating π of Liu Hui [劉徽] and Zu Chongzhi [祖沖之] (429–500) of China has long been lost in Japan, the records of Zu Chongzhi's achievements in Book of Music and Calendar [律暦志 Lulizhi] and History of the Sui Dynasty [隋書 Suishu] stimulated mathematicians to seek for the method of evaluating π. In Platter of Mathematics [算組 Sanso] (1663), Muramatsu Shigekiyo [村松茂清] (1608–1695) started from square he took the circumference of 2^{15} polygon to approach the circumference of circle, and $\pi = 3$ was obtained. It seems that Seki wanted to restore Zu Chongzhi's method, he adopted Muramatsu's method, and started cutting the circle from inscribed square quadrilateral of circle, and evaluated circumference C_n of 2^{n+1} regular polygon inscribed in a circle successively, when evaluating circumference C_{14}, C_{15}, C_{16} of $2^{15}, 2^{16}, 2^{17}$ regular polygons, conduct accelerated processing with (incremental) method of accelerated approximation [増約術 zōyakujutsu] (i.e., the Aitken method), so the formula of circle ratio [定周 teishū] presented was obtained.

$$C = C_{15} + \frac{(C_{15} - C_{14})(C_{16} - C_{15})}{(C_{15} - C_{14}) - (C_{16} - C_{15})}.$$

Thus $\pi = 3.14159265358979323\,86$ [4, p. 2] was obtained, which is accurate to the 18th place of decimals; however, Seki was not sure of its accuracy, he just confirmed the first 11 digits as correct.

Takebe Katahiro made creative reformation for Seki's method, and established the method of successive accelerated approximation [累遍増約術 ruihen zōyaku jutsu] on the basis of method of one-time accelerated approximation [一遍増約術 ippen zōyaku jutsu]. He took the circumference power of inscribed regular polygon of circle to approach circumference power of circle, set circle diameter $d = 1$, denoted the side length of n regular polygon inscribed in a circle by a_n, the circumference is p_n, and its circumference power is p_n^2, for unity, denoted by $T_n^{(0)}$, so as to represent initial approximate value series of π^2, the circumference is denoted by p, obviously, $T_n^{(0)} = p_n^2 \to p^2 = \pi^2$.

He evaluated $\{T_i^{(0)}\}$ $(i = 1, 2, 3, \cdots, n)$ sequence with cyclotomic method, and observed its first difference, the composed common ratio is the geometric progression of $1/4$, and as the first method of accelerated approximation, that is:

$$\Delta_i^{(1)} = T_{i+1}^{(0)} - T_i^{(0)}, \quad r_i = \frac{\Delta_{i+1}^{(1)}}{\Delta_i^{(1)}} \approx \frac{1}{4},$$

$$T_i^{(1)} = T_i^{(0)} + (1 + r_1 + r_1^2 + r_1^3 + \cdots)\Delta_i^{(1)} = T_i^{(0)} + \frac{4\Delta_i^{(1)}}{4-1}.$$

As for $\{T_i^{(1)}\}$ $(i = 1,2,3,\cdots,n-1)$, he observed its difference, the composed common ratio is the geometric progression of $1/4^2$, and as the the second method of accelerated approximation:

$$\Delta_i^{(2)} = T_{i+1}^{(1)} - T_i^{(1)}, \quad r_2 = \frac{\Delta_{i+1}^{(2)}}{\Delta_i^{(2)}} \approx \frac{1}{4^2},$$

$$T_i^{(2)} = T_i^{(1)} + (1 + r_2 + r_2^2 + r_2^3 + \cdots)\Delta_i^{(2)} = T_i^{(1)} + \frac{4\Delta_i^{(2)}}{4^2-1}.$$

As for $\{T_i^{(2)}\}$ $(i = 1,2,3,\cdots,n-2)$, he observed its difference, the composed common ratio is the geometric progression of $1/4^3$, and as the third method of accelerated approximation. Evaluate continuously in this way

$$\Delta_i^{(m)} = T_{i+1}^{(m-1)} - T_i^{(m-1)}, \tag{1}$$

$$r_m = \frac{\Delta_{i+1}^{(m)}}{\Delta_i^{(m)}} \approx \frac{1}{4^m}, \tag{2}$$

$$T_i^{(m)} = T_i^{(m-1)} + (1 + r_m + r_m^2 + r_m^3 + \cdots)\Delta_i^{(m)} = T_i^{(m-1)} + \frac{4^m\Delta_i^{(m)}}{4^m-1}. \tag{3}$$

Takebe evaluated $T_i^{(8)}$ $(m = 8, i = 1,2,3,\cdots,9)$ as approximate value of π^2, and obtained

$$\pi = 3.1415\ 9265\ 3589\ 7932\ 3846\ 2643\ 3832\ 7950\ 2884\ 1971\ 2$$

is accurate to the 41st place of decimals.

In fact,

$$\begin{aligned} T_i^{(m)} &= T_i^{(m-1)} + \frac{4^m\Delta_i^{(m)}}{4^m-1} \\ &= T_i^{(m-1)} + \frac{4^m(T_{i+1}^{(m-1)} - T_i^{(m-1)})}{4^m-1} \\ &= \frac{4^m T_{i+1}^{(m-1)} - T_i^{(m-1)}}{4^m-1}. \end{aligned}$$

Obviously, Takebe's method of successive accelerated approximation and the Romberg method in numerical integration are completely consistent [9].

Romberg algorithm, which adopted Richardson Extrapolation Method, was proposed by Romberg, an American calculation mathematician in 1955 [6], while Richardson Extrapolation Method was proposed by Lewis Fry Richardson in 1910 [5]. Although Takebe's method of successive accelerated approximation was recorded on Mathematical Treatise on the Technique of Linkage of 1722, it was appeared in Volume 12 of the Complete Book of Mathematics written by him during the period of Genroku. The advantage of this algorithm is: reduce the cyclotomic frequency can increase the convergence speed instead, and high-precision value can be obtained very quickly. The process from Seki's method of one-time accelerated approximation to Takebe's method of successive accelerated approximation is a qualitative leap in mathematical method, and a glorious model of the spirit of Wasan algorithm.

2.2.2 Exploitation study of expansion of infinite series: from Seki's Konton Method to Takebe's Tan-kojutsu.

One direct cause for the generation of calculus algorithm is the problem of evaluating the length of curve, and the original integral form was used to represent curve with expansion of infinite series. The breakthrough of *wasan* circle theory [円理 enri] algorithm also displayed in the calculation of arc length, Seki, the first one who made creative contributions on this aspect, applied rule of finding differences [招差法 zhaochafa or shōsahō] creatively on the calculation of arc length, and composed the following interpolation polynomials with Newtonian form:

$$
\begin{aligned}
p^2 = \left\{\lambda h^2 + 4h(d-h)\right\} - \frac{h^2(d-2h)}{(d-h)^5}\left\{k_1(d-h)^4 - k_2(d-h)^3(h-h_1)\right. \\
+ k_3(d-h)^2(h-h_1)(h-h_2) - k_4(d-h)(h-h_1)(h-h_2)(h-h_3) \\
\left. + k_5(h-h_1)(h-h_2)(h-h_3)(h-h_3)\right\}.
\end{aligned}
$$

Among them, $\lambda = (355^2 - 4 \times 113^2)/113^2$, p is arc length, h is vector length, d is diameter of circular arc, and h_i is interpolation nodes.

The error of taking interpolation polynomials with more than 3 times to approach trigonometric function is huge, so Seki's formula is just accurate to the 6th place of decimals. Seki's method provides enlightenment to Takebe Katahiro, and he realized that:

> In the search of the form and attribute of the back arc, the true number is hidden if it is close to the half circle and the true number appears if it is close to the side. If it is close to the half circle, it belongs to the latitude and its curve is rapid; if it is close to the side, it belongs to the longitude and its curve is slow. Therefore, taking the sagitta to be extremely small, we should search for the number and seek the procedure. [8].

He pointed that Seki's failure lies in that the essential characteristics of arc were not found: when evaluate arc with vector, the closer the arc approaches the semicircle, the larger the error of arc length formula is; the thinner the arc, the smaller the error of the arc length formula is. Thus, Takebe started from the situation that h/d as the minimum to explore the formula of arc length. His method made breakthroughs from two aspects, one is the application of method of successive accelerated approximation to evaluate the approximate value of arch length with high precision, another is the application of method of approximate fractions [零約術 reiyaku jutsu] to calculate coefficients of infinite series, and represent arc length as power series of vector. The deduction processes are as follows:

If the diameter of a circle $d = 1$ feet, vector $h = 10^{-6}$ feet, inscribe two chords a_2 in the arc, each arc is corresponding to half of the original arc, then inscribe another two chords a_4, inscribe two chords in the arc of chord many correspondence a_8, till it was inscribed to 2^n chord a_{2^n}, evaluate a_{2^n} with pythagorean theorem, thus, calculate the circumference power inscribed polygon with arch, that is, the approximate value of arc length power s_k^2. If the corresponding chord of 2^k arc was recorded as a_k, then evaluate:

$$\left(\frac{s_k}{2}\right)^2 = \left(\frac{2^k a_k}{2}\right)^2 = \{p_k\}, \quad (k = 1, 2, 3, \cdots, n).$$

Take this as initial approximate value, then evaluate with method of successive accelerated approximation:

$$\left(\frac{s}{2}\right)^2 = 1.000000333333511111225396906666728234776947959587 5\cdots \times 10^{-6}.$$

Then apply the method of approximate fractions to construct power series, and obtained expansion of power series of $(s/2)^2$ concerned with h/d:

$$\left(\frac{s}{2}\right)^2 = dh + \frac{1}{3}h^2 + \frac{8}{45}\left(\frac{h^3}{d}\right) + \frac{4}{35}\left(\frac{h^4}{d^2}\right) + \frac{128}{1575}\left(\frac{h^5}{d^3}\right) + \frac{128}{2079}\left(\frac{h^6}{d^4}\right) + \cdots.$$

It equals expansion of power series of inverse sine function. Takebe Katahiro's work, which is the starting point for Wasan circle theory research to the analysis method, paved a new road for infinitesimal algorithm of Wasan, and it is comparable with western modern mathematics.

2.2.3 Evaluation of Surface Area and Volume of Sphere: Calculus Thought

Evaluation of the surface area and the volume of a sphere belongs to circle theory, which was first seen in Record of Jugai [竪亥録 Jugairoku] (1639) of Imamura Tomoaki [今村知商]. The book offers the approximate formula $S_m \approx \pi^2 d^2/4$ of surface area of sphere. Elucidation of Mathematics with Author's Comment [増補算法闕疑抄 Zōho Sanpō Ketsugishō] (1684) of Isomura Yoshinori [礒村吉徳] (?—1710) offers accurate formula $S_m = \pi d^2$. The Platter of Mathematics of Muramatsu Shigekiyo begins to apply segmentation method to compute the volume of a sphere [1, p. 7]. Seki cuts 50, 100 and 200 thin sections by using parallel sections, by which the first approximate value V_1, second approximate value V_2 and third approximate value V_3 have been evaluated successively, and speeds up the approximation by method of accelerated approximation, then there is:

$$\overline{V} = \frac{(V_3 - V_2)(V_2 - V_1)}{(V_2 - V_1) - (V_3 - V_2)} + V_2 = 666\frac{2}{3}.$$

And thus the approximate formula of the volume of a sphere is obtained as:

$$V = \overline{V} \cdot \frac{\pi}{4} = 666\frac{2}{3} \cdot \frac{\pi}{4}.$$

In the Mathematical Treatise on the Technique of Linkage, Takebe applies Seki's method to derive the formula of sphere surface area. He first evaluates the volume of the sphere V_i (the diameter is $d_i = 1 + \varepsilon_i$ $(i = 1, 2, 3)$) and the volume of the sphere V_0 (the diameter is $d_0 = 1$), and then obtains the corresponding difference of volumes $W_i = V_i - V_0$ $(i = 1, 2, 3)$ that is then divided by thickness difference $(d_i - d_0)/2$ of the two homocentric spheres, and the volume of thin sections is given by

$$m_i = \frac{W_i}{h} = \frac{V_i - V_0}{(d_i - d_0)/2}.$$

That is, regard surface area of sphere as differential of sphere volume to radius. And then apply Seki's decremental method of accelerated approximation [損約術 sonyaku jutsu] to evaluate surface area of sphere, that is:

$$S_m = m_2 - \frac{(m_1 - m_2)(m_2 - m_3)}{(m_1 - m_2) - (m_2 - m_3)} = 3.14159265359 \approx \pi.$$

Then the surface area of sphere $S_m = \pi d^2$ is obtained. Takebe's such method can be called the peeling method [薄皮饅頭法 usukawa manjū hō] which is expressed in differential form as

$$\begin{aligned} S_m &= \lim_{d_{外} - d_{内}} \frac{V_{外} - V_{内}}{(d_{外} - d_{内})/2} \\ &= \lim_{r_1 \to r_2} \frac{4}{3}\pi(r_2^2 + r_2 r_1 + r_1^2) = \frac{4}{3}\pi(3r_1^2) = 4\pi r_1^2 = \pi d_1^2 = \pi d_2^2. \end{aligned}$$

In addition, Takebe also put forward another method, that is, regarding that the sphere is constructed by infinite number of small cones, of which the apex is the center of the sphere, the undersurface of cones constitutes surface S_m of the sphere, the height is radius of the sphere r, so the volume of the sphere $V = S_m r/3 = \pi d^3/6$ is the sum of volumes of small cones. Therefore, there is $S_m/3 \cdot d/2 = \pi d^3/6$ and then $S_m = \pi d^2$. Such thought is completely consistent with that of John Kepler (1571-1630) in the 17th century, which indicates that the West and East have common knowledge on mathematics. However, Takebe Katahiro's differential thought has not been further developed in Wasan.

2.3 Establishment of Extremum Algorithm

The Chapter Six of Mathematical Treatise on the Technique of Linkage offers the maximum problem regarding evaluation of polynomial function $f(x) = abx + (b-a)x^2 - x^3$. Takebe first applied Seki' exactly vanishing condition of modulus class [適尽方級法 tekijin hokyū hō] to evaluate the stable point of the function, that is, in case of $a = 7$ and $b = 8$, evaluate the function's derived function $V'(x)$ and its zero point $x = 14/3$, and then determine the maximum $f(14/3)$ of the function. The so-called exactly vanishing condition of modulus class equals the methods of deriving $f'(x) = 0$ from $f(x) = 0$. Seki and Takebe didn't involve the concept of derivatives; however, it was consistent with the Fermat method for evaluating function extremum in differential calculus [11, p. 64]. Takebe's such efforts opened up a new field of Wasan regarding research of theory of function extremum.

2.4 Diophantine approximation：From Seki's Jōichi jutsu to Takebe's Method of Residues

Before the Works and Days Calendar [授時暦 Shoushili], in the light of no adopting decimal fractions in Chinese ancient calendar, astronomical constants are usually expressed only by fractions. Generally speaking, astronomical constants x and y are both real numbers, so astronauts need to choose the natural number a and b which satisfy the inequality $|a/b - y/x| < \varepsilon$ (ε is an arbitrarily small and positive number) according to x and y. Then rational approximation becomes a fundamental issue of Chinese ancient astronomical calculation. In Han dynasty this kind of calculation had emerged so as to obtain the meeting periodic of many planets. He Chengtian [何承天] of Southern-Northern dynasties constructed a method of day adjustment [調日法 tiaorifa] for getting the value of lunar month as expected, the approximate rate [約率 yakuritsu, $22/7$] value of π and the close rate [密率 mitsuritsu, $355/113$] value of π which were given by Zu Chongzhi in the same dynasty maybe drive from the method of day adjustment.

To find out the source of a calendar, Chinese astronaut founded a method to arrive at the solution of sets of congruence expressions, Qin Jiushao [秦九韶] (1202-1261) of Song dynasty viewed it as the generalized algorithm program and called it great art of general remainders [大衍總数術 dayan zongshushu]. This method was called art of cutting bamboo [剪管術 jianguan shu] appeared in Yang Hui's [楊輝] works and its core is great art on the remainder one [大衍求一術 dayan qiuyishu], that is to realize the algorithm of $ax = 1 \pmod b$, which is equivalent to solve $ax - by = 1$. Seki called it the art to leave remainder one [剰一術 jōichi jutsu]. Without the background of real number theory, East Asian mathematicians deem the algorithm of $|a/b - y/x| < \varepsilon$ and $ax - by = 1$ as the same thing.

Seki generalized Yang Hui's art of cutting bamboo, obtained an algorithm similar to Qin Jiushao's great art of general remainders and found the method of approximate fractions which is similar to the method of day adjustmentin order to deal with rational approximation of real numbers. In Method of Residues [累約術 Ruiyaku-jutsu] which is revised by Nakane Genkei, Katahiro gave a generalization of Seki's method of approximate fractions to create Method of Residues, that is to get the algorithm of continued fraction expressions of integer x and y which satisfy $ax - by = c$. If we generalize the coefficients of an indeterminate equation of the first degree to irrational numbers, the rational approximation problem naturally appears which satisfy $|a/b - y/x| < \varepsilon$. He unified the method of approximate fractions and the art to leave remainder one. In west, this kind of issue firstly appeared in the posthumous manuscripts of Jacobi [Crelle Jouanal 69, 1869] while Takebe Katahiro discussed it 140 years before Jacobi.

Furthermore, Takebe Katahiro generalized the method of approximate fractions to get the method of repeated fractions [重約術 Jūyaku jutsu]. As a matter of fact, he gave out the simultaneous rational approximation problem [12].

2.5 *The improvement of Shōsahō: From Seki's Ruisai-Shōsahō to Takebe's Hōtei-Shōsahō*

Without knowing the equation of planet movement, for the sake of its position, Chinese ancient calendar commonly used the method of interpolation whose polynomial had reached 3 degrees in Works and Days Calendar from the start of Sui-Tang dynasties. It was called the rule of finding differences by Chinese mathematicians. After that, Seki gave a generalization of rule of finding differences and successfully applied it in the calculation of arc length. Then he constructed two types of algorithms: repeated rule of finding differences [累裁招差法 ruisai shōsa hō] and mixed rule of finding differences [混沌招差法 konton shōsa hō]. The repeated rule of finding differences is a kind of finite difference algorithm which has a significance of the general algebraic and programmed property. The mixed rule of finding differences, which is the Newton interpolation formula, is widely used in the calculation of arc length. It paved a new road for research of Wasan circle theory to the analysis method. Takebe Katahiro constructed the equation rule for finding differences [方程招差法 hōtei shōsa hō] in Volume 5 of the Complete Book of Mathematics.

As to the interpolation polynomial $f(x) = a_1x + a_2x^2 + a_3x^3 + \cdots + a_nx^n$, Takebe substitutes the values of the interpolation nodes $(x_i, f(x_i))$ into it in turn. So we could get a group of linear equations regarding its coefficient a_i, that is to say construct an augmented matrix of its coefficients.

$$\begin{pmatrix} x_n & x_{n-1} & \cdots & x_2 & x_1 \\ x_n^2 & x_{n-1}^2 & \cdots & x_2^2 & x_1^2 \\ x_n^3 & x_{n-1}^3 & \cdots & x_2^3 & x_1^3 \\ \vdots & \vdots & & \vdots & \vdots \\ x_n^n & x_{n-1}^n & \cdots & x_2^n & x_1^n \\ -f(x_n) & -f(x_{n-1}) & \cdots & -f(x_2) & -f(x_1) \end{pmatrix}$$

By the traditional procedure of equations [方程術 hōtei jutsu], we could obtain the solution after changing the matrix into triangular one through elementary transformation.

There are two sources of algorithms in Chinese rule of finding differences: one is the issue of summation of high-order arithmetic sequence in the theory of integers. The other is numerical approximation problem of continuous function. East Asian mathematicians make no distinction between them all the time for lack of theory basis of real numbers. The generalization of Seki's method towards Guo Shoujing's [郭守敬] (1231–1316) resulted in the connection between rule of finding differences and ancient procedure of equations made by Takebe and strengthened the significance of rule for finding differences as a method of undetermined coefficients. From this sense, the contrast among repeated rule for finding differences, mixed rule for finding differences, equation rule of finding differences, only exists in the different elementary transformation of the coefficient matrix above.

3 The contributions of Seki, Takebe Katahiro, Takebe Kataakira in the Complete Book of Mathematics

The Complete Book of Mathematics was compiled by Seki, Takebe Katahiro and Takebe Kataakira, who collaborated to finish this huge encyclopedic mathematical works for 28 years, but never published it. Until now there are about 20 editions [2]. These are the Description of Kataakira in the Biography of The Takebe [建部氏伝記]:

> >From the third year of the *Tenna* period [天和] (1683), three gentlemen projected, as Katahiro as leader, to write all the new ideas which each one can produce and all the old rules of mathematics and compile them till the middle of the *Genroku* period [around 1696]. Total 12 volumes were named the Complete Book of Mathematical Methods [算法大成 Sanpō Taisei], for which a rough draft was prepared. As he [Katahiro] became a busy administrative officer and could not examine the details, while Takakazu reached an advanced age and, because of chronic illness, could not concentrate himself on careful consideration. Therefore, from winter of the 14th year of the same period [1701], Kataakira himself began to investigate for eleven years, to think extensively, to annotate minutely and to expand the book into twenty volumes, which was renamed the Complete Book of Mathematics [大成算経 Taisei Sankei]. He himself transcribe it by hand and completed it. (The compilation of the book started in the *Tenna* period and finished in the last year of the *Hōei* period [1711]. Each volume was revised more than ten times. Because of this hard work, the compilation required 28 whole years. [1683-1711]) [7].

It describes the course of compilation of the Complete Book of Mathematics. The three men commenced to plan and compile it in 1683. At first the task was carried out only by Takebe Katahiro himself. Around 1695, 12 volumes had been completed. At this time, Takebe Katahiro was busy with public affairs and Seki got older. Then only Takebe Kataakira could undertake the task. After all, it had been accomplished until 1710.

But what kinds of achievements in mathematics are involved in this book which belongs to the three parts respectively? In order to clear up this issue, we have to find out the relations between the Complete Book of Mathematics and the other works of Seki and Takebe Katahiro.

The book is composed of 3 Parts: first Part consists of the first 3 volumes, which describe the basic knowledge of arithmetic. The materials are mainly chosen from Systematic Treatise on Arithmetic [算法統宗 Suanfa Tongzong] (1592), Introduction to Mathematics and Jinkōki [塵劫記, 1627] as the representative of Wasan books. Volume 3 is named Various Techniques [変技 Hengi] and contains the method of extracting roots [開方 kaihō], which is almost the arrangement of Seki's books Methods of Equation Modifications [開方翻変之法 Kaihō honhen no Hō] and Extraction of Roots from Equations [開方算式 Kaihō Sanshiki]. It contains five paragraphs (general method of extraction [開出總法 kaishutsu sōhō], three formulas [三式 sanshiki], ten quotients [十商 jūshō], exactly vanishing condition of modulus class, and replacement of numbers [替数 taisū]), amends some errors of the Seki's books, and presents detailed explanations about some brief issues. Seki Takakazu classified the extraction formulas [開方式 kaihō shiki] into 4 types: full quotient formula [全商式 zenshō shiki], alternative quotient formula [交商式 kōshō shiki], variable quotient formula [変商式 henshō shiki] and null quotient formula [無商式 mushō shiki]. Meanwhile, Takebe concludes only in 3 types considering the alternative quotient formula as a kind of the variable quotient formula.

The second Part is from Volume 4 to Volume 15. Volume 4 is named Three Essentials [三要 Sanyō], which is the core of authors' mathematical ideas and the center of this book. Their ideas are made up of symbol and figure [象形 shō–kei], flow and ebb [満干 man–kan] and number [数 sū], under the background thought of mathematics of symbols [象数学 shō sūgaku]. Then mathematical problems are divided into two types, symbol [象 shō] and figure [形 kei]. The variation of symbol and figure are classified into three conditions: ordinary [全 zen], excessive [背 hai], and extreme [極 kyoku]. They are applied to numbers. The number comprises integer [全 zen], rational [繁 han], algebraic [畸 ki], and irrational [零 rei]. First two are finite numbers, the other are infinite numbers, while the other volumes are written around these contents [10, p. 233].

The contents discussed in Volumes 5–7 belong to Method of Symbols [象法 Shō hō]. It contains mutual multiplication [互乗 gojō], polynomial fitting [畳乗 Jōjō], and pile sum [朶積 daseki]. Its last two chapters share similar contents with the Pile Sum and Rule of Finding Differences [朶積招差 Daseki Shōsa], which is the first volume of the Concise Collection of Mathematical Methods [括要算法 Katsuyō Sanpō], Seki Takakazu's posthumous publication in 1712. Volume 6 consists of three chapters: fractions [之分 shibun], various divisions [諸約 shoyaku], and art of cutting bamboo [剪管 senkan]. The same situation also happens between the last two chapters and various divisions and art of cutting bamboo, which are in the second volume of the Concise Collection of Mathematical Methods. Volume 7 consists of three chapters: magic numbers [聚数 shusū], Josephus problems [計子 keishi], and check of sign [験符 kenpu]. The content of this volume is also found in Seki's books, Method of Magic Squares, Method of Magic Circles [方陣之法・円攢之法 Hōjin no Hō, Ensan no Hō] (1683) and Method of Solving Josephus Problem, Method of the Check of Sign [算脱之法・験符之法 Sandatsu no Hō, Kenpu no Hō] (1683).

Volumes 8 and 9 discuss algorithms in daily use. Its contents derive from the daily mathematical issues appearing in diverse arithmetic books of Japan and China. It contains the issues of grain, gold, silver, money, apparel, consumption, taxation, quantity, movement, profit, transportation, conversion, difference, equilibrium, multiple, surplus, equation, and accumulation.

The contents of Volumes 10–15 belong to Geometry [形法 Kei hō]. Angles [角 Kaku] and Method of Regular Polygons [角術 Kaku jutsu] in Volume 11 coincide with the contents of Volume 3 of the Concise Collection of Mathematical Methods. Geometrical constants [形率 keiritsu] in Volume 12 mainly concern with the calculation of π, arc length, volumes of sphere and spherical crown, which are similar treated in circle theory, i.e., Volume 4 of the Concise Collection of Mathematical Methods. Its operational methods are mainly the method of successive accelerated approximation and the method of approximate fractions which is mentioned in Takebe's Mathematical Treatise on the Technique of Linkage.

Measurements [求積 Kyūseki] in Volume 13 is divided into two types of plane and solid. Applied Measurements [形巧 Keikō] in Volume 14 is classified into "section, tangency and consistency." Two kinds of load and coiling exist in Applied Measurements in Volume 15. It is mainly about the calculation problems of various geometric figures. A part of these problems scatter in Seki's Answers to the Elucidation of Mathematics [闕疑抄答術 Ketsugishō Tōjutsu], Answers to the Clarification of Mathematics [勿憚改答術 Futsudankai Tōjutsu], Measurements, Methods of Solving Explicit Problems [解見題之法 Kaikendai no Hō], and Mathematical Methods for Exploring Subtle Points.

The last Part is from Volume 16 to Volume 20. two meanings [両義 ryōgi] in volume 16 origin from Seki's Critical Studies of Problems [題術辯議之法 Daijutsu Bengi no Hō] (1685). It discusses whether a mathematical problem has a solution or not and its condition respectively. Solution of Whole Problems [全題解 Zendaikai] in Volume 17 divides the regular issues with right question set into four types of explicit problem [見題 kendai], implicit problem [隱題 indai], concealed problem [伏題 fukudai], covered problem [潜題 sendai] whose contents are from Seki's Methods for Solving Explicit Problems, Methods of Solving Implicit Problems [解隱題之法 Kaiindai no Hō] (1685), Methods of Solving Concealed Problems [解伏題之法 Kaifukudai no Hō] (1683). On Defective Problems [病題議 Byōdaigi] in Volume 18 discusses the remodification of defective problems whose contents are from Seki's Method for Restoring Defective Problems [病題明致之法 Byōdai Meichi no

Hō] (1685). Volumes 19–20 are examples of explicit problems, implicit problems, concealed problems, and covered problems.

3.1 Source of knowledge of the Complete Book of Mathematics

Vol. 1–3, is basic knowledge of mathematics from Introduction to Mathematics, Systematic Treatise on Arithmetic, Yang Hui's Methods of Mathematics [楊輝算法 Yang Hui Suanfa], Jinkōki etc. , and part from Seki's Methods of Equation Modifications and Extraction of Roots from Equations.

Vol. 4, its main content is Three Essentials, which is referred in Takebe's Mathematical Treatise on the Technique of Linkage.

Vol. 5, it contains the pile sum and the rule for finding differences which is also recorded in Volume 1 of the Concise Collection of Mathematical Methods of Seki Takakazu.

Vol. 6, it contains procedure of various divisions and art of cutting bamboo, which is also recorded in Volume 2 of the Seki's Concise Collection of Mathematical Methods.

Vol. 7, its Main content is analysis of combination, which is recorded in Seki's book Method of Magic Squares, Method of Magic Circles and Method of Solving Josephus Problem, Method of the Check of Sign

Vol. 8–9, is common algorithm which is selected from Introduction to Mathematics, Systematic Treatise on Arithmetic, Yang Hui's Methods of Mathematics, Jinkōki, etc.

Vol. 10, is Pythagorean problem which is selected from Systematic Treatise on Arithmetic, Yang Hui's Methods of Mathematics, Jinkōki, etc.

Vol. 11, its content is Method of Regular Polygons which is consistent with the content in Volume 3 of the Concise Collection of Mathematical Methods.

Vol. 12, its content is circle theory which is also recorded in Volume 4 of the Concise Collection of Mathematical Methods. and Takebe's Mathematical Treatise on the Technique of Linkage.

Vol. 13, it contains Integral formula of geometry which is also recorded in Seki's Measurements.

Vol. 14–15 is deal with computation of graph which is also recorded in Seki's books Answers to the Elucidation of Mathematics, Answers to the Clarification of Mathematics, Methods for Solving Explicit Problems, and Mathematical Methods for Exploring Subtle Points.

Vol. 16, Math problems and mathematical methods are discussed in this part, which is identical with Seki's Critical Studies of Problems Exactly.

Vol. 17, its main content is elimination method, Determinant and equations which are identical with Seki's Trilogy [三部抄 Sanbushō] in principle.

Vol. 18, its content is Restoration of Defective Problems which is identical with Seki's Method for Restoring Defective Problems

In a contrast, The first 12 volumes of Takebe Katahiro and the late 8 volumes of Kataakira are considered separately, we could easily reach at the conclusion that the Complete Book of Mathematics is mainly the adaptation of Seki's works. In the first 12 volumes, not only mathematical ideas but also its methods both have breakthrough on the basis of Seki's work. At first, three essentials never appeared in Seki's book. Second, circle theory in Volume 12 is the same as Searching Circle Number [探円数 tan-ensu], Searching Arc Length [探弧術 tan-ko jutsu] in the Mathematical Treatise on the Technique of Linkage. In addition, the equation rule for finding differences of Volume 5 were not appeared in Seki's book. But the last 8 volumes generally inherited Seki's traditions. Volumes 13–15 contains new classification of the geometric problems of Seki. No innovation appeared in the method. The only thing that merits to mention is the improvement on the dealing of Seki's expansion of determinant. The type of problem [題 dai] and algorithm [術 jutsu] of Critical Studies of Problems was distinguished again in Volume 16 and 18, and the Transforming way of Defective Problems in Restoration of Defective Problems was revised.

Above all, the new thoughts and innovative methods in the Complete Book of Mathematics contribute to Takebe which are all reflected in the first 12 volumes. It shows that some of Takebe's work had exceeded Seki's before 1695. But Kataakira said like this in Biography of The Takebe:

> However, I am originally of hermitic character and do not want to become famous in the world. As I intend to keep myself incognito and hide all of my good act, I transfer my achievement to Katahiro and consider myself as a foolish person. [7].

He means he had contributed more than his younger brother Takebe on the completion of this book. The book has been never published yet, so profit never arrives at Katahiro. But why did Kataakira once said "I have given all my accomplishments to Katahiro. I declared myself completely an idiot." From this point of view, Kataakira seemed not so indifferent to fame and wealth as he claimed himself. And he perhaps had some ambition to surpass his younger brother.

4 Takebe Katahiro's respect for Seki Takakazu

As the most complacent student and intimate collaborator of Seki, Takebe went through an extraordinary relations with his master. He showed great homage and respect to Seki. And we could find such emotions for Seki in his works.

When most of Wasan mathematicians blamed on the new discovery of Seki, Takebe Katahiro wrote Mathematical Methods for Clarifying Slight Signs to explain his teacher's new method and defended Seki's academic reputation.

Even when Takebe became aged and distinguished, he showed a deep reverence for his teacher Seki. His admiration is beyond description. Seki was referred for 14 times by Takebe in his Mathematical Treatise on the Technique of Linkage. Though Takebe had made a very outstanding achievement in mathematics, even surpassing his teacher, he still held Seki in high esteem.

In Takebe's view, the gift of a man could be divided into wholly pure and partially pure. At the same time, mathematics owns this kind of essence. After all, we could make it if we follow this trend. The wholly pure kind would much more easily stand out without any doubt. He was complaining all the time that his gift was less than his teacher's. Once he said:

> Because originally I am of foolish attribute, if I want to understand, by principles, the true rule only by observation, although it may be very easy if we encounter a procedure like this, which is simple in principle, I cannot always attain a solution when a given procedure is not based on a simple principle. In such a case, we investigate repeatedly, relying exclusively on numbers, to understand there is some basis, on which we can establish the true rule. For this reason, I do not dare to consider the former procedure second rate. Probably, is it because of my distorted attribute that it is difficult for me to understand without any investigation? If I were straight, without distinguishing the bases of numbers and principles, I would be able to understand everything immediately without any investigation. But because I am of distorted attribute, even though I study deeply, I will not be able to attain such a state. Generally speaking, in numbers, in procedures, and in rules, everything is originally natural. He who understands this does not trod on a new path; his path merges with the natural path to attain understanding. If this is the case, it is also appropriate to attain understanding after investigation. I strongly recognize that Master Seki's natural intelligence is without parallel in the world. He always said that problems on the circular "measure" were very difficult to solve. Alas, this is because he [chose to] operate in a relaxed manner, but I dare say that even problems on the circular "measure" can certainly be solved by tenacity. This is only because I work in a painstaking manner. The reason why Master Seki said that he could not solve this type of problem was that he operated in a relaxed manner to find a quick and easy solution, endeavoring to solve problems immediately without any investigation. It was not because he could not solve them. Perhaps, he did not like to go into the matters thoroughly. Because natively I am of foolish attribute, I cannot reach a quick and easy solution operating in a relaxed manner. I have found a way to be peaceful even operating always in a painstaking manner. Therefore, if I investigate [in this way], I know I will certainly obtain the solution. Reflecting on this, I know that my native attribute is one [part] out of ten less than that of Takakazu. [8].

As a matter of fact, his gift on maths compares favorably with Seki's. High appraisal has been given out in Biography of The Takebe:

> He [Katahiro] started to work on mathematics at 13 years. In the same way as his elder brother Kataakira, he studied all day long. As he was very intelligent, he understood the whole theory of mathematics and mastered also the art of calendar and astronomy. His mathematical talent was not less than [Seki] Takakazu; in particular, he was better than his master in calculating numerical values and in processing various operations. [7].

Takebe also gives high appreciation on Seki's method and makes a comparison between his new ways and Seki's. For example:

> Master Seki said that, in order to understand thousands of rules, it is most essential to observe the form and to establish the path [of reasoning]. His hidden purpose was to understand the true procedure from the beginning without any investigation. Thus, in the latter procedure, he observed the form of a sphere and considered it as a cone and its center as the apex. In this way, observing the form and establishing the path [of reasoning], he understood the true procedure immediately without any investigation. Therefore, he considered the former procedure second rate. [8].

He thinks it is Seki who constructs operations of celestial element [天元演段 tengen endan]. He views it as an unparalleled method in solving maths problems. Full of praise is also provided on Seki's circle theory. Although Seki and Zu Chongzhi are aliens, their comprehension accords.

It is rare and commendable that Takebe "learns from conventions but not restrained by them." He didn't follow Seki blindly nor drag his feet, and many mathematics methods he used were improved on the basis of those of Seki. The solutions to each problem are provided with reasons for tenability and evaluated by comparing with former methods with their innovations being figured out. Only by his rational analysis, he can point out his superiority over his own teacher. In deed, Takebe never has a complex of self-inferiority. Although he claims himself stupid, the words in Mathematical Treatise on the Technique of Linkage also shows his confidence of the new methods that he ever found.

5 Seki and Takebe in the history of mathematics in "culture circle of Chinese characters"

In the middle of the sixteenth century most countries in East Asia began to keep in touch with western Christian culture. After that, East Asian countries' policy turned rapidly from prohibition to seclusion. Under this historical milieu, the development of East Asian traditional science diverged. Qing's mathematics integrated the features of the Occident and China guided ideally by traditional Chinese values aided with modern Western science and technology. On the other hand, Wasan was basically not influenced by Western mathematics while it developed only by itself inheriting Chinese mathematical traditions. Seki and Takebe have laid the foundation of Wasan.

Chinese traditional mathematics takes practical problem solution as its objective and algorithm construction as its kernel. Mathematics in Han-Tang dynasties are primarily concerned with arithmetic knowledge while in Song-Yuan dynasties it turned to algebraic knowledge as its main part which are most in lost during Ming dynasty. Wasan is an arithmetic knowledge system full of Edo cultural features which is built on the basis of Song-Yuan's algebraic traditions. It enlarged the content of algebra and geometry of Song-Yuan and developed the infinitesimal algorithm. Wasan mathematicians furthered the cause of traditional mathematics in "culture circle of Chinese characters" into the level of modern maths just before Newton. Takebe has left an outstanding achievements in algebra, infinitesimal analysis, Diophantine approximation, especially in infinitesimal analysis. It reveals that East Asian mathematicians' achievements on calculus in the seventeenth century compares favorably with those of Europe at the same time. The soul of algorithm is the key feature of Takebe's maths. His construction of method of successive accelerated approximation and method of residues is a paradigm of procedural algorithm.

East Asian mathematics usually pays much attention on creating procedure [術 shu] so as to solve practical problems. Meanwhile, it neglects the theory underlying in "procedure." But it doesn't mean procedure is only accumulation of experience or lack of theoretical recognition. For example, in Commentary on the Nine Chapters on the Mathematical Art [九章算術注 Jiuzhang Suanshu Zhu, 263], Liu Hui elucidated each theory of procedure respectively. Takebe carried forward the tradition of Liu Hui and emphasized the important significance of induction full of oriental characteristics in the field of mathematical innovation and algorithm construction. He rebuilt the knowledge system of East Asian mathematics through the basic framework of five techniques [五技 gogi], three essentials. two meanings. He uncovered the essence of East Asian mathematics from four stages: mathematical method, mathematical objects, mathematical precision, mathematical truth. This deed bears a special significance in the world's mathematical history.

References

1. Akira Hirayama [平山諦]: *People in History of WASAN* [和算史上の人々], Fuji Junior College [富士短期大学出版部] (1966).
2. Hikosaburo Komatsu [小松彦三郎]: A Report on the Revised Edition of the Complete Book of Mathematics [大成算経校訂本の現状報告], *RIMS Kokyuroku*, **1546**, Kyoto Univ., pp. 140–156 (2007).
3. Tsukane Ogawa, Ken'ichi Sato, Osamu Takenouchi and Mitsuo Morimoto: *The Mathematics of Takebe Katahiro* (in Japanese), Kyōritsu Press [共立出版] (2008).
4. Tsukane Ogawa [小川束]: History of methods of mathematical analysis in modern Japan [近世日本数学における解析的算法の歴史], *Special Lecture at Mathematical Society of Japan* [企画特別講演, 日本数学会企画特別講演録 1392], at Gakushūin University [学習院大学] (1999).
5. L. F. Richardson: The approximate arithmetical solution by finite differences of physical problems including differential equations, with an application to the stresses in a masonry dam, *Philosophical Transactions of the Royal Society of London*, Series A 210, pp. 307–357 (1910).
6. W. Romberg: Vereinfachte numerische Integration, *Det Kongelige Norske Videnskabers Selskab Forhandlinger*, 28:7, Trondheim (1955).
7. Takebe Kataakira [建部賢明]: *Biography of the Takebe* [建部氏伝記抄録], Tohoku University Library [東北大学附属図書館], Code [登録番号] ws000106.
8. Takebe Katahiro: Mathematical Treatise on the Technique of Linkage [綴術算経 *Tetsujutsu Sankei*], hand-copied book collected in Cabinet Library, No. 23581.
9. Xu Zelin: Takebe Katahiro and Romberg Algorithm, *Historia Scientiarum*, vol. 9, no. 2 (1999).
10. Xu Zelin: Takebe Katahiro's Epistemology of Mathematics, *Studies in the History of Natural Sciences*, vol. 21, no. 3, p. 233 (2001),
11. Xu Zelin: *Kyokusujutsu* of *Wasan* and Bud of Extremum Concept in Traditional Mathematics of China, *Journal of Dialectics of Nature*, vol. 24, p. 37 (2002).
12. Xu Zelin: *Shoyakujutsu* of *Wasan* and Diophantine Approximation and their headwaters in Chinese traditional mathematics, *Journal of History of Mathematics*, Japan, vol. 180, pp. 1-15 (2004).

Takebe Katahiro's Algorithms for Finding the Circular Arc Length

Mitsuo Morimoto

Abstract In the 17th century, Japanese mathematicians could calculate the arc length numerically at any accuracy once the diameter and the sagitta were given numerically. But they could not find any formulas to express the arc length in terms of the diameter and the sagitta; only polynomials or fractions of polynomials were considered as formulas. Finally in 1722, Takebe Katahiro (1664–1739) succeeded in expressing the arc length in terms of the diameter and the sagitta in the form of an infinite series expression.

Introduction

Let an arc with sagitta c be given on a circle of diameter d. In modern mathematics, the arc length s is given by one of the inverse trigonometric functions:

$$s = d \arcsin\left(2\sqrt{\frac{cd-c^2}{d}}\right) = d \arccos\left(1-\frac{2c}{d}\right) = 2d \arcsin\left(\sqrt{\frac{c}{d}}\right). \quad (1)$$

In the 17th century, Japanese mathematicians could calculate the arc length s numerically at any accuracy once the diameter d and the sagitta c were given numerically. But they could not find any formulas to express the arc length s in terms of d and c; only polynomials or fractions of polynomials were considered as formulas. Finally, Takebe Katahiro [建部賢弘] (1664–1739) published in the Mathematical Treatise of the Technique of Linkage [綴術算経 Tetsujutsu Sankei] (1722) [10] the following infinite series expression of s in d and c:

Mitsuo Morimoto
1-64-5, Wada, Suginami-ku, Tokyo 166-0012, Japan
e-mail: mormormor@jcom.home.ne.jp

$$\left(\frac{s}{2}\right)^2 = cd + \frac{1}{3}c^2 + \frac{1}{3}\frac{8}{15}\frac{c^3}{d} + \frac{1}{3}\frac{8}{15}\frac{9}{14}\frac{c^4}{d^2} + \frac{1}{3}\frac{8}{15}\frac{9}{14}\frac{32}{45}\frac{c^5}{d^3} + \frac{1}{3}\frac{8}{15}\frac{9}{14}\frac{32}{45}\frac{25}{33}\frac{c^6}{d^4} + \frac{1}{3}\frac{8}{15}\frac{9}{14}\frac{32}{45}\frac{25}{33}\frac{72}{91}\frac{c^7}{d^5} + \cdots . \tag{2}$$

Takebe rejoiced at this discovery considering himself to have surpassed his master Seki Takakazu [関孝和] (ca. 1642–1708) : indeed (2) was the first infinite series ever found in the history of Japanese mathematics. (See [2], [4] and [6].) But he was not quite well satisfied with his discovery; once truncated the Taylor series expansion (2) gave him very accurate approximation of the arc length s if the sagitta c is small but did not give him good approximation as expected if c approaches to $d/2$, i.e., if the arc is close to the half circle. He strove for rational approximation formulas which ensure good uniform approximation of the arc length.

Following [10], we shall review approximation formulas of the arc known to Seki Takakazu and Takebe Katahiro. (An outline of this paper was published in [3].)

1 Numerical value of arc lengths

At early stage of the *Edo* period (1603–1868), Japanese mathematicians were interested in the calculation of the length of the regular 2^n-gons inscribed to a circle of diameter $d = 10$ for $n = 2, 3, 4, \cdots$ and obtained the approximate numerical value of the circular length 10π. For example, in Platter of Mathematics [算俎 Sanso] (1663), Muramatsu Shigekiyo [村松茂清] (1608–1695) presented the calculation of the length of 2^n-gons for $n = 2, 3, \cdots$, up to 15. Seki Takakazu invented an acceleration method, equivalent to the Aitken method, to find the limit of a sequence and obtained 12 digits: $\pi = 3.14159265359\cdots$. He also calculated the length of the arc with sagitta $c = 1, 2, 3, 4$, and 4.5. Seki's results were published posthumously in the Concise Collection of Mathematical Methods [括要算法 Katsuyō Sanpō] (1712) [7] as follows:

$$\begin{aligned} s(10,1) &= 6.4350116, & s(10,2) &= 9.272953, \\ s(10,3) &= 11.5927958, & s(10,4) &= 13.6943852, \\ s(10,4.5) &= 14.70629030, \end{aligned} \tag{3}$$

where $s(d,c)$ denotes the length of the arc with diameter d and sagitta c. Note that $s(10,5) = 5\pi$. Takebe Katahiro found that the repeated applications of Seki's acceleration method could increase the degree of approximation and obtained the 43 digits of π. (See sheet 37 verso of [10].)[1] Takebe thought that he had exceeded his master Seki in the numerical calculation of π. As quoted below, Takebe also calculated $s(10,1)$, $s(10,2)$, $s(10,3)$, $s(10,4)$, $s(10,4.5)$, and $s(10,4.9)$ with his newly ameliorated method and examined the values.

[1] Pages of [10] are counted according to the sheet. The sheet is folded; the front part is called recto and the back part is called verso.

Takebe and his contemporaries could calculate the value of arc length as accurate as they wished once a value was given to the sagitta c . But they were not satisfied with the numerical calculation and tried to find a polynomial or a rational formula (fraction of polynomials) of diameter d and sagitta c to represent s or s^2. But they could not find any satisfactory answer. As the inverse trigonometric functions (1) are not rational, we know that any such attempts should fail. There were, nevertheless, several attempts to find the formula, as Takebe states at sheet 41 recto in [10]:

> At the beginning, assuming the diameter to be 1 *shaku* [i.e., 10 *sun*] and the sagitta to be 1 *sun*, 2 *sun*, 3 *sun*, or 4 *sun*, we searched for the definite back arc [弧背 kohai][2] by procedures of decomposition and of incremental divisor. Further, we continued to determine the definite back arc for the sagitta of 4 *sun* 5 *fun* [i.e., 4.5 *sun*], 4 *sun* 9 *fun* [i.e., 4.9 *sun*], etc. We examined these numbers but could not understand the underlying rule when the back arc was close to the half circle. Therefore, although Master Seki formed and revised the coefficient formula of the back arc twice and I myself also formed and revised once, we abandoned these procedures because all the formulas were not accurate.[3]

In [10] Takebe formulated three formulas for the squared arc length s^2. In his first formula (2) the coefficients followed a simple rule and he almost thought he had solved the problem. He rejoiced at this discovery considering himself to have exceeded his master Seki for the second time.[4] Then he realized that the first formula (2) gave him very precise approximation of the squared arc length while the sagitta c was small but did not give him the precision when c became larger, i.e., good approximation uniformly for all the value c with $0 \leq c \leq d/2$. He continued his search for the rational formula which would give him better uniform approximation. He elaborated the second and the third formulas for the squared arc length in the rational formula.

Including these three formulas, Takebe cited eight algorithms (procedures, or formulas) to express the back arc in [10]:

1. (sheet 42 verso in [10]) "the old method" (See (4) and (5) below.),
2. (sheet 45 recto in [10]) "This original procedure": Takebe's first formula (2), which coincides with the truncated Taylor expansion,
3. (sheet 46 recto in [10]) "Master Seki's 4-multiplication procedure for finding the back arc"[5] (See (7) and (9) below.),
4. (sheet 49 recto in [10]) "To search the differences by the devision of the difference of the diameter and the sagitta": Takebe's second formula (12),
5. (sheet 49 verso in [10]) "the 6-multiplication original procedure for finding the back arc, which we [Takebe] established earlier" (See (8) below.),

---

[2] This means an arc in the sense of a segment of a circle. On the other hand, an arc [弧 ko] in the Edo period meant a disk segment (editor).

[3] *shaku*, *sun*, *fun* are units of length. 1 *sun* is about 3 cm. 1 *shaku* = 10 *sun*, 1 *sun* = 10 *fun*.

[4] In the spring of 2005 Mr. Yokotsuka found a manuscript titled the Circular Arc and its multisections [弧背截約集 Kohai Setsuyaku Shū], in which it was stated that Takebe Katahiro found the formula unexpectedly on January 13, 1722. See [12].

[5] Let n be a natural number. A "n-multiplication procedure" means a procedure which contains polynomials of degree less than or equal to $n+1$.

6. (sheet 49 verso in [10]) "This old procedure": the revised old method (See (6) below.),

7. (sheet 53 verso in [10]) "This main procedure": Takebe's third formula (13),

8. (sheet 54 verso in [10]) "the general procedure. All details are recorded in the Arc Rates [弧率 Koritsu]".

Although Takebe's three formulas were formulated explicitly, other algorithms were not clearly mentioned in [10]. We shall examine mathematics books in the early *Edo* period to find what were indeed these algorithms.

2 The Register of Jugai

In the Register of Jugai [竪亥録 Jugai roku] (1639), Imamura Tomoaki [今村知商] gave the following formula in a chant form (See [5], vol.1, page 221.):

$$s^2 = 4cd + 2c^2. \tag{4}$$

As we have $s^2 = (5/2)d^2$ for $c = d/2$ (half circle), we know that π was assumed to be $\sqrt{10}$ in this book. Takebe stated at sheet 42 verso in [10]:

> It corresponds with the old method where the squared sagitta multiplied by the coefficient 5.8696 (strong) is added to the square of the chord to find the squared back arc.

As the square of the chord is equal to $4(cd - c^2)$, the old method referred here is the following approximation formula:

$$s^2 = 4(cd - c^2) + 5.8696c^2 = 4cd + 1.8696c^2. \tag{5}$$

As Takebe assumed π to be $355/113$, which he called the precise value [密率 mitsuritsu], the formula (4) was revised as (5), which was quoted as the "old method" at sheet 42 verso in [10]. In fact, when $d = 10$ and $c = 5$, the formula (5) gives

$$s(10,5)^2 = \left(4 \times \frac{1}{2} + 1.8696 \times \frac{1}{4}\right) \times 100 = 246.74$$

and $\left(5 \times \dfrac{355}{113}\right)^2 = 246.740\cdots$, whence the coefficient 5.8696 appeared.

There is another mention of the "old method" at sheet 49 verso in [10]:

> In an old method, multiplying the sagitta by itself, multiply by the coefficient of the squared sagitta [i.e., 5.8696], add it to the square of the chord. The sum is called the square of the approximate back arc. Subtract the double of the sagitta from the diameter. Multiply the remainder by the squared sagitta, divide it by the difference of the sagitta and the diameter and halve it. Subtract the obtained number from the squared approximate back arc and call it the square of the definite back arc. This old procedure corresponds naturally to the previous main procedure with two differences.

Here, the square of the approximate back arc means the right-hand side of (5) given by the "old method" and s^2 is claimed to be better approximated by the square of the definite back arc as given in the following:

$$s^2 = 4(cd - c^2) + 5.8696c^2 - \frac{(d-2c)c^2}{2(d-c)}. \tag{6}$$

This is the "revised old method" given at sheet 49 verso in [10]. Note that the correction term in (6) becomes 0 when the sagitta is null or $d/2$.

3 The Mathematical Methods for Clarifying Slight Signs

Takebe Katahiro published the Mathematical Methods for Clarifying Slight Signs [研幾算法 Kenki Sanpō] [9] when he was 19 years old. (For the life of Takebe, we refer the reader to [11], which contains English translation of Takebe's three mathematical monographs and his short biography in English.) In this book, he gave solutions to the 49 problems proposed by Ikeda Masaoki [池田正意] in the Mathematical Textbook on Multiplication and Division [数学乗除往来 Sūgaku Jōjo Ōrai] (1674). The first problem is to ask the area of the part of the plane surrounded by the arc and the chord. But the area can be expressed by means of the arc length s, the solution contained the following approximation formula of the squared arc s^2:

$$K(d,c) = \frac{-A_0d^5 + A_1cd^4 + A_2c^2d^3 + A_3c^3d^2 - A_4c^4d + A_5c^5}{A_6d^2} \tag{7}$$

where the coefficients are given

$$A_0 = 81,\ A_1 = 18393267,\ A_2 = 6021104,\ A_3 = 4081524,$$

$$A_4 = 715920,\ A_5 = 5500232,\ A_6 = 4596840.$$

This formula is a polynomial of degree 5, hence, is a 4-multiplication procedure. In [9] π being assumed to be $355/113$, the formula (7) gives good approximation for $d = 10$, $c = 1$, $c = 2$, $c = 3$, $c = 4$, and $c = 4.5$ but gives negative values if the sagitta c becomes small : $K(10,0) = -45/25538$. This shows that Takebe and his master Seki had no idea, at least when they were writing this book, to consider cases with very small sagittas.

According to the Supplementary Guide [凡例 hanrei] in [9] Takebe admitted that the problems related with the arc length (Problems 1 and 48) were instructed by the master. If we respect this passage, we may ascribe (7) to Seki Takakazu and consider it as the "Master Seki's 4-multiplication procedure" in [10].

4 The Concise Collection of Mathematical Methods

The Concise Collection of Mathematical Methods [7] contains a formula in d and c to approximate the squared arc length s^2. Although the book was published in 1712 posthumously, we can guess the result had been obtained around 1685, Seki's most productive years, and that it would be little later than [9].

In volume 4 of [7] we find the following approximation formula $Y(d,c)$ for s^2:

$$Y(d,c) = \frac{B_1 cd^6 - B_2 c^2 d^5 + B_3 c^3 d^4 - B_4 c^4 d^3 + B_5 c^5 d^2 - B_6 c^6 d - B_7 c^7}{B_8 (d-c)^5}. \tag{8}$$

where the coefficients are given by

$$B_1 = 5107600,\ B_2 = 23835413,\ B_3 = 43470240,\ B_4 = 37997429,$$

$$B_5 = 15047062,\ B_6 = 1501025,\ B_7 = 281290,\ B_8 = 1276900.$$

Because $B_1 = 4B_8$, $Y(d,c) = 4cd + \cdots$ when c is very small. Because $Y(d,0) = 0$, (8) is better than (7) when the sagitta is very small.

As the numerator of (8) is a polynomial of degree 7, this is a 6-multiplication procedure. But it is not certain that it corresponds to the "original 6-multiplication procedure" cited in [10] which was ascribed to Takebe himself. Only the 4-multiplication procedure was ascribed to master Seki in [10].

Assuming that the circular coefficient $\pi = 355/113$, the formula (8) is constructed by an interpolation of the values (3) at $c = 1$, $c = 2$, $c = 3$, $c = 4$, $c = 4.5$, $c = 5$ with $d = 10$. This method is equivalent to Newton's interpolation formula. (See [1], English part, page 63.)

5 The Complete Book of Mathematics

Seki Takakazu, Takebe Katahiro, and Takebe Kataakira (Katahiro's elder brother) started editing the Complete Book of Mathematics [大成算経 Taisei Sankei] [8] in 1683 and compiled a version in 12 volumes around 1695. The whole 20 volumes were completed in 1711 by the effort of Kataakira. It took 28 years to accomplish the editing of the encyclopedic collection of mathematics of Seki and Takebe brothers.

The circular coefficient π, the arc length and related topics are treated in volume 12 of [8]. The formula for the arc length in [8] improves that of [7] it can be concluded the former is later than the latter. The formula in [8] is reproduced in [5] (vol. 2, page 428):

$$T(d,c) = \frac{H_1 cd^4 - H_2 c^2 d^3 + H_3 c^3 d^2 - H_4 c^4 d - H_5 c^5}{K_0 d^3 - K_1 cd^2 + K_2 c^2 d - K_3 c^3} \tag{9}$$

where the coefficients are given by the following:

$$H_1 = 39020125496,\ H_2 = 61434714678,\ H_3 = 25918266069,$$

$$H_4 = 1828448393,\ H_5 = 102756994,\ K_0 = 9755031374,$$

$$K_1 = 18610356125,\ K_2 = 10948798854,\ K_3 = 1913138432.$$

In [8], the circular coefficient π is more precise than $355/113$. The formula (9) is also a 4-multiplication procedure and earlier than [8]. It is another candidate for the "Seki's 4-multiplication procedure" referred in [10].

6 Takebe's First formula

We already discussed in details the Takebe's three formulas in [4]. They are usually called "formulas" but in the context of the Seki and Takebe mathematics it is better to consider them as algorithms or procedures. This is the reason why the author chose the term "algorithm" in the title of this paper.

In *wasan*, Japanese traditional mathematics, a polynomial of numerical coefficients was represented by a configuration (column vector of polynomial's coefficients) on the counting board. This is the procedure of celestial element [天元術 tengen jutsu]. If a value is given to the invisible unknown x, the value of the polynomial was calculated on the counting board. There was no polynomial notation of modern mathematics but only operations on configurations on the counting board. In this regard, it is hard to reproduce Seki and Takebe's mathematics in modern mathematical notation. Having said so, we try to reproduce their mathematical ideas using modern terminology.

Takebe's first formula is usually written as (2) and considered as an infinite series which coincides with the Taylor series expansion. But the following form resembles more the standard notation of the *wasan*:

$$\begin{aligned} s^2 = 4cd\Bigg(1+\frac{1}{3}\frac{c}{d}\Bigg(1+\frac{8}{15}\frac{c}{d}\Bigg(1+\frac{9}{14}\frac{c}{d}(1+ \\ +\frac{32}{45}\frac{c}{d}\Bigg(1+\frac{25}{33}\frac{c}{d}\Bigg(1+\frac{72}{91}\frac{c}{d}(1+\cdots)\Bigg)\Bigg)\Bigg)\Bigg)\Bigg)\Bigg) \end{aligned} \tag{10}$$

The more faithful reproduction of the *wasan* notation looks as follows:

$$D_1(d,c) = 4\times\frac{c^2}{3},\quad S_1(d,c) = 4cd + D_1(d,c),$$

$$D_2(d,c) = D_1(d,c)\times\frac{c}{d}\times\frac{8}{15},\quad S_2(d,c) = S_1(d,c)+D_2(d,c),$$

$$D_3(d,c) = D_2(d,c)\times\frac{c}{d}\times\frac{9}{14},\quad S_3(d,c) = S_2(d,c)+D_3(d,c),$$

$$D_4(d,c) = D_3(d,c) \times \frac{c}{d} \times \frac{32}{45}, \quad S_4(d,c) = S_3(d,c) + D_4(d,c),$$

$$D_5(d,c) = D_4(d,c) \times \frac{c}{d} \times \frac{25}{33}, \quad S_5(d,c) = S_4(d,c) + D_5(d,c),$$

$$D_6(d,c) = D_5(d,c) \times \frac{c}{d} \times \frac{72}{91}, \quad S_6(d,c) = S_5(d,c) + D_6(d,c),$$

Starting from $4cd$, adding the differences D_n, we approximate the arc length s^2 by S_n, $n = 1,2,3,\cdots$.

Takebe observed carefully the denominators and the numerators separately and deduced recursively that the fraction which should be multiplied to the $(i-1)$-th term to obtain the i-th term ($i \geq 2$) was given by $\dfrac{2i^2}{(2i+1)(i+1)}$ when i is even and by $\dfrac{i^2}{(2i+1)(i+1)/2}$ when i is odd.[6]

Takebe recognized that the calculation could be continued for as many steps as wished using the following algorithm:

$D := 4c^2/3;$
$S := 4cd + D;$
for $i := 2$ **to** N **do begin**
  if i **mod** $2 = 0$ **then**
    begin $P := (2i+1)(i+1); Q := 2i^2$ **end**
  else
    begin $P := (2i+1)(i+1)/2; Q := i^2$ **end**;
  $D := D \cdot \dfrac{Q}{P} \cdot \dfrac{c}{d};$
  $S := S + D;$
end;

The formula (2) was later reformulated as

$$\left(\frac{s}{2}\right)^2 = cd\left\{1 + \frac{2^2}{3\cdot 4}\left(\frac{c}{d}\right) + \frac{2^2\cdot 4^2}{3\cdot 4\cdot 5\cdot 6}\left(\frac{c}{d}\right)^2 + \frac{2^2\cdot 4^2\cdot 6^2}{3\cdot 4\cdot 5\cdot 6\cdot 7\cdot 8}\left(\frac{c}{d}\right)^3 + \cdot\right\} \quad (11)$$

in the Circle Theory and the Circular Arc Length [円理弧背術 Enri Kohai Jutsu], where (11) was derived by an algebraic method. (See [6].)

Note that Takebe could not recognize the identity of the two forms of fraction in the coefficients of (11) and wrote the rule of coefficients for odd and even terms separately in (2).

At sheet 48 recto in [10], Takebe commented his first formula as follows:

> The original procedure is a natural method which follows the attribute of the arc. If we seek the squared half back arc for an extremely small sagitta, the successive differences decrease

---

6 This fact was shown by numerical data in the table at sheet 47 recto in [10]. Seki and Takebe could not represent general i.

> more rapidly and the truer number can be achieved quickly. But if the sagitta is getting larger in the case of a half circle, the successive differences decrease slowly and more and more differences must be calculated. It cannot be considered as the definite coefficient.

He was not satisfied with this procedure (2) because the approximation is not uniform for c, $0 \leq c \leq d/2$. He then started another quest using the form of fraction.

7 Takebe's Second formula

Takebe's second formula was the following:

$$\begin{aligned}\left(\frac{s}{2}\right)^2 = cd + \frac{1}{3}c^2 + \frac{1}{3}\frac{8}{15}\frac{c^3}{d-c} - \frac{1}{3}\frac{8}{15}\frac{5}{14}\frac{c^4}{(d-c)^2} + \frac{1}{3}\frac{8}{15}\frac{5}{14}\frac{12}{25}\frac{c^5}{(d-c)^3} \\ - \frac{1}{3}\frac{8}{15}\frac{5}{14}\frac{12}{25}\frac{223}{396}\frac{c^6}{(d-c)^4} + \cdots,\end{aligned} \tag{12}$$

where the last denominator was erroneously stated as 398 in [10]. This formula is the transcription of the following procedure:

$$D_1(d,c) = 4c^2/3, \quad S_1(d,c) = 4cd + D_1(d,c),$$

$$D_2(d,c) = D_1(d,c) \times \frac{c}{d-c} \times \frac{8}{15}, \quad S_2(d,c) = S_1(d,c) + D_2(d,c),$$

$$D_3(d,c) = D_2(d,c) \times \frac{c}{d-c} \times \frac{5}{14}, \quad S_3(d,c) = S_2(d,c) - D_3(d,c),$$

$$D_4(d,c) = D_3(d,c) \times \frac{c}{d-c} \times \frac{12}{25}, \quad S_4(d,c) = S_3(d,c) + D_4(d,c),$$

$$D_5(d,c) = D_4(d,c) \times \frac{c}{d-c} \times \frac{223}{396}, \quad S_5(d,c) = S_4(d,c) - D_5(d,c).$$

Same as before, first we approximate s^2 by $4cd$, and we arrange (add or subtract) the differences with fraction whose denominator is $d-c$. The sums S_n will approximate the arc length s^2 successively.

Takebe also abandoned the second formula, saying its precision did not increase very much even with the increased number of multiplications. See the following passage at sheet 49 recto and verso in [8]:

> Applying this procedure to the half circle, the sagitta being large, we find 3 orders by using two differences, 4 orders using three differences, and 5 orders by using four differences. We find one order more if we use one more difference. That is, this coincides with the 6-multiplication original procedure for finding the back arc, which we established earlier. Originally, expecting to find 7 orders using 6-multiplication we established the method, which turned out not to be accurate even using multi-multiplication. Therefore, we did not employ that procedure and abandoned it.

8 Takebe's Third formula

Takebe's third formula have more complicated form with polynomials of higher degree in the denominator:

$$\left(\frac{s}{2}\right)^2 = cd + \frac{1}{3}c^2 + \frac{1}{3} \times \frac{c^3}{d - \frac{9}{14}c} \times \frac{8}{15} + \frac{1}{3} \times \frac{c^5}{d - \frac{9}{14}c} \times \frac{1}{d^2 - \frac{1696}{1419}cd + \frac{6743008}{26176293}c^2} \times \frac{8}{15} \times \frac{43}{980}. \tag{13}$$

The formula (13) is also a transcription of the following procedure:

$$D_1(d,c) = \frac{4c^2}{3}, \quad S_1(d,c) = 4cd + D_1(d,c),$$

$$D_2(d,c) = D_1(d,c) \times \frac{c}{d - \frac{9c}{14}} \times \frac{8}{15}, \quad S_2(d,c) = S_1(d,c) + D_2(d,c),$$

$$D_3(d,c) = D_2(d,c) \times \frac{c^2}{d^2 - \frac{1696cd}{1419} + \frac{6743008c^2}{26176293}} \times \frac{43}{980},$$

$$S_3(d,c) = S_2(d,c) + D_3(d,c).$$

After explaining how to proceed, Takebe stops calculating the fourth difference D_4 saying it would be very complicated.

The closing remark of Chapter 12 of [10] reads as follows:

> In the above investigation of numbers of the arc, the determination of the coefficients of multipliers and divisors at each difference is the investigation of numbers by numbers. The procedure to determine the back arc is the investigation of rules by numbers. Truly, in the investigation of the circular circumference and the back arc, neither numbers nor procedures can be obtained by investigation by principles. We can obtain them through investigation by numbers. This is because of the attribute of the arc and the circle.

Takebe stresses that the numerical research is more important than the theoretical research in the study of circle and arc. He claims that it is due to the nature of the circle and the arc.

9 The Arc Rates

I have a copy of the manuscript (No. 1427) named the Arc Rates [弧率 Koritsu] of the Japan Academy. Although there are many discussion, on the manuscript Takebe Katahiro is referred to as the author. In this manuscript we find the following formula:

$$T_2(d,c) = \frac{L_1 cd^4 - L_2 c^2 d^3 + L_3 c^3 d^2 - L_4 c^4 d - L_5 c^5}{M_0 d^3 - M_1 cd^2 + M_2 c^2 d - M_3 c^3} \tag{14}$$

$$L_1 = 17243148700,\ L_2 = 27148244837,\ L_3 = 11453384892,$$

$$L_4 = 807998619,\ L_5 = 45408726,\ M_0 = 4310787175,$$

$$M_1 = 8223990414,\ M_2 = 4838317774,\ M_3 = 845423484.$$

This is a 4-multiplication procedure. It has a form similar to that of [8] with different coefficients. This is one of the trials to simplify the coefficients of (9). There are many manuscripts named the Arc Rates and this book does not seem to be the Arc Rates mentioned at sheet 54 verso.

10 Summary

We list the maximum error of the formulas which appeared in this note:

1. The algorithm (4) consists of a polynomial of degree 2. The maximum error is 1.7×10^{-2}. Note $\pi = \sqrt{10}$.
2. The algorithm (7) consists of a polynomial of degree 5. The maximum error is 4×10^{-5}. Note $\pi = 355/113$ and $s(d,0) < 0$.
3. The algorithm (8) consists of a fraction of which numerator and denominator are degree 7 and 5, respectively. The maximum error is 1.5×10^{-7}. Note $\pi = 355/113$.
4. The algorithm (9) consists of a fraction whose numerator and denominator are polynomials of degree 5 and 3, respectively. The maximum error is 1.5×10^{-10}.
5. Takebe's first formula (2) is a polynomial of degree 7. (Truncated Taylor series expansion.) The maximum error is 4×10^{-4}.
6. Takebe's second formula (12) is a fraction of polynomials whose numerator and denominator are of degree 6 and 4 respectively. The maximum error is 2.5×10^{-3}.
7. Takebe's third formula (13) is a fraction of polynomials whose numerator and denominator are of degree 5 and 3 respectively. The maximum error is 5×10^{-4}.
8. The procedure (14) consists of a fraction of polynomials whose numerator and denominator are of degree 5 and 3 respectively. The maximum error is 5×10^{-10}.

These maximum errors are calculated by the computer algebra system Mathematica. We used Takebe's three formulas as given in [10] for the evaluation of error. As explained in [6], we can easily increase the degree of Takebe's three formulas higher and increase their accuracy. Although Takebe did not appreciate his first formula (truncated Taylor series expansion), this is an efficient method to calculate the arc length numerically.

References

1. Akira Hirayama [平山諦], Kazuo Shimodaira [下平和夫] and Hideo Hirose [広瀬秀雄] ed.: *Takakazu Seki's Collected Works Edited with Explanations* [関孝和全集], Osaka Kyōiku Tosho [大阪教育図書] (1974).
2. Annick Horiuchi: *Les mathématiques japonaises à l'époque d'Edo*, Vrin, Paris (1994).
3. Mitsuo Morimoto [森本光生]: On the "old method", the "4-multiplication procedure for finding the back arc length", and the "6-multiplication original procedure for finding the back arc length [古法、四乗求背の術、六乗求背の元術について Kohō, shijō kyūhai no jutsu, rokujō kyūhai no motojutsu ni tsuite], *RIMS Kôkyûroku*, **1546**, Kyoto Univ., 175–180 (2007).
4. Mitsuo Morimoto [森本光生] and Tsukane Ogawa [小川束]: The mathematics of Takebe Katahiro: his three formulas of an inverse trigonometric function, *Sugaku Exposition*, **20** (2), 237–252 (2007); This article originally appeared in Japanese in *Sūgaku*, **56** (3), 308–319 (2004).
5. Nihon Gakushiin [日本学士院 Japan Academy]: History of Pre-Meiji Japanese Mathematics [明治前日本数学史 Meijizen Nihon Sūgaku Shi], 5 vols., Iwanami Shoten [岩波書店], Tokyo [東京] (1954–1960).
6. Tsukane Ogawa [小川束], Ken'ichi Sato [佐藤健一], Osamu Takenouchi [竹之内脩] and Mitsuo Morimoto [森本光生]: Takebe Katahiro's Mathematics [建部賢弘の数学 Takebe Katahiro no Sūgaku], Kyōritsu Shuppan [共立出版], Tokyo [東京] (2008).
7. Seki Takakazu: The Concise Collection of Mathematical Methods [括要算法 Katsuyō Sanpō], pp. 269–370 (English page 38–67) in [1].
8. Seki Takakazu [関孝和], Takebe Kataakira [建部賢明], and Takebe Katahiro [建部賢弘]: The Complete Book of Mathematics [大成算経 Taisei Sankei] (1711); The image of the book is available at Tohoku University digital collection (http://dbr.library.tohoku.ac.jp/infolib/meta_pub/G9200001CROSS).
9. Takebe Katahiro [建部賢弘]: Mathematical Methods for Clarifying Slight Signs [研幾算法 Kenki Sanpō] (1683); The image of the book is available at Tohoku University digital collection (http://dbr.library.tohoku.ac.jp/infolib/meta_pub/G9200001CROSS);
   An English translation can be found in [11].
10. Takebe Katahiro [建部賢弘]: Mathematical Treatise on the Technique of Linkage [綴術算経 Tetsujutsu Sankei] (1722), manuscript preserved in the National Archives of Japan; The book is being translated and commented by the author and T. Ogawa and will be published in near future.
11. Osamu Takenouchi and Mitsuo Morimoto: *Selected Mathematical Works of Takebe Katahiro (1664–1739)*, Wasan Institute (2004).
12. Hiroyuki Yokotsuka [横塚啓之]: On the Kohai-setsuyaku-shū regarded as the work of Katahiro Takebe, *Sūgakushi Kenkyū* (Journal of History of Mathematics, Japan), no. 182, 1-39 (2004).

The Method of Successive Divisions by Takebe Katahiro and Nakane Genkei*

Tamotsu Tsuchikura

Abstract The method of successive divisions developed by Wasan mathematicians was considered as a kind of Diophantine approximations. Nevertheless their main aim is not to find the close approximating fractions of irrational numbers but to obtain the simple fractions which enter the assigned range. We shall here reproduce their crucial algorithm. About fifty years later another Wasan mathematician made an alternate treatment for the same problem which is rather troublesome but gives systematical and complete solutions.

1 The Problem

In the research book the method of successive divisions [累約術 Rui-yaku-jutsu] published in 1728, Takebe Katahiro [建部賢弘][3, 4, 5], (1664–1739) and his colleague Nakane Genkei [中根元圭] (1662–1733) treated the following indefinite inequality problems, which are originated in the calendrical determination of the changing dates of the seasons.

After some normalizations, the problem is written in the modern form:

Let $a > 0, b > 0, c$, and $\gamma > 0$ be given real numbers. Find the pairs of integers x and y such that

$$|ax - by + c| \leq \gamma.$$

Tamotsu Tsuchikura
Professor emeritus, Graduate School of Science, Tohoku University,
6-3 Aramaki, Aoba, Aoba-ku, Sendai 980-8578, Japan
e-mail: o-tsuchikura@nifty.com

* This note was published in Japanese with some additional remarks. "Ruiyaku-jutsu, An algorithm for the solution to the indefinite inequality of Diophantine type", The Bulletin of Wasan Institute, No.8, 2008, pp.3–14.

In the book mentioned above, they obtained the solution pairs x,y by constructing a skillful algorithm. Prof. Matsusaburo Fujiwara [藤原松三郎][1] (1881–1946) first explained this theory in the modern style, and pointed out that this is of the Diophantine approximation. Indeed, we get the difference $b/a - x/y$ from the above inequality, and if b/a is an irrational, the inequality is of the Diophantine type, but the examples they adopted are not in the case. The main aim of Takebe and Nakane is not to determine the degree of approximation, but to find the smallest x;y and the following simple pairs.

They made a crucial use of the theory of continued fractions more than one hundred years before the famous Jacobi treatment of the Diophantine approximation.

2 Takebe-Nakane's Algorithm

In the problem we can suppose $b \geq c$ without loss of generality. First, we make a calculation of the continued fractions, and get the sequences of natural numbers $a_1, a_2, \cdots$, and of the rests $r_1, r_2, \cdots$:

$$\frac{b}{a} = a_1 + \frac{r_1}{a}, \quad \frac{a}{r_1} = a_2 + \frac{r_2}{r_1}, \quad \frac{r_1}{r_2} = a_3 + \frac{r_3}{r_2}, \quad \cdots.$$

Let $\frac{p_n}{q_n}$ be the n-th partial fraction of this continued fraction, then we have the well known relations:

$$\begin{cases} p_n = a_n p_{n-1} + p_{n-2} \\ q_n = a_n q_{n-1} + q_{n-2} \end{cases} \quad (n = 1, 2, \cdots),$$

where $p_1 = a_1, p_0 = 1, p_{-1} = 0, q_1 = 1, q_0 = 0, q_{-1} = 1$.

$$r_n = (-1)^n (a p_n - b q_n) \quad (n = 1, 2, \cdots; r_0 = a).$$

Next, we shall make an expansion of the fraction $\frac{b-c}{a}$ into a version of decimal or binary systems, namely, its own different system. The sequence of non-negative integers $b_1, b_2, \cdots$ and of the rests $s_1, s_2, \cdots$ will be obtained:

$$\frac{b-c}{a} = b_1 + \frac{s_1}{a} = b'_1 - \frac{s'_1}{a} \quad (b'_1 = b_1 + 1,\ s_1 + s'_1 = a),$$

$$\frac{s'_1}{r_1} = b_2 + \frac{s_2}{r_1} = b'_2 - \frac{s'_2}{r_1} \quad (b'_2 = b_2 + 1,\ s_2 + s'_2 = r_1),$$

$$\cdots$$

$$\frac{s'_{n-1}}{r_{n-1}} = b_n + \frac{s_n}{r_{n-1}} = b'_n - \frac{s'_n}{r_{n-1}} \quad (b'_n = b_n + 1, s_n + s'_n = r_{n-1})$$

$$(n = 1, 2, \cdots; \text{ where } \ s'_0 = b - c, r_0 = a).$$

We can see easily the following formulas by the mathematical induction.

$$s_n = (-1)^n (au_n - bv_n + c) ,$$

$$\begin{cases} u_n = u'_{n-1} + b_n p_{n-1} \\ v_n = v'_{n-1} + b_n q_{n-1} , \end{cases}$$

$$s'_n = (-1)^{n+1} (au'_n - bv'_n + c) ,$$

$$\begin{cases} u'_n = u'_{n-1} + b'_n p_{n-1} = u_n + p_{n-1} \\ v'_n = v'_{n-1} + b'_n q_{n-1} = v_n + q_{n-1} , \end{cases}$$

for $n = 2, 3, \cdots$ and

$$u_1 = b_1, v_1 = 1, u'_1 = b'_1, v'_1 = 1 .$$

If we find an index n such as $|s_n| < \gamma$, then the integral pair $x = u_n$, $y = v_n$ satisfies the required inequality.

We shall treat an example of Takebe and Nakane, [3, 4, 5], for the sake of convenience we adopt \$ as the money unit.

Having in hand the principal \$240.02, we add \$75.36 everyday, and just when the total is over \$501.63, we save this amount in the saving box. How many days the money in hand will be less than or equal to \$1?

This problem is to find the pair of integers x, y such that

$$0 \leqq 240.02 + 75.36x - 501.63y < 1 ,$$

x being the number of days and y the saving times.

Multiply both sides by 100, and divide by the common divider 3 of the coefficients of x and y. Then, we get

$$-\frac{24002}{3} \leq 2512x - 16721y < -\frac{23902}{3} . \quad (1)$$

The extreme sides will be replaced by -8000 and -7968 respectively, and we get

$$|2512x - 16721y + 7984| \leq 16 .$$

Hereafter we put

$$a = 2512, \quad b = 16721, \quad c = 7984 .$$

Following the above mentioned process, we get:

$$\frac{b}{a}=\frac{16721}{2512}=6\,(a_1)+\frac{1649}{2512}^{(r_1)} \qquad (a_1=6, r_1=1649 \text{ and so on})$$
$$\frac{a}{r_1}=\frac{2512}{1649}=1\,(a_2)+\frac{863}{1649}^{(r_2)}$$
$$\frac{r_1}{r_2}=\frac{1649}{863}=1\,(a_3)+\frac{786}{863}^{(r_3)}$$
$$\frac{r_2}{r_3}=\frac{863}{786}=1\,(a_4)+\frac{77}{786}^{(r_4)}$$
$$\frac{r_3}{r_4}=\frac{786}{77}=10\,(a_5)+\frac{16}{77}^{(r_5)}$$
$$\frac{r_4}{r_5}=\frac{77}{16}=4\,(a_6)+\frac{13}{16}^{(r_6)}$$
$$\frac{r_5}{r_6}=\frac{16}{13}=1\,(a_7)+\frac{3}{13}^{(r_7)}$$
$$\frac{r_6}{r_7}=\frac{13}{3}=4\,(a_8)+\frac{1}{3}^{(r_8)}$$
$$\frac{r_7}{r_8}=\frac{3}{1}=3\,(a_9)+\frac{0}{1}^{(r_9)}\;,$$

$$\frac{b-c}{a}=\frac{8737}{2512}=3\,(b_1)+\frac{1201}{2512}^{(s_1)}=4\,(b'_1)-\frac{1311}{2512}^{(s'_1)}$$
$$\frac{s'_1}{r_1}=\frac{1311}{1649}=0\,(b_2)+\frac{1311}{1649}^{(s_2)}=1\,(b'_2)-\frac{338}{1649}^{(s'_2)}$$
$$\frac{s'_2}{r_2}=\frac{338}{863}=0\,(b_3)+\frac{338}{863}^{(s_3)}=1\,(b'_3)-\frac{525}{863}^{(s'_3)}$$
$$\frac{s'_3}{r_3}=\frac{525}{786}=0\,(b_4)+\frac{525}{786}^{(s_4)}=1\,(b'_4)-\frac{261}{786}^{(s'_4)}$$
$$\frac{s'_4}{r_4}=\frac{261}{77}=3\,(b_5)+\frac{30}{77}^{(s_5)}=4\,(b'_5)-\frac{47}{77}^{(s'_5)}$$
$$\frac{s'_5}{r_5}=\frac{47}{16}=2\,(b_6)+\frac{15}{16}^{(s_6)}=3\,(b'_6)-\frac{1}{16}^{(s'_6)}$$
$$\frac{s'_6}{r_6}=\frac{1}{13}=0\,(b_7)+\frac{1}{13}^{(s_7)}=1\,(b'_7)-\frac{12}{13}^{(s'_7)}$$
$$\frac{s'_7}{r_7}=\frac{12}{3}=4\,(b_8)+\frac{0}{3}^{(s_8)}=5\,(b'_8)-\frac{3}{3}^{(s'_8)}\;,$$

$$\frac{p_1}{q_1} = a_1 = \frac{6}{1} \quad (p_1 = 6, q_1 = 1 \quad \text{and so on})$$
$$\frac{p_2}{q_2} = a_1 + \frac{1}{a_2} = \frac{a_2 p_1 + 1}{a_2} = \frac{1 \times 6 + 1}{1} = \frac{7}{1}$$
$$\frac{p_3}{q_3} = \frac{a_3 p_2 + p_1}{a_3 q_2 + q_1} = \frac{1 \times 7 + 6}{1 \times 1 + 1} = \frac{13}{2}$$
$$\frac{p_4}{q_4} = \frac{a_4 p_3 + p_2}{a_4 q_3 + q_2} = \frac{1 \times 13 + 7}{1 \times 2 + 1} = \frac{20}{3}$$
$$\frac{p_5}{q_5} = \frac{a_5 p_4 + p_3}{a_5 q_4 + q_3} = \frac{10 \times 20 + 13}{10 \times 3 + 2} = \frac{213}{32}$$
$$\frac{p_6}{q_6} = \frac{a_6 p_5 + p_4}{a_6 q_5 + q_4} = \frac{4 \times 213 + 20}{4 \times 32 + 3} = \frac{872}{131}$$
$$\frac{p_7}{q_7} = \frac{a_7 p_6 + p_5}{a_7 q_6 + q_5} = \frac{1 \times 872 + 213}{1 \times 131 + 32} = \frac{1085}{163}$$
$$\frac{p_8}{q_8} = \frac{a_8 p_7 + p_6}{a_8 q_7 + q_6} = \frac{4 \times 1085 + 872}{4 \times 163 + 131} = \frac{5212}{783},$$

$$\begin{aligned}
u_1 &= b_1 = 3 & u_1' &= b_1 + 1 = 4 \\
u_2 &= u_1' + b_2 p_1 = 4 + 0 \times 6 = 4 & u_2' &= u_2 + p_1 = 4 + 6 = 10 \\
u_3 &= u_2' + b_3 p_2 = 10 + 0 \times 7 = 10 & u_3' &= u_3 + p_2 = 10 + 7 = 17 \\
u_4 &= u_3' + b_4 p_3 = 17 + 0 \times 13 = 17 & u_4' &= u_4 + p_3 = 17 + 13 = 30 \\
u_5 &= u_4' + b_5 p_4 = 30 + 3 \times 20 = 90 & u_5' &= u_5 + p_4 = 90 + 20 = 110 \\
u_6 &= u_5' + b_6 p_5 = 110 + 2 \times 213 = 536 & u_6' &= u_6 + p_5 = 536 + 213 = 749 \\
u_7 &= u_6' + b_7 p_6 = 749 + 0 \times 872 = 749 & u_7' &= u_7 + p_6 = 749 + 872 = 1621 \\
u_8 &= u_7' + b_8 p_7 = 1621 + 4 \times 1085 = 5961 & u_8' &= u_8 + p_7 = 5961 + 1085 = 7046,
\end{aligned}$$

$$\begin{aligned}
v_1 &= 1 & v_1' &= 1 \\
v_2 &= v_1' + b_2 q_1 = 1 + 0 \times 1 = 1 & v_2' &= v_2 + q_1 = 1 + 1 = 2 \\
v_3 &= v_2' + b_3 q_2 = 2 + 0 \times 1 = 2 & v_3' &= v_3 + q_2 = 2 + 1 = 3 \\
v_4 &= v_3' + b_4 q_3 = 3 + 0 \times 2 = 3 & v_4' &= v_4 + q_3 = 3 + 2 = 5 \\
v_5 &= v_4' + b_5 q_4 = 5 + 3 \times 3 = 14 & v_5' &= v_5 + q_4 = 14 + 3 = 17 \\
v_6 &= v_5' + b_6 q_5 = 17 + 2 \times 32 = 81 & v_6' &= v_6 + q_5 = 81 + 32 = 113 \\
v_7 &= v_6' + b_7 q_6 = 113 + 0 \times 131 = 113 & v_7' &= v_7 + q_6 = 113 + 131 = 244 \\
v_8 &= v_7' + b_8 q_7 = 244 + 4 \times 163 = 896 & v_8' &= v_8 + q_7 = 896 + 163 = 1059.
\end{aligned}$$

Among this table we pick up s_n and s_n' whose values are less than $\gamma = 16$. We get

$$s_6 = 15, s_6' = 1, s_7 = 1, s_7' = 12, s_8 = 0, s_8' = 3.$$

From $s_6 = au_6 - bv_6 + c = 15$, we have $x = u_6 = 536$, $y = v_6 = 81$, the smallest solution. From $s'_6, (x,y) = (u'_6, v'_6) = (749, 113)$, from $s_7, (u_7, v_7) = (749, 113)$ the same with the above; from $s'_7, (u'_7, v'_7) = (1621, 244)$; from $s_8, (u_8, v_8) = (5961, 896)$, and from $s'_8, (u'_8, v'_8) = (7046, 1059)$.

We shall remark that, by some combinations of s_n or r_n, we may obtain another solutions. Using the formulas

$$(-1)^n s_n = u_n a - v_n b + c\,, \quad (-1)^m r_m = p_m a - q_m b\,,$$

we get, h being a non-zero integer,

$$(-1)^n s_n + h(-1)^m r_m = (u_n + hp_m)a - (v_n + hq_m)b + c\,.$$

If the absolute value of the second-hand side is not over the bound γ, then $(u_n + hp_m, v_n + hq_m)$ is a pair of solution. In the Takebe-Nakane book[3, 4, 5], only the case $m = n-1$ is treated. We may use s'_n instead of s_n, or further additional r_ℓ . If (x_0, y_0) is a pair of solution, then, as we see easily, the pair $(x_0 + kb, y_0 + ka)$ is also a solution for any integer k.

For the purpose to get the smaller solution, as we see in the formula for u_n, v_n it is favorable to use smaller b_n. Hence, in the case when $s_n < \gamma$, if we have an integer N such that $b_n \geq N, s_n + Nr_{n-1} < \gamma$, then we put $\bar{b}_n = b_n - N, \bar{s}_n = s_n + Nr_{n-1}$. we have

$$b_n + \frac{s_n}{r_{n-1}} = \bar{b}_n + \frac{\bar{s}_n}{r_{n-1}} \quad (\bar{b}_n \geq 0,\ \bar{s}_n < \gamma)\,,$$
$$\bar{s}_n = (-1)^n(a\bar{u}_n - b\bar{v}_n + c)\,,$$
$$\bar{u}_n = u'_{n-1} + \bar{b}_n p_{n-1}\,, \quad \bar{v}_n = v'_{n-1} + \bar{b}_n q_{n-1}\,,$$

the pair $(\bar{u}_n, \bar{v}_n)$ is smaller than (u_n, v_n). We must say this discussion a very careful step. In our example we consider s_8. For this case $b_8 = 4$, $s_8 = 0$, $r_7 = 3$, we put $N = 4$, then

$$\bar{b}_8 = b_8 - 4 = 0,\ \bar{s}_8 = s_8 + 4r_7 = 12 < \gamma\,(= 16)\,,$$

and $\bar{u}_8 = u'_7 + \bar{b}_8 p_7 = u'_7, \bar{v}_8 = v'_7 + \bar{b}_8 q_7 = v'_7$, the same as s'_7.

We shall make some examples, put $a = 2512$, $b = 16721$, $c = 7984$.

Example 1.

$$s_6(=15) = au_6 - bv_6 + c = 536a - 81b + c,$$
$$r_8(=1) = ap_8 - bq_8 = 5212a - 783b\,.$$

By adding $16 = 5748a - 864b + c$, hence $(x,y) = (5748, 864)$ is a solution.

Example 2.

$$s_8 + r_6 - 2r_7 = 0 + 13 - 2 \times 3 = 7 < 16,$$

hence

$$(au_8 - bv_8 + c) + (ap_6 - bq_6) + 2(ap_7 - bq_7)$$
$$= (5961 + 872 + 2 \times 1085)a - (896 + 131 + 2 \times 163)b + c = 9003a - 1353b + c\,.$$

and $(x, y) = (9003, 1353)$ is also another solution.

Example 3.

$$\begin{aligned} &\text{By} - s'_8 - r_7\,, \quad -6 = 8131a - 1222b + c\,,\\ &\text{By} - s'_8 - 2r_7\,, \quad -9 = 9216a - 1385b + c\,,\\ &\text{By} - s'_8 - 3r_7\,, \quad -12 = 10301a - 1548b + c\,,\\ &\text{By} - s'_8 - 4r_7\,, \quad -15 = 11386a - 1711b + c\,, \end{aligned}$$

hence $(8131, 1222)$,$(9216, 1385)$,$(10301, 1548)$,$(11386, 1711)$ are solutions.

Example 4.

$$\begin{aligned} s'_5 - 3r_6 &= 47 - 3 \times 13 = 8\\ &= (u'_5 a - v'_5 b + c) - 3(p_6 a - q_6 b)\\ &= (110a - 17b + c) - 3(872a - 131b)\\ &= -2506a + 376b + c\,. \end{aligned}$$

We add $0 = 16721a - 2512b$, then the last side is equal to $14215a - 2136b + c$, hence $(14215, 2136)$ is a positive solution.

3 Approach To The Problem

About fifty years later, Aida Yasuaki [會田安明][6] (1747–1817) introduced this theory in his text book adopting the original examples, Several methods on mathematics, 3 vols. [算法諸約術, 上中下], and for the last example, which is the one we concerned here itself, he gave the complete solutions, that is, determined all the pairs x, y satisfying the required inequality, using the elementary theory of numbers. We shall repeat his process. The coefficients $a = 2512$ and $b = 16721$ are prime with each other, we can find a pair of integers x, y, such that

$$2512x - 16721y = -1\,.$$

The method to find such pair is called ken-ichi-jutsu or jiku-ichi-jutsu in the old Japanese mathematics, and it depends on the continued fraction theory. In our case $x = 11509$, $y = 1729$ is a pair of solution, i.e.,

$$2512 \times 11509 - 16721 \times 1729 = -1\,. \tag{2}$$

As we remarked on the inequality (1), the pair x, y satisfies the following thirty three equations:

$$2512x - 16721y = A\,, \tag{3}$$

where $A = -8000, -7999, -7998, \cdots, -7968$. Make $(2) \times A + 3$, we get easily

$$2512(11509A + x) = 16721(1729A + y)\,, \tag{4}$$

so we can write $1729A + y = 2512k$ (k, integer), as 2512 and 16721 are prime with each other. Substitute this into (4), and we get easily,

$$\begin{cases} x = 16721k - 11509A \\ y = 2512k - 1729A\,. \end{cases}$$

For example, put $A = -8000$, we get

$$\begin{cases} x = 16721k + 11509 \times 8000 \\ y = 2512k + 1729 \times 8000\,. \end{cases}$$

To find the smallest $x, y > 0$, we put $k = -5506$, and get $x = 6174$, $y = 928$.

After finding such positive pairs for all the values A, we can get the smallest pair. (Cf. Following table)

Table of the smallest pairs (x, y) for each F or G.

$$F = 240.02 + 75.36x - 501.63y,\ G = 2512x - 16721y + 7984,\ F = \frac{1}{100}(3G+50)\,.$$

F	G	x	y	Examples required
0.98	16	5748	864	$s_6 + r_8 = s_8 + 16r_8$
0.95	15	536	81	$s_6 = s_8 + 15r_8$
0.92	14	12045	1810	$s'_7 + 2r_8$
0.89	13	6833	1027	$s'_7 + r_8$
0.86	12	1621	244	s'_7
0.83	11	13130	1973	$s_8 + r_6 - r_7 + r_8$
0.80	10	7918	1190	$s'_7 - r_7 + r_8 = -s'_8 + r_6$
0.77	9	2706	407	$s'_7 - r_7$
0.74	8	14215	2136	$s_8 + 2r_7 + 2r_8$
0.71	7	9003	1353	$s_8 - r_6 - 2r_7$
0.68	6	3791	570	$s'_7 - 2r_7$
0.65	5	15300	2299	$s'_7 - 3r_7 + 2r_8$
0.62	4	10088	1516	$s'_7 - 3r_7 + r_8$
0.59	3	4876	733	$s'_7 - 3r_7$
0.56	2	16385	2462	$s_8 + 2r_8$
0.53	1	11173	1679	$s_8 + r_8$
0.50	0	5961	896	s_8
0.47	−1	749	113	$-s'_6 = -s_7 = s'_7 - r_6$
0.44	−2	12258	1842	$-s'_8 + r_8$
0.41	−3	7046	1059	$-s'_8$
0.38	−4	1834	276	$-s'_8 - r_7 - r_8$
0.35	−5	13343	2005	$-s'_8 - r_7 + r_8$
0.32	−6	8131	1222	$-s'_8 - r_7$
0.29	−7	2919	439	$-s'_8 - r_7 - r_8$
0.26	−8	14428	2168	$-s'_8 - 2r_7 + r_8$
0.23	−9	9216	1385	$-s'_8 - 2r_7$
0.20	−10	4004	602	$-s'_8 - 2r_7 - r_8$
0.17	−11	15513	2331	$-s'_8 - 3r_7 + r_8$
0.14	−12	10301	1548	$-s'_8 - 3r_7$
0.11	−13	5089	765	$-s'_8 - 3r_7 - r_8$
0.08	−14	16598	2494	$-s'_8 - 4r_7 + r_8$
0.05	−15	11386	1711	$-s'_8 - 4r_7$
0.02	−16	6174	928	$-s'_8 - 4r_7 - r_8$

N.B. The pairs $x + 16721k$, $y + 2512k$ are also solutions, $k = 1, 2, \cdots$.

References

1. Matsusaburo Fujiwara [藤原松三郎]: Miscellaneous notes on the history of Wazan, II [和算史の研究 II], Diophantine approximation [建部賢弘の累約術], *Tohoku Math. J.* [東北数学雑誌], vol. 46, pp. 135–144 (1940); *Invitation to history of mathematics in east Asia* [東洋数学史への招待], Tohoku University Press [東北大学出版会], Sendai (2007).
2. Japan Academy ed.: *History of pre-Meiji Japanese mathematics* [明治前日本数学史], vol. 5, Japan Academy [日本学士院日本科学史刊行会] (1979).
3. Takebe Katahiro [建部賢弘] writes [著] & extracted by [删定] Nakane Genkei [中根元圭] : *the method of successive divisions* [累約術 Rui-yaku-jutsu], MS., Okamoto Collection 304, Tohoku University Library [写本, 東北大学図書館 岡本文庫 写 304].
4. Takebe Katahiro [建部賢弘]: *new methods of successive multiplications and divisions* [累乗累約新術], MS., Hayashi Collection 740, Tohoku University Library [写本, 東北大学図書館林集書 740].
5. Takebe Katahiro [建部賢弘]: *new methods of successive multiplications and divisions* [累乗累約新術], MS., Sekisan zenden 119, Miyagi Prefectural Library [写本, 宮城県図書館 関算前傳 119].
6. Aida Yasuaki ed. [會田安明 編]: *Several methods on mathematics, 3 vols.* [算法諸約術 上中下], MS., Hayashi Collection 1191, Tohoku University Library (写本, 東北大学図書館 林集書 1191).

Manuscripts in the Edo Period: Preliminary Study on Manuscripts Written by Seki Takakazu

Tomomi Nishida

Abstract Hatsubi-Sanpō written by Seki Takakazu (ca.1642–1708) was published in 1674 in the form of woodblock printing. The rest of his works are known only in the form of hand-copied books. His case was not exceptional, however. A famous Confucianist Arai Hakuseki (1657–1725) was Seki's colleague as a government official. Arai also wrote a lot of works as manuscripts but did not publish them in woodblock printings; his works spread widely as hand-copied books. Why they did not publish printed books? We can see some suggestion through recent studies on publications in the Edo period.

1 Studies on Manuscripts

According to former studies on the history of books during the Edo period (1603–1867), the woodblock printing was regarded as the most popular style of publications, while manuscripts and hand-copied printings were less popular and recognized as private publications.

On the contrary, recent research has proven the expansion of the hand-copied book market and advantages of manuscripts. In 2007, an epoch-making book Sequel to *Wahon* for Beginners [2] was published by Kōnosuke Hashiguchi [橋口侯之介], where *Wahon* [和本] means books in Japanese style. A lot of manuscripts and hand-copied printings were shown in it and we can see various ways of publication business in the Edo period. (For general information about books in the Edo period, see [1] and [3].)

In some cases, manuscripts and hand-copied books were more convenient for bookstores and at the same time for readers. It was easier to make hand-copied books than we think of, and less expensive for readers in those days. Those advantages had

Tomomi Nishida
3-7-10, Mama, Ichikawa, Chiba 272-0826, Japan
e-mail: sancan-cufu@goo.jp

encouraged the wide spread of manuscripts and hand-copied books. That was one of the characteristics of culture in the Edo period.

Hand-copied books were handled just in the same way as printed books at any bookstores, and were one of the principal commodities in those days. They were also handled just in the same way as printed books at rental bookstores. Furthermore, there were many book copiers [書本屋 kakihon-ya] that received requests from customers and copied books with their hands.

Most of book copiers were young samurai who could not afford to buy enough commodities including books and stationaries. Manuscript business kept alive until the beginning of the Meiji period (1868–1911). Details are written in the autobiography of Fukuzawa Yukichi [福沢諭吉], the founder of Keio University.

Most of experts in Wasan (Japanese traditional mathematics) could not afford to make woodblock printings because of lack of money. Some experts, however, might have had enough money to publish printed books, but they did not always sell well in those days.

On the other hand, hand-copied books were helpful for experts and readers. Experts could make them less expensively than woodblock printings. Because it did not cost much, some could make many hand-copied books in his or her life.

Another crucial advantage was the speed of publication, which was important to claim priorities. In general, it took a considerable time to publish a woodblock printing book. In a case, it took several years to publish a book because of prudent preparation. In another case, some mistakes found in manuscript delayed the publication. A lot of new mathematical ideas might be published during the period of preparation. A delay of publication sometimes caused a serious trouble.

Isomura Yoshinori [礒村吉徳] (?–1710), a famous expert of Wasan in the middle of the 17th century, published a book on Wasan entitled Mathematical Methods without Doubts [算法闕疑抄 Sanpō Ketsugishō] in 1659. In the preface, he blamed some pupils on the ground that they had published books on Wasan that included new mathematical ideas given by Isomura, and moreover, the presentation of the mathematical ideas written in those books were not satisfactory for Isomura. So he pointed out some mistakes in those books of pupils. Though he did not write their names clearly, researches on the books have revealed the names of two authors.

From the historical point of view, this trouble was caused in part by the delay of publication of Mathematical Methods without Doubts. In the end, some pupils could not wait for years; they might have wanted to be far ahead of the time in Wasan. However, hand-copied books were not yet popular in 1650's, and hence they could not make use of advantages of non-printing books developed later.

2 Hand-copied Books

In the 18th century, hand-copied book made an astonishing progress. A famous Wasan expert Aida Yasuaki [会田安明] (1747–1817) made thousands of manuscripts and some of them were spread as hand-copied books. Some were entitled

Commentary [評林 Hyōrin]. Aida wrote a lot of commentaries of books by other Wasan experts and compared with his own. In this way, he could publish up-to-date researches. It is quite similar to research articles in modern time.

Seki Takakazu [關孝和] also made a lot of manuscripts in the latter half of the 17th century. We can read them as hand-copied books, which may tell us how he always created new ideas in mathematics. Hand-copied books might be a suitable way for showing his new ideas.

Although hand-copied book business was not very popular until the beginning of the 18th century, some hand-copied books became famous and were widely circulated. For example, books written by Arai Hakuseki [新井白石] were among the most influential works throughout the Edo period. People in Edo period did not distinguish the difference between the two types of publications. We should reconsider books published in Edo period.

Former studies on the history of Wasan have conjectured that there were only closed or even exclusive groups of experts. One of the reasons was the existence of manuscripts and hand-copied books, which were considered as circulated in each small group. Taking into account the circulation of hand-copied books in the Edo period, we should also reconsider this conjecture.

References

1. Kōnosuke Hashiguchi [橋口侯之介]: *Wahon for Beginners* [和本入門], Heibonsha [平凡社], Tokyo (2005).
2. Kōnosuke Hashiguchi [橋口侯之介]: *Sequel to Wahon for Beginners* [続和本入門], Heibonsha, Tokyo (2007).
3. Chiyoji Nagatomo [長友千代治]: *Books and Readers in Edo Period* [江戸時代の書物と読書], Tokyōdō Shoten [東京堂書店], Tokyo (2001).

Influence of European Mathematics on Pre-Meiji Japan

Tatsuhiko Kobayashi*

Abstract The mathematics developed properly in Japan during the Edo period (1603–1867) is called Wasan (Japanese Mathematics). Its roots are in the ancient Chinese mathematics which flourished in the Song and Yuan dynasties (962–1368), in particular, but the later developments in Japan have been thought to be independent of China or Europe. The author claims, however, that Wasan was influenced by Europe first through Chinese translations of European books starting with the Fundamentals for Astronomy (1629), which was purchased by a son of Tokugawa Ieyasu as early as 1632, and from the end of the eighteenth century on directly through books in Dutch.

1 Introduction

Wasan (Japanese Mathematics) is the mathematics independently developed in Japan during the Edo period (1603–1867) while it was originated in ancient Chinese mathematics.

In the Edo period, the Tokugawa shogunate government adopted the seclusion policy and strictly controlled foreign trade. In particular, there was Dutch monopoly of European trade, which was allowed only in Dejima [出島], a small island in Nagasaki. The shogunate government had also kept trade relation with China only in Dejima. From these facts, it is naturally believed that Western mathematics did not make any impact on the development of Wasan. Although the influence in mathematics of its own may have been small, we cannot deny the fact that Wasan'ka (Japanese native mathematician) got a significant impression on their thought for-

Tatsuhiko Kobayashi
Maebashi Institute of Technology, 460-1 Kamisadori-machi, Maebashi, Gunma 371-0816, Japan
e-mail: koba@maebashi-it.ac.jp

* This work was supported by Grant-in-Aid for Scientific Research (C) 8500760.

mation. In fact, they knew several aspects of Western academic circles by reading mathematics and astronomy books imported from the both countries.

In the beginning of the eighteenth century, Western mathematics and astronomy were transmitted to Tokugawa Japan by way of Qing China. While foreign books had been strictly prohibited since 1630, the policy was relaxed in 1720 by the Tokugawa shogunate government. This started an influx of Western scientific books written in Chinese by Jesuits missionaries. This liberalization policy was adopted so that the Shogunate could reform calendar system, and it also allowed eventually the introduction of Western astronomy and mathematics, in particular, trigonometry and logarithm. Wasan'ka made use of them for the study of surveying, navigation, etc.

In the latter half of the nineteenth century, Western higher mathematics was transmitted into Japan through Chinese books on Western mathematics. Around 1860 Introduction to Algebra and Calculus [代微積拾級 Dai wei ji shi ji] was brought in Japan. Wasan'ka came to understand Western differential and integral calculus by this book. Uchida Gokan [内田五観] (1805–1882) may have been the first mathematician who read Introduction to Algebra and Calculus and used terminology algebra [代数 daisū], differential [微分 bibun] and integral [積分 sekibun].

In this paper, the author discusses, first, the influence of Western mathematics through Chinese books on Western calendrical calculations and Dutch scientific books, and then will introduce how Western higher mathematics was accepted in the end of Edo period.

2 Introductory Astronomy and the Book Prohibition Policy

At the beginning of Edo period, the Tokugawa shogunate government gave permission to Japanese merchants to trade with European countries. Accordingly not only merchants but also Christianity missionaries came to Japan from Europe. The Jesuit missionaries, who established places of activity in Japan, built many elementary schools for propagandism and education. They taught reading, writing, religion, songs and manners there. Moreover there were schools where the Latin, arithmetic, natural sciences and geography were taught. The educational system and the curriculum are, however, not well known in detail [10, pp. 61–62 and p. 228].

Before leaving Europe for mission, the Jesuit missionaries took courses in natural sciences at Christian colleges in Europe. The training was also performed at the Jesuit colleges of Goa in India and of Macao in China [1, pp. 33–79]. For example, Matteo Ricci (1552–1610), who was the most important Jesuit priest from our point of view, came to Ming China in 1582. He took courses in mathematics, astronomy, geography and so on at Collegio Romano. His teacher was the most famous mathematician Christoph Clavius (1538–1612) in those days. The main purpose of the education was to get many believers in Christianity in east Asia by making use of the knowledge of natural sciences in their propagation activities. But it is not known how much mathematics and natural sciences influenced Japanese intelligencers in those days.

In 1630, as a part of suppression of Christianity, the Tokugawa shogunate government banned the influx of all books related to Christianity, including Chinese scientific books translated by the Jesuit missionaries in China. This policy was introduced on the watch for Christianity. In fact, the number of catholic converts was increasing, which was perceived as posing a threat to Tokugawa Japan. This book prohibition policy is called Ban of books in Kan'ei [寛永の禁書令 Kan'ei no kinsho rei]. While it was proclaimed in Japan, one great book was published in China in 1629, which can be called an encyclopedia on the Western mathematics, astronomy, calendar, surveying, technology, geography, literature and Christianity doctrine. The title of the book is Fundamentals for Astronomy [天学初函 Tianxue Chuhan]. It was clearly an object of the ban. Fundamentals for Astronomy, however, was purchased in 1632 by Owari feudal lord Tokugawa Yoshinao [徳川義直] (1600–1650) who was the ninth son of the first Shogun Tokugawa Ieyasu [徳川家康] (1542–1616) [9, p. 58].

This incident has prompted many questions, such as how Fundamentals for Astronomy was brought in Japan, how it evaded the mesh of a net of the ban, why Tokugawa Yoshinao was able to purchase it and so on. These remain, however, a mystery in history of cultural exchange between Japan and China.

Fig. 1
Title page of Fundamentals for Astronomy (Nagoya-shi Hōsa Bunko Library)

Fundamentals for Astronomy was compiled by Chinese scholar Li Zhizao [李之藻] (1565–1630) with the support of a Jesuit missionary Sabbathin de Ursis (Italy, 1575–1620) and is composed of twenty four books and in sixtieth volumes. It included Elements of Geometry [幾何原本 Jihe Yuanben] and Elements of Surveying [測量法義 Celiang Fayi] that Matteo Ricci translated. These complete works were divided into two chapters of natural sciences and technology [器編 Qibian], and literature and theology [理編 Libian]. Since the chapter of natural sciences and technology had nothing to do with Christianity doctrine, it seems that Japanese scholar could read it without caring about the Shogunate's eye. In fact, a member of the Shogun's Council of Elders Itakura Shigenori [板倉重矩] (1617–1673) obtained a copy of Book of Western Watersupply [泰西水法 Taixi Shuifa] in Fundamentals for Astronomy, and a historian Matsushita Kenrin [松下見林] (1637–1704) attached, in 1665, grammatical marks for rendering Chinese into Japanese on the copy that Itakura had [2, pp. 366–367].

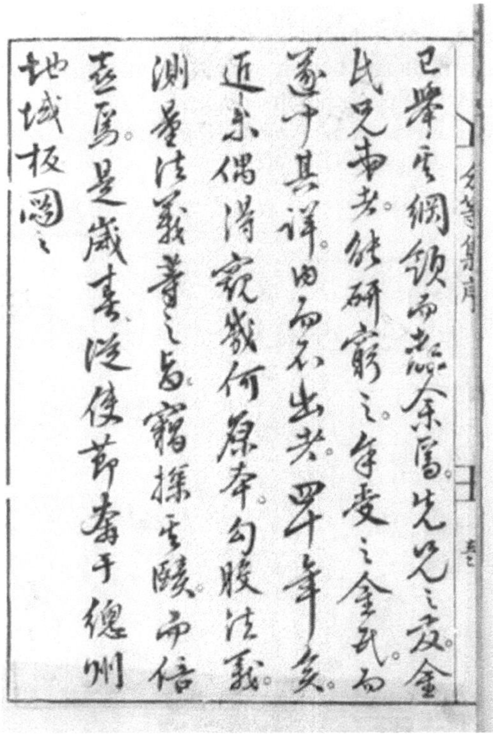

Fig. 2
Preface by Hosoi Kōtaku in Dividing Method by Compasses in 1722

Hosoi Kōtaku [細井広澤], who was known as calligrapher or surveyor, talked about his joy of reading Elements of Geometry, Elements of Right-angled Triangles [勾股法義 Cougu Fayi], Elements of Surveying in the preface of Dividing Method by Compasses [規矩分等集 Kiku buntō shū] which was published by Mao Tokiharu

[万尾時春] (1683–1755) in 1722. Hosoi's impression on these books was as follows (see *ibid.*, p. 1):

> In those days, I obtained by chance Elements of Geometry, Elements of Right-angled Triangles and Elements of Surveying, and was able to read them. When I knew an abstruse principle of the surveying, my joy doubled.

The book prohibition policy was relaxed in 1720 by the eighth Shōgun Tokugawa Yoshimune [徳川吉宗] (1684–1751). He was interested in reforming the out-dated Japanese calendar system, which urged him strongly to get up-to-date astronomical knowledge from Chinese books. Thereby his interest induced the liberalization policy. Although it is unknown where Hosoi Kōtaku read these Chinese books on Western mathematics and surveying, we must stress that his remark was only two years after the relaxation by Tokugawa Yoshimune.

3 Transmission and Influence of the Complete Treatise on Calendar and Computation

In 1726, the second edition of Mei Wending's work Complete Treatise on Calendar and Computation [暦算全書 Li Suan Quan Shu] was imported into Japan from China. Mei Wending [梅文鼎] (1663–1721) is recognized as one of the most influential mathematicians and astronomers in China in the eighteenth century. He devoted his scholarly life to the integration and assimilation of Western sciences into traditional Chinese mathematics and scientific know-how. The study of Western mathematics and astronomy is considered to have started from the introduction of Complete Treatise on Calendar and Computation into Japan by overcoming the obstacles imposed by the Tokugawa Shogunate government.

Soon after the transmission of Complete Treatise on Calendar and Computation, Takebe Katahiro [建部賢弘] (1664–1739) and Nakane Genkei [中根元圭] (1662–1733) began to translate it into Japanese and completed it in 1728 and in 1733, respectively. Takebe Katahiro presented the Japanese version to the Shōgun Tokugawa Yoshimune under the title New Japanese version of Complete Treatise on Calendar and Computation [新写訳本暦算全書 Shinsha Yakuhon Rekisan Zensho], which contained a preface by Takebe. In this preface Takebe stressed the importance of a careful study of Mei Wending's works (see *ibid.*, pp. 1–2):

> Mei Wending's compendium is composed of thirty volumes. The editor named his work Complete Treatise on Calendar and Computation. This compendium mainly explains Western calendar; there are included other subjects such as Manual of Calculation, Napier's Rods, Geometry and Trigonometry. Trigonometry is the newest one and particularly is so much useful for astronomy and in their astronomical calculation.

In the year when Complete Treatise on Calendar and Computation was imported, the Tokugawa government ordered Chinese traders to bring tables of trigonometric functions. It is because the editor of Complete Treatise on Calendar and

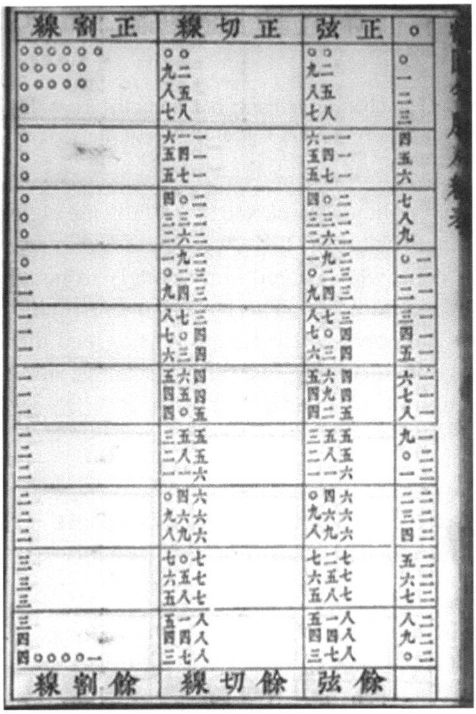

Fig. 3
First page of the trigonometric function table in Table for the eight lines cutting a circle

Computation did not print the trigonometric function tables in this compendium, although they actually compiled two chapters for it: Elements of Plane Trigonometry [平三角挙要 Ping Sanjiao Juyao] and Elements of Spherical Trigonometry [弧三角挙要 Hu Sanjiao Juyao]. One year later, the following five books on trigonometric function tables and their applications arrived:

(1) Table for the eight lines cutting a circle [八線表 Ba xian biao],
(2) Method of computation for the eight lines cutting a circle and its application [八線互求法 Ba xian hu qiu fa],
(3) Table for the eight lines cutting a circle [割圓八線之表 Ge yuan ba xian zhi biao],
(4) Method of computation for the eight lines cutting a circle and its application [割圓八線互求法 Ge yuan ba xian hu qui fa],
(5) Table for the eight lines cutting a circle [割圓勾股八線之表 Ge yuan gou go ba xian zhi biao].

While (2) and (4) contained similar content concerning plane trigonometry and its applications to land surveying, (1), (3) and (5) were related to trigonometric function tables, which were introduced for the first time from China to Japan. Soon after these books were brought to Japan, one of them, Table for the eight lines cutting a circle was passed on to Takebe Katahiro. He and his disciple, Nakane Genkei, were just translating Mei Wending's Complete Treatise on Calendar and Computation. Based on our research, we can confirm that Table for the eight lines cutting a circle is reproduced from Book of Chongzhen Calendar [崇禎暦書 Chongzhen Li Shu] and is actually identical to the Japanese version included in New Japanese version of Complete Treatise on Calendar and Computation. Book of Chongzhen Calendar thirty volumes, were compiled by the Jesuit missionaries Jacques Rho (Italy, 1593–1638) and Johann Adam Shall von Bell (Germany, 1591–1666) with support of a Chinese bureaucrat of high rank, Xu Guangqi [徐光啓] (1562–1633).

After this, the trigonometry and trigonometric function tables began to prevail among Japanese mathematicians and astronomers. The first Japanese mathematicians who made use of the trigonometry were Nakane Genkei and his pupils. Nakane has left two manuscripts: Measuring of height of the sun and the moon [日月高測 Nichigetsu kōsoku] written in 1732 and Calculation for using trigonometric function table [八線表算法解義 Hassenhyō Sanpō Kaigi]. These are the earliest examples of the use of the trigonometry by Japanese mathematicians [3, pp. 5–7].

From the middle of the eighteenth century, Wasan'ka were devoted to calculating accurately the volume of regular and semi-regular polyhedra and finding correct formulas for them. In these processes, they discovered a unique solid: a kind of Stella polyhedron, Sixtieth semi-regular polyhedra [六十等面 Rokujū tōmen] constructed of as regular dodecahedron and twelve regular pyramids with a regular pentagonal base. The first mathematician who calculated the volume of Sixtieth semi-regular polyhedra was Matsunaga Yoshisuke [松永良弼] (?–1744). Matsunaga's linking of Sixtieth semi-regular polyhedra to the family of regular polyhedra was to be criticized later by a Wasan'ka Fujita Teishi [藤田貞資] (1734–1807). In the manuscript titled by Mensuration of regular polyhedra [等面求積 Tōmen kyūseki], Fujita seriously criticized Matsunaga's idea using the description of regular polyhedra in Complete Treatise on Calendar and Computation. He pointed out as follows (see *ibid.*, pp. 13–14):

> It is possible to make arbitrarily various type of Stella polyhedra, if we employ same manner such as making Sixtieth semi-regular polyhedra. Even the author Mei Wending of Complete Treatise on Calendar and Computation did not count Stella types as regular polyhedra. Now due to his definition the number of regular polyhedra should be limited only five kinds. Hence we have to omit Sixtieth semi-regular polyhedra from regular polyhedra.

Fujita's opinion quoted above might be a typical case based on Mei Wending's thought. It is, however, true that Fujita defined the number of regular polyhedra based on the description in Complete Treatise on Calendar and Computation. This case entails a further evidence that Mei Wending's thought strongly influenced on Japanese mathematicians. At the same time, we may consider Fujita's opinion as one explicit example that Western way of thinking gave an impact on Japanese mathematics.

4 The Compendium of Calendrical Science and Astronomy, The Sequel of Compendium of Calendrical Science and Astronomy and Trigonometry

The transmission of Compendium of Calendrical Science and Astronomy [曆象考成 Li xiang Kao cheng], published in 1723, and Sequel of Compendium of Calendrical Science and Astronomy [曆象考成後編 Li xiang Kao cheng hou bian], published in 1742 are of significant importance in the history of Japanese mathematics, because Wasan'ka carefully studied these Chinese books on Western astronomy and quoted astronomical examples from them.

In 1761, about thirty-five years after the import of Complete Treatise on Calendar and Computation, Chinese traders brought in Japan Compendium of Calendrical Science and Astronomy [曆象考成 Li xiang Kao cheng] which contained a French astronomical new observation result. This astronomy book was divided into two volumes: First volume [上編 Shang bian] and Second volume [下編 Xia bian]. First volume is composed of sixteen chapters and Second volume of ten chapters. The two volumes mainly explained the motion of the planets based on the system of Tycho Brahe (1544–1601).

The second and third chapters of First volume of Compendium of Calendrical Science and Astronomy was, in particular, devoted to the explanation of the spherical trigonometry. Because of the importance of astronomical calculation, the methods of the spherical astronomy were presented in a distinct way. The compilers of this volume not only presented the basic formulas of the spherical trigonometry, but also gave many applied examples of astronomical computation, corresponding to the various phenomena in the celestial sphere. Thereby Compendium of Calendrical Science and Astronomy became a suitable text to learn the spherical trigonometry for Japanese mathematicians like Complete Treatise on Calendar and Computation. Even if the Meiji period had begun at this time, Compendium of Calendrical Science and Astronomy would have been still studied among local Japanese mathematicians [3, pp. 10–12].

About twenty years later, a new astronomical theorem was complemented as the sequel of Compendium of Calendrical Science and Astronomy. Then this reedited book was published with the title of Sequel of Compendium of Calendrical Science and Astronomy in 1742. The editor of the book was a German Jesuit missionary Ignatjus Köngker (1680–1746). The contents of Sequel of Compendium of Calendrical Science and Astronomy, which focused on the elliptical orbits theorem of Johannes Kepler (1571–1630), were very difficult to understand for conservative astronomers[1]. Aoki Konyō [青木昆陽] (1698–1769) may have been the first Japanese who read Sequel of Compendium of Calendrical Science and Astronomy in a relatively early time after this astronomical book arrived. We can find it from his writing

[1] The method of elliptic drawing is given in the illustration of Explanation on Construction of Astronomical Observatory and Instrument for Astronomical Observation [靈臺儀象志 Ling tai Yi xiang zhi] which was compiled by Ferdinand Verbiest (Belgium, 1623–1688) in 1673.

Fig. 4

Explanation of drawing method of ellipse in the first volume of Sequel of Compendium of Calendrical Science and Astronomy

with the title of Sequel of desultory essay by Konyō [続昆陽漫録補 Zoku konyō manroku ho] in 1761 [2, p. 369].

After the transmission of Sequel of Compendium of Calendrical Science and Astronomy to Japan, the studies of drawing method or on the focus of ellipse started among Wasan'ka. In the latter half of the eighteenth century, Aida Yasuaki [会田安明] (1747–1817), Ishiguro Nobuyoshi [石黒信由] (1760–1836) and Ono Eijū [小野栄重] (1763–1831) were Wasan'ka devoted to the study of ellipse. Aida Yasuaki wrote a manuscript of six volumes with the title of Complete book of ellipse problems [算法側円集 Sanpo sokuen shū] in 1799. In this manuscript, he referred to himself as the first Japanese mathematician in his method of elliptic drawing.

It was the most basic one using a string with the both ends fixed at two needles. One needle could move freely so that one could adjust the scale of ellipse. Aida Yasuaki called this wooden compasses compasses for drawing ellipse [側円規 Sokuen ki], which was invented probably by himself. Ishiguro Nobuyoshi wrote Drawing method of ellipse [側円周規法 Sokuen shūki hō] in 1815, and Ono Eijū wrote Commentary book of ellipse [側円規矩例弁解 Sokuen kikurei benkai] in 1828.

There is an interesting episode related to the method of elliptic drawing. In 1823, four Japanese astronomers, including Shibukawa Kagesuke [渋川景佑] (1787–1856) and Yamaji Yukitaka [山路諧孝] (1777–1861), visited the Dutchmen who came over to Edo to have an audience with the Tokugawa Shōgun. They met three Dutchmen, Jan Cock Blomhoff (Kapitein, 42 years old), Jan Vreserik over meer

Vijfer (Secretie, 22 years old) and Nicollaf Tulling (Surgeon, 38 years old) in a Japanese hotel in Edo and asked about the trend of the latest Western astronomical studies, among which the method of elliptic drawing was included. A Japanese astronomer asked to the Dutchmen about the accurate drawing of ellipse and about tools for it. The Dutchmen, however, taught an approximate drawing method as none of them could answer the question. The Dutchman's idea was a combination of the two circles with the sectors.[2]

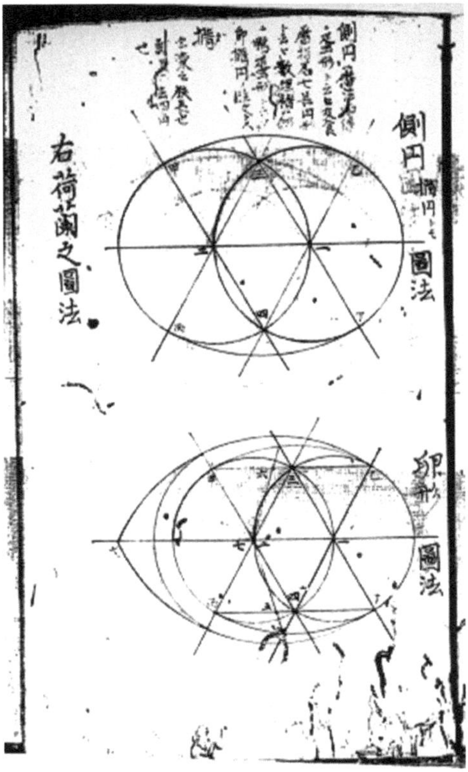

Fig. 5

Two approximate of drawing methods of ellipse that Dutchmen taught (picture was taken from Record of calendar experience)

Hazama Shigetoshi [間重新] (1786–1838), a son of Hazama Shigetomi [間重富] (1756–1816), was a civilian astronomer and a merchant in Ōsaka prefecture, wrote a commentary book on mechanical instrument for elliptic drawing in 1828 with the title of Origin of ellipse [楕円起源 Daen kigen]. In this commentary, he mentioned a unique curve that a mechanical instrument draws. From today's viewpoint, the

[2] See Record of calender experience [暦学見聞録 Rekigaku kenbunrokou], vol. 4. This note was preserved in the Library of National Astronomical Observatory of Japan.

curve that Hazama Shigetoshi discovered was equivalent to a nephroid, a kind of epicycloids. He, however, believed that it would be an elliptic curve (see *ibid.*, pp. 13–15).

5 Acceptances of Dutch Scientific Books

Toward the end of the eighteenth century, Shiduki Tadao [志筑忠雄] (1760–1806) started a great work. He was a Dutch interpreter in Nagasaki in his early years, resigned it later, and then devoted himself to a study of Dutch scientific books. Shiduki wrote Guide book of new natural science and astronomy [暦象新書 Rekishō shinsho], composed of three volumes in 1798–1802, which was not published. He tried to comprehend Western natural sciences systematically by studying Dutch scientific books. Thereby he studied *Inleidinge tot de ware Natuur-en Sterrekunde of de Natuur-en Sterrekundige Lessen*, Leiden, in 1741, which was written by a Dutch mathematician Johan Lulofs (1711–1768) and introduced physics and astronomy of Newton. A cycloid problem was taken up in this book. Shiduki called it Locus of

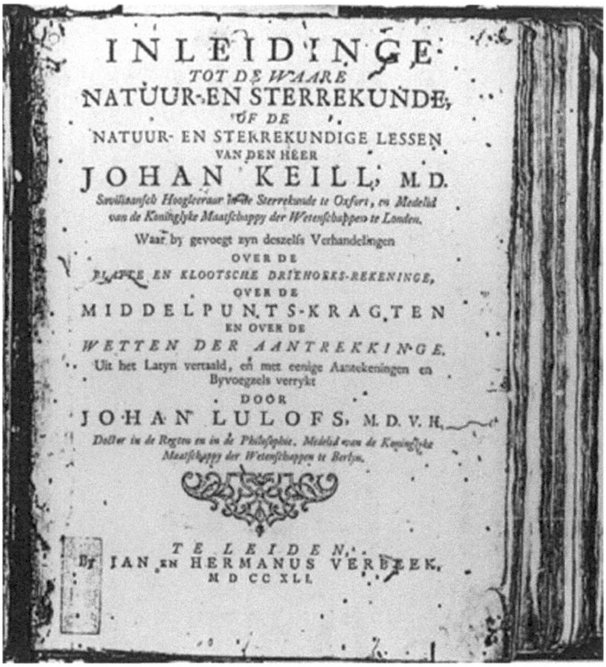
INLEIDINGE
TOT DE WAARE
NATUUR-EN STERREKUNDE,
OF DE
NATUUR- EN STERREKUNDIGE LESSEN
VAN DEN HEER
JOHAN KEILL, M.D.
Savilliaansch Hoogleeraar in de Sterrekunde te Oxfort, en Medelid van de Koninglyke Maatschappy der Wetenschappen te Londen.
Waar by gevoegt zyn deszelfs Verhandelingen
OVER DE
PLATTE EN KLOOTSCHE DRIEHOEKS-REKENINGE,
OVER DE
MIDDELPUNTS-KRAGTEN
EN OVER DE
WETTEN DER AANTREKKINGE.
Uit het Latyn vertaald, en met eenige Aantekeningen en Byvoegzels verrykt
DOOR
JOHAN LULOFS, M.D. V. H.
Doctor in de Regten en in de Philosophie, Medelid van de Koninglyke Maatschappy der Wetenschappen te Berlyn.
TE LEIDEN,
By JAN en HERMANUS VERBEEK,
M D CC XLI.

Fig. 6

Title page of *Inleidinge tot de ware Natuur-en Sterrekunde of de Natuur-en Sterrekundige Lessen*, Leiden, in 1741. This book is preserved in the Library of Dōhō University, Nagoya, Japan. This picture was taken from it.

the dust on wheel [塵跡線 Jinseki sen].

While Shiduki Tadao was translating J. Lulofs' book into Japanese, he made several partial manuscripts. The following three manuscripts from (1) to (3) and a draft paper (4) are about the study of trigonometry.

(1) New chapters for plane trigonometry [勾股新編 Kōko shinpen],
(2) Origin of Dutch trigonometry [法蘭三角形起元 Hōran sankakukei kigen],
(3) Secret treatise on trigonometric calculation [三角提要秘算 Sankaku teiyō hisan],
(4) Secret treatise on trigonometry [三角算秘伝 Sankakusan hiden].

The first manuscript New chapters for plane trigonometry was a translation of the chapter of Platte Driehoeks Rekeninge (The plane trigonometry) in J. Lulofs' book. Origin of Dutch trigonometry and Secret treatise on trigonometric calculation focused, in particular, on translation of *In Schuynhoekige Driehoeken* (The spherical trigonometry).

The proofreader of Secret treatise on trigonometric calculation is Ōtsuki Seijun [大槻清準] (1773–1850), a clansman of the Sendai feudal clan [仙台藩]. He studied Dutch arts and sciences at Shiduki's school in 1803, and then revision work was performed by him only in a month. From his Japanese translation, we find that Shiduki comprehended the principles of plane and spherical trigonometry, including Napier's rule for the right-angled spherical triangle.

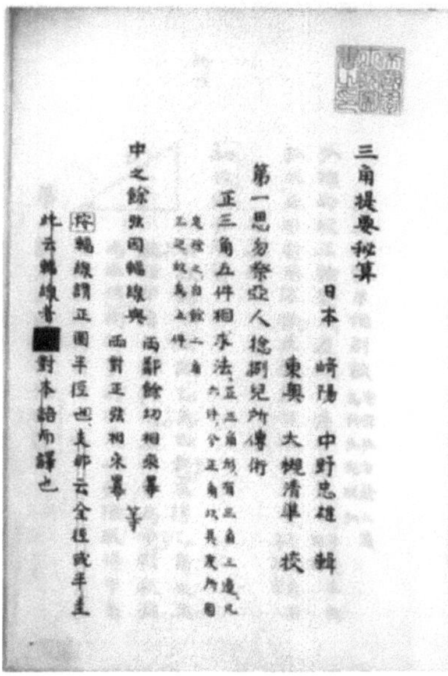

Fig. 7

Title page of Secret treatise on trigonometric calculation (Library of Japan Academy). Compiler of this manuscript is Nakano (or Shiduki) Tadao and proofreader is Ōtsuki Seijun

Ordinary Napier's rules are explained by the following expression:

Rule 1. *The cosine of any middle part is equal to the product of the cotangents of adjacent parts.*
Rule 2. *The cosine of any middle part is equal to the product of the sines of opposite parts.*

These rules can be easily memorized by the expressions of *otan. ad.* and *sin. op.* He completely understood these rules from J. Lulofs' book. It can be said that this was the first transmission of Napier's rules into Japan through a Dutch book. Secret treatise on trigonometry was only a draft paper which was made of six sheets, in which he explained spherical trigonometry, Napier's rules and logarithms. In calculating by logarithms, he used the following formulas [4]:

$$\log_a(M \times N) = \log_a M + \log_a N,$$
$$\log_a(M \div N) = \log_a M - \log_a N,$$
$$\log_a \sqrt[r]{N} = (\log_a N)/r.$$

Shiduki also obtained these formulas from J. Lulofs' book.

On the other hand, Wasan'ka Honda Toshiaki [本多利明] (1743–1820) and Sakabe Kōhan [坂部廣胖] (1759–1824) have emphasized the utility of the calculation by logarithms. Sakabe, a disciple of Honda, published fifteen-volume mathematical textbook series titled by Record to guidebook in computational procedures [算法點竄指南録 Sanpo tenzan shinanroku] in 1815, in which he gave, by using trigonometry and logarithms, some extraordinary solutions to problems. For example, with the aid of trigonometry, he solved a problem to find angel of revolution of the sun on elliptic orbit, and by using logarithms, gave numerals one to 300 to seven decimal places for it (see *ibid.*, vol.4, pp. 11–14). Both Honda and Sakabe knew that these problems came from Dutch mathematics.

In the beginning of the nineteenth century, there appeared Japanese mathematicians who tried to read Dutch scientific books. Ichino Shigetaka [市野茂喬] is one of such mathematicians. Although his career is unidentified, Ichino is known as a mathematician who belonged to the Saijō-ryū faction while working as an official of the astronomical position of the Edo Shogunate government. In this circumstance, he was strongly attracted by Western astronomy and mathematics. We can find it from Dutch equivalent notes of all five volumes by Ichino Shigetaka, which are conserved in Kobe Municipal Institution Museum [6, p. 2]. Since he did not give, however, any title to Dutch equivalent notes[3], it is not yet identified what Dutch book he read. But judging from the places and their latitudes of all over the world written down in four volumes of all five volumes, it seems that he read Dutch geography books. Here we will discuss only the contents of Vol. 5, because it attracts our interest best among all equivalents in his work.

[3] In the first volume of this note Ichino wrote down the title of the each chapters in original book by Dutch as follows: "Geographische Tafel van de graaden Der Breedte of Poolshoogte en lengte of Middagcirkel, van de meet bekende Steeden; Rivieren, caapen, Baayen, en Inzonderheid Der voornaamste Zee-haavens Des Aardryks."

In fact, Ichino Shigetaka wrote down, in this volume, so many Dutch terms with Japanese equivalents for astronomy, mathematics, calendar, etc.[4] As for mathematical terminology, he translated about 160 terms into Japanese, from which we show only some typical terms in the following table.

Dutch	Japanese	English
	算詞	Mathematical term
Bytellen, Bygeteld, meer	＋加	plus
aftrekken, min, minder	－減	minus
addeerf, som	相加, 相併	add
Substraheern	相減	subtract
redden	比例	ratio
eenhaid	天元之一	unknown
Cercle, rond, kring, ring	圓	circle
maal, omtrek	圓周	circumference
Diameter	圓径	diameter
hoekmaat, Sin·Sinuffen	正弦	sine
Schilboogs-hoekmaat, cos·cosinuffen	余弦	cosine
angle, hoek	角, 三角ノ角也	angle
Cube	四角六面 	cube
cilindre	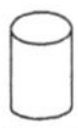	cylinder

Fig. 8

Comparison list for Dutch terms and its Japanese equivalents. The author gave the English equivalents for the sake of convenience

Let us explain a particularly interesting word, eenhaid which we showed in the table of Fig. 8. The term eenhaid generally means a unit in Dutch. But Ichino Shigetaka could not find a correct equivalent for it. It may have been the case that he did not know the true meaning for eenhaid. He employed Tengen no ichi [天元之一] for it. This phrase had been usually used for a mathematical term expressing an unknown in the East Asia since the thirteenth century. Needless to say, the word eenhaid does not mean unknown in Dutch. Japanese equivalents for term of the trigonometry and of the geometry were given correctly. On the other hand, he gave Japanese equivalents for polyhedra in some explicit figures as we indicate two examples in the table of Fig. 8. For example, the term "cylinder" was the one for which Japanese mathematicians already had had the exact corresponding Entō [円壔] in those days. He would think that visual description was simple to understand complicated polyhedron.

[4] Ichino has been assigned to duty of Nagasaki in 1810; it seems that at this time he made these notes.

The Translation works by Ichino Shigetaka can be considered as a case of mathematicians of the eighteenth century who were accepting Western mathematics positively. Moreover many Japanese mathematicians at that time had or at least began to have the interest as Ichino had.

6 Transmission of Western Higher Mathematics from China

Around 1860, a Chinese book on Western mathematics was brought in Japan, which was Introduction to Algebra and Calculus [代微積拾級 Dai wei ji Shi ji] published in 1859 in Shanghai [上海]. This book was a joint translation by British propagator Alexander Wylie [偉烈亞力] (1815–1887) and Chinese mathematician Li Shanlan [李善蘭] (1811–1882), and the original was *Elements of Analytical Geometry and of the Differential and Integral Calculus* (1851, New York) by Elias Loomis (1811–1889). The transmission of Introduction to Algebra and Calculus had brought, for the first time, an impact of Western higher mathematics to the traditional Japanese mathematics, though in an indirect way. It is not known how Introduction to Algebra and Calculus was brought to Japan around 1860. Uchida Gokan [内田五観] (1805–1882) may be, however, the first mathematician who read it and used terminology [代数 daisū], [微分 bibun] and [積分 sekibun] for *algebra*, *differential* and *integral*. Uchida Gokan was a samurai of lower rank in the Edo Shogunate while he was well-known as a leading mathematician at that time. What is more, he had deep interchanges with Japanese scholars in Dutch. Thus it is naturally considered that Uchida got a chance to come across Introduction to Algebra and Calculus.

Ara Shijū [荒至重] (1826–1909), who was a clansman of Nakamura feudal clan [中村藩], published, in 1865, a book about directions to use the surveying tools and their applications. The book is titled by Three methods for land surveying [量地三略 Ryōchi sanryaku] and composed of two volumes. Ara Shijū was a pupil of Uchida Gokan, who wrote preface for Ara's book. In this preface Uchida pointed out that 自少好数学, 而代数諸術, 微分積分諸法, 皆修其奥 (see *ibid.*, vol. 1, p. 2), which means that Ara has enjoyed mathematics since he was young and masters an abstruse principle of mathematics such as algebra, differential and integral now. Here Uchida used Chinese characters 代数, 微分, and 積分 for algebra, differential and integral. Wasan'ka of the Edo period had never used these mathematical terminologies before. The concept of differential, in particular, had not grown up in their studies. Since Uchida Gokan, who read Introduction to Algebra and Calculus, pointed out the importance of this book, he may have understood that the substantial part of Western mathematics lied in algebra, differential and integral.

Kanda Takahira [神田孝平] (1830–1898) was a mathematics professor at Kaisei-sho Institute [開成所], which was established in 1862 by the Tokugawa shogunate government [7, pp. 252–253]. Kaisei-sho Institute was installed to study the Western arts and sciences and to educate young people. Kanda also realized the importance of Introduction to Algebra and Calculus and tried to translate it into Japanese. He finished writing a manuscript entitled 代微積拾級 from 1864 to 1865, which fi-

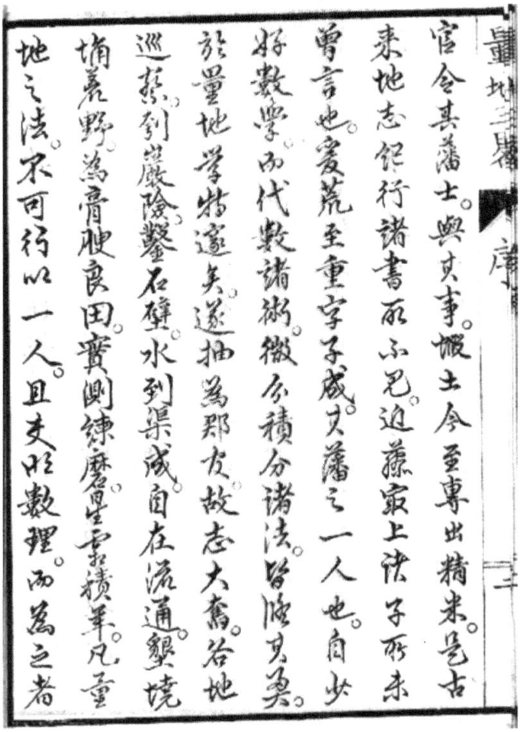
量地三略 序

官令其藩士。與其事。假士令至專出精米。是古
來地志紀行諸書所不見。近藤家上法子所求
曾言也。爰荒至重字子成。其藩之一人也。自少
好數學。而代數諸術。微分積分諸法。皆除其奧。
於量地学特邃矣。遂插為郡吏。故志大奮。若地
巡察。到巖險鑿石壁。水到渠成。自在流通。磽境
埔荒野。為膏腴良田。實則練磨星象積筭。凡量
地之法。不可行以一人。且支於數理。而為之者

Fig. 9

Preface of Three methods for land surveying which was written by Uchida Gokan. He used Chinese characters 代数, 微分 and 積分 for algebra, differential and integral here.

nally remained unfinished. He stopped translating the book though it seems that Kanda Katahira could understand calculus. We would not assume unclear reason why Kanda gave up the work, but rather pay our attention to the fact that he wrote down some remarkable descriptions and corrected a few mistakes of Introduction to Algebra and Calculus. We emphasize, first of all, that he comprehended meaning of calculus, and that he could make calculations by using Chinese notation in Introduction to Algebra and Calculus. While understanding calculus, he depended on Dutch mathematical books [8, pp.15–21]. It is remarkable that they still needed Dutch books even at that time to understand the Western arts and sciences.

7 Conclusion

In order to study how European mathematics left its mark on Japanese mathematics before modern times, it is necessary to trace it in two directions the indirect influence from China and the direct influence from the Netherlands. In this paper the author has mainly discussed the indirect influence from China. In fact, many Chinese books on Western Calendrical Calculations were imported in the Edo period, and mathematicians and astronomers tried to accept European arts and sciences by reading them. After these books became, in 1720, not conflicting with the foreign book prohibition policy of the Edo Shogunate, they could devote themselves to those studies without any fear of being arrested. Among these books, Complete Treatise on Calendar and Computation, Compendium of Calendrical Science and Astronomy and Sequel of Compendium of Calendrical Science and Astronomy should be worthy of special attention as the Chinese book which had a far-reaching ramification to the Japanese in the Edo period.

The impact by Chinese books did not remain to mathematicians, Wasan'ka, and scholars of ancient Japanese thought and culture at that time could not be unaffected. A recent research tells us that Motoori Norinaga [本居宣長] (1730–1801) made copies of trigonometric function table in 1799, and that Hirata Atsutane [平田篤胤] (1776–1843) possessed copies of Book of Chongzhen Calender, 96 volumes[5] and read them. In addition to these facts, a person of Confucianism Yamagata Bantō [山片蟠桃] (1748–1821) studied Chinese books on Western calendrical calculations.[6]Moreover, according to Takeo Suzuki's recent research, the Shogun's Council of Primary Elder Matsudaira Sadanobu [松平定信] (1759–1829) had many Chinese books on Western calendrical calculations including Complete Treatise on Calendar and Computation, Compendium of Calendrical Science and Astronomy, Sequel of Compendium of Calendrical Science and Astronomy, Elements of Geometry, etc. [11]. Matsudaira probably read these books, but he may not have understood its contents. From these facts, we may say that the Chinese book had a deep influence to the Japanese of every hierarchy in the latter period of the Edo period.

On the other hand, we must promote a more study on the direct influence from Dutch scientific books. For example, it is believed that the first Japanese scholar in Dutch who employed the term Western arithmetic [洋算 Yōsan] as a word for Western mathematics is Yanagawa Shunsan [柳川春三] (1832–1870), which may go against the fact. He published a book which introduced Western mathematics with the title of Arithmetic by Western numeral and notation [洋算用法 Yōsan yōhō] in 1857, in which he certainly used a term Western arithmetic to bring in Dutch arithmetic. We should, however, not give him honor to be the first person who used it. Another Japanese scholar in Dutch Satō Masayasu [佐藤政養] (1821–1877) should be the first person, who employed the term Western arithmetic in contrast with a term Wasan [和算] that expressed Japanese mathematics in the Edo period. In 1856, he wrote a manuscript with the title of Land surveying by trigonometry [三角惑問 Sankaku wakumon],[7] and the term Western arithmetic appeared in the explanatory notes of this manuscript [5]. Satō Masayasu was a disciple of Katsu Kaishū [勝海舟] (1823–1899), one of the most influential politicians in the end of Edo period, and

[5] These copies are conserved at Akita Prefecture Library.

[6] See his book, Dream world [夢ノ代 Yume no shiro] in 1820.

[7] This manuscript was conserved at Library of Kyōto University.

Satō learned Dutch arts and sciences in Katsu's school. In other words, the Japanese scholars in Dutch at that time came to understand European mathematics.

It is expected that we can develop a new standpoint in the field of the history of Japanese mathematics by investigating the intellectual activities of Japanese scholars in Dutch.

References

1. Ugo Baldini: The Jesuit College in Macao as a Meeting Point of the European, Chinese and Japanese Mathematical Traditions. Some Remarks on the Present State of Research, Mainly Concerning Sources (16th–17th centuries), *The Jesuits, The Padroado and East Asian Science (1552-1773)* (Luis Saraiva and Catherine Jami edited), pp. 33–79 (2008).
2. Tatsuhiko Kobayashi [小林龍彦]: Chinese Books on Western Calendrical Calculation and Japanese Calendrical Calculators in Edo Era [漢訳西洋暦算書と近世日本の暦算家], *The Astronomical Herald* [天文月報], vol. 98, no. 6, pp. 366–372 (2005).
3. Tatsuhiko Kobayashi: What Kind of Mathematics and Terminology was Transmitted into 18th-Century Japan form China?, *Historia Scientiarum*, vol. 12, no. 1, pp. 1–17 (2002).
4. Tatsuhiko kobayashi: On Secret treatise on trigonometry by Tadao Nakano [中野忠雄輯「三角算秘傳」について], *Bulletin of Narutaki* [鳴滝紀要], vol. 10, pp. 1–13 (2000).
5. Tatsuhiko Kobayashi: Measurements by trigonometry, questions and answers by Masayasu Satō [佐藤政養著「測量三角惑問」] and Dutch mathematics [蘭算], *Journal of History of Mathematics* [数学史研究], Japan, vol. 189, pp. 1–17 (2006).
6. Kōbe City Museum [神戸市立博物館]: *Preservation Catalogue*, Documents V (2003).
7. Tsuyoshi Kurasawa [倉沢剛]: *Study of the History of Education in the Late Tokugawa Period*, [幕末教育史の研究], vol. 1, Yoshikawa Kōbunkan [吉川弘文館] (1983).
8. Feng Lisheng [馮立昇] : On the spread and the influence of *Daibiseki Shūkyū* [代微積拾級の日本への伝播と影響について], *Journal of History of Mathematics*, Japan, vol. 162, pp. 15–28 (1999).
9. The Nagoya-city Board of Education [名古屋市教育委員会] ed.: *The Classified List of Chinese Books in Nagoya-shi Hōsa Bunko Library* [名古屋市蓬左文庫漢籍分類目録] (1975).
10. Dorotheus Schiling: *Das Schulwesen der Jesuiten in Japan* (1551–1614), Münster (1931); Yoshitomo Okamoto [岡本良知] translated, [日本に於ける耶蘇会の学校制度], Tōyōdō [東洋堂], pp. 61–62 and p. 228 (1943).
11. Takeo Suzuki [鈴木武雄]: Matsudaira Sadanobu and Elements of Geometry [松平定信と幾何原本], *Ōsaka Journal of Mathematics Education* [数学教育研究], vol. 37, pp. 81–90 (2007).

On Contemporary Mathematics in Vietnam

Ha Huy Khoai

Abstract We give a brief survey of the development of mathematics in Vietnam since 1947, when the first mathematical research paper written by a Vietnamese mathematician was published in an international journal. We describe how mathematics in Vietnam developed under very special conditions: the anti-French resistance, the struggle for the reunification of the country, the American war, the economic crisis, and the change toward a market economy.

Introduction

In this talk I like to give a brief survey on the development of contemporary mathematics in Vietnam. Here by *contemporary* I mean the period since 1947, when the first mathematical research paper by a Vietnamese mathematician was published in an international journal. Moreover, Vietnam declared its independence from French colonialism in September 1945, so the contemporary history of mathematics in Vietnam is the history after the colonial period. I should say that until now there is no research paper on this subject. My talk could be considered as a story told by a Vietnamese mathematician, who was born in November 1946 and who was a student at Hanoi University during the time of the American war, when the University was evacuated to the jungle, rather than a research paper on the history of mathematics.

1 Le Van Thiem—the founder of contemporary mathematics in Vietnam

The contemporary history of mathematics in Vietnam dates from 60 years ago, when a Vietnamese mathematician, Le Van Thiem, published a paper [3] in an interna-

Ha Huy Khoai
Institute of Mathematics, 18 Hoang Quoc Viet, 10307 Hanoi, Vietnam
e-mail: hhkhoai@math.ac.vn

tional journal.Le Van Thiem was born in 1918 in Ha Tinh, Vietnam, into an intellectual family. He was the youngest of 13 brothers and sisters. Le Van Thiem's oldest brother earned his "doctoral" degree (Tien si) after having passed the last Confucian traditional examination (1919, Nguyen Dynasty), whereas Le Van Thiem was the first Vietnamese to earn a "modern" doctoral degree. In 1939, after passing the final term examination with excellent marks, Le Van Thiem was offered a scholarship to study at the École Normale Supérieure in Paris. His education was interrupted by the outbreak of the Second World War and did not resume until 1941. He graduated with a Bachelor's Degree in Mathematics within a year rather than the conventional 3-year time. In 1942, under the supervision of George Valiron, he began his research on the value distribution theory of meromorphic functions (Nevanlinna Theory). It was in this period that he made important contributions to the solution of the inverse problem of the Nevanlinna theory that constituted the core of his doctoral dissertation (1945, Göttingen) and Docteur d'État (1949, Paris) and placed him among the best young researchers in the field at that time (See Drasin [1]).[1]

Meanwhile, in Vietnam, the resistance war against French colonialists was at its height. Despite his great passion for mathematics and the bright prospect of his scientific career, Le Van Thiem made in 1949 a dramatic decision which would not only drastically change his life but which would exert a profound influence on many generations of students in Vietnam to come—abandoning his academic position at the prestigious Zurich University, he returned to Vietnam to actively take part in Vietnam's struggle for independence.

To return to Vietnam, Le Van Thiem first flew to Bangkok and then headed for the liberated region in the far south of Vietnam. A few months later, following a narrow footpath through the mountains, which later during the American war became the famous Ho Chi Minh trail, he made the long trek to Viet Bac, in the far north of Vietnam and which used to be the headquarters of the Resistance. It was in Viet Bac that Le Van Thiem met other intellectuals, most of them educated in France—Ta Quang Buu (a mathematician, former Minister of Defense (1947), Minister of Higher Education, and President of the National Committee of Science and Technology), Tran Dai Nghia (a former *polytechnicien* and President of the Vietnamese Academy of Science). Convinced of the importance of education and science in this fight, Le Van Thiem founded, in the liberated zone, a teacher training college and a college of fundamental sciences with the aim of providing the country with qualified teachers and technicians, of which the resistance was in dire need. These colleges functioned until the end of the French war in 1954. The later development of science and research in Vietnam highlighted the essential contribution of these colleges to upgrading and sustaining the education system at a satisfactory level, even in complete isolation from the outside world during the French and then the American wars. Furthermore, these colleges formed the foundation for the immediate reopening in 1955 of Hanoi University with a strictly Vietnamese teaching staff, which at that time was a remarkable accomplishment in this region of Asia. Le Van Thiem,

[1] In this paper Drasin commented on Le Van Thiem's paper: *Using an important principle of Teichmuller, Le Van Thiem first applied this principle to the inverse problem, and the method was further exploited by Goldberg.*

together with other mathematicians (Hoang Tuy, Ta Quang Buu) founded two Vietnamese research mathematical journals in foreign languages (English, French, and Russian): *Acta Mathematica Vietnamica* and the *Vietnam Journal of Mathematics*. He also was a founder of the journal *Mathematics and Youth*, a friend of many generations of secondary school students. The appearance of these three journals during the American war in Vietnam was an important and hardly believable event.

Le Van Thiem passed away on June 3rd, 1991 in Ho Chi Minh city. He was the first modern Vietnamese mathematician to be commemorated by having a street (located in Hanoi) named after him.

2 Mathematics in Vietnam during the resistance war against French colonialists (1946–1954)

On 19 December 1946, a little more than one year after its declaration of independence, Vietnam began the resistance war against French colonialists. On that morning, all the government organizations received the order to evacuate Hanoi and move to the liberated zones, mostly again to Viet Bac. However, some high ranking officers only learned of the order later, among them mathematics Professor Nguyen Thuc Hao, who, like Le Van Thiem, was educated in France and returned to Vietnam in 1935. In December, Nguyen Thuc Hao left Hanoi for his native land in Nghe An, a province in the IV liberated zone. Some months later Nguyen Thuc Hao was appointed by the Ministry of Education to organize a mathematical school, more precisely, a class on mathematics at the university level. The unique professor at this "university" was Nguyen Thuc Hao himself. Although Hao's university was small in size, it was not small in importance. The first students from this mathematics class later became leading scientists of Vietnam. Nguyen Thuc Hao's mathematics class marked the beginning of the history of higher education in Vietnam after the colonial period.

A great landmark in the development of mathematics and mathematical education in Vietnam was the return of Le Van Thiem from France. At that time he was an idol for young Vietnamese. The return of Le Van Thiem attracted many talented young people to Viet Bac. The first students of the University of Sciences, founded in Viet Bac by Le Van Thiem, later became leading scientists of Vietnam. In the first peaceful years after the resistance war, some students of the University of Sciences founded by Le Van Thiem were sent to Russia to follow a postgraduate program. Almost all of them received the *candidate of sciences* (Ph. D.) degree after just 2 or 3 years of study. In particular, after only one year, Hoang Tuy had written a candidate dissertation in real analysis under the supervision of Menshov, and he had published 5 papers in leading Russian journals during the 20 months of his stay in Moscow for the Ph. D. program. Some years later, Hoang Tuy became the "founding father" of global optimization, with the famous *Tuy's cut* in non-convex programming theory. Another student, Nguyen Canh Toan, successfully defended his Dr. Sci. thesis in Russia with important results in projective geometry, which he had obtained during his years at the University of Sciences. We can say that the University of Sciences had an important role not only in training students at the university level, but also in building the first mathematical research group in Vietnam after the colonial years.

3 The years after the resistance war and during the American war (1954–1975)

The resistance war was finished in 1954, and the university reopened in Hanoi in 1955. Le Van Thiem was the rector of the university. The students who graduated from the university in Viet Bac now had an opportunity to do research in mathematics. Many of them were sent abroad, mostly to the USSR, to Eastern Europe, or to China.

During this period, many mathematicians, after receiving their Ph. D. degrees abroad, changed their interests to applied mathematics, following the scientific policy of the government. Note that this tendency appeared also in China at this time—for example, Hua Lo-Keng was actively promoting operations research at this time. In Vietnam one can see this tendency in the following examples:

—Hoang Tuy, who successfully defended his thesis on real analysis, became the first one to introduce operations research and optimization in Vietnam in 1961. From the very beginning, Vietnamese mathematicians have made many efforts to use mathematics to solve practical problems. In 1961–1962, Hoang Tuy and his group worked on a problem in transportation—reorganizing the logistics of trucking so as to reduce the distance that trucks have to travel with an empty load. I would like to mention that the Soviet mathematicians ventured into this applied problem later, in about 1963. Of course, they were able to carry it through much better than their Vietnamese colleagues could. Hoang Tuy said that after his visit with Kantorovich (a Soviet mathematician and Nobel laureate in economics) in Novosibirsk in 1962, he fully changed from real analysis to operations research. In 1964 Hoang Tuy obtained an outstanding result on concave minimization, which brought him international recognition. Hoang Tuy proposed a new kind of cutting plane, a notion introduced in integer programming by Gomory in the 1950's for use in convex programming. Hoang Tuy suggested a new type of cut, which would enable one to carry out a concave minimization algorithm. His cutting plane now is known as the *Tuy's cut,* and Hoang Tuy is sometimes called *the father of global optimization.*

—Phan Dinh Dieu, who obtained the Doctor of Sciences degree in Moscow with a thesis on constructive mathematics, redirected his interests to computer science. Later he became the first director of the Institute of Information Technology at the Vietnamese Academy of Science and Technology.

—Le Van Thiem, a famous expert in function theory with pioneering results in Nevanlinna theory, began to study the theory of groundwater movement and its applications in Vietnam. In this new area for him, Le Van Thiem obtained a remarkable result—he was the first who solved explicitly the problem of filtration via two ground layers [4]. Le Van Thiem and his students also applied the methods of complex analysis in orienting explosives during the time of the American war.

In 1964, the U. S. army began to bombard the North of Vietnam, including Hanoi and other cities. All universities were evacuated to the jungle. Many of them were again moved to Viet Bac, the former headquarters of the anti-French resistance. However, even during the war, the Vietnamese mathematical community continued its activities.

The mathematical society, which was founded in 1965 by Le Van Thiem, organized joint seminars in optimization, probability, functional analysis, complex analysis, algebra, and numerical analysis. People from Hanoi University, the Pedagogical Institute, and the Polytechnic Institute participated. Since the three Institutions had been evacuated in different directions from Hanoi, the seminars were held in Hanoi. They met twice a month, and one should say that people were very diligent about attending.

During the war, some mathematicians from abroad visited Vietnam and gave lectures for students and researchers. Among the visitors were Alexandre Grothendieck, Chandler Davis, Laurent Schwartz, André Martineau, Bernard Malgrange, and Alain Chenciner. To better understand the experiences of Vietnamese mathematicians, the experiences of the foreign mathematicians who visited Vietnam, and the mathematical life in Vietnam at the time, I would like to recall some parts from Alexandre Grothendieck's report of his visit to Vietnam in November of 1967, a report widely distributed among universities in the world in 1968. The first few days of Grothendieck's lectures took place in Hanoi. But one day a missile exploded only 100–200 meters away from the lecture hall. As a result, the Higher Education Minister, Ta Quang Buu, ordered us to be evacuated. Grothendieck was delighted with the news that we were being evacuated and approached the unusual situation with a spirit of adventure.

Grothendieck lectured on abstract algebraic geometry for four hours a day and met with students and colleagues during the afternoons. After the visit, Grothendieck wrote a very interesting and famous report, which gives the reader an overview about the mathematical life in Vietnam during the war. Here is his description of lecturing in Hanoi during the bombing:

Like most more or less public activities, the lectures were scheduled between about 6 and 10 a.m. During most of my stay the sky was cloud-covered and consequently there were few bombing raids. The first serious bombardments had been anticipated—they took place on Friday 17 November, two days before we left for the countryside. Three times my talk was interrupted by alarms, during which we took refuge in shelters. Something which is at first very striking to the newcomers is the great calm, almost indifference, with which the population reacts to the alarms, which have become a daily routine....

During one of the air raids in that Friday morning, a delayed-action cluster bomb fell right in the courtyard of the Hanoi Polytechnic Institute, and (after the alarm was over) killed two mathematics instructors at the Institute. Ta Quang Buu, who is a mathematician as well as the Minister of Higher Education (and who attended the lectures that I gave while in Hanoi), was discreetly informed of this during the lecture. He left at once—the rest of the audience continued to follow the lecture while waiting for the next alert. The next day's lecture had to be rescheduled for the following week in the university in evacuation, so as not to have large group of cadres in the city during the period of bombardment.

Grothendieck made some comments about the scientific as well as practical difficulties that an aspiring Vietnamese mathematician had to endure in such an isolated part of the world:

Life is very primitive. Everyone—university administrators, teaching staff, and students—live in the same type of straw huts made of bamboo with mud walls, window open to the wind, and the sun baking the earth. Since there is no electric lighting, they use kerosene lamps.... Very often when the weather is clear enemy planes fly over the university, occasionally dropping their bombs—haphazardly, so as to get rid of them before returning to base—sometimes wounding or killing some civilians.

In a country which, by force of circumstance, has few relations with the outside (unless one counts the cluster bombs as a form of relation), it is particularly difficult for an inexperienced mathematician to orient himself among the multitude of possible directions, to distinguish what is interesting from what is not.

He explained that he was astonished to find an active community of research mathematicians in Hanoi:

The first statement to make—a rather extraordinary statement in view of the circumstances—is that there is in fact a mathematical life worthy of the name in North Vietnam. To properly appreciate this 'existence theorem,' first of all one must keep in mind that in 1954, after the eight-year war of liberation against French colonial occupation (i.e. thirteen years ago), higher education was practically nonexistent in North Vietnam. During the extremely brutal war of 1946–1954, the main effort in education was directed toward achieving literacy for the large masses of peasants, an effort which was carried through to its final goal in subsequent years, until about 1958, at which time illiteracy was practically eradicated in the lowlands.

...The method followed (undoubtedly the only one possible) was to send young people to universities in the socialist countries, especially in the USSR. Among the hundred or so mathematics instructors at Hanoi University and the Pedagogical Institute, about thirty have gone abroad for four or six years of training. They have generally reached the level of a soviet 'candidate' thesis.

Finally, Grothendieck concluded on an optimistic note:

I can attest that both the political leaders and the senior academic people are convinced that scientific research—including theoretical research having no immediate practical applications—is not a luxury, and that it is necessary to promote theoretical scientific research (as well as the development of instruction and the applied sciences) starting now, without waiting for a better future.

... And through an effort undoubtedly without precedent in history, in spite of everything, they are succeeding in increasing the cultural and professional level of their citizens, even as their country is to a great extent being devastated by the largest industrial power in the world. They know that once the war ends, there will be people with the professional and moral qualities needed to reconstruct the country.

The university was in evacuation for four years. It reopened in Hanoi in September 1969. Then again from 1972–73, there was another evacuation, when the U.S Army used B-52 bombers to carpet-bomb Hanoi and some other cities of Vietnam.

During the American war, every year Vietnam sent about 100–150 students to the mathematics departments of universities in the USSR and in Eastern European countries. Also, every year about 20 mathematics instructors from Vietnamese universities were sent to follow Ph. D. programs in these countries. Returning to Vietnam, these scholars became the leaders of research groups in Vietnam's universities. The main difficulty at the time was our relative isolation from the mathematical community in the world. Even communication between new Ph. D.'s, who just returned from abroad, and their former supervisors was not easy. Fortunately, at the time, Vietnam was able to receive from China almost all the main international mathematical periodicals (and in fact usually within 1–2 years of their publication). At this time China had not yet signed the Berne Convention on copyright, and they regularly made copies of journals and gave some copies to Vietnam for the National Library. Other materials, like journals and books in Russian, could be found in book stores at very low prices (for example, a Russian translation of S. Lang's *Algebra* sold for about 20 cents.).

In the period from 1955 to 1975, mathematics in North Vietnam made significant progress. Some strong research groups were established—optimization (headed by Hoang Tuy), singularity theory (with the guidance of Vietnamese overseas mathematicians Frédéric Pham and Le Dung Trang), complex analysis (Le Van Thiem and his students), P.D.E, etc. The creation in 1966 of the Mathematics Section in the National Committee for Science and Technology (later, in 1970 it became the Institute of Mathematics, directed by Le Van Thiem) further stimulated mathematical research in Vietnam. Even during the most difficult years of the American war, the Mathematics Section (and then the Institute of Mathematics) organized annual scientific conferences and printed the proceedings under the name *Toan hoc – Ket qua nghien cuu (Mathematics – research results)*. Many people published their results in leading Soviet journals, like *Doklady of the Academy of Science of the USSR, Mathematics Sbornik, Functional analysis and its applications,* etc. The mathematicians established in this period are still the leading mathematicians in Vietnam at the present time.

Before the reunification of the country (in 1975), in South Vietnam, there was almost only one research group, namely the group researching P.D.E lead by Professor Dang Dinh Ang, a mathematician educated at the California Institute of Technology (U.S.A.). Other mathematicians, such as Nguyen Dinh Ngoc (a topologist who returned from France), were teaching at Saigon university, but not doing research. I would like to mention that the group of Dang Dinh Ang and his students up to the present day is still the strongest group researching analysis and P.D.E in Vietnam.

4 Mathematics in Vietnam after the reunification of the country

After the reunification of the country in 1975, mathematics in Vietnam finally experienced favorable conditions for development. In particular, cooperation with the mathematical community in the world became much easier. Many young people obtained fellowships to go study abroad, and not only to the socialist countries, but also to other countries: France, West Germany, Italy, Japan, etc. For example, from our Institute of Mathematics, 16 members received Alexander-von-Humboldt fellowships—about 20 people were named associate members of the Mathematics Section of the International Centre for Theoretical Physics (ICTP) in Italy, etc.

Only a few years after the reunification of the country, the number of mathematicians who held the Ph. D. degree grew quickly to the point where in 1980, Vietnam had about 300 mathematicians with Ph. D. degrees. The establishment of several new universities, almost all of which included mathematics in their program, promoted the development of mathematicians, and especially, mathematicians with a Ph. D. degree.

However, during the period from 1980 to 1995, mathematics in Vietnam faced a serious difficulty. Vietnam experienced an economic crisis during the 1980's, and in the beginning of the 1990's, Vietnam began its transition toward a market economy, the so called *Doi moi*. Many mathematicians had to leave mathematics because the salary of a mathematics lecturer was very low, only about 3–4 USD per month. Almost all had to do a "second job", which usually required much more time and effort than the first one – doing mathematics! If, for many years, mathematics was the first choice for the best high school students, then in the early 90's, an opposite tendency appeared. It even happened one year that there was not a single student who entered the mathematics department of Hanoi University. At that time, some mathematicians predicted that mathematics in Vietnam was at risk to become extinct in only 15 years! [5]

Fortunately, mathematics in Vietnam survived this difficult period. The first and the most important reason was that during this period many Vietnamese mathematicians continued their mathematical research despite the extremely hard conditions. On the other hand, it is worth mentioning valuable help from the mathematical community around the world, especially from France, Italy, Germany, and Japan. I would like to mention here the role of the program 'ForMathVietnam' from France and the importance of fellowships like Alexander-von-Humboldt (Germany), JSPS (Japan), and ICTP (Italy and UNESCO). Vietnamese mathematicians also appreciate the assistance offered to them from mathematicians abroad during this difficult time for them. Here are two examples. There was a time when almost all the new books coming to the library of the Institute of Mathematics were donated by foreign and overseas-Vietnamese colleagues. The guest house of the Institute of Mathematics was built with money donated by mathematicians from Japan, the U.S., and other countries.

Beginning in the mid-1990's, Vietnam step–by–step got out of the economic crisis, and Vietnamese mathematics returned again to a normal development. Young Vietnamese now can go to study abroad not only with fellowships from foreign institutions, but also with financial support from the Vietnamese government (Project '322'). Good students with a passion for mathematical research now do not hesitate to choose mathematics as their future career. Some of the best students continue their study in famous universities throughout the world, such as Harvard University, Princeton University, École Normale supérieure, École Polytechnique, Trinity College....

On top of the fast growth in the number of mathematicians with a Ph. D. degree (at the present time, it is around 700), Vietnamese mathematicians have contributed outstanding results and solved fundamental problems in mathematics. I would like to highlight here Ngo Bao Chau's proof of the *Fundamental Lemma*, a famous problem in modern mathematics that is a cornerstone in the Langlands program, one of the most important and active areas of the 21st century mathematical research. Notices that, for the proof of this lemma in a partial case [2], he was awarded (jointly with Gérard Laumon) the prestigious Clay Research Award in 2004 from the Clay Mathematics Institute.[2]

5 Some remarks

During the 60 years since Vietnam's colonial period, a period not so long, mathematics in Vietnam developed under very special and difficult conditions—the anti-French resistance, the struggle for the reunification of the country, the American war, the economic crisis, and the transition toward a market economy. I would like to conclude by suggesting the following topics for more detailed study:

—The higher mathematical education and research during the wars.

—The impact of the assistance and cooperation received from abroad on contemporary mathematics in Vietnam.

—The influence of the change toward a market economy on the development of mathematics in Vietnam, and other former socialist countries.

References

1. D. Drasin: A meromorphic function with assigned Nevanlinna deficiencies, *Bulletin of the American Mathematical Society*, vol. 80, pp. 766-768 (1974).
2. Gérard Laumon et Ngô Báo Châu, Le lemme fondamental pour les groupes unitaires, *Annals of Mathematics*, vol. 168, no. 2, pp. 477–573 (2008).
3. Le Van Thiem: Beitrag zum Typenproblem der Riemannschen Flachen, *Commentarii Mathematici Helvertici*, vol. 20, pp. 270–287 (1947).
4. P. Ya. Palubarinova-Kochina: *Theory of Groundwater Movements* (in Russian), Nauka, Moscow (1977).
5. F. Pham: Y aura-t-il toujours des mathématiciens au Vietnam l'an 2000 ?, *Gazète de Mathématiques*, vol. 64, pp. 61–63 (1995).

[2] Added in Proof: In the December 8, 2009 issue, the American weekly journal "Time" selected Ngo Bao Chau's proof as one of The Top 10 of Scientific Discoveries of 2009.
http://www.time.com/time/specials/packages/0,28757,1945379,00.html

Part II
Supplements

Notes on Complete Book of Mathematics Vol. 4: Three Essentials

Mitsuo Morimoto

Abstract The Complete Book of Mathematics is the most comprehensive treatise of mathematics in the Edo Period of Japan. The 20 volume book is about 900 sheets or 1800 pages long. Seki Takakazu (1642?–1708), Takebe Kataakira (1661–1721) and Takebe Katahiro (1664–1739) spent 28 years (1683–1711) in writing it. Unfortunately, the Book has never been published as a whole. We reproduce here Volume 4 Three Essentials, which is a rare exposition of mathematical philosophy during the Edo period.

1 The Complete Book of Mathematics

The Complete Book of Mathematics [大成算経] was complied by Seki Takakazu [関孝和] and the Takebe [建部] brothers (Kataakira [賢明] and Katahiro [賢弘]) during the period of 1683–1711 (28 years) according to the Biography of the Takebe . (See [2, p. 270].) Along with the tradition of Chinese mathematics, Takebe Katahiro recognized mathematics as a bunch of mathematical problems. He tried to classify mathematics (i.e., mathematical problems) and to organize the Book as an encyclopedia of mathematics at the beginning of the 18th century in Japan.

The Book is preceded by Introduction [首篇], which includes Origin of Arithmetic [算数論], Basic Numbers [基数], Large Numbers [大数], Decimal Fractions [小数], Counting Numbers [度数], Quantity [量数], Weight [衡数], Money Counting [鈔数], Counting Boards [縦横], Positive and Negative Numbers [正負], Operations on Counting Boards [上退], and Technical Terms [用字例].

The 20 volumes are divided into four parts. Part 1 is composed of the first 3 volumes and treats elementary arithmetic ending with an introduction to discriminants. Volume 1 is entitled Five Techniques [五技] and treats Addition [加], Subtraction

Mitsuo Morimoto
1-64-5, Wada, Suginami-ku, Tokyo 166-0012, Japan
e-mail: mormormor@jcom.home.ne.jp

[減], Multiplication [因乗], Division [帰徐], and Extraction of Roots [開方]; Volume 2 is entitled Miscellaneous Techniques [雑技] and surveys Addition and Subtraction [加減], Multiplication and Division [乗除], and Extraction of Roots [開方] in classical textbooks; and Volume 3 is entitled Various Techniques [変技] and treats advanced aspects of the contents of the previous volumes.

Part 2 is composed of 6 volumes and treats traditional mathematics and games. Volumes 4 is named Three Essentials [三要] and includes Symbols and Figures [象形], Flow and Ebb [満干], and Numbers [数]; Symbols and Figures are the classification of mathematical objects and hence, problems. Volumes 5 to 9 are named Methods of Symbols Methods [象法] and discuss problems on "symbols". Volume 5 treats Mutual Multiplication [互乗], Polynomial Fitting [畳乗], and Pile Sums [垜積]; Volume 6 treats Fractions [之分], Several Methods of Fractions [諸約], and Arts of Cutting Bamboos [翦管]; Volume 7 treats Magic Squares, Magic Circles [聚数], Joseph's Problems [計子, 算脱], Coding Problems [験符]; and Volumes 8 and 9 treat Daily Mathematics [日用術].

Part 3 is composed of 6 volumes and named Methods of Figures [形法]. It treats various problems in geometry and measurements. Volume 10 treats Squares [方], Rectangles [直], and Rectangular Triangles [勾股], Polygons [斜 (三斜、四斜、五斜)]; Volume 11 discusses Regular Polygons [角法]; Volume 12 is concerned with Ratios of Figures [形率], i.e., Circle Theory [円理], and treats Length of the Circular Circumference [円率], Length of Arcs [弧率], Volume of a Ball [立円率], and Volume of Spherical Segments [球闕率]. Volume 13 is the same as Seki Takakazu's monograph Measurements [求積]. Volumes 14 and 15 are concerned with Complicated Configuration of Figures [形巧].

Part 4 is composed of 5 volumes and treats the theory of equations. Volume 16 is named Discussion on Problems [題術辨] and similar to Seki's Critical Studies of Problems [題術辨議之法]. Volume 17 is named Solutions of Whole Problems [全題解], contains Seki's Trilogy [三部抄], and composed of Explicit Problems (i.e., direct calculation) [見題], Implicit Problems (i.e., equations of one variable) [隠題], Concealed Problems (i.e., equations of several variables) [伏題], and Submerged Problems (i.e., non algebraic equations) [潜題]. Volume 18 is similar to Seki's Restoring Defective Problems [病題擬]; and Volumes 19 and 20 are named Examples of Operations [演段例] and contain 23 examples of algebraic and non algebraic equations.

In [2, p.385] Matsusaburo Fujiwara wrote that Volume Four was "very strange and meaningless as mathematical theory." Because of this negative comments almost no researches on Volume 4 had been done until Xu Zelin published [8]. In this important article he examined the Three Essentials and understood it in the context of traditional Chinese culture. Recently, appeared a few papers like Fumiaki Ozaki [5] and Hikosaburo Komatsu [1].

2 English translation in part

Here we translate general statements into English. 67 problems are not translated. Three Essentials [三要, san'yō] starts with the following statements:

Symbols and figures are the most fundamental of all, and appear at the beginning of every problem. Usually there are determined formulas but there are also exceptional cases. Under a name of Symbols or Figures there are the variations represented by flow and ebb in which numbers play very useful roles. These three form a focus which the theories should exactly investigate to their extreme ends. In fact, from the techniques of solving [mathematical] problems to the movements of the heaven and the earth, and also the substance, its movement and transformations of everything, everything is equipped with its theory involving its numbers. Therefore, the student should observe throughly all the changes of matters to investigate their theories to their extremes.

1. Section 1 is named Symbols and Figures [象形, shō-kei] and starts with the following statements:

A symbol is something not yet embodied, while a figure is what is embodied. Each is composed of two kinds, respectively. That which gives rise to the reason why Spring and Autumn come and the moon waxes and wanes, and why the heaven appears as a circle and the earth as a square, and so on, is originally equipped by the nature. That which makes the market price function in daily life, and the decision of the shape of containers and other articles for daily life are all done by the human beings.

The theories distinguish each symbol or each figure from tens of thousands of matters first by its name and then by its numbers; by measuring its length with a rule, its weight with a scale, its volume with a container or by its number by counting; all measuring [主, hakaru, see Ozaki [5]] its numbers naturally according to the matter.

There are two kinds of symbols. Those which have originally no shape or those which have a shape but it is of no use to express it by a geometrical figure are called [abstract] symbols; those which can be compared with a figure in length or those which are represented by a picture of procession are called [concrete] symbols.

There are two kinds of figures. Those with length and breadth are called planar figures and those with length, breadth and height are called solid figures.[1]

As symbols with a given name have only the total numbers as their numbers, they are of no use for the identification. Therefore, they are used along with other

[1] Geometrical figures are easy to understand but there are other mathematical objects, which he called symbols. Symbols [象, shō] were classified into two subcategories: ◯ symbols [◯象, [ ]shō] and ☐ symbols [☐象, [ ]shō], while Figures [形, kei] were classified into two subcategories: planar figures [平形, heikei] and solid figures [立形, ritsukei]. In the original text ◯ and ☐ are hiatuses as the authors could not find suitable characters to express their idea. Following Hikosaburo Komatsu, we propose to read these hiatuses as (abstract) symbols [(抽) 象, (chū)shō] and (concrete) symbols [(表) 象, (hyō)shō], respectively.

things or by being applied to other things. Therefore, there are the total sums, the numbers in a unit and the numbers in a cluster. (The numbers in a cluster and the global numbers are defined in a similar way but they may be different according to the time of evaluations.) The definitions are provided naturally but the different results may occur by artful schemes.

Figures have shapes for each name but there are differences according to their width, length and so on. Therefore, they are equipped with the names of lengths, widths, diagonals and circumferences etc. and their measured numbers. Moreover, if we cut one, or connect one to another, or inscribe one into another, or put one on another, or let one wind another, then we have complicated configurations according to our skill. That is the reason why we start with symbols and figures as the heading of a problem; there is a multitude of variations.

2. Section 2 is named Flow and Ebb [満干, man-kan] and starts with the following statements:

The flow and ebb [満干, man-kan] are originally associated with the symbols and figures [象形, shōkei] and are classified into three phases: ordinary [全, zen], extreme [極, kyoku] and excessive [背, hai].

The flow is to increase and continues infinitely. The ebb is to decrease and goes to an end. The ordinary phase, which is realized in the real world, exhausts in the extreme phase; the excessive phase is contrary to the real world. Symbols and Figures always assert paired notions to a matter like long and short, many and few, expensive or cheap, heavy and light, etc.. Only the total number has no pairing, but there can be the difference of old or new ones. If a parameter is given in the problem, it is called old; if the parameter has no theoretical definition but is allowed between two limits, it is called new. (A symbol is associated with a thing. If there are two symbols, the main symbol is always compared with its pair. A length is compared with another length, a capacity with another capacity, and a weight with another weight. These are two numbers associated with the thing and its class and name are the same. Therefore, they are paired and complete. If a length is compared with a capacity or with a weight, two parameters are of different classes, the theory cannot be completed; we consider a new limit for the parameter. In this way, the parameters can be paired.) Therefore, according to the origin of symbol and figure and conditions in the problem, (If in the problem like in the case of symbol, local parameters are the diagonal of a square, the circumference of a circle, diameters of angular figure, and the height of a triangle, they are assigned numerically but determined late by a procedure. Therefore, we do not take them as the basis of increase and decrease.) we observe all the parameters can be compared. (Some parameters may be given by the problem; some parameters may be calculated by multiplication or by division. After that they are paired.)

If it is paired with the large, the parameter increases and determines the flow. If it is paired with the small, the parameter decreases and determines the ebb. If it is paired with none, the parameter increases or increases and determines both phases. (But it happens that, according to the condition in the problem, it increases but determines the ebb or it decreases but determines the flow.) If parameters are repeatedly paired with the large, they determine the flow. Therefore, we use the fewest of them. If parameters are repeatedly paired with the small, they determine the ebb. Therefore, we use all of them. We examine all parameters to find where are limits of their increase or decrease. One by one, we study three statuses (ordinary, extreme and excessive) of all parameters.

Certainly, limits of each flow and ebb and conditions in a problem will correspond with each other. Therefore, limits appear according to the species of a symbol and the picture of figure. (If the equality of numbers or the correspondence are mentioned in a problem, there will be confusion in the theory and some will not correspond with a limit.) Its variable itself rotates and has a constant number. (Although a symbol is a mathematical object, it can be useful accompanied with parameters. Therefore, we cannot decide its character. A figure has, from the beginning, its shape, which is large or small, slant or straight. Therefore, even if a figure is named, its shape cannot be determined. Taking symbols and figures, we consider them together as a limit. It is the limit of the problem. Considering its flow and ebb, we make a variable. Considering further three statuses, ordinary, extreme and excessive, we have a general view of the variable.) According to symbols and figures or according to the condition in a problem, there are pairs of properties. Although there are many ways of consideration, if we investigate the theory, all are attributed to one determined limit.

Notes: There are several parameters in a mathematical problem. Parameters wax and wane. It is very important to understand the range of a parameter and limits of the range. In some cases, it is interesting to consider the case where the parameter goes beyond the limit.

In sum, the authors claimes there are the following six statuses: Ordinary Flow [満全], Ordinary Ebb [干全]; Extreme Flow [満極], Extreme Ebb [干極]; and Excessive Flow [満背], Excessive Ebb [干背].

3. Section 3 is named Numbers [数, sū] and starts with the following statements:

Numbers which are employed for symbols and figures have two categories; provisional [動] and stationary [静]. As each symbol is identified from tens of thousands symbols with the same name by its total number or its number in a unit, each figure with the same name has its length, width, length of the diagonal or the circumference as its character which is determined naturally and measurable as a number, which is called stationary. The number obtained by addition is called a sum; the number obtained by subtraction is called a difference; the number obtained by multiplication is called a product; the number sought for and obtained by division or by root extraction is called a quotient. They are all obtained by procedures applied [to known numbers]. Therefore, each of these numbers is provisional. If a number is obtained by formulas or by the root extraction to retrieve it form the known, then such a number looks like provisional but we have to understand the situation and the reason to distinguish it [from the provisional].

If numbers are confined, they are well-posed [整]; if not confined, inexhaustible [不尽]. There are two kinds of well-posed number: if it has no (non-decimal) denominator, it is called whole [全] and used in many usual cases. If it has (non-decimal) denominator, it is called complicated [繁] and used in several special cases. There are two kinds of inexhaustible number: if it becomes well-posed [整] after operations of multiplication and division, it is called residual [畸]; if it never becomes well-posed, it is called degraded [零]. The [degraded] number loses its true value after painful operations of multiplication and division. But consulting symbol and figure and considering the procedure, we can use it by taking the original number, reducing fractions to a common denominator, selecting or making coefficients. Taking into account the real meaning, we should select the numbers.

Notes: First, the authors classified numbers into two categories: provisional numbers [動, dō] and stationary numbers [静, sei]. The numbers originated from a problem of symbol and figure are called stationary, while those obtained by addition/subtraction, multiplication, division or root extraction are called provisional.

Second, the authors proposed another type of classification of numbers. There are well-posed numbers [整] (i.e., rational numbers) and inexhaustible numbers [不尽] (i.e., irrational numbers). Well-posed numbers are classified into whole numbers [全] (i.e., finite decimal fractions, not necessarily integers) and complicated numbers [繁] (i.e., general fractions), while inexhaustible numbers are classified into residual number [畸] (i.e., algebraic numbers) and degraded numbers [零] (i.e., numbers with errors). A whole number can be placed on one row of a counting board [算盤]; a complicated number can be expressed using two rows; and a residual number can be expressed using a finite number of rows.

Although the integers, the rational numbers and the algebraic numbers were recognized by Takebe Katahiro, he did not recognize that the algebraic numbers are closed under arithmetic operations. As we see this classification, we are tempted to consider the "degraded" numbers as transcendental numbers; looking at five examples of "degraded" numbers given as problems, we know he was regarding them as inexact numbers with error term.

References

1. Hikosaburo Komatsu: On Volume Four (Three Essentials) of the Complete Book of Mathematics (in Japanese), *Abstract in the Section of History of Mathematics, Annual Meeting of Mathematical Society of Japan, March 2007, Saitama University.*
2. Japan Academy: *History of Mathematics before the Meiji Restoration* (in Japanese), volume 2, Iwanami Shoten [岩波書店] (1956).
3. Mitsuo Morimoto and Tsukane Ogawa: The Mathematics of Takebe Katahiro: His Three Formulas of an Inverse Trigonometric Function, *Sugaku Expositions*, volume 20, no. 2, 237–252 (2007); original paper (in Japanese), *Sugaku*, vol. 56, no. 3, 308–319 (2004).
4. Tsukane Ogawa, Ken'ichi Sato, Osamu Takenouchi and Mitsuo Morimoto: *Mathematics of Takebe Katahiro* (monograph in Japanese), Kyōritsu Shuppan [共立出版] (2008).
5. Fumiaki Ozaki: Questions of Volume Four of the Complete Book of Mathematics, Three Essentials (in Japanese), *RIMS Kôkyûroku*, **1392**, Kyoto Univ., 186–196 (2004).
6. Takebe Katahiro: *Mathematical Treatise on the Technique of Linkage*; translated by Mitsuo Morimoto and Tsukane Ogawa with commentary (in English), to be published with a copy of the manuscript conserved at the National Archives (Naikaku Bunko), in preparation.
7. Osamu Takenouchi and Mitsuo Morimoto (translated and Commented): *Selected Mathematical Works of Takebe Katahiro (1664–1739)*, Wasan Institute (July 2004, a tentative edition).
8. Xu Zelin: Takebe Katahiro's Epistemology of Mathematics (in Chinese), *Research of History of Natural Science*, vol. 21, no. 3, 232–243 (2002); Japanese translation was prepared by Mitsuo Morimoto: *Journal of History of Mathematics*, Japan, vol. 206, 30–47 (2010).

Complete Book of Mathematics Vol. 4: Three Essentials, by Seki Takakazu, Takebe Kataakira and Takebe Katahiro, collated by Fumiaki Ozaki and Hikosaburo Komatsu

Seki Takakazu, Takebe Kataakira and Takebe Katahiro

Abstract Volume 4 is named Three Essentials [三要, san'yō] and composed of three sections. Each section starts with a general statement followed by problems, which serve as examples for the general statements. Section 1 is named Symbols and Figures [象形, shō-kei] and divided into (abstract) symbols [(抽) 象, (chū)shō] (Problems 1 – 6), (concrete) symbols [(表) 象, (hyō)shō] (Problems 7 – 11), planar figures [平形, heikei] (Problems 12 – 16), and solid figures [立形, ryūkei] (Problems 17 – 21). Section 2 is named Flow and Ebb [満干, man-kan] (Problems 22 – 37). Section 3 is named Numbers [数, sū] and divided into two subsections: the first subsection is named Provisional and Stationary [動静] numbers (Problems 38 – 47). The second subsection deals with two kinds of finitely presented numbers [整数], i.e., Whole [全] numbers (≒ finite decimal fractions) (Problems 48 – 52) and Complicated [繁] numbers (≒ fractions) (Problems 53 – 57), and two kinds of Inexhaustible numbers [不尽], i.e., Residual [畸] numbers (≒ algebraic numbers) (Problems 58 – 62) and Degraded [零] numbers (≒ transcendentals and observed constants) (Problems 63 – 67).

大成算經 巻之四 三要

巻之四 中集 三要

關孝和
建部賢明 編
建部賢弘

二〇〇四年三月十九日 尾崎文秋校
二〇〇八年八月二十日 小松彦三郎再校

大成算經卷之四　中集

三要

夫象形者萬事之本爲題問之首而常有定法之式亦有臨場之機然滿干變化之道備而數能致其用矣此三者爲衆理當窮之要也蓋自問題答術之技以至天地之運萬物之氣與動作云爲之事悉莫不以具其理包其數焉是以學者宜盡物變而窮其理矣

象形第一

象者未顯之稱形者已顯之稱其所成各有二焉如生春秋盈虧之理顯天地方圓之狀者本自然而所具也如成商價日用之功制器用什物之狀者皆人

一

爲之所定也衆理萬物之所分一象一形各其名具而度長短秤輕重量容受計名目者皆應物而自主其數也象有二義焉本無狀者雖有狀不用畫圖者謂之抽比長短之形成行伍之圖者謂之表也形有二義焉縱横二畫謂之平縱横高三畫謂之立也凡象者每名皆一偏之總數而不能自爲用是以或托事而特爲用或宛物而相爲用故有通計及屬一與屬衆之數乃屬衆者與總數雖其理相同或題中言之或術中得之則各其數自有多少而新舊之意異矣其理各本自具而唯依所言之巧異象生焉形者每名有狀據其廣狹長短自爲用故縱横斜圓之號及計積之數相具然或截之或接之或容之或繞之則隨其巧奇形生焉是此所以象形爲

題首而其變化無窮也

抽象

假如有物不知總數幾數剩若干幾數剩若干問總數

是不言名以物喩之故無狀而不能畫之唯計箇數之理自然所主也然總數一號具而自不能爲用故宛幾數而爲用也

假如有三乘方積若干問每面

是有名而無狀故不能畫之雖然主度長之定理故面一號本具而自有計積之用也

假如有酒斛若干每斗若干價錢文若干問該錢

是酒本一氣渾然而無定狀故唯量其容數錢雖有狀不關于其事故以畫不論之唯計其緡

二

數是皆自然所主也若別各爲一物則皆總計之一數而不能自爲用是以酒與錢相宛而爲用故酒本總斛與屬錢一之數相具錢本總緡與屬酒一之數相具今題中言屬衆酒之錢而問屬總酒之錢故雖二屬衆與總之理相似兩數異而其意亦不同也

假如有銀錢若干買羅綾共尺若干羅尺價錢若干綾尺價錢若干問羅綾數

是銀與羅綾三物雖有狀各不據畫圖其所主銀秤重羅綾各度長是皆自然之理也今以銀聯宛二物而爲用故銀有屬總羅與羅一之重又有屬總綾與綾一之重羅有屬該銀與銀一

之長綾有屬該銀與銀一之長若羅綾相宛爲用則復有屬羅一之綾與屬綾一之羅也

假如有元米若干斛每若干月利若干斗今借米若干斛經若干月問該利

是米所稟之狀本不據畫唯量容數經月本無狀而計其名然今分米一名而相宛元利之二物又宛于月數爲三名且別其新古而爲用故有三本利經月及總與屬一之數各新舊二色是皆自相具也

假如有織工若干人每若干日織絹若干匹布若干匹今工若干人經若干日問織絹布

是絹布及人皆雖有狀不據之日數本無狀乃

三

所主絹布各度長人與日各以名計之今相聯絹布二物而以工與日又相宛有四名且分新舊而爲其用故有四絹布與工及經日總與屬一之數二品皆本所具也

表象

假如有樹高若干尺春生嫩枝至秋長若干尺問該高

是本雖有狀主株根數而宛物則不用其畫今主長而托事增枝之爲用故釋題意而寫一根之稟狀唯原高與通高及杪長相具也

假如有紅絲若干斤每斤價銀若干兩問計銀

是絲銀二名本有狀而不用其畫皆主重而相宛爲用故各總重與屬一之重相具是故題中固雖無借狀之意依術釋其乘除之理則摸直形也

假如有金毬一隻徑若干尺問重

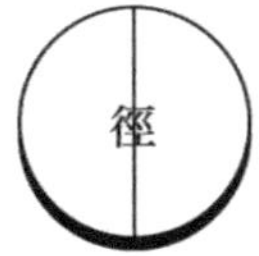

四

是常主秤而宛物相爲用故雖以畫不論之題中借立圓之形問之故摸其狀而釋題意也

假如有幾方陣縱横角斜各等數備之問備圖

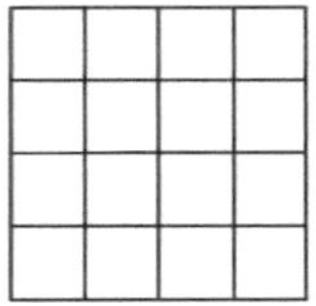

是本聚數之法借形而自爲用故畫方于每一面而證其配圖是以唯方與一遍之總兩數相具也

假如有圓陣若干隊分騎步而備之隔若干隊順擊及餘步一隊却自其逆擊則騎隊亡而止步一隊問備圖

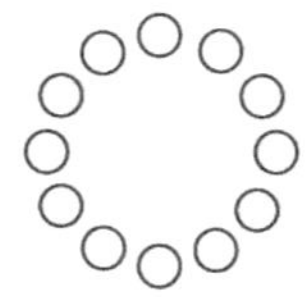

是本計數之法借形而托事爲用故畫小圈而
證配列是以總與計及順逆限三數相具也

平形

假如有平方面若干問積

是固有狀故常摸其圖勢乃諸形畫皆以度計之故積亦冒一位之度名而計其數是皆自然所主也凡積者縱橫高下相通之總數故每一形專以之爲要也

五

本雖爲縱橫二畫四旁相等而唯以每面與外
圍之一畫自爲用是故角斜之畫本自具也

假如有直長若干闊若干問積

是本縱多橫少之狀長闊二號相宛而爲用故
斜畫自具也

假如有梭長若干闊若干從右旁截長若干問截闊

是亦縱多橫少之狀長闊相宛而爲用故外四面畫自具今斷形之巧其勢繩直故舊號長闊與截長一畫共爲其用是以截闊截斜二畫新具也

假如有梯大頭若干小頭若干長若干應準而接作圭問接長

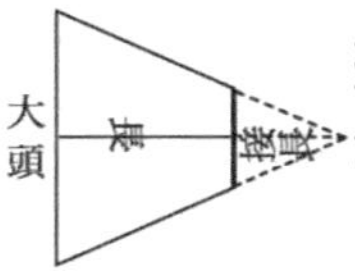

是本橫有廣狹之狀以兩頭及長三號爲用故內外二斜畫自具今雖成補闕之巧以外斜與長相會者爲限故據舊三號爲其用是以新接長與斜二畫具也

六

假如有三斜大斜若干中斜若干小斜若干內容圓問圓徑

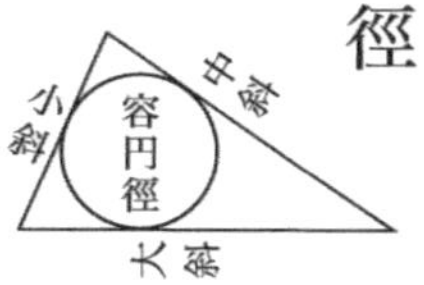

是三條長皆轉折之狀以大中小之三號互爲用故每斜之中股及左右闊三畫各相具今雖成容罅之巧而新徑圍之畫具周各有所交故依舊三號爲其用也

立形

假如有方墻方若干高若干問積

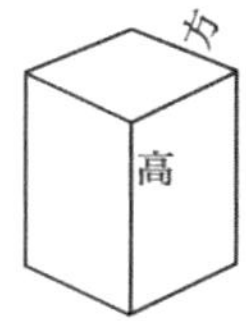

是立起之方上下同狀以方高二號爲用故上
下方面斜四旁直面斜內四稜斜各三畫相具
也

假如有方臺上方若干下方若干高若干問積

是方上小下大之狀以上下方及高三號爲用
故上下方面斜四旁梯面長及兩斜內四稜斜

七

總六畫相具也

假如有直錐下闊若干下長若干高若干問積

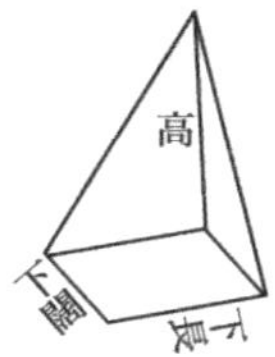

是上鋭下直之狀以長闊高三號爲用故下直
面斜四旁圭面長及斜總三畫相具也

假如有甲乙丙丁圓球各一甲徑若干乙徑若干丙徑
若干丁徑若干下敷乙丙丁三球上載甲球問中高

是本四球四徑之狀唯周圍之畫各具今雖成
數載之巧每一周互有所交會故皆依舊四徑爲
其用是以新中高一畫具也
假如有圓臺上徑若干下徑若干高若干以絲繞之每繞
隙廣各若干問絲長

是圓上小下大之狀以上下徑及高三號爲用
故外圍之斜高與形內之稜斜二畫相具今成
周旋委蛇之巧與隙廣相共爲其用是以繞長
畫新具也

八

滿干第二

滿干者本屬象形而有全極背三科矣所謂滿者增也其所至遂無窮干者損也其所至已有盡全者物理之常所用極者所窮背者相反也凡象形者必對物而論長短多少貴賤輕重之理不對則唯總計之一數耳然其所對有新舊之異矣蓋其數本有多少之際者舊具也雖本無其理言相減之餘而互有限者新爲也象者每宛于一物其所主依二象之等類而屬有自對者所謂以度宛度以量宛量以秤宛秤之類皆所屬之兩數位名相同故相對自具若以度宛量宛秤之類兩數位名異而雖其理本不具新爲限而言之則其數却相對有之矣是故先依象形之原與題辭若題辭如象屬衆之數形方斜圓周及諸角之中徑三斜之中股者各雖其號本具其數皆由技而後得之故多不以爲增損之所據也悉察其相對之理依問旨與題數或相乘或歸除之後

有相對者也而後對多者其數增之故爲滿對少者其數
損之故爲干無對者自增損而包兩理或依問旨增而反得干理
損而反得滿理者亦有之矣若對累而多者皆得滿理故用最少
數對累而少者皆得干理故用最多數各隨其多少
而視增損之所窮每一品一畫如此而究三科之變
化也蓋滿干各一科之所化與題問之辭兩數相通
故隨象品形畫而有限若題中言或等數或應準之辭則其理相混故却有不應
限者也其變亦循而有定數也象者本雖爲一品數據兩屬數互成巧故依所
言其品不定形者本有大小斜正之勢故依形名其畫不定是以象形各品數畫數相幷爲一科化限數
是即其題辭限數也以之乘滿干二名爲一科變數亦乘全極背三科爲總變數也是以或依
象形或隨題辭有相對之同異而雖所據之道多理
所窮悉歸于其定限也

九

假如有錢若干貫文買綿每斤價錢若干文問計綿
是以二品錢綿爲一科化限又爲題辭限錢本無
多少之論又綿重與價錢緡二類異而各相對
不具故自增損而得滿干之理也
有錢無對物故自增而雖有滿理無窮故其極不具又自損而有干理以所盡爲極
干全有錢最少　干極有錢空　干背有錢負
滿全有錢最多　滿極無　滿背無
綿無對物故自損而有干理以所盡爲極又自增而雖有滿理無窮故極不具
干全綿最少　干極綿空　干背綿負
滿全綿最多　滿極無　滿背無
右二品變每一科各有四條全者有錢滿干
兩數多少雖異其理同綿滿干二數雖異其

理各同　極者錢干一數綿干一數　背亦
準之皆隨限數二化爲二數也

假如有金（若干兩）每金（幾兩）換銀（若干兩）問該銀
是本以（有金及該銀）二品爲一科化限亦爲題辭限
（題中雖有三數所言者二辭也）有金本無多少而相對不具每
金與屬銀各主重（二類相同）而多少之限自具故據
屬一之數相對之也

有金（無對物故自損而有干理以所盡爲極自增而雖有滿理無窮故極不具）
干全（有金最少）　滿全（有金最多）
干極（有金空）　滿極無
干背（有金負）　滿背無
屬銀（對一金之少損之故有干理而極具又無對多之物故自增而雖有滿理無窮）（而極不具）

干全（屬銀與一金相似）　滿全（屬銀最多）
干極（屬銀與一金相等）　滿極無
干背（屬銀少一金多）　滿背無

右二品變每一科各四條全者有金滿干屬
銀滿干數雖異理各同屬銀干一數極者
有金干一數屬銀干一數　背亦準此皆隨
限數化爲二數也

假如有米（若干斛）換豆（若干斛）麥（若干斛）每米一斗豆不
及麥（若干斛）問二直米
是本（豆麥及二直米）四品故即以四爲一科化限又爲
題辭限豆麥各無多少而相對不具屬豆與每
米皆主量數故有限而自相對具屬麥與每米
又有限而相對自具也（乃題中雖言屬豆與屬麥之多少其理本自具）

故不用之

豆無對物故自損而有干極自增而雖有滿理極不具

干全豆最少　干極空豆　干背負豆

滿全豆最多　滿極無　滿背無

麥增損之理如前

干全麥最少　干極空麥　干背負麥

滿全麥最多　滿極無　滿背無

屬豆對每米之少而干極具對屬麥之多而滿極具

干全屬豆每米相似　干極屬豆每米相等　干背屬豆少每米多

滿全屬豆屬麥相似　滿極屬豆屬麥相等　滿背屬豆多屬麥少

屬麥對屬豆之少而干極具又累對每米之少而雖干極具最少數故不用之亦無對多之物自雖有滿理極不具

十一

干全屬麥屬豆相似　干極屬麥屬豆相等　干背屬麥少屬豆多

滿全屬麥最多　滿極無　滿背無

右四品變每一科各八條全者豆滿干數雖異理各同麥滿干數雖異理各同屬豆干屬麥滿數雖異理各同屬豆滿屬麥干數理相同　極者豆干麥干屬豆干各一數屬豆滿屬麥干數理相同　背亦準之皆隨限數化爲四數也

假如有買馬牛共若干隻馬價金若干兩牛價金若干兩每馬幾隻價不及每牛幾隻價若干兩問馬牛數

是本兩獸及其價四品故以四爲一科化限亦爲題辭限雖題中有六數所言者四辭也買馬牛本多少不定而相

對不具又據幾隻之價則貴賤相反故以屬一
之數別高下而相對之也

買馬 無對物故自損而有干極自增而雖有滿理極不具
干全 馬最少　干極 馬空　干背 馬負
滿全 馬最多　滿極無　滿背無

買牛 理如前
干全 牛最少　干極 牛空　干背 牛負
滿全 牛最多　滿極無　滿背無

馬一價 對牛一價之少而干極具無對多之物故自雖有滿理極不具
干全 馬牛價相似　干極 馬牛價相等　干背 馬價少牛價多
滿全 馬價最少　滿極無　滿背無

牛一價 對馬一價之多而滿極具無對少之物故自有干極

十二

干全 牛價最少　干極 牛價空　干背 牛價負
滿全 牛馬價相似　滿極 牛馬價相等　滿背 牛價多馬價少

右四品變每一科各八條全者買馬滿干數
雖異理各同買牛滿干數雖異理各同馬一
價干牛一價滿數理相同馬一價滿牛一價
干數理相同　極者買馬干買牛干各一數
馬一價干牛一價滿數理相同牛一價干一
數　背亦準此皆隨限數化爲四數也

假如有銀若干錢買羅綾絹羅尺價若干綾尺價若干絹
尺價若干綾不及絹若干尺卻多羅若干尺問羅綾絹
是本羅綾絹及三價六品故以六爲化限又爲題辭限
羅綾絹雖各無定數而多少不具於題中言過

不及之差故有相對之限又尺價本有高下而
自相對具也
羅
無對少之物故自損而有干極
對綾之多有滿理而其極具
羅最羅羅
干全干極干背
少空負
羅綾羅綾羅多
滿全滿極滿背
相似相等綾少
綾
對羅之少而干極具
對絹之多而滿極具
綾羅綾羅綾少
干全干極干背
相似相等羅多
綾絹綾絹綾多
滿全滿極滿背
相似相等絹少
絹
對綾之少而干極具無對多
之物故雖自有滿理極不具
絹綾絹綾絹少
干全干極干背
相似相等綾多
絹最
滿全滿極無滿背無
少
羅尺價
對綾尺價之多而干極具無對
多之物故雖自有滿理極不具

羅綾尺羅綾尺羅尺價少
干全干極干背
價相似價相等綾尺價多
羅尺價
滿全滿極無滿背無
最多
綾尺價
對絹尺價之少而干極具
對羅尺價之多而滿極具
綾絹尺綾絹尺綾尺價少
干全干極干背
價相似價相等絹尺價多
綾羅尺綾羅尺綾尺價多
滿全滿極滿背
價相似價相等羅尺價少
絹尺價
對綾尺價之多而滿極具
無對少之物故自有干極
絹尺價絹尺絹尺
干全干極干背
最少價空價負
絹綾尺絹綾尺絹尺價多
滿全滿極滿背
價相似價相等綾尺價少
右六品變每一科各一十二條全者羅干絹
滿數雖異理各同羅滿綾干數理相同綾滿
絹干數理相同羅尺價干綾尺價滿數理相
同羅尺價滿絹尺價干數雖異理各同綾尺

價干絹尺價滿數理相同　極者羅干一數
羅滿綾干數理相同綾滿絹干數理相同羅
尺價干綾尺價滿數理相同綾尺價干絹尺
價滿數理相同絹尺價干一數　背亦準此
皆隨限數化爲六數也

假如有人出米若干斛換金若干兩銀若干兩銅若干兩鐵
若干兩每金銀各一兩換米和共若干每銀銅各一兩
換米和共若干每銅鐵各一兩換米和共若干問四色
直米

是本四金及直米八品故以八爲一科化限又爲題
辭限四物互無多少之際換米亦準其數而相
對不具故據屬一屬諸金一之米屬米一之諸金之數則有增

十四

損之衰差而相對具也

金無對物故自損而有干極自增而雖有滿理極不具
干全金最少　干極金空　干背金負
滿全金最多　滿極無　滿背無
銀理如前
干全銀最少　干極銀空　干背銀負
滿全銀最多　滿極無　滿背無
銅理如前
干全銅最少　干極銅空　干背銅負
滿全銅最多　滿極無　滿背無
鐵理如前
干全鐵最少　干極鐵空　干背鐵負

滿全鐵最多
屬金米對屬銀米之少而干極具無對多之物故雖自有滿理極不具
干全與屬銀米相似
滿全屬金米最多
屬銀米對屬金米之多而滿極具對屬銅米之少而干極具
干全與屬銅米相似
滿全與屬金米相似
屬銅米對屬銀米之多而滿極具對屬鐵米之少而干極具
干全與屬鐵米相似
滿全與屬銀米相似
屬鐵米對屬銅米之多而滿極具無對少之物故自有干極
干全屬鐵米最少

滿極無
干極與屬銀米相等
滿極無
干極與屬銅米相等
滿極與屬金米相等
干極與屬鐵米相等
滿極與屬銀米相等
干極屬鐵米空

滿背無
干背屬金米少屬銀米多
滿背無
干背屬銀米少屬銅米多
滿背屬銀米多屬金米少
干背屬銅米少屬鐵米多
滿背屬銅米多屬銀米少
干背屬鐵米負

十五

滿全與屬銅米相似
滿極與屬銅米相等
滿背屬鐵米多屬銅米少

右八品變每科各一十六條全者四金共滿
干數雖異理各同屬金米干與屬銀米滿數
理相同屬金米滿與屬鐵米干數雖異理各
同屬銀米干與屬銅米滿數理相同屬銅米
干與屬鐵米滿數理相同　極者四金共干
各一數屬金米干與屬銀米滿數理相同屬
銀米干與屬銅米滿數理相同屬銅米干與
屬鐵米滿數理相同屬鐵米干一數　背亦
準此皆隨限數化爲八數也

假如有平方圜若干問斜
是縱橫等而混爲面一畫以之卽爲一科化限

又爲題辭限本無長短而相對不具故自增損
而究滿干之理也乃題中雖言圖而問斜皆由技求之故不爲所據也
面無對物故自損有干理以所盡爲極又自增雖有滿理無窮而極不具

右一畫變每科各二條全者方滿干圖勢大
小雖異理同　極者方干一圖　背亦準此
如此隨限數一化皆爲一圖也

十六

假如有直積若干平不及長若干問長平
是本以二畫爲一科化限又爲題辭限長短之
狀自相對具故每一畫互據多少而增損之也乃題中雖言差之多少其理本具故不用之
平無對少之物故自損有干理以所盡爲極對長之多故有滿理而極自具

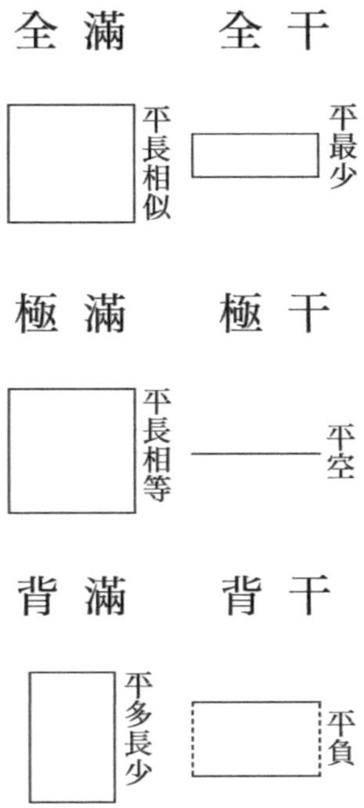

長對平之少有干理而極具又無對多之物故雖有滿理極不具

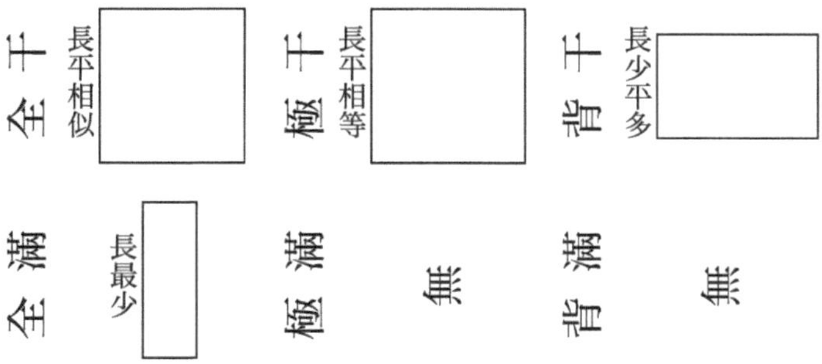

右二畫變每一科各四條全者平滿長干二
圖理相同長滿平干二圖理相同 極者平
滿長干二圖理相同平干一圖 背亦準此
皆隨限數二化各爲二圖也
假如有三斜大斜若干中斜若干小斜若干問中股
是以三畫爲一科化限亦爲題辭限本其狀有

十七

屈伸故依小斜之多少相對亦有異也
小斜對大中斜差之少干極具
又對中斜之多滿極具

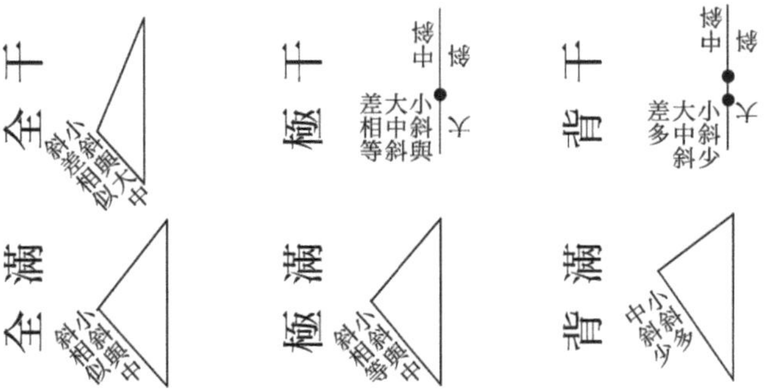

中斜若小斜多於半大斜者對小斜之
少干極具又對大斜之多滿極具

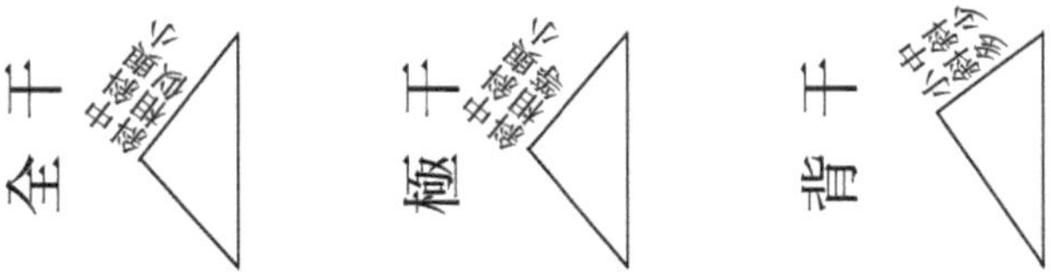

全滿

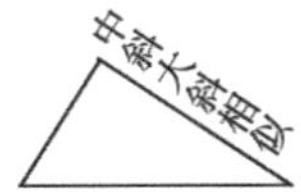

極滿

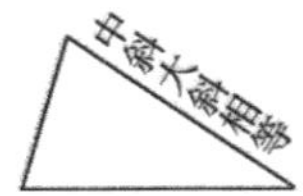

背滿

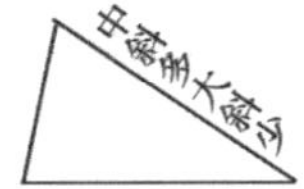

若小斜少於半大斜者以中斜對大小斜差之少干極具又對大斜之多滿極如前故再不圖于茲

全干

極干

背干

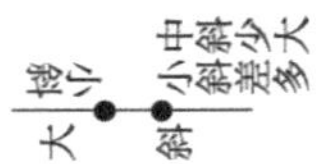

大斜對中斜之少干極具又對中小斜之多而滿極具

全干

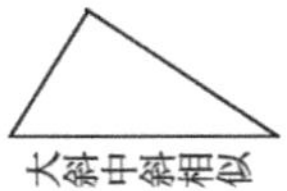

極干

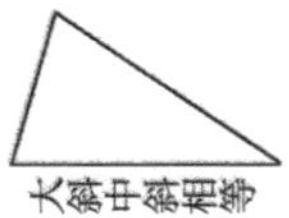

背干

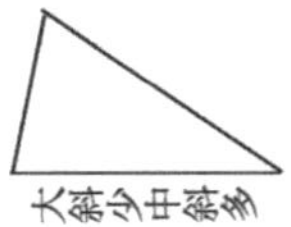

十八

全滿

極滿

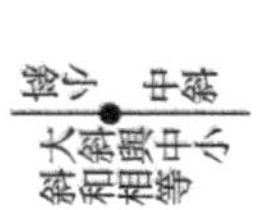

背滿

右三畫變每科各六條全小斜多半大斜少者小斜干大斜滿二圖理相同小斜滿中斜干二圖理相同中斜滿大斜干二圖理相同小斜少半大斜多者小斜干中斜干大斜滿三圖理相同中斜滿大斜干二圖理相同小斜滿一圖　極背各準此皆隨限數三化爲三條也

假如有三廣積干若上下廣和干若下廣多於中廣干若却少於長干若問上中下廣及長

是以四畫爲一科化限又爲題辭限本上下大

小相對又中最小而與上相對各自具長本雖
無多少之際因言題中下廣差有相對也

上廣
對中廣之少干極具
對下廣之多滿極具

全干
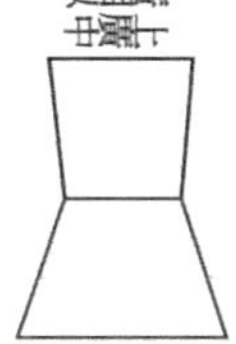

極干
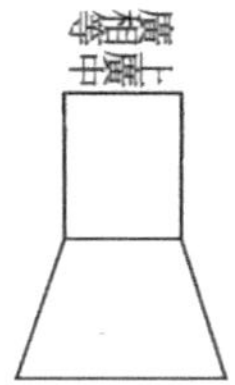

背干
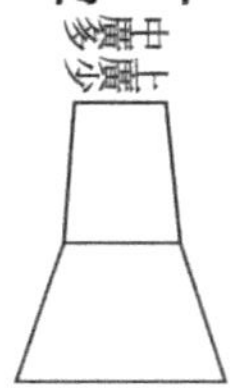

全滿
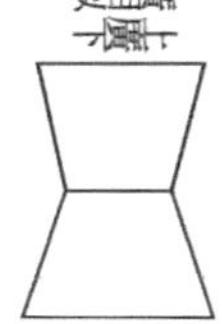

極滿
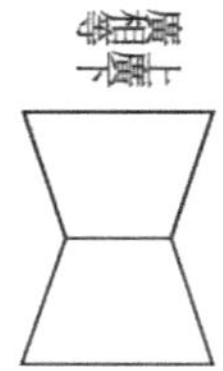

背滿
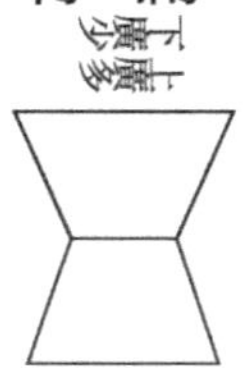

中廣
對上廣之多滿極具又無
對少之物故自有干極

全干
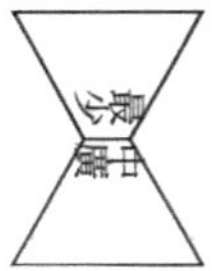

極干

背干

十九

全滿
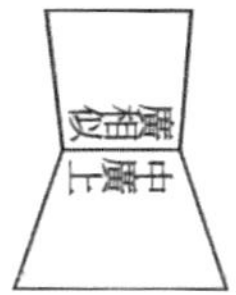

極滿
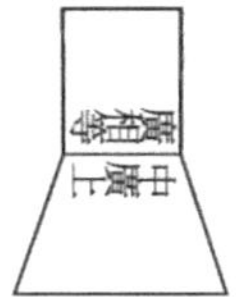

背滿
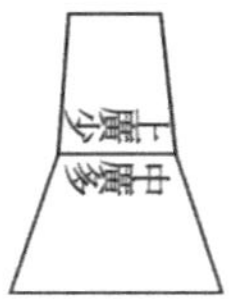

下廣
對上廣之少干極具題言中廣之差故
別雖有干極最少數而不用之又依言
與長之差對
多而有滿極

全干
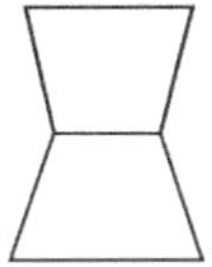

極干
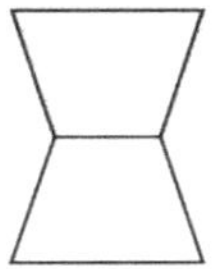

背干
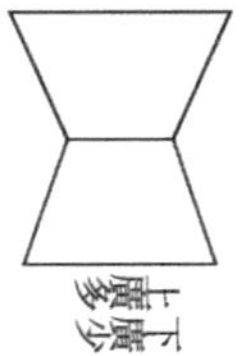

全滿

極滿
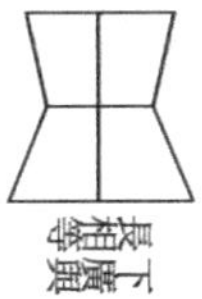

背滿
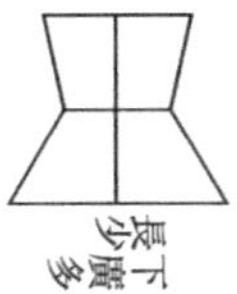

長
對下廣之少有干極又無對多
之物故雖自有滿理極不具

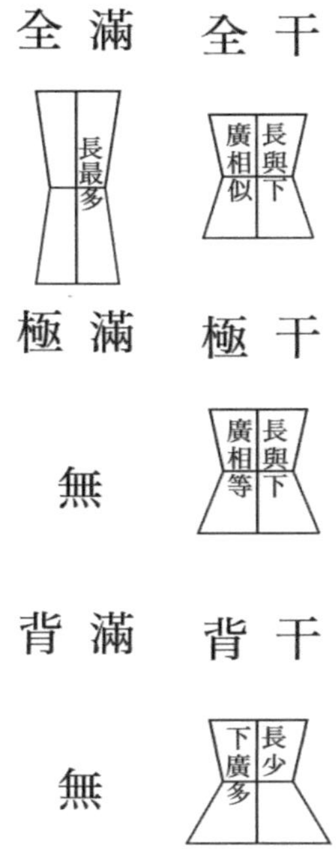

右四畫變每一科各八條全者上廣干中廣滿二圖理相同上廣滿下廣干二圖理相同中廣干長滿二勢雖異理同下廣滿長干二圖理相同　極者上廣干中廣滿二圖理相同上廣滿下廣干二圖理相同中廣干一圖下廣滿長干二圖理相同　背亦準之皆隨

二十

限數四化爲四圖也

假如有四斜甲干若乙干若丙干若丁干若戊干若問積

是以五畫爲化限亦爲題辭限本無定形而諸斜大小之所在不必論上下左右其號常隨長短而分次序故依每斜多少有相對之同異也

戊對甲乙差之少干極具對丁之多而滿極具

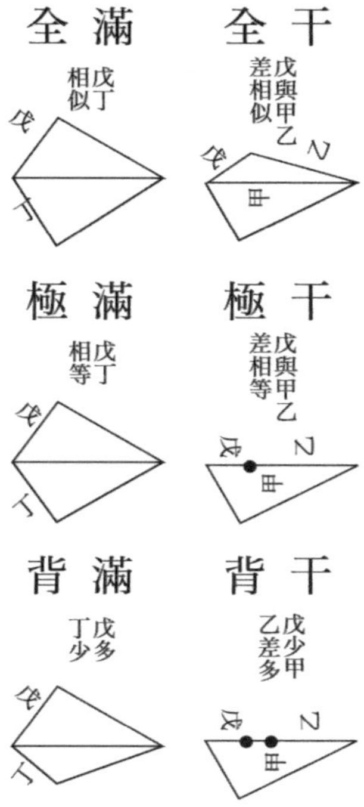

丁
若戊多於甲丙差者對戊之
少干極具對丙之多滿極具

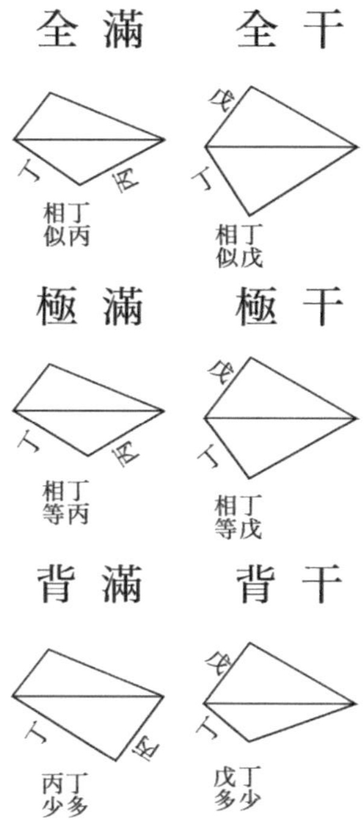

若戊少於甲丙差者丁與甲丙差之少
相對干極具又對丙之多滿極如前

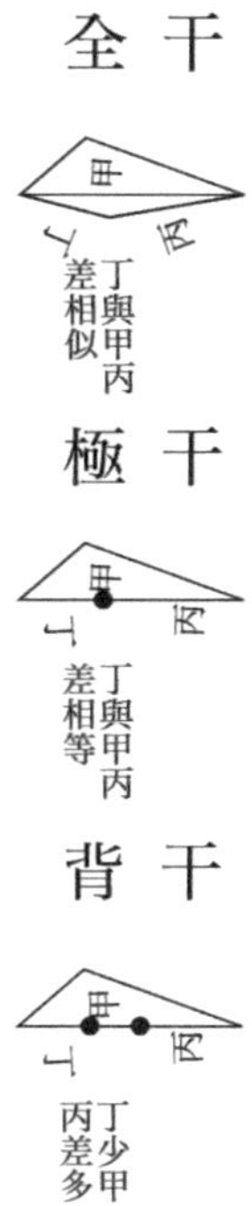

丙
若丁多於半甲者對丁之少
干極具對乙之多滿極具

二十一

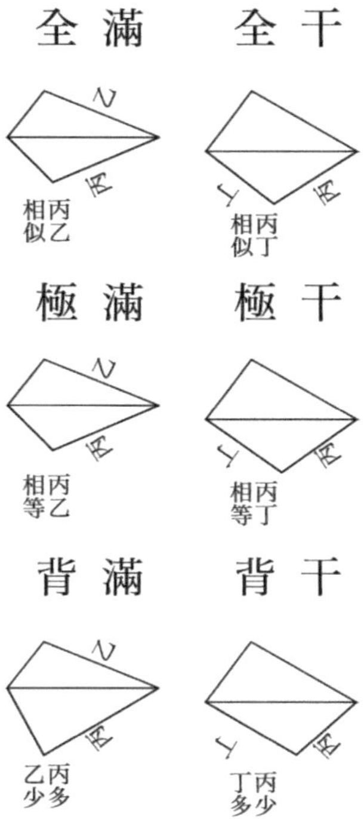

若丁少於半甲者丙與甲丁差之
少相對干極具對乙而滿極如前

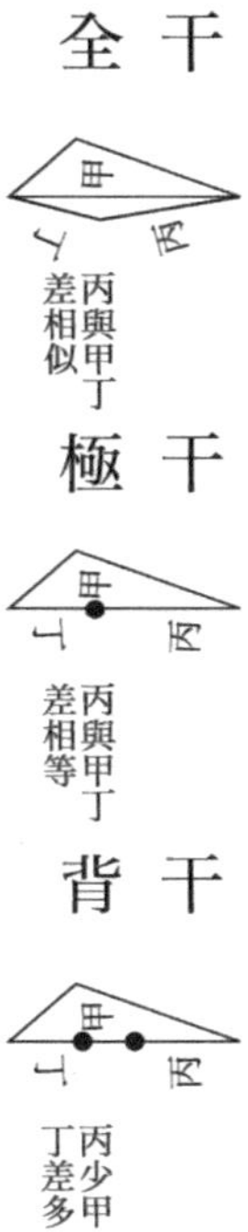

乙
若戊多於半甲者對丙之少
干極具對甲之多滿極具

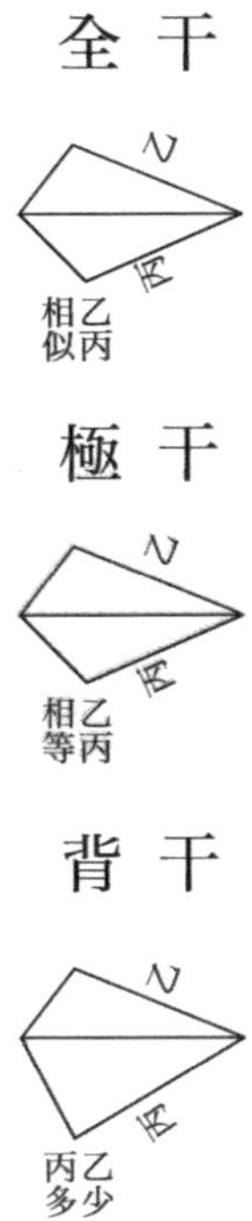

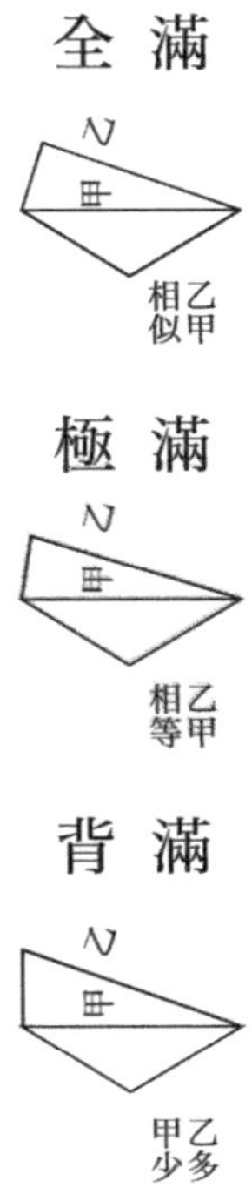

若戊少於半甲者乙與甲戊差之少干極具又對甲之多滿極如前

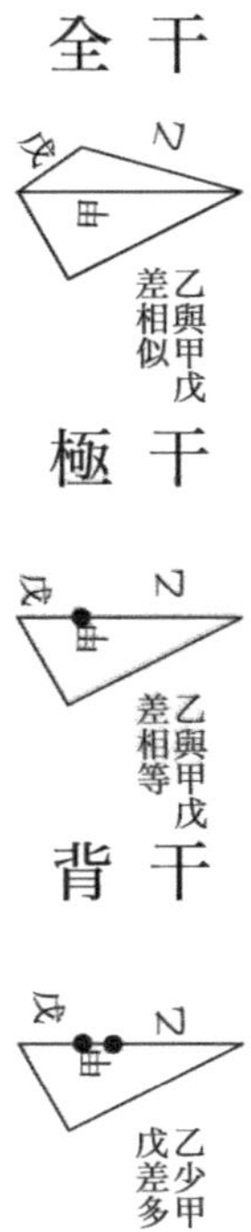

甲若丙丁和多於乙戊和者對乙戊和之多滿極具對乙之少干極具

二十二

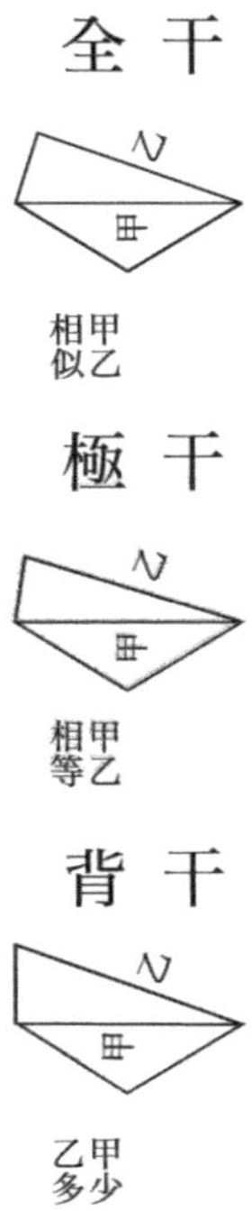

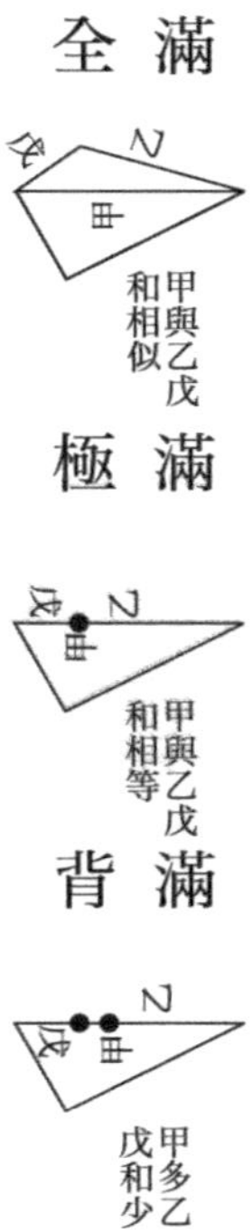

若丙丁和少於乙戊和者甲與丙丁和之多相對滿極具對乙干極如前

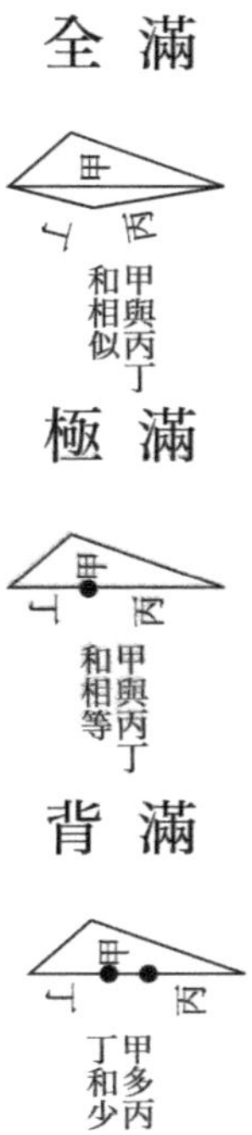

右五畫變每一科各一十條全多於者戊干甲

滿二圖理相同戊滿丁干二圖理相同丁滿
丙干二圖理相同丙滿乙干二圖理相同乙
滿甲干二圖理相同於少者戊干乙干二圖理
相同戊滿丙滿二圖雖異理同丁干丙干甲
滿三圖理相同乙滿甲干二圖理相同丁滿
一圖　極背亦準此皆隨限數五而化爲五
圖也

假如有方壔積干若方不及高干若問方高
是本雖爲縱橫高三畫方面等而相混故以二
畫爲一科化限亦爲題辭限方高本無大小之
際而雖相對不具今題中依言多少有相對也
方無對少物故自損有干理以所盡爲極對高之多有滿理而極具

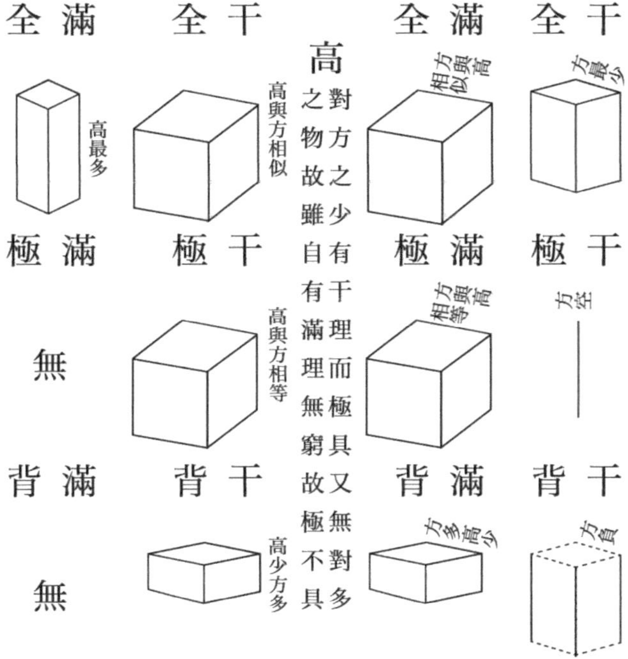

右二畫變每一科四條全者方干高滿二圖
理相同方滿高干二圖理相同　極者方滿
高干二圖理相同方干一圖　背亦準此皆
隨限數二化爲二圖也
假如有直錐長若干闊若干高若干問積
是以三畫爲一科化限又爲題辭限上鋭而下
直故長闊自具高本無多少之際題中亦不言
其差故相對無之也

闊 無對少之物故以自所盡爲干極又對長之多滿極具

二十四

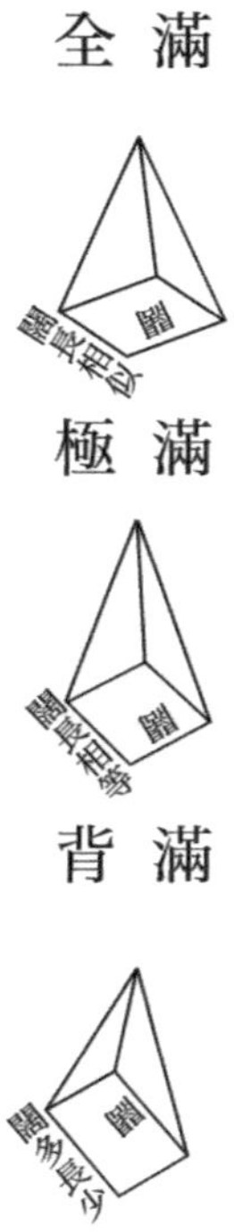

長 對闊之少干極具又無對多之物而滿極不具

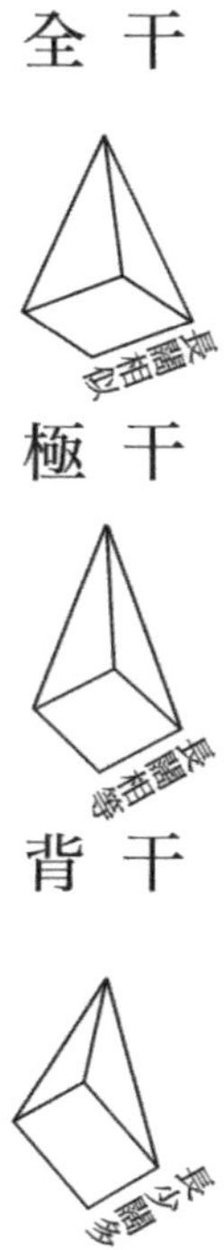

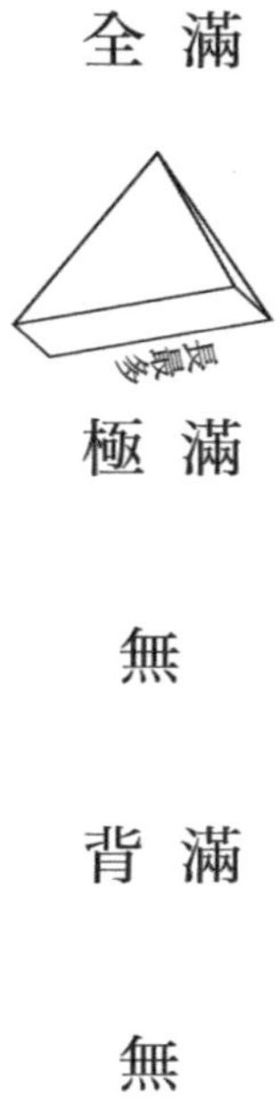

高 無對少之物以自所盡爲干極又無對多之物而雖有滿理極不具

全干
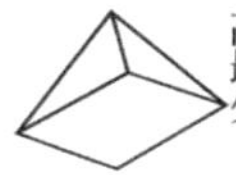

極干
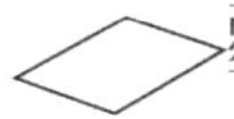

背干
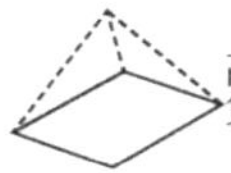

全滿

極滿　無

背滿　無

右三畫變每一科各六條全者闊滿長干二
圖理相同高滿干雖圖勢理同闊干長滿二
圖理相同　極者闊滿長干二圖理相同闊
干高干各一圖　背亦準此皆隨限數三化
各爲三圖也

假如有方臺積干若上下方和干若下方與高和干若問

二十五

上下方及高

是以三畫爲一科化限亦爲題辭限其狀固上
小下大而相對自具高本與上下方互無長短
之際而相對不具題中亦不言多少故無其理
也

上方無對少之物故以自所盡爲極又對下方之多滿極具

全干

極干
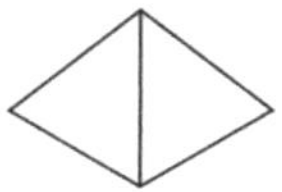

背干
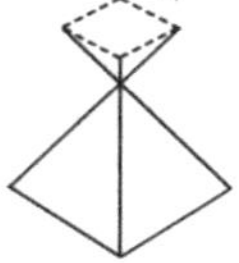

全滿
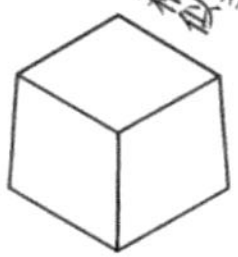

極滿
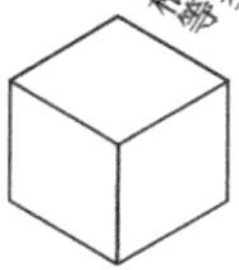

背滿
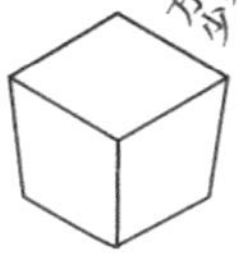

不具

下方對上方之少故有干理而其極自具又無對之物故雖自有滿理無窮而其極

全干

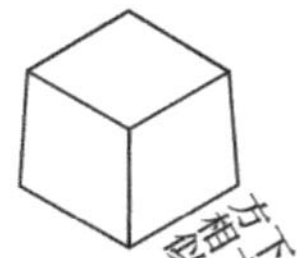

極干

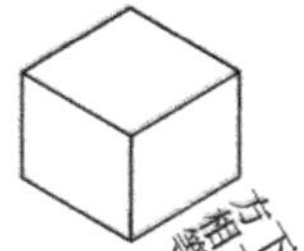

背干

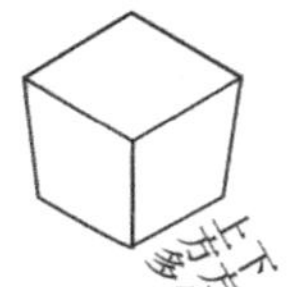

全滿

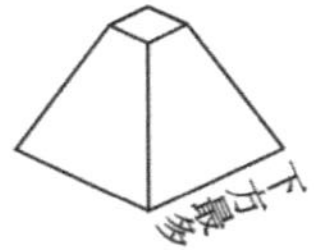

極滿

無

背滿

無

高無對物故自損而以所盡爲干極又自增雖有滿理無窮而其極不具

全干

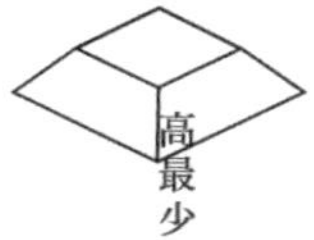

極干

背干

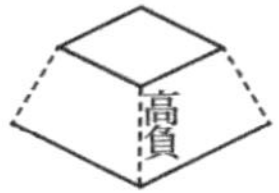

二十六

全滿

極滿

無

背滿

無

右三畫變每一科各六條全者上方滿下方
干二圖理相同上方干下方滿二圖理相同
高滿干圖勢雖異理同　極者上方滿下方
干二圖理相同上方干高干各一圖　背亦
準此皆隨限數三化爲三圖也

假如有楔積若干縱橫差若干刃與橫和若干縱與長和若干
問縱橫刃及長

是以四畫爲一科化限亦爲題辭限本縱橫相
對具縱亦與刃相對具故依橫與刃之廣狹各

有異長本無多少之際故相對不具也

横 若多於刃者對刃之少干極具對縱之多滿極具

全干

極干

背干

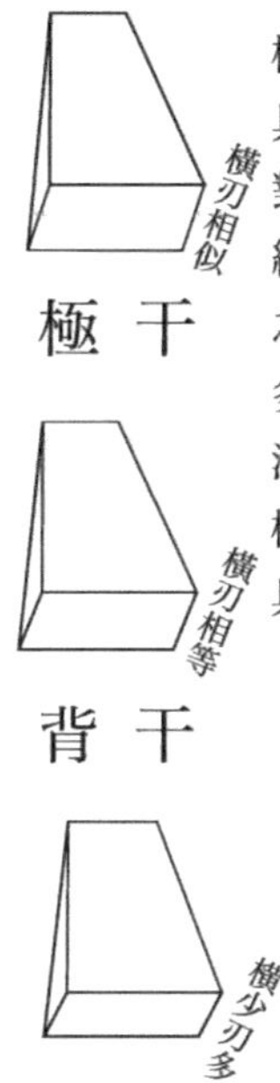

全滿

極滿

背滿

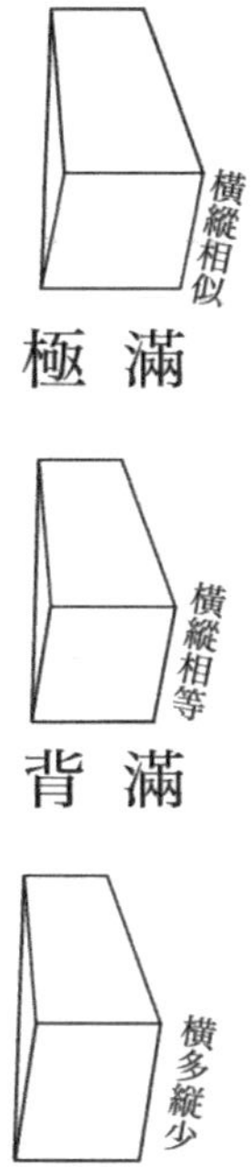

若横少刃多者對刃之多滿極具無對少之物以自盡爲干極

全干

極干

背干

全滿

極滿

背滿

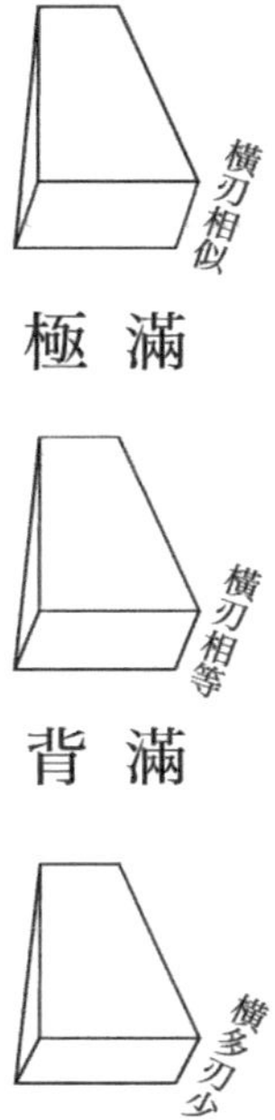

縱 具不 若多刃少者對横之少而干極具又無對多之物故雖自有滿理遂無其窮而極

全干

極干

背干

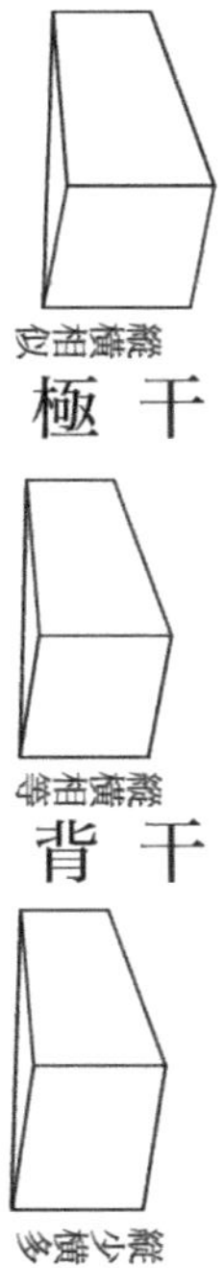

全滿

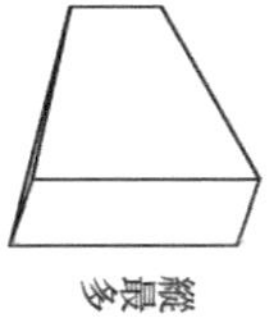

極滿

無

背滿

無

若横少刃多者對刃之少干極具又無對多之物而滿極不具如前

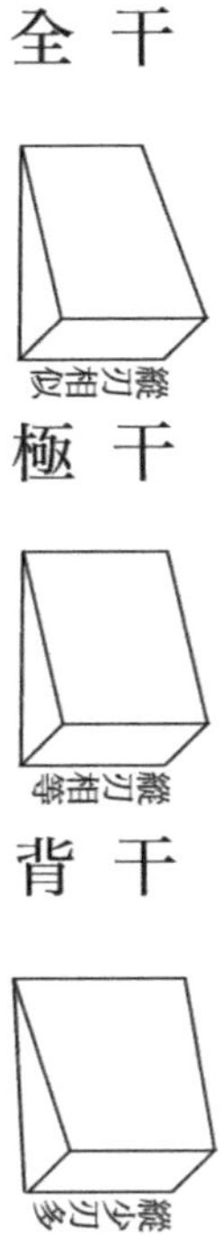

刃

若横多刃少者對横之多滿極具又無對少之物故以自盡爲干極

若横少刃多者對横之少干極具對縱之多而滿極具

二十八

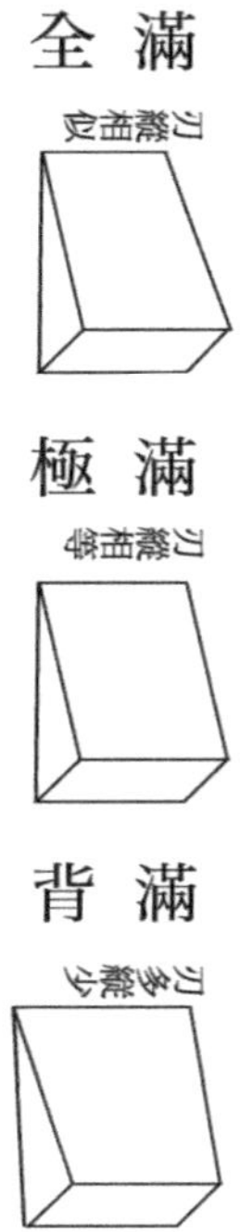

長

本無對物故自損以所盡爲干極又自增雖有滿理極不具

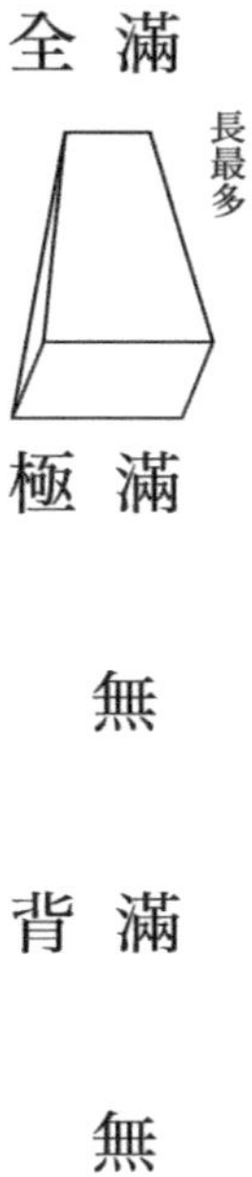

右四畫變每一科各八條全横多刃少者横干刃
滿二圖理相同横滿縱干二圖理相同縱滿
刃干二圖理相同長滿干圖勢雖異理同横少
刃多者横干縱滿二圖理相同横滿刃干二圖
理相同縱干刃滿二圖理相同長滿干如前
極横多刃少者縱干刃滿二圖理相同横滿縱
干二圖理相同刃干長干各一圖横少刃多者横
滿刃干二圖理相同縱干刃滿二圖理相同
横干長干各一圖　背亦準此皆隨限數四
化各爲四圖也

假如有直臺積加入上下長共若干上下闊差若干上
長多於下闊若干高多如上長若干却少下長若干問上

二十九

下長闊及高

是以五畫爲一科化限又爲題辭限本上下大
小之直故各長闊相對亦上長對干下長上闊
對干下闊是皆本自具也然上長與下闊其廣
狹不定高亦無多少之際而雖相對各不具依
題中言過不及之數有新舊之相對而分所用
之同異也

上闊無對少之物故自損而以所盡爲干極對上長之多滿極具又對下闊之多滿極具是皆舊所對也於題中無新對之理故視兩舊滿極下闊對少故用之

全干

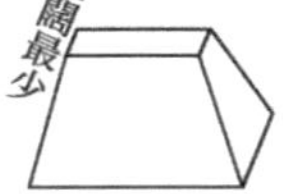

極干

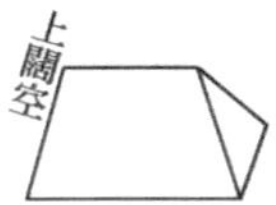

背干

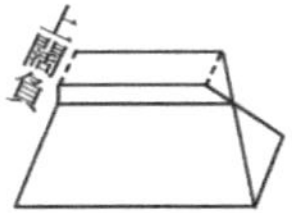

上長
故兩高
用干之
之極多極舊
下有具對
闊滿又上
對極據闊
多是題之
故皆辭少
用新對干
之所下極
視具闊具
兩也之對
滿於少下
極是有長
高視干之
對新極多
少舊對滿

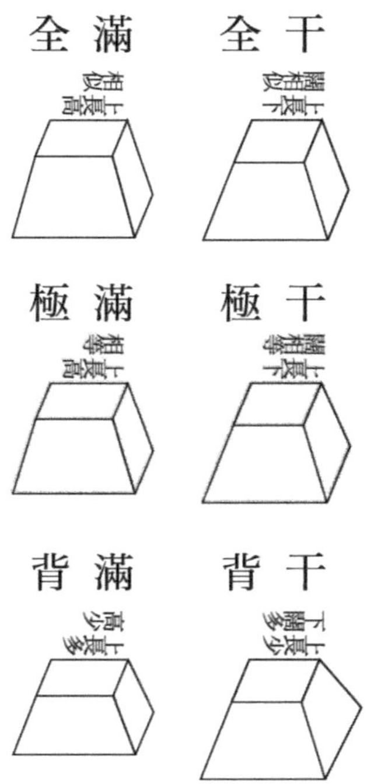

三十

下闊
少於
故是
用視極舊
之新具對
干舊又上
極兩據闊
依滿題之
舊極辭少
而上新干
用長對極
之對上具
長對
之下
多長
有之
滿多
極滿

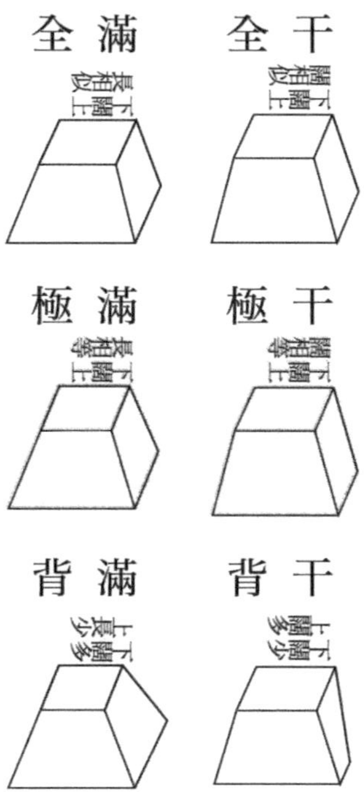

下長
故少理
用復本
之有無而舊
干其干對
極際極下
於而具闊
是極無之
視不對少
新具多干
舊又之極
三據物具
干題故對
極辭自上
高新增長
對對雖之
最高有少
少之滿累

全干

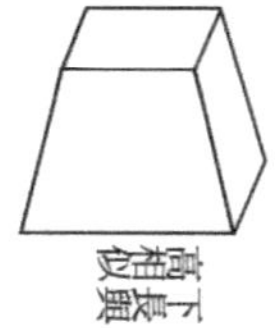

極干

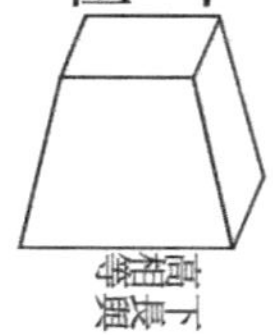

背干

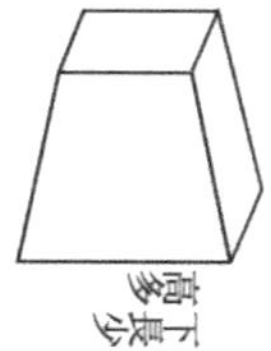

全滿

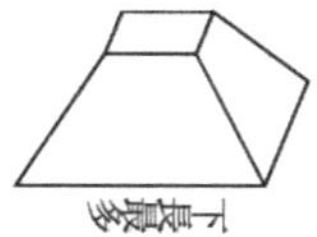

極滿

無

背滿

無

高本無對物故自增損而雖相對不具據題
辭新對上長之少有干極對下長之多有
滿極故
各用之

全干

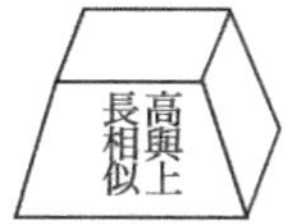

極干

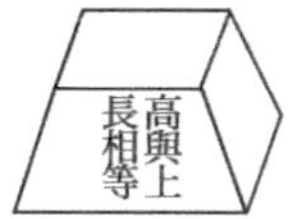

背干

全滿

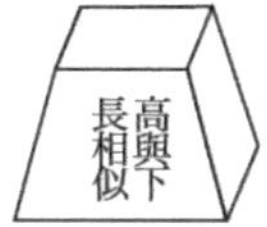

極滿

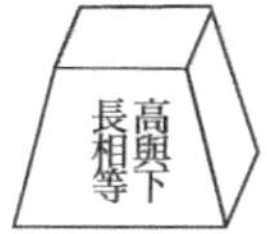

背滿

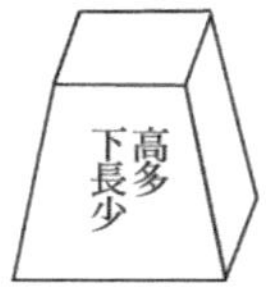

右五畫變每一科各一十條全者上闊干下長滿二圖勢雖異其理同上闊滿下闊干二圖理相同上長干下闊滿二圖理相同上長滿高干二圖理相同下長干高滿二圖理相同　極者上闊干一圖上闊滿下闊干二圖理相同上長干下闊滿二圖理相同上長滿高干二圖理相同下長干高滿二圖理相同　背亦準之如此皆隨限數五化各爲五圖也

數第三

數者象形之用而有動靜之異矣凡如萬象分總計與屬幾之物衆形有縱横斜圜之畫者是皆無爲而本其數具故所主各靜也加幷而共得者號和減損而相較者號差因乘而量總者號積開除而攷求者號商是皆由技而變得其數故所主各動也若所求數或據法或得式開除之則雖其數皆似動能察其具與變之理而別之也夫數窮者謂之整無窮者謂之不盡也整有二等不帶約數者曰全是常所用也遍帶約數者曰繁用之成諸技則徒有過數之患也不盡有二等遭乘除之後整者曰畸遂不得整者曰零各循其數則專成乘除之勞且失其眞然照象形

與術式之所得或依舊或通約或收棄或作率而用之是以隨物理之變其數各有取捨矣

動靜

假如有出金若干兩買米若干斛問每斛價

是金米兩總皆無爲而具數故各靜也求每斛價則逢歸除之技而雖似變本自其象具數故亦靜也

假如有絹若干尺每錢若干文換絹若干尺問價錢

是有絹與每錢及換絹各其數無爲而自具故皆靜也價錢者雖由乘除之技而求之其象本自具故不爲變也

假如有蜜蠟共若干斤蜜價銀若干兩蠟價銀若干兩蜜

斤價不及蠟斤價若干兩問蜜蠟數

是共數者相幷之和逢加技而變故其數動也二價者其數本自具故靜也少如者本相減之差亦由技而變故動數也所問數者本無爲之象雖逢技其數靜也

假如有銀若干錢買瓜若干箇桃若干箇每錢瓜不及桃若干箇問二價

是有銀者所言雖似無爲而具本兩價和而逢加技之變故動數也兩果者本自具數故靜也不及者由相減之技而變故動數也求兩價則二象本其數具故雖逢開方之技而似變其數靜也

假如有欠錢若干文歷若干日利錢若干文今歷若干日共還錢若干文問元利

是欠錢與兩歷日及利錢皆無爲之具數故靜也共還錢者元利幷數由加技而變故動也所問者雖逢技而求之皆其象本自具故靜也

假如有方圜若干問斜

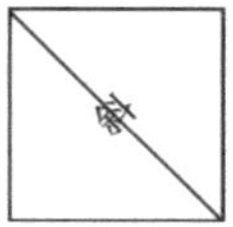

是圜本幷四面之形雖變而似動其畫自具故靜也斜者雖由開方之技而求之亦本具故靜也

假如有勾股積若干股若干問勾弦

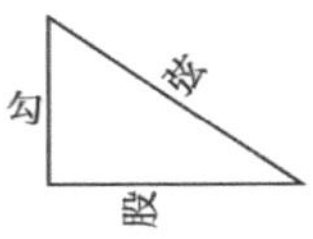

是積者形中之總數由相乘之技而變故動數
也股者本畫具故靜也求勾者歸除求弦者開
方各雖逢其技自具而無爲之數故是亦靜也

假如有直墻闊若干長若干高若干問積

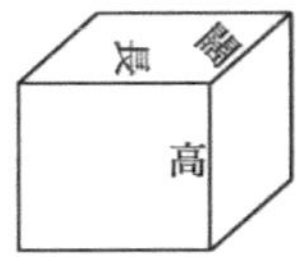

是長闊及高皆無爲之數具故靜也所問之積

三十四

者逢因乘之技而變故動數也

假如有梯大小頭和若干長多如大頭若干以大頭除
斜長得若干問大小頭及二長

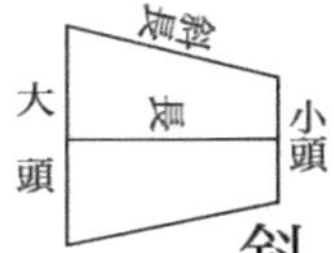

是和者相和多如者相減得者歸除皆由技而
變故各動數也所問之四數者雖由開方而各
求之其形本自具故皆無爲之靜數也

假如有四不等上斜若干旁斜若干闊爲實平方開之
得數與長再自乘數相幷共若干長多於闊
若干問長闊

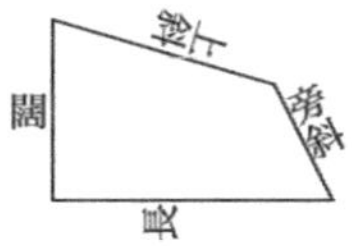

是兩斜者本自具數故靜也共數者開方與再
乘兩技相變而復有加技之再變故其數累而
動也多於者逢相減之變技故動也所問數者
雖由技而求之皆靜也

整數二等

全

假如有平方積四百尺問方

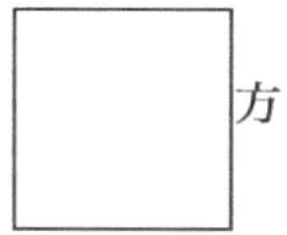

是本來一畫之形而無當約之對數也

假如有紅羅二十一尺換白絹八十四尺問每尺
絹數

三十五

是羅絹各有約數而法實雖似繁不爲患也

假如有直積八十四寸長闊差五寸問長闊

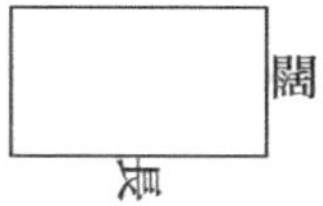

是兩數固不帶約數且及得式諸級亦無遍約
之數也

假如有芳茗四十八斤每六斤價銀七錢問總價

是三數相對而無遍約數作式則雖似有約 六
數不爲繁

假如有鼓左右廣五寸中廣八寸長一尺問積

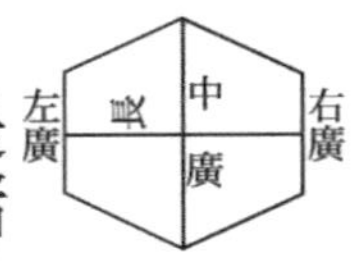

是諸數旁無相約之數也
如此等之數者皆不帶約法而不成術中之繁冗
答數亦整故依舊而用之
繁
假如有梭積一百寸長闊和三尺問長闊

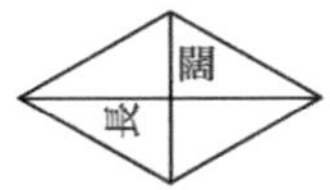

是兩數各帶約法雖不成術式之煩及得答數

三十六

各有約數也
假如有小麥二十五斛每八斗磨成麵六斗問計
麵
是三數雖似無遍約之法舊有數者不相對故
每與麵相對而帶二約也
假如有簫上廣九寸下廣一尺二寸長二尺四寸
問積

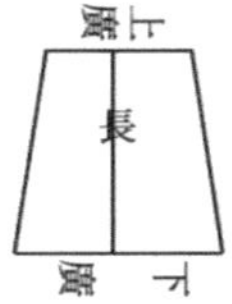

是三數遞相對而各有三約之數也
假如有馬六隻牛八隻共價金四十兩馬隻價多
於牛隻價二兩問各價

是兩獸與共價及多如各有約法得答數而後亦帶二約也

假如有方田自方一百五十四丈七尺六寸中開廣九丈一尺四寸之曲尺道二條餘積等三段配之問截長闊

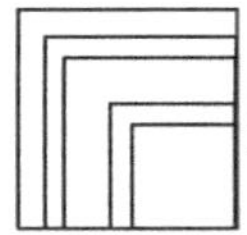

是兩數各帶二約數也原式數位繁故不約則增散漫之患也

如此等之諸數者各帶約數而徒致繁擾之弊雖然依術式有强不成勞者是以或依舊或約而後用之

不盡二等

畸

假如有立方每面三寸六分八釐四毫二絲一忽微强問積

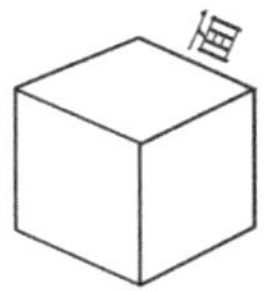

是以一十九分乘之則爲整數也

假如有古據方三角各一方斜一尺四寸中徑八寸五分七釐一毫四絲二忽八五强問各面

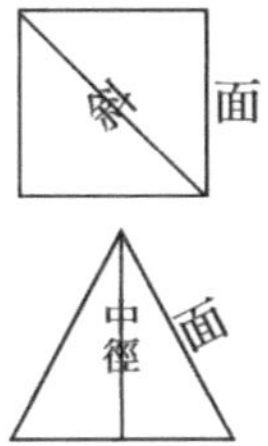

是皆古法而最雖疎若用之則斜以五因之中徑以七因之各相整也若以此數爲定率者分乘除率而用之

假如有米五斛三斗八升四合六勺一抄五撮强
每斛舂成糲七斗六升九合二勺三抄微强問該糲
是各以一十三分乘之則爲整數也
假如有方臺上方五尺下方八尺高四尺問積

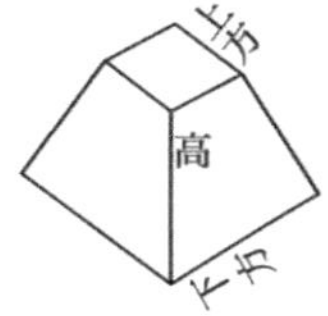

是皆數整而雖似全及求答數而有歸除之不
盡故各以三因之則無畸餘之數也
假如有健二十人怯三十七人八分共運米二百
一十一斛七斗六升四合七勺微强健一人運米多
於怯一人運米五斛〇九升八合〇三抄九撮强

三十八

問每人運米
是各以二百五十五分乘之則爲整數也
如此等之諸數者皆依舊則位太繁亂而成乘除
之勞故通而可用之
零
假如有圓周三尺一寸四分一釐五毫九絲二忽
六半强問徑

是以若干數或通或約雖屢累其技而驗之遂
不得整也
假如有上米一十四斛下米一十五斛支一十一

人上一人支米不及下一人支米一斛問各一人
支米
是題數雖整而似全及求答數有不盡雖累乘
除之數遂不整也
假如有周天三百六十五度二十五分强太陽日
行一度太陰日行一十三度三十六分太强問一周
歷日
是行度素不整也
假如有圭積二十八寸三分〇七毫六絲一忽强
長闊差七寸五分九釐三毫〇八忽弱問
長闊

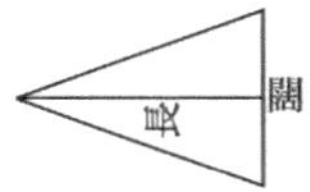

三十九

是兩數雖成幾乘除各不得整也
假如有大中小平方各一共方和一尺八寸八分
二釐三毫五絲三忽弱大中方差二寸
六分〇八毫四絲一忽强中小方差三
寸二分六釐三毫〇八忽弱問三方

是方和以一十七通之則雖整兩差數遂不得
整也
如此等之諸數者依舊則術式散漫而且失眞故
或作乘除率或收棄尾位而後用之

大成算經卷之四終

Notes on Collation of Complete Book of Mathematics Vol. 4: Three Essentials

The following are the call numbers of manuscripts made use of in these notes:

DT: Date Library [伊達文庫] 1682 KD090-セ 5 / The Later Bestowals of Seki's School [關算後伝] no. 45 in Miyagi Prefectural Library [宮城県図書館].

NK: T20 / 70 in the Main Library of University of Tokyo. It used to be in the Nan-Ki Library [南葵文庫] of Kii Tokugawa family [紀伊徳川家].

NS: 618 / 64 in the Nakanoshima Library of Osaka Prefecture [大阪府立中之島図書館].

1r. l. 9 : 象形第一 is written 象形 in NK.
1v. l. 4 : 抽 is missing in DT, NK and NS.
1v. l. 4 : 表 is missing in DT, NK and NS.
2r. l. 2 : 抽象 is written 象 in DT, NK and NS.
2v. l. 7 : Lower 若干銭 is written 若干尺 in NK.
3. all : 經 is written 徑 in DT, NK and NS.
3v. l. 5 : 表象 is written 象 in DT, NK and NS.
5v. ll.10–12 figure : 截闊 is written 残闊 in NS.
6r. ll. 8–10 figure : 接長 is written 長 in DT and NS.
7v. ll. 3–5 : 下長 in figure is written 下闊 in DT.
8v. l. 1 : 満干第二 is written 満干 in NK.
8v. l. 2 : 科 is written 斜 in NK.
8v. l. 3 and many other places in NK : 已 is written 己.
9r. l. 7 : 等 is written 著 in DT.
9v. l. 6 : 少 is written 多 in NS.
10v. l. 3 : 有金 is written 金有 in NS.
13v. l. 4 : 綾絹尺價相等 is missing in NS.
13v. l. 5 : 綾羅尺價相等 is missing in NS.
14v. l. 6 : 負 is written 空 in DT and NK.
15v. l. 1 : 米相等 is written 相等 in DT, NK and NS.
17r. ll. 1–3 : 干背 is written 干極 in NK.
18v. ll. 1–3 figure 2 : 小斜 and 中斜 are missing in NK, and 中斜 is missing in DT and NS.
18v. ll. 1–3 figure 3 : 中斜 is missing in DT.
18v. l. 11 : 於 is missing in DT, NK and NS.
20v. ll. 7–9 figure 1 : 乙 is missing in NS.
21r. ll. 9–11 figure 2 : 丁與甲丙差相等 is missing in NK.
21v. ll. 4–6 : figure 3 : 丙多乙少 is missing in NS.
21v. ll. 9–11 : 干極 is written 満極 in NK.
22r. ll. 4–6 figure 3 : 満背 is written 干背 in DT.
23v. ll. 1–3 figure 3 : dotted lines are written straight lines in DT.
23v. ll. 8–10 figure 2 : 高與方 is written 方與高 in NK.
24r. l. 4 : 隨 is missing in DT, NK and NS.
24r. ll. 10–12 figure 3 : dotted lines are written straight lines in DT.

24v. ll. 1–3 figure 2 : 闊 is missing in NS.
24v. ll. 5–7 figure 3 : 長少闊多 is written 長多闊少 in NK and 長少闊少 in DT.
25v. l. 3 : 互 is written 反 in DT.
26r. l. 9 : 以 is missing in NK.
27r. ll. 10–12 figure 3 : dotted lines are written straight lines in DT.
28r. l. 5 : 物 is written 而 in NS.
29r. l. 4 : 横満 is written 満満 in NK.
29r. l. 9 : 四 is missing in DT, NK and NS.
29v. l. 5 : 各 is written 無 in NS.
29v. l. 8 : 少 is written 分 in NS.
30r. ll. 7–9 figure 2 : 干極 is written 干全 in DT.
31r. ll. 10–12 figure 1 : 高與上 is written 高與下 in NK.
32r. l. 1 : 數第三 is written 數 in NK.
32v. l. 10 : 本 is written 各 in NS.
33r. l. 9 : 果 is written 異 in NS.
33v. ll. 7–9 figure : 斜 is missing in DT .
34r. l. 6 : 本 is written 太 in NS.
36r. l. 5 : 冗 is written 宛 in NS.
36v. l. 6 : 籥 is written 蕭 in DT, NK and NS.
38r. l. 2 : 斛 is written 斗 in DT, NK and NS.
38r. ll. 5–7 figure : 上方, 高, 下方 are missing in NS.
38r. l. 9 : 数 is written 故 in NS.
38v. l. 8 figure : 周 is missing in NS.
38v. l. 10 : 約 is written 終 in NS.
39v. l. 1 : 整 is missing in DT, NK and NS.
39v. l. 8 : 且 is written 旦 in NS.

Seki's Trilogy: Methods of Solving Explicit Problems, Methods of Solving Implicit Problems and Methods of Solving Concealed Problems

Hikosaburo Komatsu*

Abstract Seki Takakazu (1642?–1708) classified mathematical problems into three categories: explicit problems, which can be solved by arithmetic, implicit problems, which can be solved by an algebraic equation of one unknown, and concealed problems, which needs simultaneous algebraic equations with more than one unknowns. He is believed to have written for each category of problems a book of their solutions. The three books: Methods of Solving Explicit Problems, Methods of Solving Implicit Problems and Methods of Solving Concealed Problems are called Seki's Trilogy and used as the standard textbooks in the later Seki School of Mathematics. If a student masters a book of them, he is given a license which certificates his degree of understanding in mathematics.

Each part of these books is without doubt Seki's writing but we don't think that the books as they are now are the same as what Seki wrote. The Methods of Solving Implicit Problems may be his original but the others seem to be the results of edition by Yamaji Nushizumi (1704–1772), who instituted Seki's School of Mathematics. The following are our restorations of the Yamaji editions around 1726 or later.

Hikosaburo Komatsu
Graduate School of Mathematical Sciences, University of Tokyo,
3-8-1 Komaba, Tokyo 153-8914, Japan
e-mail: komatsu@ms.u-tokyo.ac.jp

* Supported by JSPS Grant-in-Aid Scientific Research (C) 16540119 (2004–2006), MEXT Grant-in-Aid on Priority Areas 17083006 (2005–2010) and JSPS Grant-in-Aid Scientific Research (C) 20540107 (2008–2011).

1 Introduction

Seki Takakazu [関孝和] classified problems in mathematics into the explicit problems [見題 kendai], which can be solved by arithmetic[1], the implicit problems [隠題 indai], which can be solved by an algebraic equation of one unknown, and the concealed problems [伏題 fukudai], which needs simultaneous algebraic equations with more than one unknowns[2].

Seki's Trilogy [三部抄 Sanbu Shō], that is, Methods of Solving Explicit Problems [解見題之法 Kaikendai no Hō], Methods of Solving Implicit Problems [解隠題之法 Kaiindai no Hō] and Methods of Solving Concealed Problems [解伏題之法 Kaihukudai no Hō] played a very important role in the later Seki School of Mathematics. If a student masters Method of Solving Explicit Problems, he is allowed to obtain the License of Explicit Problems [見題免許 kendai menkyo]. Then, if he masters Methods of Solving Implicit Problems, he gets the License of Implicit Problems [隠題免許 indai menkyo]. Finally, he is given the License of Concealed Problems [伏題免許 fukudai menkyo] after he masters Methods of Solving Concealed Problems, and he is allowed to teach his own students. In this way Seki's School of Mathematics was inherited from teachers to students. This system of education, which is similar to the other learnings and arts in Japan, seems to have been founded by Yamaji Nushizumi [山路主住]. In fact, looking at the family tree of the apprenticeship of Seki's school of mathematicians[3], we find an outburst of the number of students at Yamaji which lasts until the end of the Edo period.

Moreover, we cannot find any copies of any of these books which are surely copied before the time of Yamaji. On the other hand, the name of Yamaji appears in many copies as the person who copied them before.

Since each student copied his teacher's book, there remain many copies of each book. When we investigated for the copies of Seki's Trilogy at the Main Library of Tohoku University in February, 2004, there were 15 copies of Methods of Solving Explicit Problems, 11 of Methods of Solving Implicit Problems and 16 of Methods of Solving Concealed Problems. These numbers are exact because our visit was after Professor Tadao Oda [小田忠雄] compiled the Database of the Books of Wasan in the Library on February 23, 2001.

In this investigation we discovered that the copies by Matsunaga Sadatoki [松永貞辰] (1751–1795) were the oldest among the copies in which the dates of copies were recorded.

---

[1] Methods of Solving Explicit Problems contains a few problems which are solved by extraction of roots of algebraic equations. This happened when the arithmetic expression of the solution contained numerical coefficients with radicals. (See the note at the end of the book.) They didn't have notations to express such numbers on the counting board, so that they had to construct an algebraic equation of high degree with integral coefficients and then to solve it to get a solution.

[2] Later Seki and his pupils Takebe Katahiro [建部賢弘] and Kataakira [賢明] brothers added one more category the submerged problems [潜題 sendai] which cannot be formulated by algebraic equations in their treatise Complete Book of Mathematics [大成算経 Taisei Sankei] (1683–1711).

[3] See page 248 of these Proceedings.

At the end of his copy of Methods of Solving Explicit Problems it is written that "Finished to copy on the 22 in the 4th lunar month of Meiwa 6 (1769) [明和六年丑之四月廾二日写之畢] Matsunaga Sadatoki [松永氏貞辰]" with his Cipher [花押] following. Then the place of copying is written as "Copied this at the Mansion [of his Shinjō Fief [新庄藩]] in Edo [江戸於御屋敷写之ナリ]".

Similarly, at the end of Matsunaga's copy of Methods of Solving Implicit Problems, we find the script "Completed to copy on the 26 in the 4th lunar month of Meiwa 6 (1769) [明和六己丑年四月廾六日写之畢] Matsunaga Sadatoki [松永氏貞辰]" with his Cipher and the note "Copied the book borrowed from Master Ajima [安嶋先生より借用写之]". Master Ajima is his former teacher Ajima Naonobu [安島直圓] (1732–1798) who was from the same Shinjō Fief in Yamagata [山形] prefecture as Matsunaga. By his recommendation Matsunaga was able to study with Yamaji in Edo.

His copy of Methods of Solving Concealed Problems has no date of copy but it must have been earlier than the 8th in the 12th lunar month of Meiwa 7 [明和七] (1770) on which he was given the Licenses of Explicit Problems, Implicit Problems and Concealed Problems at the same time by Yamaji [1, p. 213].

With these copies by Matsunaga as the main source books, we have published the revised text [6] of Methods of Solving Concealed Problems and those [7] of Methods of Solving Explicit Problems and Methods of Solving Implicit Problems.

Seki's Trilogy is of course included in his Collected Works [2] but we were forced to publish more reliable texts in order to defend our papers [3] and [4]. Their Japanese version [5] had been rejected by the History of Science Society of Japan with the editor's comments saying "It has been pointed out that many of the edited contents of the Collected Works [2] are not academically reliable. When one writes an article on the history of mathematics about Seki Takakazu, it is very dangerous to make discussions depending only on [2]. No reliable arguments hold without taking extant manuscripts or prints at hand and referring to them specifying what manuscripts or prints are used and scrutinizing up to each letter and each phrase of them. This is the minimum rule to keep when one writes a historical article."

In our earlier publications [3] and [4], we referred to four to five manuscripts and Collected Works and tried to edit mathematically correct texts. The experience since then tells, however, that the errors made by the author or editors could also be important sources of investigations. Therefore, we have made use of several manuscripts in order to compile restored texts also at this time, we recorded only the comparison of them with two or three manuscripts only. As far as mathematical errors are concerned, we have revised only technical terms according to their usage in Complete Book of Mathematics [大成算經 Taisei Sankei] written in 1683–1711 by Seki, Takebe Kataakira [建部賢明] (1661–1721) and Takebe Katahiro [建部賢弘] (1664–1739).

The current texts of Seki's Trilogy are supposed to be Seki's own writings but we are dubious about this view. Probably they are the results of Yamaji Nushizumi's editing of several sketches of Seki which he prepared for the treatise Complete Book of Mathematics,

The year Kyōho 11th [享保丙午歳], which is written as the date of completion of Methods of Solving Explicit Problems, is 1726, 18 years later than the year 1708 in which Seki died. On the other hand, Methods of Solving Implicit Problems and Methods of Solving Concealed Problems are dated in 1685 and 1683, respectively. However, in Methods of Solving Concealed Problems there is a serious disorder of a table [4], which should have been caused by a careless revision. We supplemented the correction in two sheets at the end of the restored text of Methods of Solving Concealed Problems.

References

1. Matsusaburō Fujiwara [藤原松三郎], Japan Academy ed.: *History of Mathematics in Japan before Meiji* [明治前日本数学史], vol. 2, Iwanami [岩波書店], Tokyo (1957).
2. A. Hirayama [平山諦], K. Shimodaira [下平和夫] and H. Hirose [広瀬秀雄]: *Takazu Seki's Collected Works Edited with Explanations* [關孝和全集], Osaka Kyōiku Tosho [大阪教育図書], Osaka (1974).
3. Takefumi Goto–Hikosaburo Komatsu: Determinants, resultants and discriminants in Japan in the seventeenth century and in Europe in the eighteenth and nineteenth centuries, *J. Northwest Univ., Natural Science Edition*, vol. 33, pp. 363–367 (2003); these Proceedings, pp. 553–559.
4. Takefumi Goto–Hikosaburo Komatsu: A correction of Seki's error in the expansion of determinants, *J. Northwest Univ., Natural Science Edition*, vol. 33, pp. 376–380 (2003); these Proceedings, pp. 559–565.
5. Takefumi Goto [後藤武史]–Hikosaburo Komatsu [小松彦三郎]: Determinants, resultants and discriminants in Japan in the seventeenth century and in Europe in the eighteenth and nineteenth centuries [17 世紀日本と 18–19 世紀西洋の行列式, 終結式及び判別式], *Researches on History of Mathematics* [数学史の研究], *Colloquium of Research Institute for Mathematical Sciences* [数理解析研究所講究録], vol. 1392, pp. 117–129 (2004).
6. Hikosaburō Komatsu [小松彦三郎]: The restoration of the Yamaji Nushizumi edition of "Methods of Solving Concealed Problems" and its comparison with the Collected Works of Seki Takakazu [『解伏題之法』山路主住本の復元と「關孝和全集」との比較], *Researches on History of Mathematics* [数学史の研究], *Colloquium of Research Institute for Mathematical Sciences* [数理解析研究所講究録], vol. 1392, pp. 225–245 (2004).
7. Hikosaburō Komatsu [小松彦三郎]: The restoration of the Yamaji Nushizumi edition of Seki Takakazu's "Trilogy" [關孝和著『三部抄』山路主住本の復元], *Researches on History of Mathematics* [数学史の研究], *Colloquium of Research Institute for Mathematical Sciences* [数理解析研究所講究録], vol. 1444, pp. 169–202 (2005).

Methods of Solving Explicit Problems, by Seki Takakazu, collated by Hikosaburo Komatsu

Seki Takakazu

Abstract This is a collated text of the first of Seki's Trilogy. The book is composed of four chapters: Chapter 1 Addition and Subtraction [加減 kagen], Chapter 2 Partition and Joining [分合 bungō], Chapter 3 Whole Products [全乗 zenjō] and Chapter 4 Partial Products [折乗 setsujō]. In Chapter 1 easiest examples are given to show how a mathematical problem is interpreted as an algorithm on a counting board. In Chapter 2 the byscript method [傍書法 bōsho hō] of Seki is introduced by which we can express a polynomial as a sum of monomials represented by a numerical coefficient of counting rods and literal factors written on its right side. The heading means the distributive law of sums and products. Chapters 3 and 4 deal with the area and the volume of geometric figures. Formulas are not given by the algebraic byscript method but as algorithms on a counting board. The Pythagorean theorem and the Eudoxos formula of the volume of a pyramid are proved by figures. The volume formula of sphere segments is correct but its proof by an illustration is hard to understand.

解見題之法

關孝和　編

二〇〇五年七月九日　小松彦三郎校

解見題之法 凡四篇

關孝和編

加減第一 附併

加減者應于題旨而兩位相從者謂加兩位相消者謂減 併者與加同

假如有直長若干平若干問和

置平加入長得和

假如有甲若干乙若干丙若干問相併共數

置甲加入乙得數又加入丙得共數

假如有直長平和若干平若干問長

一

置和減平餘得長

假如有甲乙丙相併數若干甲若干乙若干問丙

置共數減甲餘又減乙餘得丙

分合第二 附添削化

分合者依術意圖正負與段數而傍書加減相乘者名宜分之合之

假如有四不等甲若干乙若干丙若干問積

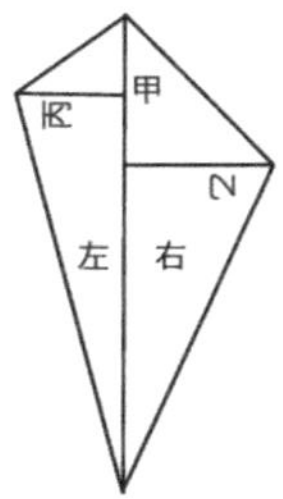

分術置甲以乙相乘得二段右積 |甲乙 置甲
以丙相乘得二段左積 |甲丙 二積相併折半
之得積

合術置乙加入丙共得數以甲相乘 |丙甲|乙甲 折
半之得積

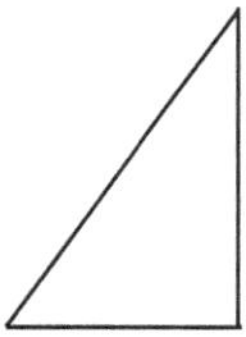

假如有勾股勾若干股若干問勾股
和冪

分術勾自乘一段 勾巾 股自乘一段 股巾 勾股
相乘二段 |股勾 三位相併得和冪

二

合術置勾加入股共得 |股勾 自乘得和冪

添
多位而正負同者添之爲寡位

假如 [|方巾 |斜巾] 添之 [|||方巾]

假如 [|勾巾 |股巾 |玄巾] 添之 [||玄巾]

刪
多位而正負異者刪之爲寡位

假如 〔⊥斜巾 ‖方巾〕 削之 〔|方巾〕

假如 〔⊥玄巾 ‖殳巾 ‖勾巾〕 削之 〔|玄巾〕

化

段數同而傍書變者謂之化

假如 〔|円径 |玄〕 化之 〔|勾 |殳〕

三

假如 〔|殳巾中勾巾 |勾巾中勾巾〕 化之 〔|殳巾勾巾〕

右添削化者雖爲分合一理意味有少差焉

全乘第三

全乘者施于正形者也長平或縱橫高相乘得積

自方

假如有平方自方若干問積

副置自方相乘之得積

假如有立方自方若干問積

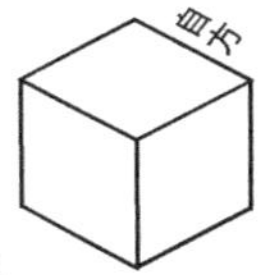

置自方再自乘之得積

圖解

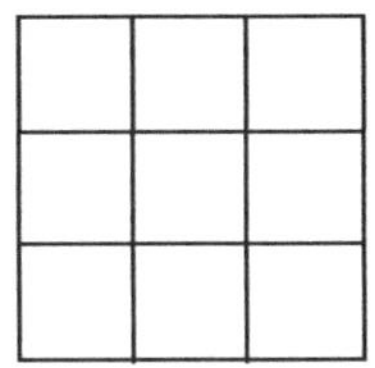

圖解

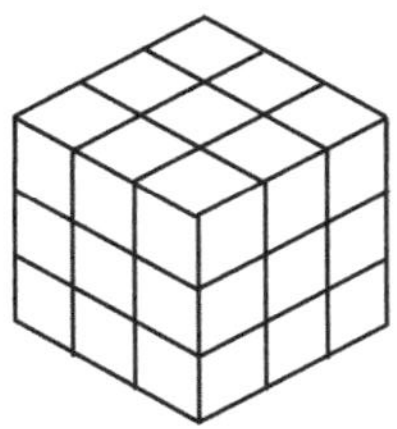

其餘直方堡壔直堡壔倣之

四

折乘第四

折乘者施于變形者也變形而方者長闊或縱橫高相乘得數隨其形之變而以其法約之得積

假如有勾股勾若干股若干問積

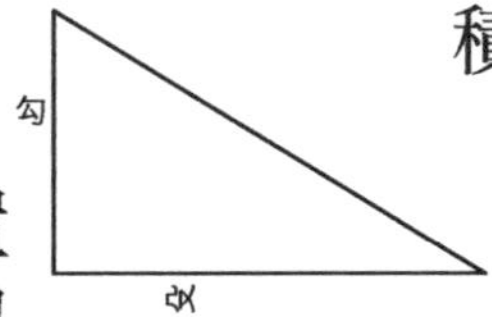

置勾以股相乘之得數折半之得積

圖解

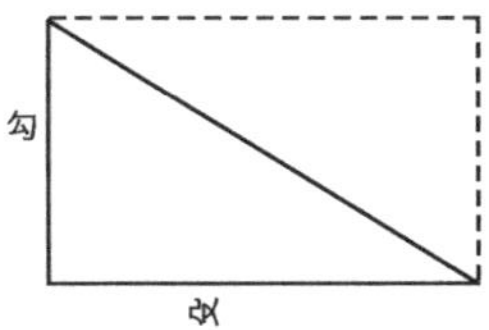

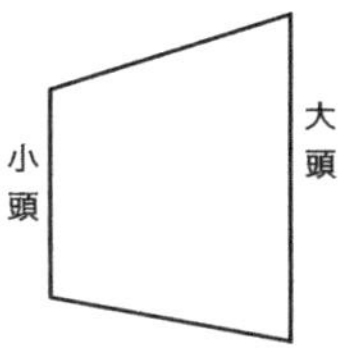

假如有梯大頭若干小頭若干長若干問積

置小頭加入大頭共得數以長相乘之得數折半之得積

解圖

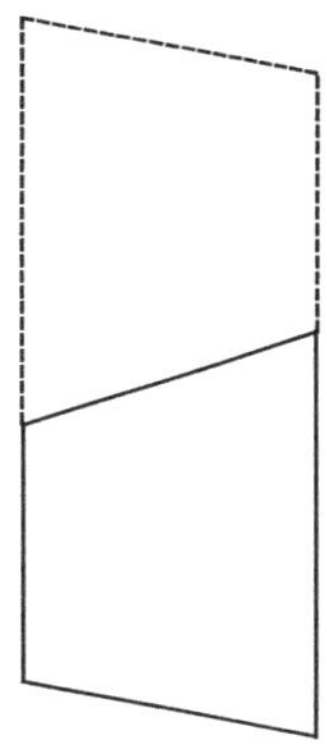

五

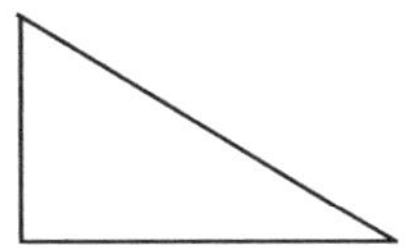

假如有勾股勾若干股若干問弦

置勾自乘之加入股冪共得數爲實開平方除之得弦

解圖

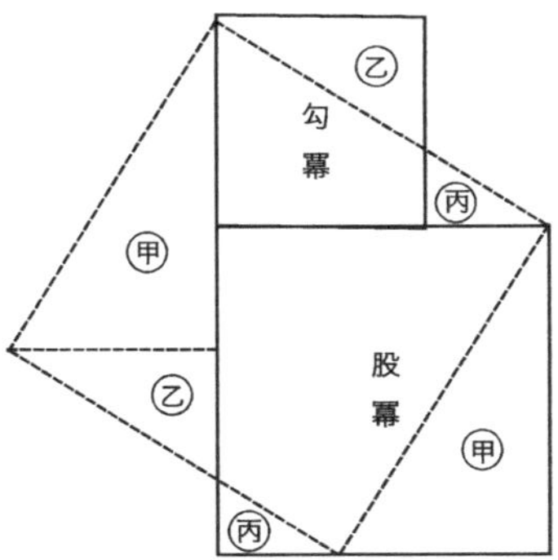

其餘圭稜斜鼓箭筈箭翎三廣腰鼓三斜曲尺
幞頭抹角四不等諸角形等皆傚之

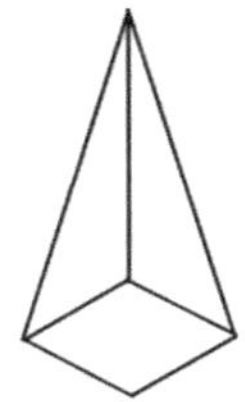

假如有方錐下方若干高若干問積
置下方自乘以高相乘之得數以三約之
得積
解術方二分之
一爲横方一箇
爲縱高二分之
一爲高三位相
乘則方冪高相乘四分之一是直堡壔
積乃四分之三方錐積也依課分術得方冪高相
乘者三段方錐積全積八分之一爲甲積全積三十二分之一爲乙積全積內減甲積一段與乙積四段餘得直堡壔積則全積四分之三也

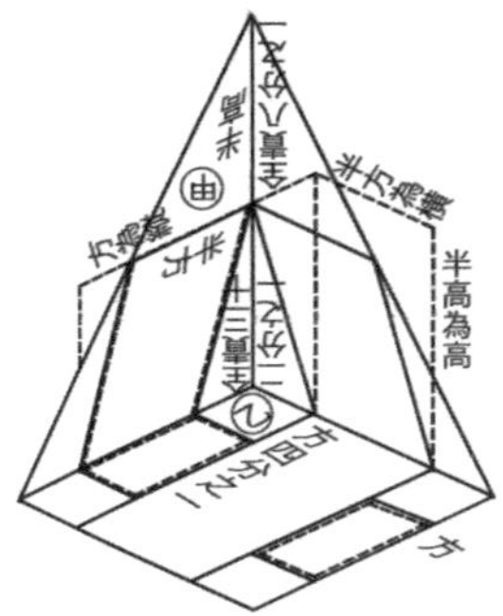

六

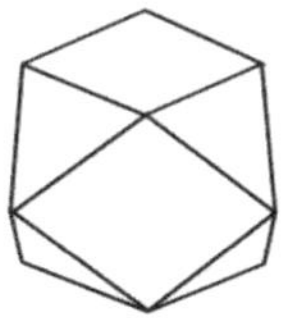

假如有方切籠每方若干問積
置方五自乘之以五十乘之得數爲實以
九爲廉法開平方除之得積
解術方堡壔一箇方爲方斜爲高直錐四箇方爲

橫斜爲縱半方爲高故方冪一
段爲橫冪方冪二十
五段爲縱冪方冪二　解
段爲高冪三位相乘
則方五乘冪五十段　圖
卽九段乃錐法冪也切籠

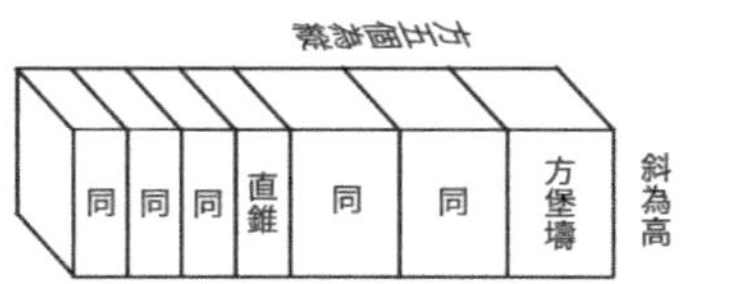

積冪也

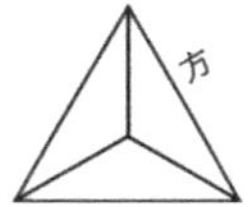

假如有蕎麥形每方若干問積
置方五自乘之得數爲實以七十二爲廉
七
法開平方除之得積
解術方冪四分之　解
三爲橫冪方冪一
段爲縱冪方冪三　圖
分之二爲高冪三

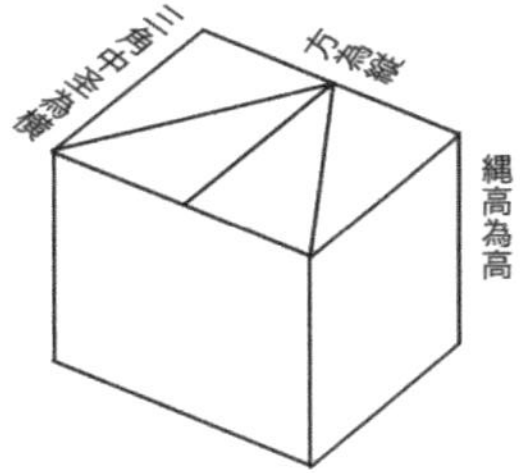

位相乘則方五乘冪一十二分之六是
直堡壔積冪也乃三十六段蕎麥形積冪也依課分
術得方五乘冪者七十二段蕎麥形積
冪倍錐法三自之得三十六
其餘直錐方臺直臺楔形等皆倣之

變形而圓者徑或徑高自乘再乘相乘得數隨
其形之變而以其法約之得積

假如有平圓周若干徑若干問積

置周以徑相乘之得數以四約之得積

解術視圭而周爲長　解
半徑爲闊相乘折半
之得積 求周徑率術載于別記　圖

假如有弧矢若干弦若干問積

八

別得 徑若干背若干 置背以徑相乘之得數寄位
置徑內減倍矢餘以弦相乘之得數以減
寄位餘以四約之得積

解術徑背相乘爲四　解
段扇積寄位　徑內
減倍矢餘爲二箇圭　圖
闊以弦爲圭長相乘
爲四段圭積以減寄位餘得四段弧積
求背術載于別記

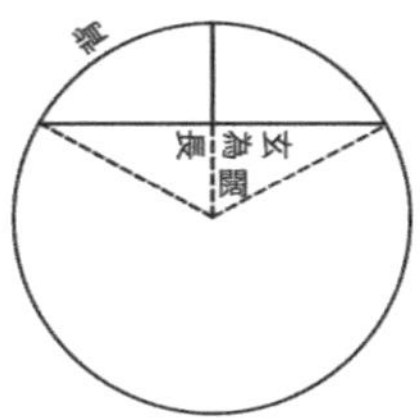

假令有側圓長徑若干短徑若干問積

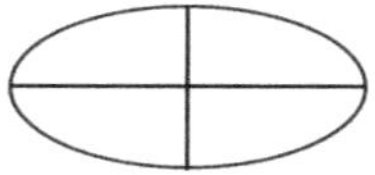

置長徑以短徑相乘之得數以圓積法乘之得積

解術視圓壔而短徑爲徑長徑爲斜又徑爲勾斜爲弦依勾股術而以所得股爲高以所得中股爲假高求

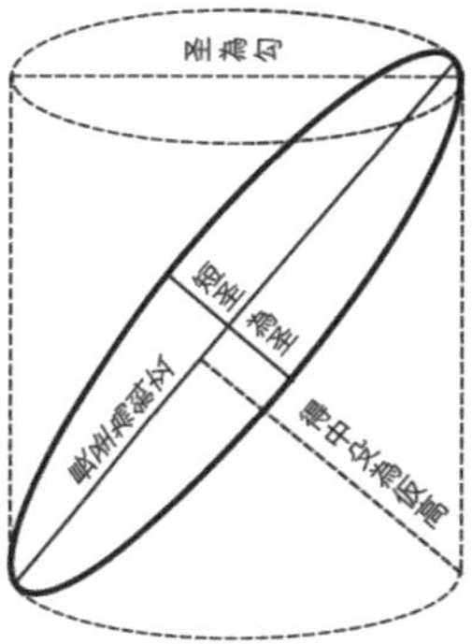

九

圓壔積以假高除之得斜截面積卽側圓積也

假如有側圓長徑若干短徑若干問周

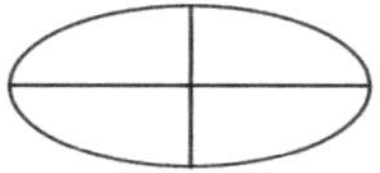

置長徑以短徑相乘之以圓周法冪乘之得數寄位置長徑內減短徑餘自乘之得數四之加入寄位共得數爲實開平方除之得周

解術正視則全圓故長短徑相乘以圓

周法冪乘之得數
傾視則二線故倍長　解
短徑差自乘之得數
二數相併得側圓周　圖
冪

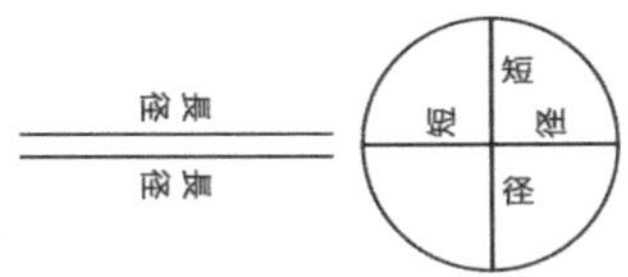

假如有半圓闕半徑若干灣若干
承背準規而週晼形問晼背

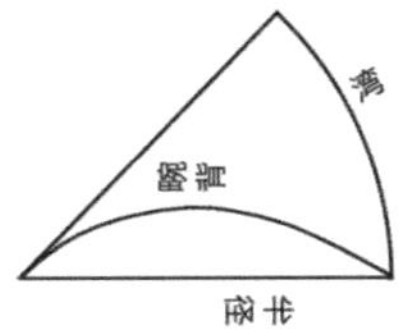

置半徑自乘三之加入灣冪共得數爲實
以三爲廉法開平方除之得背

十

解術日半灣冪依四分
之一增約術得數乃灣冪三　解
分之一也擬勾冪半徑冪擬
股冪二數相併得晼背　圖
冪

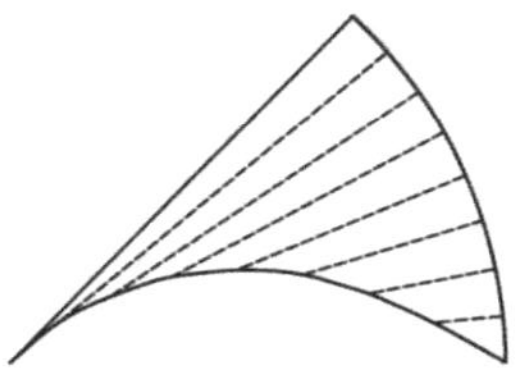

其餘環扇欖眉錠晼錢覆月車輞牛角火塘帶
直圓等皆做之
假如有立圓徑若干問積

置徑再自乘之得數以立圓積法乘之得
積求立圓積法術載于別記

假如有立圓闕矢若干弦若干問積

置矢自乘四之加入三段弦冪共得數以
矢相乘之得數以立圓積法乘之以四約
之得積

解術矢爲容立圓徑依立圓術求積得
數寄位　矢加二分之一爲錐高乃通高也

十一

弦爲錐底徑依圓錐術
求積加入寄位得立圓
闕積

解　圖

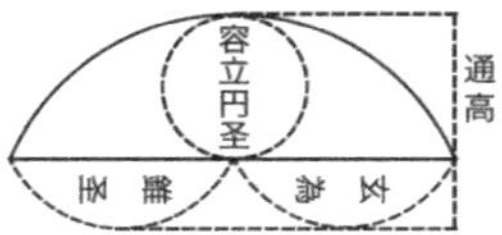

假如有立圓徑若干問覓積

置徑自乘之得數以圓周法乘之得覓積

解術視錐而半徑爲
高中心爲尖立圓積
爲錐積三之以高除

解　圖

之得錐面之覓積卽立圓覓積也
假如有立圓闕矢若干弦若干問頂覓
積

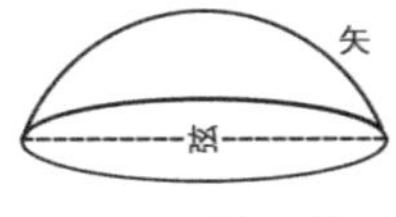

置矢自乘四之加入弦冪共得數以圓積
法乘之得頂覓積
解術別得徑半徑內減
矢餘爲錐高以弦爲
錐徑依圓錐術求積
寄位　求立圓闕積

解圖

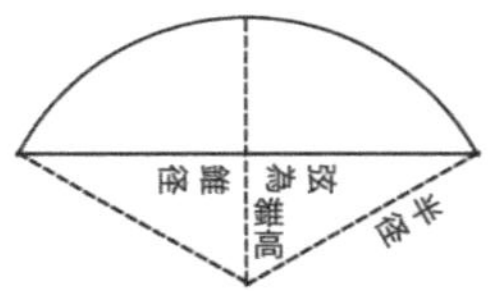

十二

加入寄位共得數三之以半徑除之得
頂覓積
其餘環圓壔圓錐圓臺環錐環臺揉立圓押立
圓帶堡圓壔圓臺斜截等之諸形甚多皆載于
其術於別記
右所錄四篇所以解見題之法也蓋此隱題
伏題皆可通用法也然見題內有似隱題者
焉學者宜熟思之其餘諸形難枚舉故標大
概而爲模範矣已

解見題之法終

享保丙午歲四月望前五日

Notes on Collation of Methods of Solving Explicit Problems

The following are the call numbers of manuscripts employed in the collation:

ME: Hayashi Collection [林集書] 649 / Matsunaga Library [松永文庫] 2491 in the Main Library of Tōhoku university = Matsunaga Book [松永本] in [1, pp. 173–174].

OE: Okamoto Library [岡本文庫] Manuscript 写 21/16944 in the Main Library of Tōhoku university = Okamoto Book [岡本本] in [1, p. 174].

SE: Kuwaki Library, Mathematics (2) [桑木文庫算学 (二)] Manuscript [写本] 679 Trilogy [三題集 Sandai Shū] in the Main Library of Kyūshū University = Gekka Book [月華本] in [1, p. 174].

1v. l. 8: 宜 is written 宣 in all of ME and OE.

2r. l. 6 & 2v. l. 1: 股 is written 段 in SE.

2v. l. 9: OE lacks the heading 削.

3r. l. 8: 段 is mistaken to be 的 in all copies including on three.

4r. right lower figure: Caption 解圖 is missing in SE.

4r. upper figure: 自方 is missing in OE.

4r. l. 10: The top character 其 is missing in ME.

4v. ll. 2&3: OE writes 面 for 而.

4v. lower figure: 勾 and 殳 are missing in OE.

5r. l. 1: 梯 is written 稊 in SE.

6r. l. 2: 抹 is written 株 in ME, and 秣 in OE.

6r. l. 2: 傚 is written 倣 in OE and SE.

6r. l. 8: In all of three 箇 is written 個.

6r. lower figure: 半高 is missing and the left edge of the prism is erroneous in ME; 半高為高 is written 半方為高 in OE and SE and 半方為横 is missing in OE. OE has the caption 解圖.

6v. l. 3: SE writes 二十二分之一 by mistake.

6v. l. 9: 開平方 is written 開方 without 平 in OE.

7r. l. 1: 半方 is 半分 in SE.

7r. l. 2: The top 段 is written 假 in SE.

7r. lower figure: 方 and 半方 are written over 方堡壔 and 直錐, resp. in ME.

7v. l. 9: 三 is mistaken to be 六 in OE and SE.

7v. l. 10: In SE 傚 is confused with 倣.

8r. l. 1: In SE 圓 is mistaken as 圖.

8v. l. 9: OE and SE don't have 于 in 于別記.

9r. & 9v. upper figures: Two diameters are dotted lines in OE.

9r. l. 3: 壔 is missing in SE.

9r. l. 6: 壔 is missing in all three manuscripts. OE lacks the following 而短, too.

9r. lower figure: 得殳為高 is missing in ME, and 得中殳為假高 is written 得中殳假為高 in ME and SE.

9v. l. 1: 面 in 面積 is 而 in SE.

10r. l. 2: 倍 is written 信 in SE.

10r. lower figure: Two 短径 are missing in OE.
10r. l. 5: 冪 is missing in OE.
11r. l. 3: 立圓 is written 平圓 in ME.
11v. lower figure: 通高 is 通玄 in SE.
11v. upper figure: No shadow in OE.
11v. l. 5: 立圓 is written 玉圓 in ME.
11v. lower figure: 尖 is written 矢 in all of the three manuscripts.
12r. l. 2: 頂覓 is written 項覓 in all.
12v. l. 1: 寄位 is written 奇位 in OE.
12v. l. 4: 帶堡圓壔 [,] 圓臺斜截 is written 帶堡圓圖壔臺斜截 in all. 帶堡圓 is an ellipse-like plane figure obtained by adding a rectangle between two half discs. 帶堡圓壔 is a cylinder with 帶堡圓 as its base. 圓臺斜截 is a conic section.
12v. ll. 6&7: 隱題 is written 隨題 in OE.
12v. l. 8: 枚擧 is written 牧擧 in OE and SE.
12v. l. 10: 享保丙午歳四月望前五日 is missing in ME and deleted by a double line in OE. This phrase denotes the date "Five days before the Full-moon of the Fourth lunar month of Kyōhō 11," which is 1726, 18 years later than the death of Seki Takakazu. Since the date at the end of a book usually meant the date of completion of the book, this shows that some one else must have completed or edited the book.

Reference

1. Hikosaburō Komatsu [小松彦三郎]: The restoration of the Yamaji Nushizumi edition of Seki Takakazu's "Trilogy" [關孝和著『三部抄』山路主住本の復元], *RIMS kôkyûroku* [数理解析研究所講究録], vol. 1444, pp. 169–202 (2005).

Methods of Solving Implicit Problems, by Seki Takakazu, collated by Hikosaburo Komatsu

Seki Takakazu

Abstract This is a collated text of the second of Seki's Trilogy. This is a very concise monograph of Celestial Element Method [天元術 tengen jutsu] and Root Extraction Method [開方術 kaihō jutsu]. In Chapter 1: Setting Element [立元], Celestial Element [天元 tengen], that is, the unknown is represented by a counting rod of 1 in the modulus [方 hō] class of a counting board. Polynomials [式 shiki] in Celestial Element with numerical coefficients and their operations [演段 endan] are discussed in Chapter 2: Addition and Subtraction [加減 kagen], and Chapter 3: Mutual Multiplications [相乗 sōjō] as manipulations of counting rods on a counting board. Two polynomials representing the same quantity define an Equation [開方式 kaihōsiki] by subtracting one from the other as shown in Chapter 4: Mutual Cancellation [相消]. The last Chapter 5: Root Extraction [開方] is devoted to the Numerical Extraction of Roots [開方術], that is the same as the so called Horner method which is erroneously attributed to W. G. Horner (1819). Japanese learnt these methods from Zhu Shijie [朱世傑] by his Introduction to Mathematics [算学啓蒙] (1299) but the methods described here are more refined. At the end of the book, Newton's method of approximation is explained very succinctly.

解隱題之法

關孝和 編

二〇〇五年七月九日　小松彦三郎校

解隱題之法 凡五篇

關孝和編

立元第一

立元者立天元一也

太極

加減第二 附併

加者單位者謂加衆位者謂併各其異名相減則同名相加正無人正之負無人負之

假右
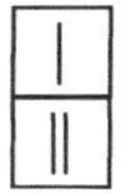

一

如左
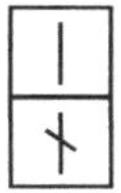

加之 右左一級數同名相加正二 ○二級數異名相減正一

得
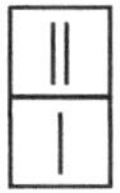

假右
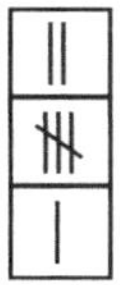

如左
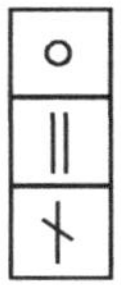

加之 右左一級數正無人故正二○二級數 異名相減負一○三級數異名相減空

得
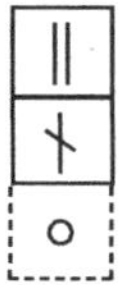

假右

如併之

中 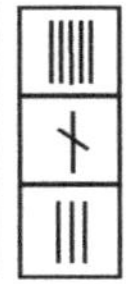

左 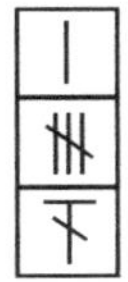

右〇八左
右中左一級數同名相加正九
〇右中左二級數同名相加負
八〇右中三級數同名相加與
左三級數異名相減正四

得 

減者其同名相減則異名相加正無人負之負
無人正之

假如減之

右 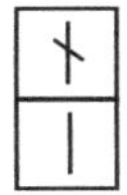

左 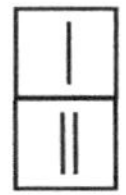

右左一級數異名相加正二
〇二級數同名相減正一

以右減左

得 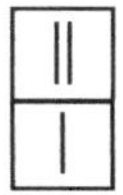

二

假如減之

右 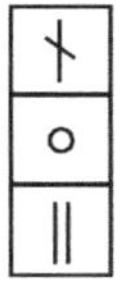

左 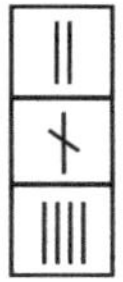

右左一級數異名相
加負三〇二級數負
無人故正一〇三級
數同名相減負二

以左減右

得 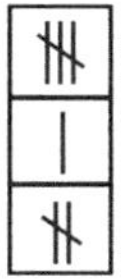

相乘第三 附見乘

相乘者置其式於左右以左自上級到下級逐

遍乘右同名相乘爲正異名相乘爲負○乃當空級而乘者爲空各相併得
式自乘者準之
見乘者置其式乘數乃歸除空平方一立方二以上做之自乘者
倍之加一再乘者三之加二三乘者四之加三
次第做之爲乘數相乘者兩式乘數相併加一爲乘
數

假如 [○|] 自乘之見乘數者歸除空加一得一爲平方式

右 ○ | [○ ○] 以左一級空遍乘右

左 ○ ∠ |

三

右 ○ | [○ |] 以左二級正一遍乘右

左 ○ |∠

二位相併
得 [○ ○ |]

假如 [||| ≠ |] 自乘之見乘數者平方一倍之加一得三爲三乘方式

右

左

以左上級正
三遍乘右

右

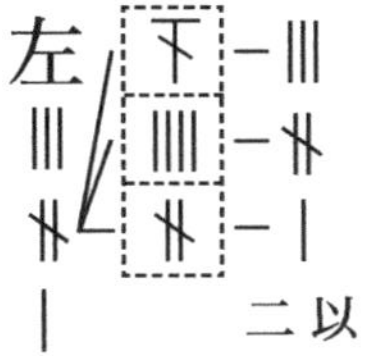

左

以左中級負
二遍乘右

右

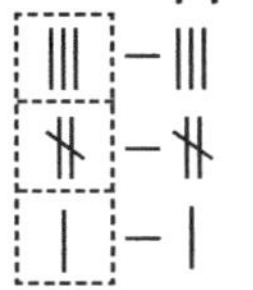

以左下級正
一遍乘右

四

三位相併

得

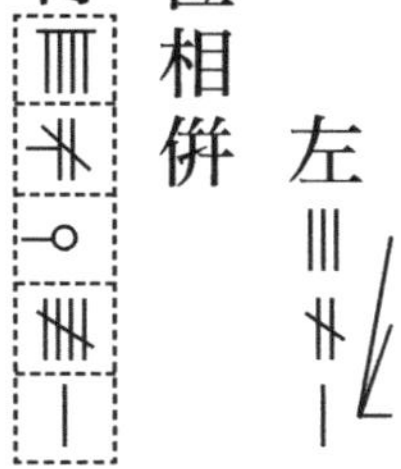

左

假如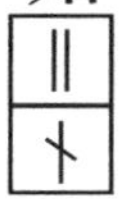再自乘之

見乘數者歸除空加二
得二爲立方式

先自乘之得

又相乘之

右

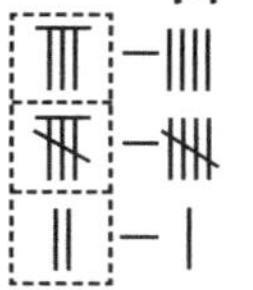

以左上級正
二遍乘右

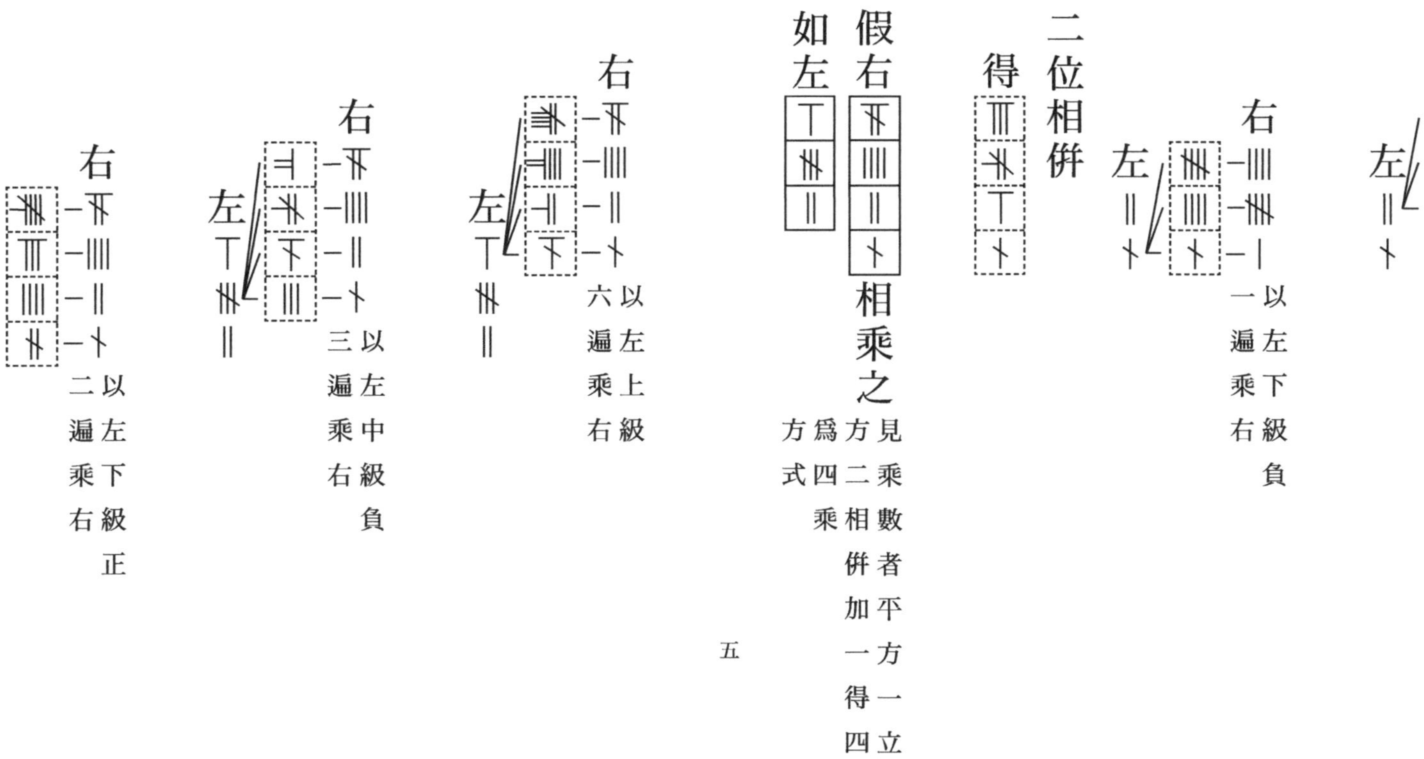
左
右
以左下級負
一遍乘右
左
二位相併
得
假右
相乘之
見乘數者平方一立
方二相併加一得四
爲四乘
方式
如左
五
右
以左上級
六遍乘右
左
右
以左中級負
三遍乘右
左
右
以左下級正
二遍乘右

左 

三位相併

得 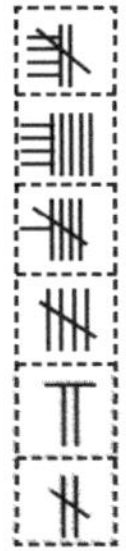

相消第四

相消者如意求之得寄左數與相消數兩數之
內任意而其同名相減則異名相加正無人負
之負無人正之得歸除及開方式

假　得數 

如　寄左 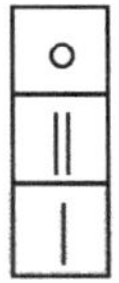　以得數消寄左

相消　一級數正無人故負八〇二
　　　級數正二〇三級數正一

得開方式 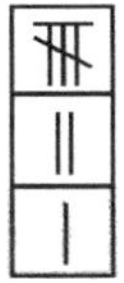

六

假　得數 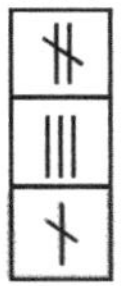

如　寄左 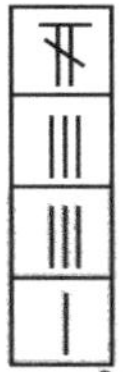　以寄左消得數

相消　一級數同名相減正五〇
　　　二級數同名相減空〇〇三
　　　級數異名相加負四〇四
　　　級數正無人故負一

得開方式 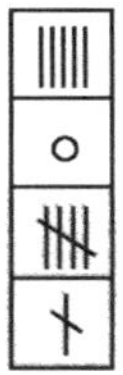

開方第五　附得商

開方者立商從隅平方式者從廉命之命之乃超位如常到實
咸同加異減而開盡之諸級中正負相反者謂之翻法也
假如開方式
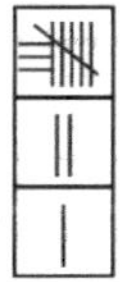
平方開之
立商五命廉同加方得方正七以商五命之異減實恰盡○又以商五命廉同加方得方正一十二
商五

假如開方式
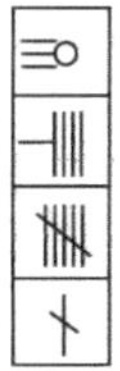
立方翻法開之
立商三命隅同加廉得廉負八以商三命之異減方得方正反負一十○以商三命之異減實恰盡○又以商三命
七
隅同加廉得廉負一十一以商三命之同加方得方負四十三○又以商三命隅同加廉得廉負一十四○是方正反爲負故爲翻法
商三
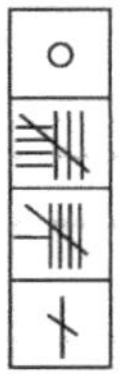
得商
先立商一自隅命之到實異減同加而實餘者
復立商一如前到實逐如此而實盡則所立商
相併爲定商
假如
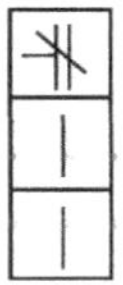
先立商一箇自廉命之到實異減同加而得

商一箇

復立商一箇如前而得

商一箇
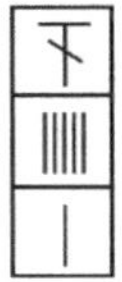
又立商一箇如前而實盡

商一箇

仍所立商相併得三爲定商

或實翻而不能盡者立負商如前到實異減同

加而實盡則前商相併内併減負商爲定商

假如
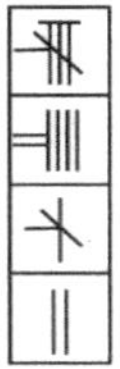
先立商一箇自隅命之到實異減同加而得

八

商一箇
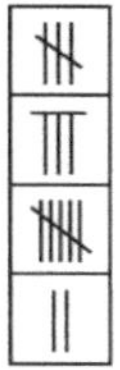
又立商一箇如前而實翻而不能盡

商一箇
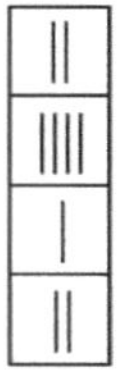
又立負商五分如前異減同加而實盡

負商五分

仍所立商相併得二箇内減負商五分餘一箇

五分爲定商

或實有不盡者以方隨開商位數除實而以所

得依正負而加減于開商爲次商以之自隅命

之到實而如前以方除實而以所得又加減于

次商也次第如此而得定商

假如

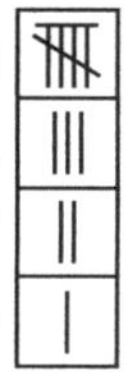

先立商一箇自隅命之
到實異減同加而得

商一箇

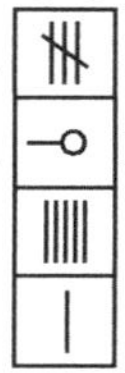

又立商二分
如前而得

商二分

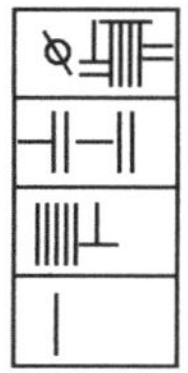

又立商六釐
如前而得

商六釐

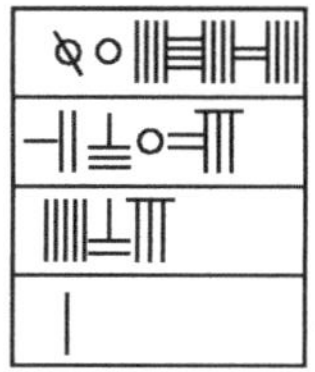

如此實有不盡故於是以方除實得正三毫四
六强加入前開商共得一箇二分六三四六强
次第如此而得定商

右所錄五篇所以解隱題之法也各々深意
有之今取捷徑誌之矣學者當研究耳

解隱題之法訖

貞享乙丑八月戊申日龔書

Notes on Collation of Methods of Solving Implicit Problems

The following are the call numbers of manuscripts employed in the collation:

MI: Hayashi Collection [林集書] 647 / Matsunaga Library [松永文庫] 2489 in the Main Library of Tōhoku university = Matsunaga Book [松永本] in [1, p. 189].

NI: Hayashi Library [林文庫] 2306 in the Main Library of Tōhoku university = Nakamura Book [中村本] in [1, p. 189]

OI: Okamoto Library [岡本文庫] 写 22/16945 in the Main Library of Tōhoku University.

1r. l. 5: 太 is written 大 in MI and OI.

1v. l. 2: 數 is written 衆 in OI.

1v. l. 8: In MI the lowest frame is not dotted and noframe in OI.

1v. l. 10: 假 is missing in NI.

2v. l. 6&10: 正一 is written 正之一 in OI, and 左 is written 方 in MI.

3v. l. 3: 二 is taken to be 一 in MI, NI and OI.

4r. l. 6: The number −2 in the medium class [廉級] is written 2 in OI.

4v. l. 2: 得 is repeated in MI, NI and OI.

5r. l. 4: The number −1 in the medium class is written 1 in OI.

6r. l. 3: The number −4 in the second medium class [次廉級] is mistaken to be 4 in almost all manuscripts including our MI, NI and OI.

6v. l. 5: The number 1 in the ultima class [隅級] is −1 in OI.

7r. l. 1&2: 到 is taken to be 列 in MI, NI, and OI, and 咸 is mistaken for 減 in OI.

7r. l. 10: The punctuation mark ○ on the right is missing in MI and NI.

7v. l. 1: 加 on the left is mistaken to be 如 in NI.

8r. l. 9: The number −18 in the res class [實級] is 18 in MI.

8v. l. 1: The number 8 in the modulus class [方級] is 3 in NI.

8v. l. 5: If we compute the modulus class as indicated in the text, we get 4.5. MI and NI write it by the arrangement of counting rods meaning 45 and set −2 and 2 in the lower classes on the left. OI tries to express this by using 4 vertical rods but there are only 3 horizontal rods for 0.5.

8v. l. 8&9: 所 and 于 are written テ in MI .

9r. l. 2: The number −9 in the res class [實級] is 9 in MI.

9r. l. 3: 一 is taken to be 二 in NI.

9v. l. 8: 申 is taken to be 中 in MI and NI.

9v. l. 9: At the end of OI there is the date 寛保癸亥四月丙午日再写之蓮貝軒, which means that Yamaji Nushizumi copied on the 丙午 day in the Fourth lunar month of Kanpō 2 (1742) for the second time, in addition to the date 貞享乙丑 (1685) 八月戊申日 on which Seki completed the manuscript.

Reference

1. Hikosaburō Komatsu [小松彦三郎]: The restoration of the Yamaji Nushizumi edition of Seki Takakazu's "Trilogy" [關孝和著『三部抄』山路主住本の復元], *RIMS kôkyûroku* [数理解析研究所講究録], vol. 1444, pp. 169–202 (2005).

Methods of Solving Concealed Problems, by Seki Takakazu, collated by Hikosaburo Komatsu

Seki Takakazu

Abstract This is the last of Seki's Trilogy. Here he gives a general procedure to eliminate a common unknown from two polynomial equations, which is the same as the elimination theory of Étienne Bézout. The book has five chapters. In Chapter 1: Real and Fictitious [真虚 shinkyo] and Chapter 2: Two Equations [兩式 ryōshiki] we learn how to formulate a system of algebraic equations as a succession of two equations in an unknown to be eliminated with polynomials of the other unknowns as coefficients. Chapter 3: Estimates of Degrees [定乘 teijō] gives an estimate of the degree of eliminated equations. The elimination is carried out in two steps. Chapter 4: Transformed Equations [換式 kanshiki] shows how to construct n equations of degree less than n out of a system of two equations of degree $\leq n$. Then the eliminated equation is obtained as the determinant of their coefficients equated to 0 as shown in Chapter 5: Creative and Annihative Terms [生尅 seikoku], where Seki gives the expansion of determinants of order up to 4. The disorder of table happens in the case of order 4, for which we append an amendment for disorder and alterations of sheets 14 and 15[1]. If Seki had stopped here, he would have been praised for having written the most concise and complete book on ellimination. His errors occurred in the expansion of determinant of order 5, for which we refer the reader to Goto–Komatsu [1].

[1] See pp. 490–491 and pp. 492–493, respectively.

解伏題之法

關孝和　編

二〇〇五年八月二十九日　小松彦三郎　校

解伏題之法 凡六篇

關孝和編

眞虛第一

隨眞術之所得而逐求虛術也

假如有勾股只云勾爲實平方開之得數與弦和若干又云勾股和若干問勾

眞術得勾

只云數有 股有 勾有

虛術見勾開方數

依只云數股勾得前式 依勾得後式

一

假如有三斜積若干只云大斜再自乘數與中斜再自乘數相併共若干又云中斜再自乘數與小斜再自乘數相併共若干問大斜

眞術得大斜

積有 中斜再自乘數有 小斜再自乘數有 大斜有

虛術見中斜

依積小斜再自乘數大斜得前式 依中斜再自乘數得後式

積有 小斜再自乘數有 大斜有 中斜有

虛術見小斜

假如有甲乙丙丁戊平方各一只云甲乙積
依積大斜中斜得前式
依小斜再自乘數得後式
差若干乙丙積差若干丙丁積差若干丁戊積差若干
又云甲乙丙丁戊方面和若干問甲方面
眞術得甲方面
乙積有丙積有丁積有戊積有乙丙丁戊
方面和有
虛術見乙方面
依丙積丁積戊積乙丙丁戊方面和
得前式○依乙積得後式
丙積有丁積有戊積有丙丁戊方面和有

二

虛術見丙方面
依丁積戊積丙丁戊方面和得前式
依丙積得後式
丁積有戊積有丁戊方面和有
虛術見丁方面
依戊積丁戊方面和得前式
依丁積得後式
右各虛術逐以次前虛術擬眞術也

兩式第二 附略省約縮
得眞虛之後求兩式也
假如有方臺積若干只云上下差與高和若干又

云下方冪與高冪相併共若干問上方
眞術得上方
積有下方與高和有又云數有上方有
虛術見高
前術曰立天元一爲高。丨以減和餘爲
下方 丨(和) 丨̸ 自乘 丨(和巾) ‖̸(和) 丨 上方自乘 丨(上方巾) 上下
方相乘 丨(和上方) 丨̸(上方) 三位相併以高乘之爲三段
積。丨(和巾) ‖̸(和) 丨 寄左 列積 ‖̸(積) 丨(和巾) ‖̸(和) 丨
丨(和上方) 丨̸(上方) 三之與寄左 丨(和上方) 丨̸(上方)
丨(上方巾) 相消得前式 丨(上方巾)

三

後術曰立天元一爲高。丨以減和餘爲
下方 丨(和) 丨̸ 自之加入高冪共得 丨(和巾) ‖̸(和) ‖ 寄
左 列又云數與 丨(和巾) ‖̸(和) ‖
寄左相消得後式 丨̸(又云)

右各以數不求式啻圖正負與段數而傍書加
減相乘者名也○各級中位傍書同而正負同
者相加之異者相減之
略 略位數也
高級式中位與卑級式同名者略之
假 前式

丨̸	子
丨	丑
丨̸	寅
丨	卯

	‖̸辰
丨辰	‖巳
‖̸巳	丨午
	卜午

如後式

卜辰
‖巳
丨午

以後式從前式上級減之又以後式從前式上第二級加之

略之

前式

卜辰	卜子
	丨丑
	卜寅
	丨卯

後式

卜辰
‖巳
丨午

或有卑級式自乘再自乘幾自乘而同名者

或有傍書段數互乘而同名者皆當依時宜

四

略之

各式之每級每位傍書遍乘同名者省之

省 省傍書也

假如

卜寅子	卜丑子
丨辰子	丨卯子
	丨子

省之 每級各省子

卜寅	卜丑
丨辰	丨卯
	丨

約 約段數也

各式之每級每位段數可遍約者約之

假如

約之 每級約二遍

縮 縮級數也

兩式空級均同者縮之

假如 前式 五乘方

如 後式 三乘方

縮之 前式縮空級而爲立方 後式縮空級而爲平方

前式 立方

後式 平方

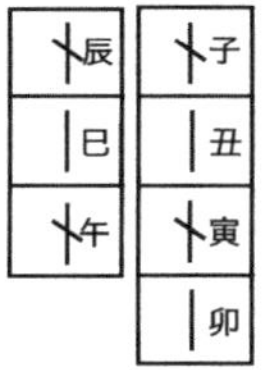

五

定乘第三 附疊括

得兩式驗略省約縮之後求定乘也

假如 前式 歸除 歸段

如 後式 立方 立平歸段

前式再自乘順行 立平歸段

同級相乘 立 立 立 立

後式直逆行 段｜歸｜平｜立

以立方爲眞術之乘數

假 前式 平方 平立歸

如 後式 立方 三立歸平

前式再自乘順行 五六七八六四立

同級相乘 九 九 十二 十四 十二 十一 十

後式自乘逆行 三立四五五六七

以一十四乘方爲眞術之乘數

六

右各毎級以眞術各位之乘數最高者記之○直自乘再自乘幾自乘者前式隨後式後式隨前式仍前式順行後式逆行也○以順逆同級相乘之乘數最高者爲眞術之乘數然換式之後遇芟者或寄消省者就而減乘數也

疊 疊級數也

卑級式之下級或上級箇數者疊之

假 前式 三乘方 

如 後式 平方 

疊之 以前式下級遍乘後式以後式下級遍乘前式相減之前式變立方 

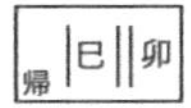

以變前式下級遍乘後式以後式下級遍乘變前式相減之變前式亦變平方

前式 平方

後式 平方

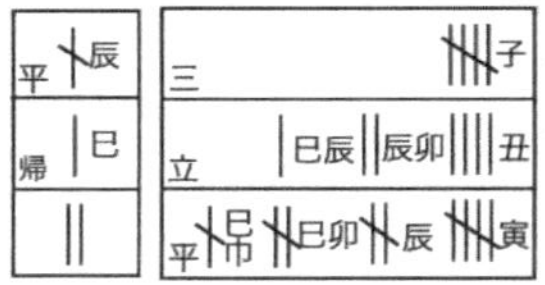

假如疊之

前式 立方

後式 平方

以前式上級遍乘後式以後式上級遍乘前式相減之前式變平方

七

前式 平方

後式 平方

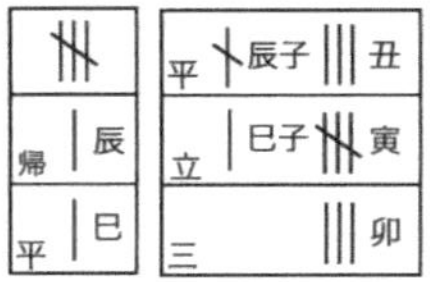

疊之每變如定乘求之而增於眞術之乘數者不疊之〇或有未括前疊之者或有已括後疊之者可依時宜〇疊之後各級中位傍書同而正負同者相加之異者相減之

括 括位數也

各級多位者括之

假

如

括之

子二箇內減丑一箇餘負爲甲○子冪三段丑冪二段寅冪一段三位相併共得負爲乙○子再自乘三段子冪丑相乘二段二位相併共得內併減丑再自乘三段丑冪寅相乘一段餘正爲丙○或有分正負而括之者可依時宜

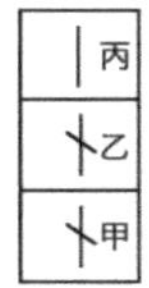

各級每位傍書遍乘同名者遍去而括之却以遍去者書之

假

八

如

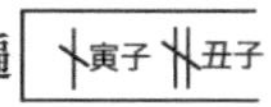

括之

遍去子○子三箇內減併丑二箇寅一箇餘正爲甲却以遍去子書之

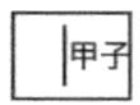

各級每位段數可遍約者遍約而括之却以遍約數圖之

假

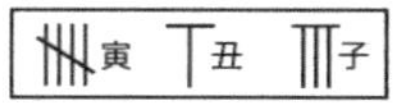

如

括

遍約二○子四箇丑三箇二位相併共得

之內減寅二箇餘正爲甲却以遍約二圖之 [‖甲]

括之位數段數同者以同名書之雖同名或

有每位段數一倍二倍幾倍者或有正負反

者皆就而圖段數正負也○各級單位者箇

數者假如者 [‖丑子] [||||] 如此之類不括之若單位者芟治之後相乘

多位者括之○或得兩式術中多位者括之

換式第四 附芟治

得定乘驗疊括之後求換式也

假如 前式 歸除
　　 後式 同

前式	後式
⧸|乙	⧸|丁
|甲	|丙

以甲正遍乘後式以丙正遍乘前式相減之得一式

一式 [⧸\|丁甲 \|丙乙]

假如 前式 平方
　　 後式 同

前式	後式
⧸|丙	⧸|己
⧸|乙	|戊
|甲	⧸|丁

以甲正遍乘後式以丁負遍乘前式相減之得一式

一 [⧸\|己甲 \|戊甲]

以乙負遍乘後式加一式以戊

式

丨̸丁丙	
丨̸丁乙	

正遍乘前式減
一式得二式

式二

丨戊丙	丨己乙
丨̸丁丙	丨̸己甲

假如
前式 立方
後式 同

前式	後式
丨̸丁	丨̸辛
丨丙	丨̸庚
丨乙	丨己
丨̸甲	丨戊

以甲負遍乘
後式以戊正
遍乘前式相
減之得一式

式一

丨戊丁	丨辛甲
丨̸戊丙	丨庚甲
丨̸戊乙	丨̸己甲

以乙正遍乘後
式加一式以己
正遍乘前式減
一式得二式

十

式二

		丨己丁	丨̸辛乙
丨̸己丙	丨̸庚乙	丨戊丁	丨辛甲
		丨̸戊丙	丨庚甲

以丙正遍
乘後式加
二式以庚
負遍乘前
式減二式
得三式

式三

丨̸庚丁	丨̸辛丙
丨己丁	丨̸辛乙
丨戊丁	丨辛甲

兩式級數有長短者借空於卑級式之下而求
換式也
乃當空級者
不及遍乘也
假 前式 三乘方
如 後式 平方

前式	後式
丨̸丁	丨̸庚
○	丨己
丨丙	丨戊
丨̸乙	○
丨甲	○

借空於後
式之下二
級而爲三
乘方○以

甲正遍乘後式去借
空一級而得一式

式一

丨̸庚甲
丨己甲
丨戊甲
○

以乙負遍乘
後式去借空
一級而加一
式得二式

式二

	丨庚乙
丨̸己乙	丨̸庚甲
丨̸戊乙	丨己甲
	丨戊甲

以丙正遍乘後
式加二式以戊
正遍乘前式減
二式得三式

式三

丨戊丁	丨̸庚丙
丨己丙	丨庚乙
丨̸己乙	丨̸庚甲
	丨己甲

以己正遍
乘前式減
三式得四
式

十一

式四

	丨己丁
丨戊丁	丨̸庚丙
	丨庚乙
	丨̸庚甲

右各從下級至上第二級同級互遍乘加減之後式爲加前式爲減得逐式也○或有未括前求換式者或有已括後求換式者可依時宜○換式之後各級中位傍書同而正負同者相加之異者相減之

芟芟傍書也

求換式而先各式之每級每位傍書遍乘同名者芟之次逐式之同級每位傍書遍乘同名者

芟之

假

一式

𝍠̸戊子再	𝍠丁子巾
𝍠丙子巾	𝍠̸乙子
	𝍠甲

二式

𝍠̸壬子三	𝍠辛子再
𝍠̸庚子再	𝍠̸己子巾
𝍠丙子巾	𝍠̸乙子

如

三式

	𝍠癸子三
𝍠̸壬子三	𝍠辛子再
𝍠̸戊子再	𝍠丁子巾

芟之先二式芟子三式芟子冪次上級芟子冪中級芟子

𝍠̸戊子	𝍠丁
𝍠丙子	𝍠̸乙
	𝍠甲

十二

𝍠̸壬子	𝍠辛
𝍠庚子	𝍠̸己
𝍠丙子	𝍠̸乙

	𝍠癸
𝍠̸壬子	𝍠辛
𝍠̸戊子	𝍠丁

治治段數也

求換式而先各式之每級每位段數可遍約者治之次逐式之同級每位段數可遍約者治之

假

一式

𝍡̸丙
𝍢乙
𝍠甲

二式

𝍦̸戊
𝍧丁
𝍢乙

如

三式

𝍧己
𝍦̸戊
𝍡̸丙

治之 先二式以三治之三式以二治之 次上級以二治之中級以三治之

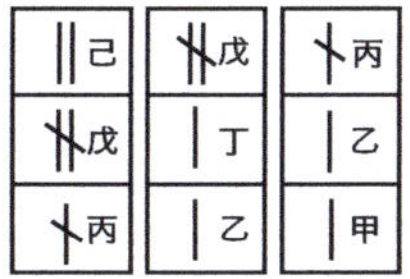

換式芟治之後或亦括之如前

生剋第五 附交式斜乘

得換式驗芟治之後求生剋也

假 一式 乙 甲

如 二式 丁 丙

十三

乙丙相乘	丁甲相乘
生	剋
○	○
一	一
甲丙	甲丙

假 一式 丙 乙 甲

二式 己 戊 丁

如 三式 壬 辛 庚

丙戊庚相乘	己辛甲相乘
生	生
○	○
一 乙戊庚	二 甲戊辛
四 甲戊庚	甲 辛

丙辛丁相乘	己乙庚相乘
剋	
○	
三 乙丁辛	乙 庚
五 甲丁辛	乙 庚

壬乙丁相乘
生
○
三 辛丁乙
六 庚丁乙

壬戊甲相乘
剋
○
二 辛戊甲
四 庚戊甲

假如

一式 房氏亢角

二式 斗箕尾心

三式 危虛女牛

四式 婁奎壁室

房箕女室相乘
生
○
一 室女箕氏
五 室女箕亢
九 室女箕角

房奎尾牛相乘
生
○
四 奎牛尾氏
十三 奎牛尾亢
十七 奎牛尾角

十四

斗虛壁角相乘	危奎亢心相乘	婁氏尾牛相乘	房奎女心相乘
剋	生	剋	剋
○	○	○	○
二 壁虛箕角	三 奎虛心亢	四 奎牛尾氏	廿一 奎女心氏
六 壁虛尾角	七 奎女心亢	八 壁牛尾氏	七 奎女心亢
十 壁虛心角	十一 奎牛心亢	十二 室牛尾氏	廿五 奎女心角

斗氏女室相乘	危箕壁角相乘	婁虛亢心相乘	房虛壁心相乘
剋	生	剋	生
○	○	○	○
一 室女箕氏	二 壁虛箕角	三 奎虛心亢	廿九 壁虛心氏
十四 室女尾氏	十五 壁女箕角	十六 壁虛心亢	十六 壁虛心亢
十八 室女心氏	十九 壁牛箕角	廿 室虛心亢	十 壁虛心角

相壁房 乘牛箕	相尾婁 乘角虛	相亢危 乘室箕	相壁斗 乘牛氏
剋	生	剋	生
○	○	○	○
廿二 壁牛箕氏	廿四 奎虛尾角	廿三 室虛箕亢	廿一 壁牛箕氏
卅三 壁牛箕亢	六 壁虛尾角	五 室女箕亢	八 壁牛尾氏
十九 壁牛箕角	廿八 室虛尾角	廿七 室牛箕亢	廿六 壁牛心氏

相尾房 乘室虛	相女婁 乘角箕	相尾危 乘室氏	相亢斗 乘牛奎
剋	剋	生	剋
○	○	○	○
卅一 室虛尾氏	卅二 奎女箕角	卅一 室虛尾氏	卅 奎牛箕亢
卅四 室虛尾亢	十五 壁女箕角	十四 室女尾氏	十三 奎牛尾亢
廿八 室虛尾角	九 室女箕角	十二 室牛尾氏	十一 奎牛心亢

十五

相女婁 乘心氏	相尾危 乘角奎	相亢斗 乘室虛
生	剋	生
○	○	○
廿一 奎女心氏	廿四 奎虛尾角	廿三 室虛箕亢
卅六 壁女心氏	卅五 奎女尾角	卅四 室虛尾亢
十八 室女心氏	十七 奎牛尾角	廿 室虛心亢

相亢婁 乘牛箕	相壁危 乘心氏	相女斗 乘角奎
生	剋	生
○	○	○
卅 奎牛箕亢	廿九 壁虛心氏	卅二 奎女箕角
卅三 壁牛箕亢	卅六 壁女心氏	卅五 奎女尾角
廿七 室牛箕亢	廿六 壁牛心氏	廿五 奎女心角

右各逐式交乘而得生剋也雖然相乘之數位
繁多而不易見故以交式斜乘代之

交式

從換三式起換四式從換四式起換五式逐如此（換二式換三式者不及交式也）○順逆共遞添一得次乃式數奇者皆順偶者順逆相交也

換三式

順	順	順
一	二	三

換四式

順
一
一
一

換五式

逆	順	逆
二	三	四
三	四	二
四	二	三

一	二	三	四	五
一	三	二	五	四
一	四	五	二	三
一	五	四	三	二
一	二	四	五	三
一	四	二	三	五
一	五	三	二	四
一	三	五	四	二
一	二	五	三	四
一	五	二	四	三
一	三	四	二	五
一	四	三	五	二

斜乘

交式各布之從左右斜乘而得生剋也若當空級者除

之○換式數奇者以左斜乘爲生以右斜乘爲

剋偶者左斜乘右斜乘共生剋相交也

換二式

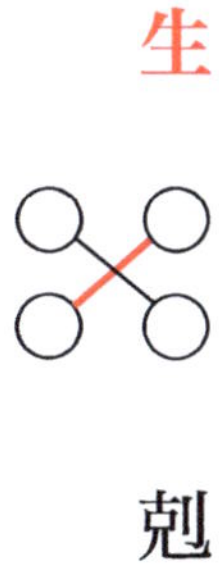

換三式

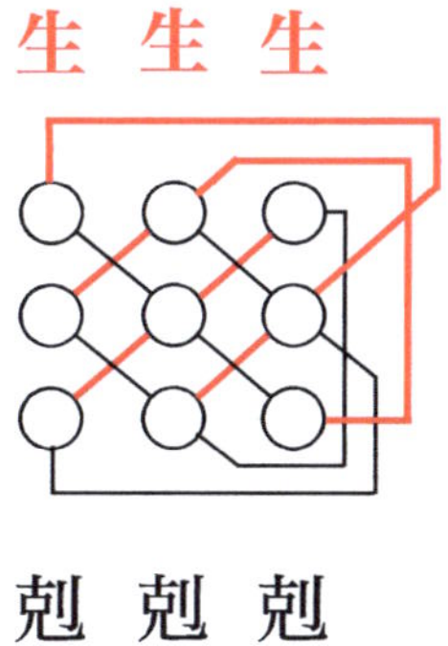

換四式

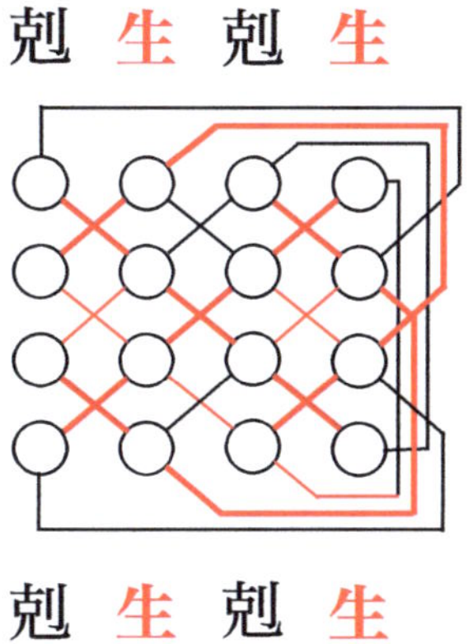

換五式

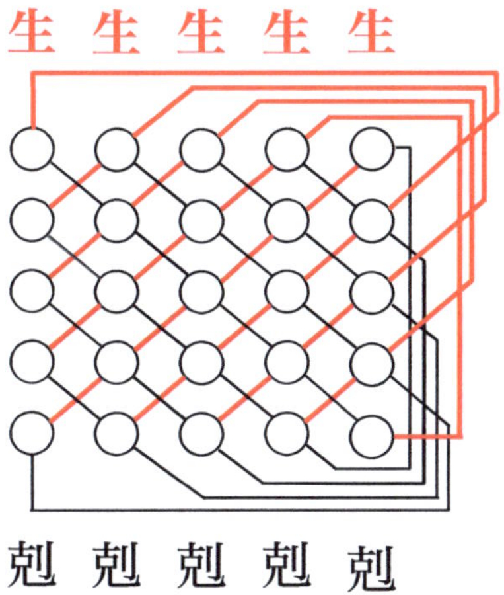

寄消第六

得生剋之後求寄消也

假　一式

　　二式

如　三式

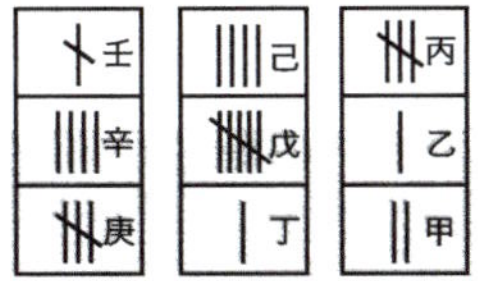

生丙戊庚相乘消

生己辛甲相乘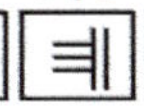寄

生壬乙丁相乘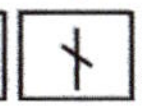消

剋丙辛丁相乘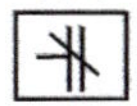寄

十八

剋己乙庚相乘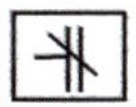寄

剋壬戊甲相乘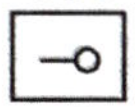消

假　一式

　　二式

如　三式

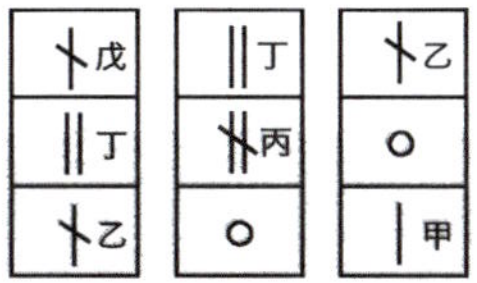

生乙丙乙相乘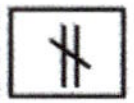消

生丁丁甲相乘寄

剋戊丙甲相乘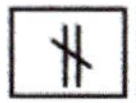消

右各生而正剋而負者相併爲寄左數生而負
剋而正者相併爲相消數也乃換一式者直以正爲寄左數以負爲相消數也○相乘同名而寄消同者相加之寄消
異者相減之○寄消或遍乘同名者省之段數
可遍約者約之如前○各起於末虛術而求到
寄消亦起次前虛術而求到寄消次第如此而
得眞術也

右所錄六篇所以解伏題之法也但舉一二
而爲之例矣學者須要分明理會得也書不

十九

盡言而已

解伏題之法畢

天和癸亥重陽日重訂書

相尾婁 乘牛氏 尅	相亢危 乘心奎 生	相壁斗 乘角虛 尅	相女房 乘室箕 生
○	○	○	○
四 奎牛尾氏	三 奎虛心亢	二 壁虛箕角	一 室女箕氏
八 壁牛尾氏	七 奎女心亢	六 壁虛尾角	五 室女箕亢
十二 室牛尾氏	十一 奎牛心亢	十 壁虛心角	九 室女箕角

相尾婁 乘角虛 生	相亢危 乘室箕 尅	相壁斗 乘牛氏 生	相女房 乘心奎 尅
○	○	○	○
十六 奎虛尾角	十五 室虛箕亢	十四 壁牛箕氏	十三 奎女心氏
六 壁虛尾角	五 室女箕亢	八 壁牛尾氏	七 奎女心亢
廿 室虛尾角	十九 室牛箕亢	十八 壁牛心氏	十七 奎女心角

補一

相亢婁 乘心虛 尅	相壁危 乘角箕 生	相女斗 乘室氏 尅	相尾房 乘牛奎 生
○	○	○	○
三 奎虛心亢	二 壁虛箕角	一 室女箕氏	四 奎牛尾氏
廿四 壁虛心亢	廿三 壁女箕角	廿二 室女尾氏	廿一 奎牛尾亢
廿八 室虛心亢	廿七 壁牛箕角	廿六 室女心氏	廿五 奎牛尾角

相女婁 乘心氏 生	相尾危 乘角奎 尅	相亢斗 乘室虛 生	相壁房 乘牛箕 尅
○	○	○	○
十三 奎女心氏	十六 奎虛尾角	十五 室虛箕亢	十四 壁牛箕氏
卅二 壁女心氏	卅一 奎女尾角	卅 室虛尾亢	廿九 壁牛箕亢
卅六 室女心氏	卅五 奎牛尾角	卅四 室虛心亢	卅三 壁牛箕角

房虛壁心相乘	斗奎亢牛相乘	危氐尾室相乘	婁箕女角相乘
生	尅	生	尅
○	○	○	○
卅三 壁虛心氐	卅四 奎牛箕亢	卅五 室虛尾氐	卅六 奎女箕角
廿四 壁虛心亢	廿一 奎牛尾亢	廿二 室女尾氐	廿三 壁女箕角
十 壁虛心角	十一 奎牛心亢	十二 室牛尾氐	九 室女箕角

房虛尾室相乘	斗奎女角相乘	危氐壁心相乘	婁箕亢牛相乘
尅	生	尅	生
○	○	○	○
卅五 室虛尾氐	卅六 奎女箕角	卅三 壁虛心氐	卅四 奎牛箕亢
卅 室虛尾亢	卅一 奎女尾角	卅二 壁女心氐	廿九 壁牛箕亢
廿 室虛尾角	十七 奎女心角	十八 壁牛心氐	十九 室牛箕亢

補二

換五式

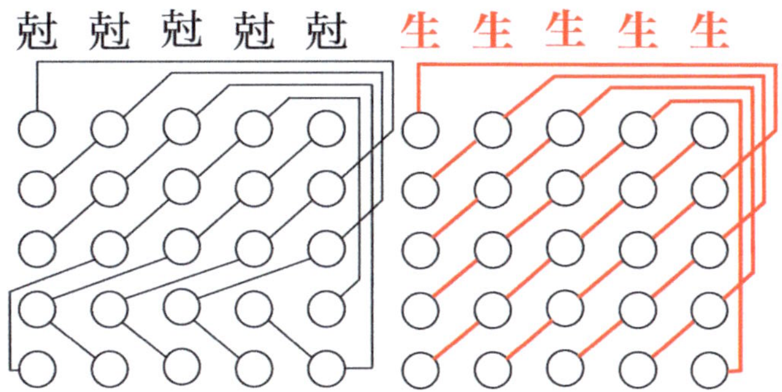

相 壬 乘 乙 丁
生
○
三 辛 丁 乙
六 庚 丁 乙

相 壬 乘 戊 甲
尅
○
二 辛 戊 甲
四 庚 戊 甲

假如

一式	房	氐	亢	角
二式	斗	箕	尾	心
三式	危	虛	女	牛
四式	婁	奎	壁	室

相 女 房 乘 室 箕	相 壁 房 乘 牛 箕
生	尅
○	○
一 室 女 箕 氐	十 壁 牛 箕 氐
四 室 女 箕 亢	十三 壁 牛 箕 亢
七 室 女 箕 角	十六 壁 牛 箕 角

改十四

相 壁 房 乘 心 虛	相 尾 房 乘 牛 奎	相 女 斗 乘 室 氐	相 壁 斗 乘 角 虛
生	生	尅	尅
○	○	○	○
二 壁 虛 心 氐	三 奎 牛 尾 氐	一 室 女 箕 氐	十九 壁 虛 箕 角
五 壁 虛 心 亢	六 奎 牛 尾 亢	廿一 室 女 尾 氐	廿二 壁 虛 尾 角
八 壁 虛 心 角	九 奎 牛 尾 角	廿三 室 女 心 氐	八 壁 虛 心 角

相 尾 房 乘 室 虛	相 女 房 乘 心 奎	相 壁 斗 乘 牛 氐	相 亢 斗 乘 室 虛
尅	尅	生	生
○	○	○	○
十一 室 虛 尾 氐	十二 奎 女 心 氐	十 壁 牛 箕 氐	廿五 室 虛 箕 亢
十四 室 虛 尾 亢	十五 奎 女 心 亢	廿七 壁 牛 尾 氐	十四 室 虛 尾 亢
十七 室 虛 尾 角	十八 奎 女 心 角	廿九 壁 牛 心 氐	卅 室 虛 心 亢

斗奎 亢牛 相乘	危氐 尾室 相乘	危箕 壁角 相乘	危奎 亢心 相乘
尅	生	生	生
○	○	○	○
廿 奎牛箕亢	十二 室虛尾氐	十九 壁虛箕角	卅二 奎虛心亢
六 奎牛尾亢	廿二 室女尾氐	卅二 壁女箕角	十五 奎女心亢
廿四 奎牛心亢	卅三 室牛尾氐	十六 壁牛箕角	廿四 奎牛心亢

斗奎 女角 相乘	危氐 壁心 相乘	危箕 亢室 相乘	危奎 尾角 相乘
生	尅	尅	尅
○	○	○	○
廿六 奎女箕角	二 壁虛心氐	廿五 室虛箕亢	卅四 奎虛尾角
廿八 奎女尾角	卅五 壁女心氐	四 室女箕亢	廿八 奎女尾角
十八 奎女心角	廿九 壁牛心氐	卅六 室牛箕亢	九 奎牛尾角

改十五

婁氐 尾牛 相乘	婁箕 女角 相乘	婁虛 亢心 相乘
尅	尅	尅
○	○	○
三 奎牛尾氐	廿六 奎女箕角	卅二 奎虛心亢
廿七 壁牛尾氐	卅二 壁女箕角	五 壁虛心亢
卅三 室牛尾氐	七 室女箕角	卅 室虛心亢

婁氐 女心 相乘	婁箕 亢牛 相乘	婁虛 尾角 相乘
生	生	生
○	○	○
十二 奎女心氐	廿 奎牛箕亢	卅四 奎虛尾角
卅五 壁女心氐	十三 壁牛箕亢	廿二 壁虛尾角
廿三 室女心氐	卅六 室牛箕亢	十七 室虛尾角

右各逐式交乘而得生尅也雖然相乘之數位繁多而不易見故以交式斜乘代之

Notes on Collation of Methods of Solving Concealed Problems

The following are the call numbers of manuscripts employed in the collation:

MC: Hayashi Collection [林集書] 648 / Matsunaga Library [松永文庫] 2490 in the Main Library of Tōhoku University = Matsunaga Book [松永本] in [2, p. 225];

NC: Hayashi Collection [林集書] 2314 in the Main Library of Tōhoku University, having the seal Nakamura [中村氏] = Nakamura Book [中村本] in [2, p. 225].

SC: Kuwaki Library, Mathematics (2) [桑木文庫算学（二）] Manuscript [写本] 679 Trilogy [三題集 Sandai Shū] in the Main Library of Kyūshū University.

2v. l. 10: A majority of the manuscripts writes 上下方 but MC and NC adopt 上下差.

3v. l. 6: 位 is 後 in MC.

6r. l. 7: 六 and 七 are written 七 and 八, respectively, in MC, and then corrected into 六 and 七.

6r. l. 8: 九 and 十二 are written 十 and 十三, respectively and then corrected into 九 and 十二 in MC.

10v. l. 10: The punctuation mark ○ is missing in MC.

11r. l. 6: 加=式 is 加-式 in NC.

12r. l. 5: The sign of the term 子^{3}庚 is minus in SC.

17r. l. 2: 生 is mistaken to be 正 in KC, and NC.

19r. l. 2: 直 is written 而 in NC.

References

1. Takefumi Gotō [後藤武史] – Hikosaburō Komatsu [小松彦三郎]: A correction of Seki's error in the expansion of determinants, *J. Northwest Univ., Natural Science Edition*, vol. 33, pp. 376–380 (2003); these proceedings, pp. 559–565.
2. Hikosaburō Komatsu [小松彦三郎]: The restoration of the Yamaji Nushizumi edition of "Methods of Solving Concealed Problems" and its comparison with the Collected Works of Seki Takakazu [『解伏題之法』山路主住本の復元と「關孝和全集」との比較], *Researches on History of Mathematics* [数学史の研究], *Colloquium of Research Institute for Mathematical Sciences* [数理解析研究所講究録], vol. 1392, pp. 225–245 (2004).

Notes on Complete Book of Mathematics Vol. 10: Geometry

Hikosaburo Komatsu*

Abstract The Complete Book of Mathematics is the most comprehensive treatise of mathematics in the Edo Period of Japan. The 20 volume book is about 900 sheets or 1800 pages long. Seki Takakazu (1642?–1708), Takebe Kataakira (1661–1721) and Takebe Katahiro (1664–1739) spent 28 years (1683–1711) in writing it. Unfortunately, the Book has never been published as a whole. Except for Volume 4 we reproduce here only one volume, Volume 10 Geometry, which is the first volume on geometry in the treatise. They develop here the basics of the plain geometry as algebraic relations among line segments which specify the geometric objects in question. This is exactly the same standpoint as René Descartes' (1596–1650) in his *Géométrie* (1637). At the end of the volume they discuss algebraic relations among the sides and diagonals in a general pentagon and hexagon by making use of Seki's theory of resultants.

1 Introduction

The Complete Book of Mathematics [大成算経 Taisei Sankei] is the most comprehensive treatise of mathematics in the Edo Period (1603–1868). Seki Takakazu [關孝和] (1642?–1708) and his pupils Takebe Kataakira [建部賢明] (1661–1721) and Takebe Katahiro [建部賢弘] (1664–1739) spent 28 years from Tenna 3 [天和 3] (1683) to Hōei 8 [宝永 8] (1711) until they completed it. The most important results Seki expounded in his Trilogy [三部抄] and Septenary [七部書] are all included in

Hikosaburo Komatsu
Graduate School of Mathematical Sciences, the University of Tokyo,
3-8-1 Komaba, Meguro-ku, Tokyo 153-9814, Japan
e-mail: komatsu@ms.u-tokyo.ac.jp

* Supported by JSPS Grant-in-Aid Scientific Research (C) 16540119 (2004–2006), MEXT Grant-in-Aid on Priority Areas 17083006 (2005–2010) and JSPS Grant-in-Aid Scientific Research (C) 20540107 (2008–2011).

the Book. The known dates of volumes in Trilogy and Septenary are concentrated at the period when Seki and his pupils started to write the Book. That will probably mean that those which are in Trilogy or Septenary are the notes Seki prepared as sketches for the Book. It consists of 20 volumes and each volume about 45 sheets or 90 pages. Unfortunately the Book has never been published as a whole. The only one exception is Volume 13 Measurements [求積], which is included in *Collected Works* [2] as a volume in Septenary. There remain more than twenty copies as manuscripts (see Komatsu [9]), but most of them are not of good quality.

2 Outlines of Volume 10

In Volume 4 Three Essentials [三要 san'yō] the authors classified the objects in mathematics into abstract Images [象 shō] and visible Appearances [形 kei]. This Volume 10 is named Algorithms for Appearances [形法 keihō] as the first of five volumes on geometry. It consists of four chapters. Chapter 1 Algorithms for Squares [方法 houhou] deals with squares. The main result is the fact that the diagonal [斜 sha] of a square [方 hō] is $\sqrt{2}$ times a side [方面 hōmen].

Chapter 2 is entitled Algorithms for Rectangles [直法 chokuhō]. The longer side of a rectangle [直 choku] is called [縦 tate] or length [長 chō] and the shorter side [横 yoko] or [平 hei] or width [闊 katsu]. Their product is the area [積 seki]. Given the product ab and the sum $a+b$ or the difference $a-b$, the difference $a-b$ or the sum $a+b$ is obtained as the square root of $(a\pm b)^2 - \pm 4ab$. In the following Chapter 3 Algorithms for Right Triangles [勾股法 kōko hō] this is used to prove the Pythagorean theorem as in the Chinese classic Mathematics of Zhou Gnomons [周髀算經 Zhoubi Suanjing]. The shorter leg hook [勾 kō] and the longer leg leg [股 ko] of a right triangle stand for the right triangle itself. The hypothenuse is called chord [弦 gen]. Seki had given another proof of the Pythagorean theorem in Methods of Solving Explicit Problems [解見題之法 Kai Kendai no Hō], which is also reproduced. Then, many problems are solved by quadratic equations.

There are no formal treatments of proportions as in Euclid's Elements but it is remarked that rectangles or right triangles with a fixed ratio of heights to widths make a straight line. They say that this fact is very useful and also that a commentary writes that this is the most valuable in arithmetic [伝曰算術之極致也].

There are two appendices to Chapter 3. The first one deals with Pythagorean triplets, i.e. the integral solutions of

$$x^2 + y^2 = z^2.$$

However, their definition of integral numbers [整数 seisū] is different from ours. In Volume 4, they classified numbers into four categories: A number is said to be entire [全 zen] or integral [整 sei] if it is represented by a finite number of counting rods, i.e. if it is a finite decimal fraction; duplex [繁 han] if it is the quotient of two entire numbers or if it is rational; multiplex [畸 ki] if it is a root of an algebraic equation represented by a finite number of counting rods or if it is algebraic; and rudimentary [零 rei] otherwise or if it is transcendental (Xu [4], Komatsu [10]).

Firstly it is shown that Pythagorean triplets with the chord $z = 1$ are obtained by a computation equivalent to the powers of complex numbers

$$\pm x + \pm iy \text{ or } \pm y + \pm ix = (0.6 + 0.8i)^n, \quad n = 1, 2, 3, \ldots.$$

There are no others, as is easily proved by the unique factorization theorem of the Gauss integers $\mathbf{Z} + i\mathbf{Z}$.

Another way is to find a rational solution x of $x^2 + n^2 = (x+m)^2$ for each pair (m,n) of integers with $0 < m < n$. Clearly any integral triple in our sense can be found in this way. Moreover, the famous formula $(n^2 - m^2, 2nm, n^2 + m^2)$ of Pythagorean triplets is obtained by multiplying the above solution by $2m$.

The second appendix is a brief introduction to the methods of survey described in Sea Islands Mathematics [海島算經 Haidao Suanjing] by Liu Hui. The last problem is taken from Nine Chapters of Mathematics [九章算術 Shizhang Suanshu].

In the last Chapter 4 Algorithms for Polygons [斜法] the authors develop a general theory of triangles [三斜 sansha], quadrilaterals [四斜 shisha], pentagons [五斜 gosha] and hexagons [六斜]. After the preparations of the cosine law and the Heron formula for triangles, they proceed to establish Algorithm for Quadrilaterals [四斜法 shishahō] which is the algebraic relation among the four sides and two diagonals of a general quadrilateral. The equation with 22 terms of total degree 8 and of degree 4 in each variable is derived from Pythagorean theorem. This became famous after Seki used it in his book Mathematical Methods that Clarify Subtleties [発微算法 Hatsubi Sanpō] (1674) in his solutions of Problems posed by Sawaguchi Kazuyuki [沢口一之] in 1670 (see Komatsu [12]).

To obtain similar results for a pentagon and a hexagon they employ Seki's theory of elimination [5]. In the case of a pentagon, it is decomposed into the sum of two quadrilaterals with a common side or diagonal. Then, they eliminate the common variable from two equations corresponding to the decomposed quadrilaterals. Since each term of the Algorithm for Quadrilaterals depends only on the squares of sides and diagonals, the elimination is actually done for two quadratic equations in the variable to be eliminated. This is the case already discussed by Zhu Shijie [朱世傑] in his Jadelike Examples of Four Unknowns [四元玉鑑 Siyuan Yujian] (1303) (see Hoe [3, pp. 133–134] and [11, p. 104]).

The principle is simple but the actual calculation of the resultant was hard. In his book Weaving Methods in Mathematics [綴術算經 Tetsujutu Sankei] Takebe Katahiro recalled of his brother Kataakira and wrote "He once tried to deparenthesize the algorithm for pentagon, which was complicated, and said that even if the number of terms was ten thousand, one could calculate it in one hundred days by calculating one hundred terms every day. He really did it in a month and a few days." A rough estimate of the number of terms of an intermediate expansion is about 4000, and we obtain an equation of degree 8 with 843 monomial terms as its complete expansion, which is not given in the text but a few main terms are calculated in order to get a formula of the algorithm for a general hexagon.

Similarly a hexagon is decomposed as the sum of a quadrilateral and a pentagon with a common segment. Thus, the algorithm for a hexagon is obtained as the resultant of a quadrilateral algorithm and a pentagonal algorithm, which is represented as a 4×4 determinant with complicated entries, is equal to 0. Its complete expansion has 273,123 terms by Kinji Kimura's calculation by a computer.

References

1. Matsusaburō Fujiwara [藤原松三郎], Japan Academy ed.: *History of Mathematics in Japan before Meiji* [明治前日本数学史], vol. 2, Iwanami [岩波書店], Tokyo (1956).
2. A. Hirayama [平山諦], K. Shimodaira [下平和夫] and H. Hirose [広瀬秀雄]: *Takakazu Seki's Collected Works Edited with Explanations* [關孝和全集], Osaka Kyōiku Tosho [大阪教育図書], Osaka (1974).
3. John Hoe: *Les Systèmes d'Équations Polynômes dans le Siyuan Yujian (1303)*, Collège de France, Institut des Hautes Études Chinois (1977).
4. Xu Zelin [徐澤林]: Takebe Katahiro's epistemology of mathematics, —A discussion on "Three Essentials" in Complete Book of Mathematics, [建部賢弘的数学認識論 —論『大成算経』中的 "三要"], *Studies in the History of Natural Sciences* [自然科学史研究], vol. 21, pp. 232–243 (2002).
5. Takefumi Goto [後藤武史] and Hikosaburo Komatsu [小松彦三郎]: Determinants, resultants and discriminants in Japan in the seventeenth century and in Europe in the eighteenth and nineteenth centuries, *J. Northwest Univ., Natural Science Edition*, vol. 33, pp. 363–367 (2003); these Proceedings, pp. 553–559.
6. Hikosaburo Komatsu: Zhu Shijie, the teacher of Seki and Takebe, *Book of Abstracts, The 22nd Congress of History of Science, Beijing, July 24–30*, p. 290 (2005).
7. Hikosaburo Komatsu: Takebe Kataakiira's deparethesization in the problem of pentagon, *Program, Abstract and List of Participants, The Sixth International Symposium on the History of Mathematics & Mathematical Education using Chinese Characters (ISHME), Tokyo, August 4–7*, p. 29 (2005); Hikosaburō Komatsu [小松彦三郎]: What is Takebe Kataakira's work "Deparethesization in the Problem of Pentagons? [建部賢明の業績『五斜の括術』は何なのか], *Abstracts, Annual Meeting of the Mathematical Society of Japan, Chūō Univ., March 26–30, Section Foundations of Mathematics and History of Mathematics*, pp. 14–15 (2006).
8. Hikosaburō Komatsu [小松彦三郎]: *History of Mathematics and, in particular, the early Wasan and Analysis in the 19th Century* [数学史の研究、特に初期の和算と 19 世紀の解析学について], The Report of the Researches Supported by JSPS Grant-in-Aid Scientific Research (C) 16540119 (2004–2006) [平成 16 年度 ∼ 平成 17 年度科学研究費補助金 (基盤研究 (C)) 研究成果報告書], Tokyo University of Science [東京理科大学] (2006); reserved in Kansai-kan of the National Diet Library [国会図書館関西館].
9. Hikosaburō Komatsu: A report on the status quo of the project of making collated text of the Complete Book of Mathematics [「大成算經」校訂本作成の現状報告], *Researches on History of Mathematics* [数学史の研究], *Colloquium of Research Institute for Mathematical Sciences* [数理解析研究所講究録], vol. 1444, pp. 140–158 (2007).
10. Hikosaburō Komatsu [小松彦三郎]: On Volume 4 Three Essentials in Complete Book of Mathematics [「大成算経」巻之四「三要」について], *Abstracts, Annual Meeting of the Mathematical Society of Japan, Saitama, March 27–30, Section Foundations of Mathematics and History of Mathematics*, pp. 20–21 (2007).
11. Jock Hoe: *The Jade Mirror of the Four Unknowns*, Mingming Bookroom, P.O. Box 29-316, Christchurch, New Zeeland (2007).
12. Hikosaburō Komatsu: Algebra, elimination and Complete Book of Mathematics, *these Proceedings*, pp. 245–274.

Complete Book of Mathematics Vol. 10: Geometry, by Seki Takakazu, Takebe Kataakira and Takebe Katahiro, collated by Hikosaburo Komatsu*

Seki Takakazu, Takebe Kataakira and Takebe Katahiro

Abstract Volume 10 is named Algorithms for Appearances [形法 keihō] as the first of five volumes on geometry. It consists of four chapters. Chapter 1 Algorithms for Squares [方法 hōhō] deals with squares. Chapter 2 is entitled Algorithms for Rectangles [直法 chokuhō]. The longer side of a rectangle [直 choku] is called [縦 tate] or length [長 chō] and the shorter side [横 yoko] or [平 hei] or width [闊 katsu]. Their product is the area [積 seki]. In the following Chapter 3 Algorithms for Right Triangles [勾股法 kōko hō] this is used to prove the Pythagorean theorem. The shorter leg hook [勾 kō] and the longer leg leg [股 ko] of a right triangle stand for the right triangle itself. The hypothenuse is called chord [弦 gen]. There are no formal treatments of proportions as in Euclid's Elements but it is remarked that rectangles or right triangle with a fixed ratio of heights to widths make a straight line. There are two appendices to Chapter 3. The first one deals with Pythagorean triplets, i. e. the integral solutions of $x^2+y^2=z^2$. The second appendix is a brief introduction to the methods of survey described in Sea Islands Mathematics [海島算經 Haidao Suanjing] by Liu Hui. In the last Chapter 4 Algorothms for Polygons [斜法 shahō] the authors develop a general theory of triangles [三斜 sansha], quadrilaterals [四斜 shisha], pentagons [五斜] and hexagons [六斜] with the use of their elimination theory.

* Supported by JSPS Grant-in-Aid Scientific Research (C) 16540119 (2004–2006), MEXT Grant-in-Aid on Priority Areas 17083006 (2005–2010) and JSPS Grant-in-Aid Scientific Research (C) 20540107 (2008–2011).

大成算經 卷之十 形法

卷之十 中集 形法

關孝和
建部賢明 編
建部賢弘

二〇〇八年十二月十一日 小松彦三郎校

大成算經卷之十　中集

形法

夫形之成也以方爲肇由其長短成直及勾股此二者其名雖異其理相同也自是之後成斜從斜成角是方之五法也然從角成圓從圓成弧及立圓球缺是圓之四率也此九者爲諸形之要也其形質之變雖無極皆本于此法而窮變理也是以別方圓之篇而解之亦立求積法與五巧術爲奇形之標準矣

方法第一

方者諸形之首也其縱横相等故曰面四面相并者曰圍自隅至隅者曰斜其求之法得四面同數之積而後以開方得其全數若含而用者卽命之曰冪諸

一

形求斜正廣狹者悉用此法也

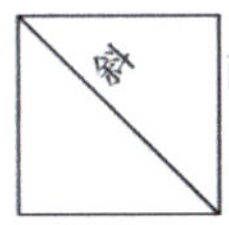

假如有平方面各一尺問斜

答曰斜一尺四寸一分四釐二毫一絲三五六強

術曰置面一尺自乘得方積一百寸倍之得斜冪二百寸爲實以一爲廉法開平方除之得方斜

解曰方爲縱亦爲横方縱自乘是乃方積也以右一半積反湊下爲黑圖積方巾横自乘如前以一半積湊下爲朱圖積方巾二積相并得四面同數之積方巾是故以方冪乃方積也二段爲斜冪也或以

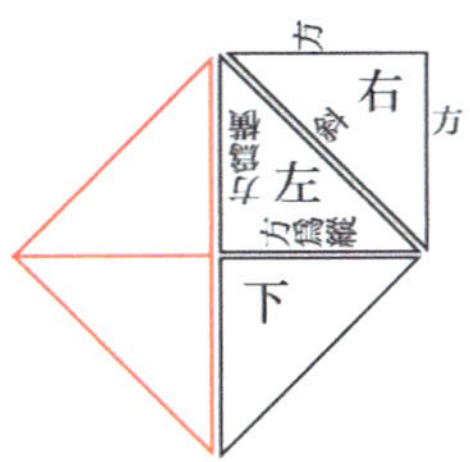

積求斜及面或以斜求面及積者皆本于此法也

直法第二

直者謂方有長短者也縱曰長横曰平曰闊自隅至隅者曰斜乃由縱横不等有長短之報也報者應準之號不論形之斜正而契符稱理之故俗謂之相應也蓋數之多少狀之大小悉承舊而求之故爲象形通用之法其所爲或除之或含而用之凡解諸象者皆假此形而證乘除之理是以古有致諸用之稱矣

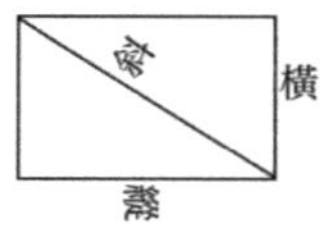

假如有直縱一尺五寸横八寸問斜

答曰斜一尺七寸

法曰縱一尺五寸自乘得二百二十五寸横八寸自乘得六十四寸二位相并得斜冪二百八十九寸爲實以一爲廉法開平方除之得斜

二

解術及演段圖具于勾股篇中

假如有直積四十寸縱横和一尺四寸問縱横差

答曰縱横差六寸

法曰縱横和一尺四寸自乘得一百九十六寸內減四因直積一百六十寸餘得縱横差冪三十六寸爲實以一爲廉法開平方除之得縱横差

假如有直積四十寸縱横差六寸問縱横和

答曰縱横和一尺四寸

法曰縱横差六寸自乘得三十六寸加入四因直積一百六十寸得縱横和冪一百九十六寸爲實以一爲廉法開平方除之得縱横和

兩術解曰縱横和自乘則直積四段與差冪一段相幷數也是故以四因直積減縱横和冪者卽得差冪加縱横差冪者卽得和冪也

假如有直積八十四寸縱一尺二寸問横

答曰横七寸

法曰置積八十四寸爲實以縱一尺二寸爲法實如法而一得横若含縱而用者以積卽命爲因縱横

三

也

假如有直積八十四寸横七寸問縱

答曰縱一尺二寸

法曰置積八十四寸爲實以横七寸爲法實如法而一得縱若含横而用者以積卽命爲因横縱也

兩術解曰直積者原縱横相乘數故以縱除之則得横以横除之則得縱也

假如有直縱六寸横三寸問報縱報横

答曰報縱二寸　報横五分

法曰置縱六寸爲實以横三寸爲法除之得報縱若不除則含横而用之卽命以縱爲因横報縱

亦置横三寸爲實以縱六寸爲法除之得報横若不
除則含縱而用之以横即命爲因縱報横也
兩術解曰報縱者屬横一寸
之縱也報横者屬縱一寸之
横也凡求形之大小數之多
寡者皆據此法而明其理也
於乘除之先後宜隨時而通變用之矣

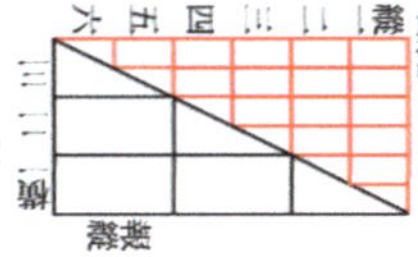

右直題七問相通于勾股法用之也

勾股法第三 附重差

勾股者謂半直也横曰勾縱曰股斜曰弦自縱横交
取縄直而其闊曰中股在左者曰小勾在右者曰小
股以勾除股者曰股報以股除勾者曰勾報俗通曰勾配

四

統直法而甚有用矣傳曰算術之極致也蓋諸形取
縄直定平正或容方圓曲直於內或量高深廣遠於
外者皆起于此法是以旁爲通變之妙法誠哉此言
也

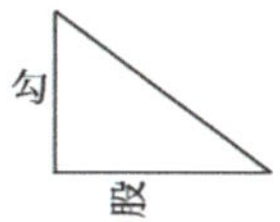

假如有勾股勾三寸股四寸問弦
答曰弦五寸
法曰勾三寸自乘得九寸股四寸自乘得一十六寸二
位相并得弦冪二十五寸爲實以一爲廉法開平方除
之得弦

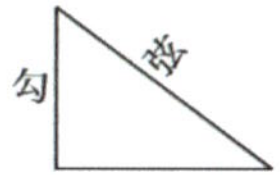

假如有勾股勾三寸弦五寸問股
答曰股四寸
法曰弦五寸自乘得二十五寸內減勾自乘九寸餘

得股冪一十六寸爲實以一爲廉法開平方除之得股
假如有勾股股四寸弦五寸問勾

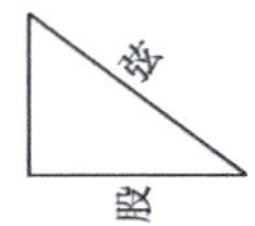

答曰勾三寸
法曰弦自乘得二十五寸內減股自乘一十六寸餘
得勾冪九寸爲實以一爲廉法開平方除之得勾
三法解曰勾股各自乘
數以右角甲積湊左上角
以上左角乙積湊左下角
以下左角丙積湊上右角

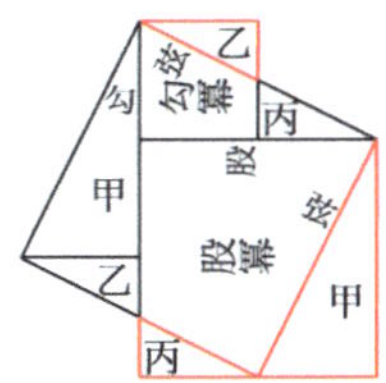

而得四面同數之方積是乃弦冪也三題互據
此圖契符而宜曉加減之理矣
假如有勾股勾八寸股弦和三尺二寸問股弦

五

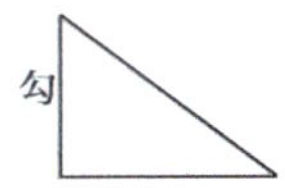

答曰股一尺五寸　弦一尺七寸
法曰先求股者股弦和三尺二寸自乘得一千〇二十四寸內減勾冪六十四寸餘九百六十寸爲實以倍
股弦和六尺四寸爲法實如法而一得股　先求弦者
股弦和自乘加入勾冪共得一千〇八十八寸爲實以倍
股弦和爲法實如法而一得弦
解曰先求股者立天元一爲股。|以減和餘
爲弦 和| |自之得內減股冪餘爲勾冪 和巾| 和‖ 與
勾冪相 和巾| 和‖
消得式 勾巾|
先求弦者立天元一爲弦。|以減和餘爲股
和| |自之得數以減弦冪餘爲勾冪 和巾| 和‖ 與勾

冪相消 |和巾 ‖和

得式 |勾巾

假如有勾股股一尺五寸勾弦和二尺五

寸問勾弦

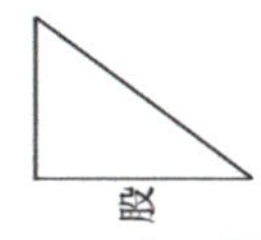

答曰勾八寸　弦一尺七寸

法曰先求勾者勾弦和二尺五寸自乘得六百二十五寸內減

股冪二百二十五寸餘四百寸爲實以倍勾弦和五尺爲法實

如法而一得勾　先求弦者勾弦和自乘加入股

冪共得八百五十寸爲實以倍勾弦和爲法實如法而

一得弦

解術與前同

假如有勾股勾五寸股弦差一寸問股弦

六

答曰股一尺二寸　弦一尺三寸

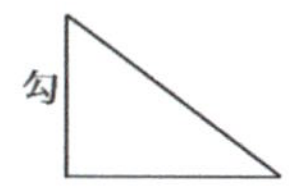

法曰先求股者勾五寸自乘得二十五寸內減股

弦差冪一寸餘二十四寸爲實以倍股弦差二寸爲

法實如法而一得股　先求弦者勾自乘加入股

弦差冪共得二十六寸爲實以倍股弦差二寸爲法實如

法而一得弦

解曰先求股者立天元一爲股。|加差爲弦

|差 |自之得內減股冪餘爲勾冪 |差巾 ‖差 與勾冪

相消 |差巾 ‖差

得式 |勾巾

先求弦者立天元一爲弦。|內減差餘爲股

|差 |自之以減弦冪餘爲勾冪 |差巾 ‖差 與勾冪相

消得 |差巾 ||差
式 |勾巾

假如有勾股股一尺二寸勾弦差八寸問
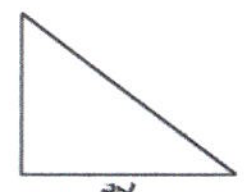

勾弦
答曰勾五寸　弦一尺三寸
法曰先求勾者股一尺二寸自乘得一百四十四寸內減勾弦
差冪六十四寸餘八十寸爲實以倍勾弦差一尺六寸爲法實
如法而一得勾　先求弦者股自乘加入勾弦差
冪共得二百八寸。爲實以倍勾弦差爲法實如法而
一得弦
解術與前同
假如有勾股勾弦和一尺八寸股弦和二尺五寸

七

問勾股弦和
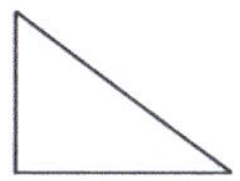
答曰勾股弦和三尺
法曰勾弦和一尺八寸以股弦和二尺五寸相乘倍
之得九百寸爲實以一爲廉法開平方除之得勾股
弦和
解曰立天元一爲勾股弦和。|內減勾弦和
餘爲股|勾弦和|自之爲股冪|勾弦和巾 ||勾弦和|寄左　列勾
股弦和內減股弦和餘爲勾|股弦和|自之加入寄
左爲|勾弦和巾 ||勾弦和||再寄　列勾弦和加入股弦|勾弦和|
弦冪|股弦和巾 ||股弦和　和內減勾股弦和餘爲弦　|股弦和
自之與再寄相消得式||勾弦和 股弦和。|
假如有勾股勾弦差九寸股弦差二寸問勾股和

與弦差
答曰勾股與弦差六寸
法曰勾弦差九寸以股弦差二寸相乘倍之得
三十六寸爲實以一爲廉法開平方除之得勾股與弦
差

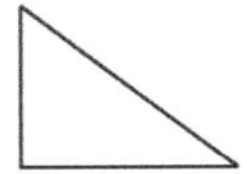

解術與前同

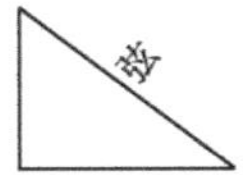

假如有勾股弦二尺五寸勾股和三尺一
寸問勾股差
答曰勾股差一尺七寸
法曰弦二尺五寸自乘倍之得一千二百五十寸内減勾股和
冪九百六十一寸餘二百八十九寸爲實以一爲廉法開平方除
之得勾股差

八

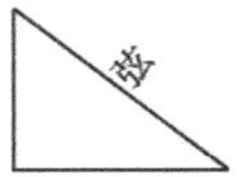

假如有勾股弦二尺五寸勾股差一尺七
寸問勾股和
答曰勾股和三尺一寸
法曰弦自乘得六百二十五寸倍之得内減勾股差冪二百
八十九寸餘一百六十一寸爲實以一爲廉法開平方除之得
勾股和

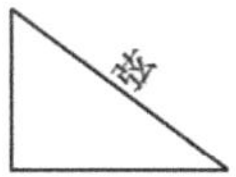

假如有勾股積二百四十寸弦三尺四寸
問勾股和及差
答曰勾股和四尺六寸　差一尺四寸
法曰求勾股和者置積二百四十寸四之得九百六十寸加
入弦冪一千一百五十六寸共得二千一百一十六寸爲實以一爲廉
法開平方除之得勾股和　求勾股差者置積四

之得數以減弦冪餘一百六十九寸爲實以一爲廉法開
平方除之得勾股差

假如有勾股積五十四寸勾股和二尺一
寸問弦及勾股差
答曰弦一尺五寸　勾股差三寸
法曰求弦者置積五十四寸四之得二百一十六寸以減勾股
和冪四百四十一寸餘二百二十五寸爲實以一爲廉法開平方
除之得弦　求勾股差者置積八之得四百三十二寸以
減勾股和冪餘九寸爲實以一爲廉法開平方除之
得勾股差

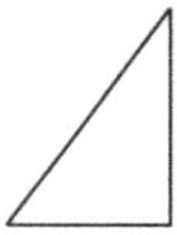

假如有勾股積五十四寸勾股差三寸問
弦及勾股和

九

答曰弦一尺五寸　勾股和二尺一寸
法曰求弦者置積五十四寸四之得二百一十六寸加入勾股
差冪九寸共得二百二十五寸爲實以一爲廉法開平方除
之得弦　求勾股和者置積八之得四百三十二寸加入
勾股差冪共得四百四十一寸爲實以一爲廉法開平方
除之得勾股和

四法解曰弦冪者四段積
與一段差冪相并數也勾
股和冪者八段積與一段
差冪相并數也是故隨題
旨察加減損益之理得所
求數也

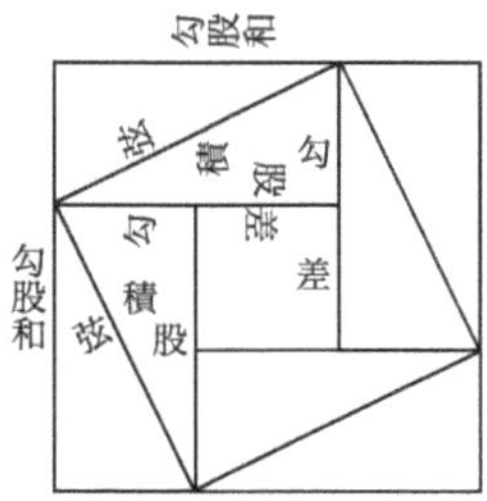

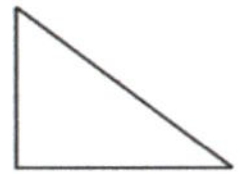

假如有勾股勾八寸股一尺問勾股報
答曰勾報一寸二五　股報八分
法曰以勾八寸除股一尺得勾報以股一尺除勾八寸得股報若不除而用者卽以股爲因勾勾報以勾爲因股股報也

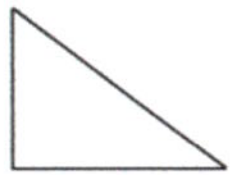

假如有勾股勾三寸弦六寸問勾弦報
答曰勾報二寸　弦報五分
法曰以勾三寸除弦六寸得勾報二寸以弦六寸除勾三寸得弦報若不除而用者卽以弦爲因勾勾報以勾爲因弦弦報

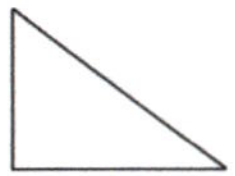

假如有勾股股八寸弦一尺問股弦報
答曰股報一寸二五　弦報八分

十

法曰以股八寸除弦一尺得股報以弦一尺除股八寸得弦報若不除而用者卽以弦爲因股股報以股爲因弦弦報
解曰三題所問皆屬法一箇之數或除之或命之者各隨法術之所施而用之也

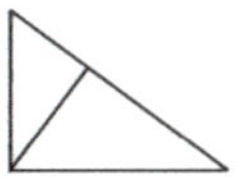

假如有勾股勾一尺五寸股二尺問中股
答曰中股一尺二寸
法曰勾一尺五寸以股二尺相乘得二段積三百寸爲實別求弦二尺五寸爲法實如法而一得中股
解曰以二段積視斜直積弦擬縱中股擬横故以弦除之則中股也或中股擬小勾股擬小弦據其報

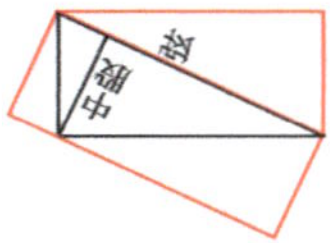

以小弦即大股乘大勾則爲因大弦小勾即中股也故
以大弦除倍積之理亦同

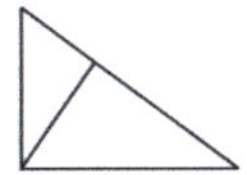

假如有勾股勾一尺五寸股二尺問小勾
股
答曰小勾九寸　小股一尺六寸
法曰求小勾者勾一尺五寸自乘得二百二十五寸爲實別求
弦二尺五寸爲法實如法而一得小勾　求小股者股
二尺自乘得四百寸爲實以弦爲法實如法而一得小
股
解曰求小勾者據勾弦報
察其理則大勾擬小弦以
大勾相乘爲因大弦小勾

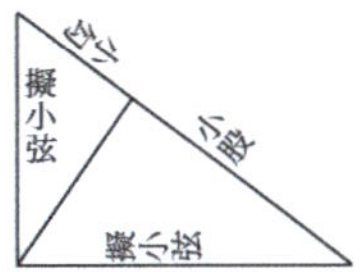

十一

故以大弦除大勾冪即小勾也求小股者據股
弦報察其理則大股擬小弦以大股相乘爲因
大弦小股故以大弦除大股冪即小股也

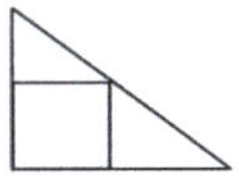

假如有勾股內容方勾二尺一寸股二尺
八寸問方面
答曰方面一尺二寸
法曰勾二尺一寸以股二尺八寸相乘得五百八十八寸爲實以勾
股和爲法實如法而一得方面
解曰以方斜爲界則其左者勾
與方相乘半段積右者股與方
相乘半段積兩數相幷倍之則
勾股相乘者即勾股和與方相

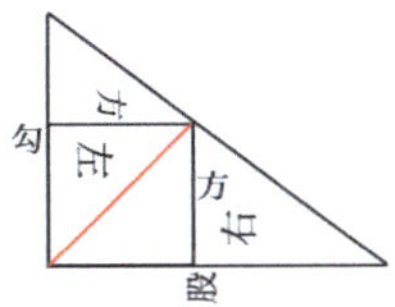

乘數也故以勾股和除倍積則方面也
假如有勾股內容圓勾八寸股一尺五寸
問圓徑

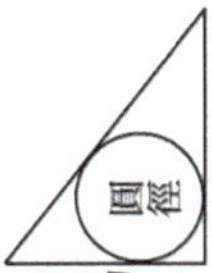

答曰圓徑六寸
法曰別得弦一尺七寸勾八寸以股一尺五寸相乘倍之得二百四十寸
爲實以勾股弦和四尺爲法實如法而一得圓徑
又置勾股和二尺三寸內減弦一尺七寸餘亦圓徑也
解曰自圓中心至三稍而界
之則上積者弦與半徑相乘
半段數左積者勾與半徑相
乘半段數下積者股與半徑
相乘半段數也三積相并四

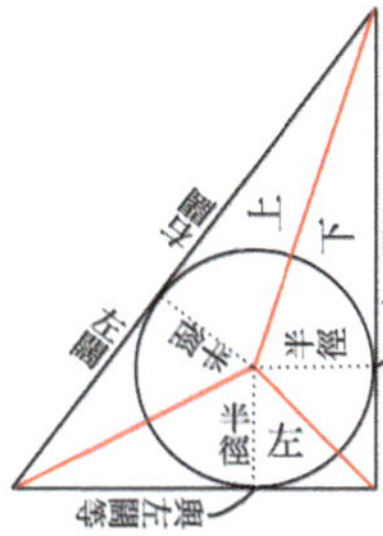

十二

之則乃勾股弦和與圓徑相乘數也故以勾股
弦和除四段積即圓徑也亦以半徑減股餘與
右闕等以半徑減勾餘與左闕等兩闕相并則
乃弦故勾股和內減弦者二箇半圓徑也其餘
勾股之法式最多且據圓徑則雖其變無極難
以一一盡述須隨時而窮之矣
假如有勾股內容直勾一尺二寸股一尺
六寸縱橫差九寸問縱橫

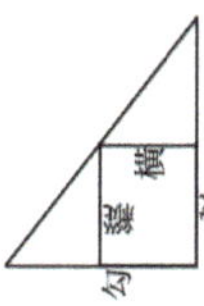

答曰橫三寸　縱一尺二寸
法曰先求橫者勾一尺二寸與股一尺六寸相乘得一百九十二寸
寄位列勾以縱橫差九寸相乘得一百八寸。以減寄位
餘八十四寸爲實以勾股和二尺八寸爲法實如法而一得

橫　先求縱者勾股相乘寄位列股以縱橫差相
乘得一百四十四寸加入寄位共得三百三十六寸爲實以勾股
和爲法實如法而一得縱若倒容者其形長短相變故以勾差相乘數加
倍積以股差相乘數減倍積各以勾股和除之少者爲橫多者爲縱也
解曰自直斜界之則右者股與
橫相乘數左者勾與縱相乘數
也求橫者以勾乘縱橫差爲朱
圖積以之減倍積餘卽因勾股
和橫也　求縱者以股乘縱橫
差爲朱圖積加倍積卽因勾股
和縱也

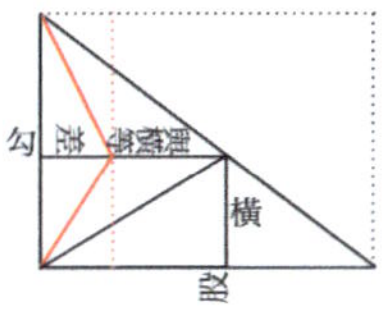

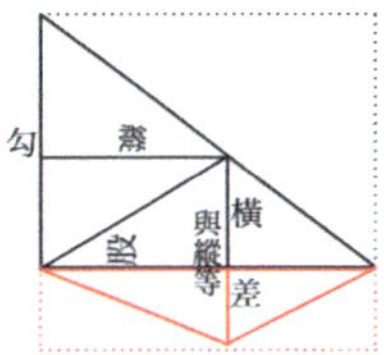

十三

求勾股弦整數

求勾股弦整數有二法矣定弦一尺者以第一勾股
報求之先起於勾一寸遞增一寸而求之則至勾六
寸得股八寸而整故以之爲第一勾股各以弦一尺
約之得勾股報乃以勾報乘勾以股報乘股其尾數
等者相幷異者相減爲勾又以勾報乘股以股報乘
勾其尾數等者相幷異者相減爲股卽得第二勾股
或有勾長股短者當相反命之逐如此求之　定勾或股者以股勾弦
差求之各置所定勾股冪內減股勾弦差冪餘以倍差除
之得數若定勾者得商少則以之爲勾以所定勾變爲股定股者得商多則以之爲股以所定股
變爲勾也不滿法者皆得等數約之得股勾加差得各弦若
欲得全數者依通分法求之也

勾

寸	二	四	四	一	七	一	四	四	二	六	○	五	三	二	六
分	二	○	八	三	○	五	七	二	○	五	七	三	五	八	
釐	五	三	九	一	○	一	二	一	六	八	五	七	二		
毫	七	八	五	五	○	二	一	九	三	九	八	六			
絲	七	八	一	八	五	四	○	七	八	四	四				
忽	五	七	四	五	一	三	三	二	七	四					
微	一	五	三	六	三	一	四	四	二						
纖	六	三	九	○	七	六	二	八							
沙	二	二	八	九	四	一	四								
塵	五	二	五	七	○	六									
埃	八	七	一	二	八										
	九	七	五	八											
	一	七	二												
	八	六													
	四														

十四

股

寸	九	九	八	九	七	九	八	九	九	七	九	八	九	九	八
分	七	一	七	九	一	八	八	○	七	五	九	四	三	六	
釐	四	四	一	一	四	八	一	六	八	二	七	三	六		
毫	一	八	九	三	○	四	五	六	四	一	一	二			
絲	七	○	九	○	九	九	四	○	七	九	二				
忽	九	八	五	四	二	六	三	八	○	二					
微	四	六	二	八	四	五	一	六	四						
纖	三	四	一	一	八	八	六	四							
沙	七	七	四	○	二	八	八								
塵	二	三	二	四	五	八									
埃	四	五	三	九	六										
	八	二	三	六											
	五	三	六												
	一	二													
	二														

法曰置弦一尺自乘得一百寸爲弦冪先視勾一寸自乘
得一寸以減弦冪餘九十九寸平方開之有不盡故不用
之　次視勾二寸自乘得四寸以減弦冪餘九十六寸平方
開之有不盡故不用之　次視勾三寸自乘得九寸以
減弦冪餘九十一寸平方開之有不盡故不用之　次
視勾四寸自乘得一十六寸以減弦冪餘八十四寸平方開之
有不盡故不用之　次視勾五寸自乘得二十五寸以減
弦冪餘七十五寸平方開之有不盡故不用之　次視
勾六寸自乘得三十六寸以減弦冪餘六十四寸平方開之得
股八寸適無不盡故以之爲第一勾股以弦一尺約之
得勾報六分股報八分
以勾報六分乘勾六寸得三寸六分以股報八分乘股八寸得六寸

十五

四分二數相減餘二寸八分爲
第二勾亦以勾報六分乘
股八寸得四寸八分以股報八分
乘勾六寸得四寸八分二數相
幷共得九寸六分爲第二股
以勾報乘第二勾二寸八分得一寸六分八鼇以股報乘第二
股九寸六分得七寸六分八鼇二數相幷得第三股亦以勾報
乘第二股得五寸七分六鼇以
股報乘第二勾得二寸二分
四鼇二數相減餘三寸五分二鼇
爲第三勾也次第如此
求之

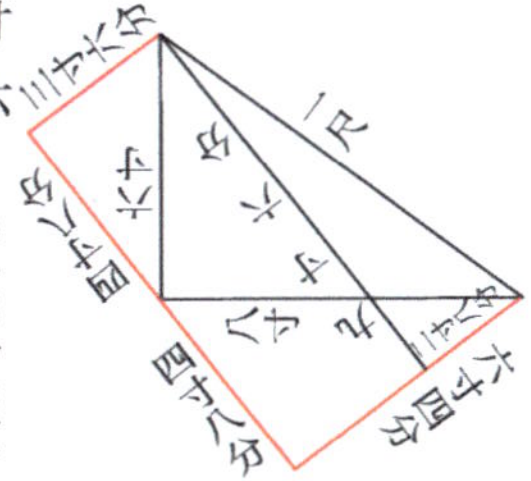

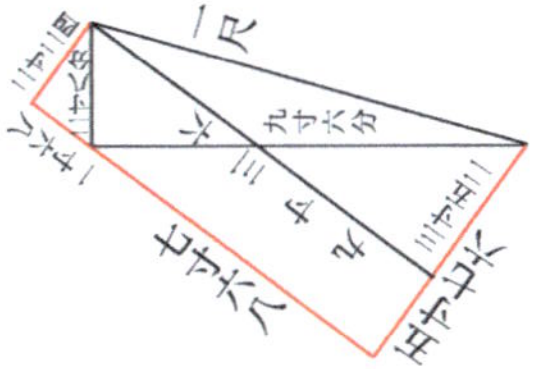

定

差一寸

勾一寸二分寸之一	股二寸	弦二寸二分寸之一

差二寸

勾一寸四分寸之一	股四寸	勾股三寸	弦五寸	弦三寸四分寸之一

差三寸

勾一寸六分寸之一	勾三寸	股七寸二分寸之一	勾股四寸	弦八寸二分寸之一	弦五寸	弦四寸六分寸之一

差四寸

勾一寸八分寸之一	勾二寸三分寸之二	股五寸四分寸之一	股一尺二寸	勾股五寸	弦一尺三寸	弦七寸四分寸之一	弦五寸三分寸之二	弦五寸八分寸之一

差五寸

十六

勾一寸十分寸之一	勾二寸二分寸之一	勾四寸二分寸之一	股八寸	股一尺七寸三分寸之一	勾股六寸	弦一尺八寸三分寸之一	弦一尺	弦七寸二分寸之一	弦六寸二分寸之一	弦六寸十分寸之一

差六寸

勾一寸十二分寸之一	勾二寸五分寸之二	勾四寸八分寸之一	勾六寸三分寸之二	股一尺一寸四分寸之一	股二尺四寸	勾股七寸	弦二尺五寸	弦一尺三寸四分寸之一	弦九寸三分寸之二	弦八寸八分寸之一	弦七寸五分寸之二	弦七寸十二分寸之一

差七寸

勾一寸十四分寸之一	勾二寸三分寸之一	勾三寸十分寸之九	勾六寸	股九寸六分寸之一	股一尺五寸	股二尺一寸二分寸之一	勾股八寸	弦二尺一寸二分寸之一	弦一尺七寸	弦一尺三寸六分寸之一	弦一尺	弦八寸十分寸之九	弦八寸三分寸之一	弦八寸十四分寸之一

差八寸

勾一寸十六分寸之一	勾二寸七分寸之二	勾三寸四分寸之三	勾五寸五分寸之三	勾八寸八分寸之一	股一尺二寸	股一尺九寸四分寸之一	股四尺	勾股九寸	弦四尺一寸	弦一尺一寸四分寸之一	弦一尺五寸	弦一尺三寸八分寸之一	弦一尺五分寸之三	弦九寸四分寸之三	弦九寸七分寸之二	弦九寸十六分寸之一

差九寸

勾一寸十八分寸之一	勾二寸四分寸之一	勾三寸十四分寸之九	勾五寸三分寸之一	勾七寸二分寸之一	股一尺二分寸之一	股二尺五寸六分寸之一	股二尺四寸	股四尺九寸二分寸之一	勾股一尺	弦五尺二分寸之一	弦二尺六寸	弦二尺八寸六分寸之一	弦二尺四寸二分寸之一	弦二尺五寸二分寸之一	弦二尺一寸三分寸之一	弦一尺十四分寸之九	弦一尺四分寸之一	弦一尺十八分寸之一

法曰定勾二寸者股弦差一寸自乘以減定勾冪四寸餘三寸爲實以倍差二寸爲法除之得股一寸二分寸之一加差一寸得弦二寸二分寸之一是得商少於定數二寸故爲勾定勾變爲股

定勾三寸者股弦差一寸自乘以減定勾冪九寸餘八寸爲實以倍差二寸除之得股四寸多於定三寸故卽爲股加差一寸得弦五寸又股弦差二寸自乘以減定勾冪九寸餘五寸爲實以倍差四寸除之得一十四分寸之一加差二寸得弦三寸四分寸之一是得商少於定數三寸故反爲勾定勾三寸變爲股也餘倣之

十七

重差

重差者度高深遠廣之法也俗謂之町見乃山岳高下城邑大小岸望谷深山望津廣古雖分重表累矩之名而解之其術皆起于勾股法故今設五題單以著測量之法云

假如有隔海樹不知其遠用方八尺板望板左前後角與樹根齊直亦從板前左角偏于右七尺九寸六分而板右後角與樹根齊直問遠

答曰遠四町二十五間二尺

法曰置偏右七尺九寸六分以方八尺相乘

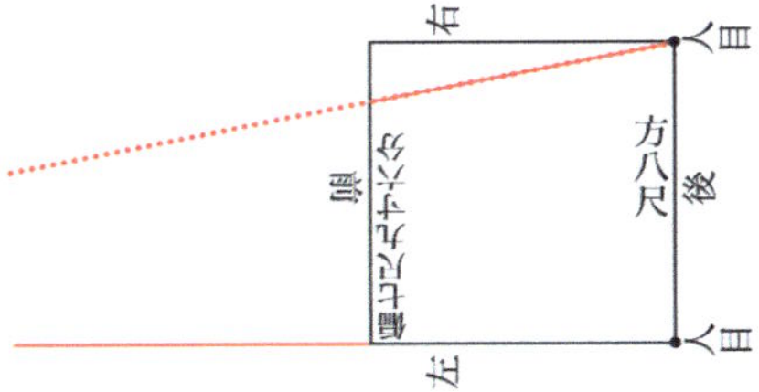

得六十三尺六寸八分爲實置方八尺內減偏得七尺九寸六分餘四分爲法實如法而一得一百五十九丈二尺以町尺率三十六丈約之得四町不滿法者以閒率六尺約之得二十五閒不滿法者命尺得遠也

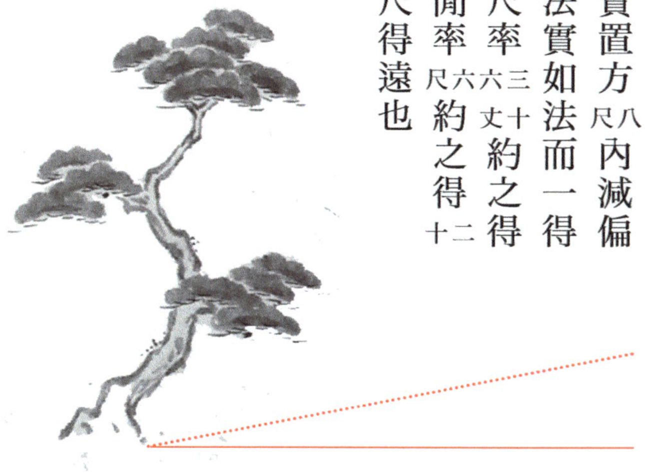

十八

假如有隔遠三町四十八閒見圓池不知其廣居頭三尺丁字杖杖頭退行二丈九尺望杖頭與池廣左右平合問池廣

答曰池廣二十五閒二尺四寸

法曰置遠三町通閒內子四十八閒得二百二十八閒以閒率六尺相乘加入退行二丈九尺共得一百三十九丈七尺以杖頭三尺相乘倍之得八千三百八十二尺爲實置退行倍之得五丈八尺內減杖頭餘五丈五尺爲法實如法而一得一十五丈二尺四寸以閒率約之不滿法者命尺得池廣

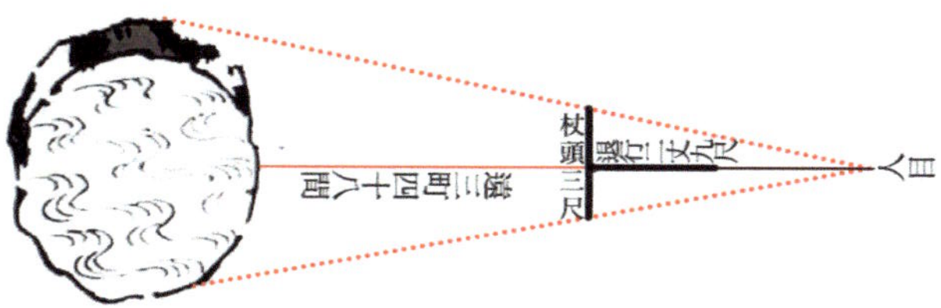

假如有竿不知其高自竿脚量遠一十六間立八尺表表後退行三丈二尺亦立三尺短表望二表倶與竿齊平問竿高

答曰竿高二丈三尺

法曰置遠一十六間通尺得九丈六尺寄位置表長八尺內減短表三尺餘五尺以寄位相乘得四百八十尺爲實以退行三丈二尺爲法實如法而一得一丈五尺加入表八尺得竿高

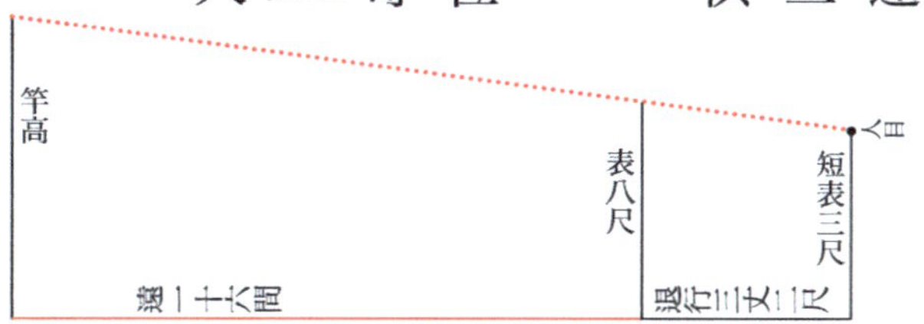

假如有海島不知其高遠立二表各高一丈五尺前後相去三百六十間從前表退行一百五間立四尺短表望二表倶與島峰參合亦從後表退行一百一十間立四尺短表望二表倶與島峰參合問島高遠

答曰
島高二町一十四間半
島遠三里一十八町

法曰先求島高者置表高一丈五尺內減短表四尺餘一丈一尺以相去三百

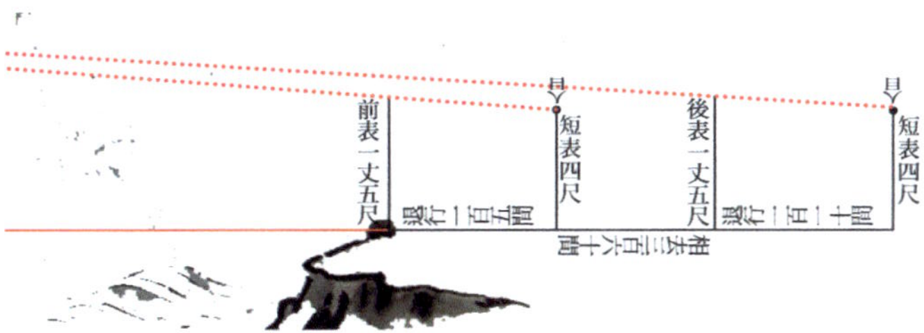

六十閒相乘得三千九百六十尺爲實置
後表退行一百一十閒內減前表退
行一百五閒。餘五閒爲法實如法而
一得七十九丈二尺加入表高得島高
八十丈七尺。以町尺率三百六十尺約之
得二町不滿法者以閒率六尺約之
得一十四閒半　求遠者置相去三百
六十閒以前表退行一百五閒。相乘
得三萬七千八百閒爲實如前法而一
得島遠七千五百六十閒以里閒率二千
一百六十閒約之得三里不滿法者以
町率六十閒約之得一十八町也

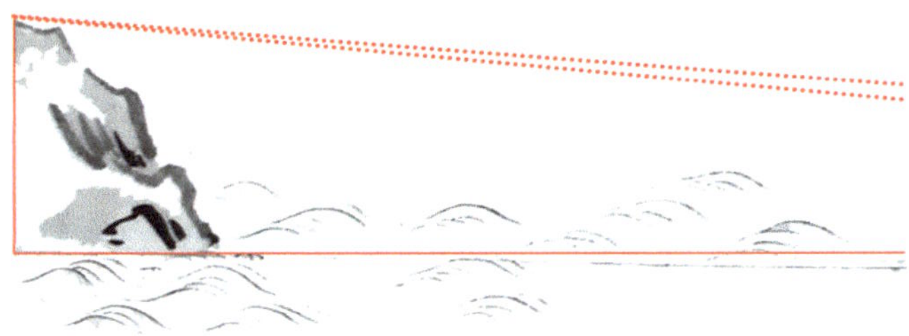

二十

假如有葭一莖生水中出水一尺引杪去五尺與
水適平問水深
　答曰水深一丈二尺
法曰置引去五尺自乘得
二十五尺寄位置出水一尺自
乘得一尺以減寄位餘二十
四尺爲實出水一尺倍之得
二尺爲法實如法而一得
水深

右所著之重差五問皆依勾股法所求而術理
已明于前故不解茲也其餘雖有遺法雜技悉
略之

斜法第四

斜者謂勾股轉折之者也是故三斜爲始本不合曲
尺故其形有屈伸也勾曰小斜股曰中斜弦曰大斜
自上稜至下取繩直而其闊曰中股右曰長股左曰
短股或自左稜至右或自右稜至左各取直而見其中闊及長短股者亦同若容方者
有方隅之交故以面爲矩容圓者有周圍之交故以
中心爲矩皆依勾股法求之乃是三斜法也四斜者
幷三斜兩段之狀五斜者幷三段六斜者幷四段七斜者幷五段也逐倣之然以外斜
數即爲其形之名一名四不等其餘隨外斜數號諸不等也依各斜之長短雖每
稜屈伸有偏正傾倒之異不論形勢之不同也係其
四稜內有縱橫兩斜故內外總六也然此形所用之
斜數五件爲限而有當求之斜一件乃置外斜數內減三餘爲內所

用斜數以外斜數相乘折半之爲其形內係總斜數減內係斜數餘爲當求斜數又置外斜數倍之內減
三餘爲其形所用總斜數是設題之辭限數也各不拘形之順倒以內斜長者準縱以短者準橫也以
內係一斜爲界或界于縱或界于橫各不爲例只隨時而可斟酌焉自兩稜至
界斜互取直各依三斜法起其術若容圓者外斜皆
有交而內斜自得故唯以外斜求之自初斜乃不拘長短任
意而爲初也隔一斜逐至末又自次斜隔一斜逐至末各
幷之其數必相等若相減則無餘而不得所交之闊
只斜一件不言而自得故代最末斜與圓交一偏之
闊而言之如六斜八斜十斜之類形數偶者如此若三斜五斜七斜九斜之類形數奇者自初
斜隔一斜逐至末又自次斜隔一斜逐至末求右交則各順幷之求左交則各逆幷之兩數相減餘折半
之得其斜與容圓所交之左右闊也皆求逐斜之交闊而後得中心至
每稜內斜冪爲內外同數之斜形起其術若容方者有四隅之

交故以方面爲縱橫之矩自四稜取繩直而依勾股法起其術也五斜已上準此五斜者係
五稜內有五斜故內外總十也所用之斜數七件爲
形限而有當求之斜三件是故以所言與所問之諸
斜視其形會四斜者卽依其法求之否則難直起術
故以虛術見別斜分兩段作四斜兩段而求之六斜者分作四斜五斜兩段七斜者分作四斜與六斜兩段八斜者分作四斜與七斜兩段也次第倣之各依四斜法得
兩式求寄消起眞術而得五斜法也六斜者內有係
九斜故內外總十五也所用之斜數九件爲形限而
有當求之斜六件是故以所言與所問之諸斜視其
形會四斜者卽以其法求之會五斜者亦以其法求
之否則又難直起術故以虛術見別斜分兩段而依
四斜法得前式依五斜法得後式求寄消起眞術而

二十二

得六斜法七斜已上皆如此而得其形之法也若形
內每稜成斜者爲內外同數之斜形故三斜內成三
斜者屬于四斜法四斜內成四斜者屬于五斜法五
斜內成五斜者屬于六斜法已上準此各據其法求之也

假如有三斜大斜二尺七寸三分中斜二
尺六寸小斜一尺六寸九分問長短股

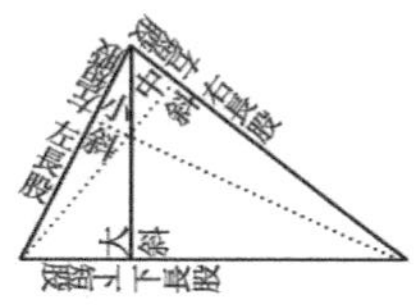

答曰
下長股二尺○八分　短股六寸五分
右長股二尺一寸八分四釐　短股四寸一分六釐
左長股一尺○五分　短股六寸四分

法曰據題數驗形之屈伸者中斜冪與小斜冪相拌共得數多於大斜冪者爲屈形少於者爲伸形也
求右下長股者列拌大斜冪七百四十五寸二九與中斜
冪六百七十六寸共得一千四百二十一寸二九內減小斜冪二百八十五寸

六一餘一千一百三十五寸六八各爲實以倍大斜五尺四寸六分除之
得下長股以倍中斜五尺二寸除之得右長股　求下短
股左長股者列幷大斜冪與小斜冪共得一千〇三十寸〇九內
減中斜冪餘三百五十四寸九各爲實以倍大斜五尺四寸六分
除之得下短股以倍小斜三尺三寸八分除之得右長股
求左右短股者列幷中斜冪與小斜冪共得九百六十
一寸六一內減大斜冪餘二百一十六寸三二各爲實以倍中斜
五尺二寸除之得右短股以倍小斜三尺三寸八分除之得左
短股若形伸短者左右短股各出于斜外故以舊斜相幷者爲長股也
解曰求下長股者立天元一爲下長股。丨自
之以減中斜冪餘爲中股冪丨中斜巾。丨寄左　列
長股以減大斜餘爲短股丨大斜丨自之以減小斜

二十三

冪餘亦丨大斜巾 ‖大斜 丨與寄左丨大斜巾 ‖大斜
爲中股丨小斜巾　相消得丨中斜巾
冪　式丨小斜巾
求下短股者立天元一爲下短股。丨自之以
減小斜冪餘爲中股冪丨小斜巾。丨寄左　列短股
以減大斜餘爲長股丨大斜丨自之以減中斜冪餘
亦爲丨大斜巾 ‖大斜 丨與寄左丨大斜巾 ‖大斜
中股丨中斜巾　相消得丨小斜巾
冪　式丨中斜巾
求左右長短股者皆如此求之故傍書式各略
之
假如有三斜大斜二尺七寸三分中斜二尺六寸

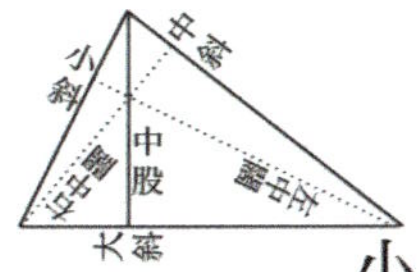

小斜一尺六寸九分問中股中闊
答曰
中股一尺五寸六分
右中闊一尺六寸三分八釐
左中闊二尺五寸二分
法曰大斜冪中斜冪相乘得五十萬〇三千八百一十六寸〇四大
斜冪小斜冪相乘得二十一萬二千八百六十二寸二七六九中斜冪
小斜冪相乘得一十九萬三千〇七十二寸三六三位相幷共得
數倍之得一百八十一萬九千五百〇一寸三五三八內減大斜三乘
冪五十五萬五千四百五十七寸一八四一中斜三乘冪四十五萬六千九百七十
六寸小斜三乘冪八萬一千五百七十三寸〇七二一餘七十二萬五千四百九十
五寸〇九七六爲各實是乃一十六段積冪也求中股者以四段大
斜冪二千九百八十一寸一六爲廉法求右中闊者以四段中

二十四

斜冪二千七百〇四寸爲廉法求左中闊者以四段小斜
冪一千一百四十二寸四四爲廉法各開平方除之得中闊若伸
矩者左右中闊各出于斜外也
解曰列幷大斜冪與中斜冪共得內減小斜冪
餘爲因 |大斜巾 自之 ||大斜巾中斜巾
大斜二 |中斜巾 以減 ||大斜巾小斜巾
箇長股 |小斜巾 大斜 ||中斜巾小斜巾
冪中斜冪相乘 |大斜三
四段餘爲因大斜 |中斜三
冪四段中股冪 |小斜三
是一十六段積冪故即爲因中斜冪四段右中
闊冪亦爲因小斜冪四段左中闊冪也

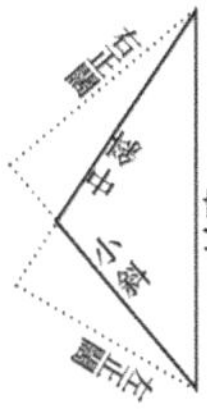

假如有三斜積二百四十二寸七分六釐中斜二尺八寸九分小斜一尺七寸問正闊

答曰 左正闊一尺六寸八分
右正闊二尺八寸五分六釐

法曰求左正闊者置倍積四百八十五寸五二為實以中斜二尺八寸九分為法除之得左正闊 求右正闊者置倍積為實以小斜為法除之得右正闊

解曰倍積擬直積中斜擬縱而除之則得直橫是左正闊又小斜擬橫而除之則得直縱是右正闊也或大斜與中股據勾股報解其理者亦同之

假如有三斜內容方大斜二尺一寸中斜一尺七

二十五

寸小斜一尺問方面

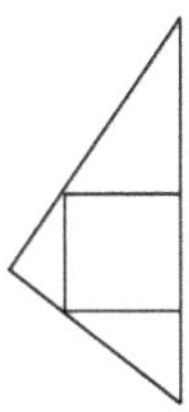

答曰方面五寸二十九分寸之二十三

法曰別得中股八寸大斜二尺一寸與中股八寸相乘得二段積一百六十八寸為實列幷大斜與中股共二尺九寸為法除之不滿法者命之得容方面

解曰立天元一為方面。|

以減中股餘 中股 | | 以大斜相

乘為因中股方面 中大股斜 | 大斜 | 寄左

列方面以中股相 中大股斜 | 大斜 |

乘與寄左相消得式 中股 |

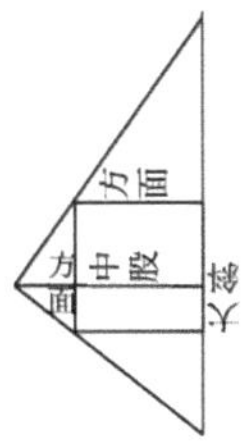

假如有三斜內容圓大斜二尺一寸中斜一尺七寸小斜一尺問圓徑

答曰

圓徑七寸

大斜交右闊一尺四寸左闊七寸

中斜交右闊一尺四寸左闊三寸

小斜交右闊三寸左闊七寸

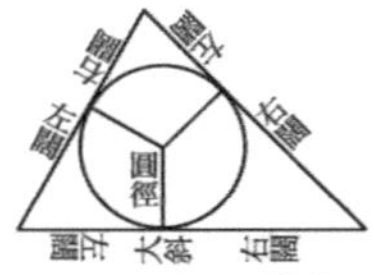

法曰求圓徑者別得中股八寸大斜二尺一寸與中股八寸相乘

倍之得四段積三百三十六寸爲實大中小斜相幷共得

四尺八寸爲法除之得圓徑　求圓與大斜交右闊者

大斜二尺一寸與中斜一尺七寸相幷共得三尺八寸內減小斜

一尺餘二尺八寸折半之得大斜交右闊中斜交右闊亦同　求

大斜交左闊者大斜與小斜相幷共得三尺一寸內減

中斜餘一尺四寸折半之得大斜交左闊小斜交左闊亦同

求中斜交左闊者中斜與小斜相幷共得二尺七寸內

減大斜二尺一寸餘六寸折半之得中斜交左闊小斜交右闊亦同

二十六

解曰容圓者詳于勾股篇中　求每斜交闊者

大斜右交與中斜右交兩闊等

大斜左交與小斜左交兩闊等

中斜左交與小斜右交兩闊等

故大中斜和內減小斜則得大

斜中斜各右交兩闊大小斜和

內減中斜則得大斜小斜各左

交兩闊中小斜和內減大斜則

得中斜左交小斜右交兩闊故

各折半之爲一遍之闊也

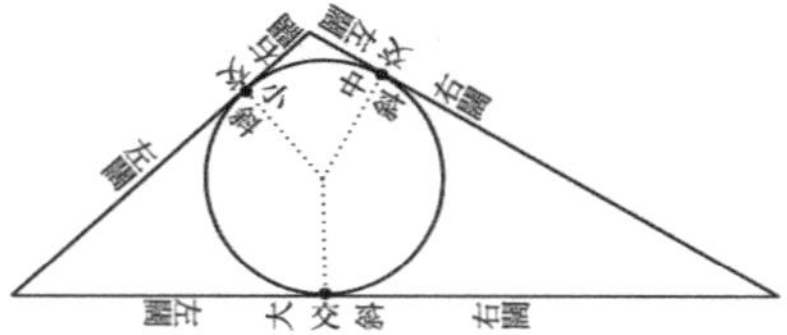

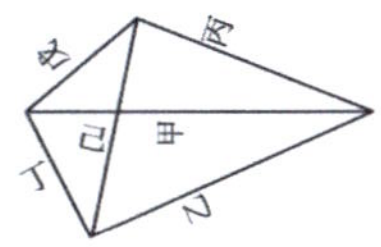

假如有四斜甲二尺一寸乙二尺丙一尺
七寸丁一尺三寸戊一尺問己
答曰己二尺〇〇二釐四毫九八四強
法曰立天元一爲己〇自之加甲冪又
以甲冪己〇〇列幷乙冪與戊冪列幷
冪相乘　以乙冪戊冪相乘丙冪
與丁冪以丙甲冪內減丁冪餘丁冪內減戊
冪丁冪相乘以丙冪戊冪相乘冪餘以甲冪
乙冪五位相〇寄左　列幷甲冪與
相乘幷共得　乙冪以丙冪丁冪相
乘列幷丙冪與丁冪列幷丁冪與戊冪〇〇
以乙冪戊冪相乘以甲冪己冪相乘

二十七

列幷甲冪與丁冪〇〇甲冪內減丙冪餘〇
以丙冪己冪相乘　以乙冪己冪相乘
〇乙冪內減丁冪餘〇〇六位相與
以戊冪己冪相乘　幷共得寄
左相消得〇〇三乘方翻法開之得己
開方式
解曰凡諸斜之名以題中所言之者必爲先故不論內外順逆隨畫之長短號之後求者各不拘形大小以先得者屬于眞數之末以又得者屬于其次也佗皆倣此以甲爲界分形於上下各依三斜法傍書而求之或以己爲界分左右者亦同其相乘畫式繁多故術中傍書皆略之列幷甲冪與丙冪

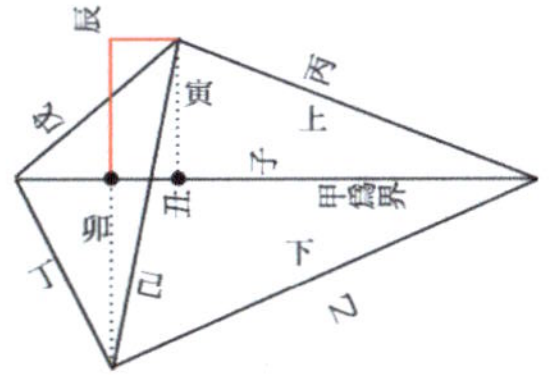

得內減戊冪餘爲因甲二箇子傍書式三位寄角位
列幷甲冪與乙冪得內減丁冪餘爲因甲二
箇丑傍書式三位寄亢位甲 甲三己巾 (一)
冪丙冪相乘四之得內減 甲巾乙巾丁巾 (五)
角位冪餘爲因甲冪四段 甲巾丙巾戊巾 (四)
寅冪傍書式六位寄氐位甲 甲巾己三 (一)
冪乙冪相乘四之得內減 乙三戊巾 (二)
亢位冪餘爲因甲冪四段 乙巾丙巾己巾 (伍)
卯冪傍書式六位寄房位亢 乙巾戊三 (二)
位內減角位餘爲因甲二 丙三丁巾 (三)
箇辰傍書式四位自之以減甲 丙巾丁三 (三)
冪己冪乃以己擬眞數而傍書之相乘 丁巾戊巾己巾 (陸)

甲赤黑丙赤黑冪之黑冪之黑冪之黑
冪符符冪符符　去符　去符　去符
乙五五戊四四　丙三　乙二　甲一
冪餘內冪餘內　冪各　冪各　冪各
去減　去減　丁幷　戊幷　己幷

二十八

四段餘爲因甲冪四段寅卯 甲巾乙巾戊巾 (五)
和冪傍書式正負總一十一位內減氐 甲巾乙巾己巾 (伍)
房位餘折半之爲因甲冪 甲巾丙巾丁巾 (壹)
因寅四箇卯傍書式正四位負六位 甲巾丙巾己巾 (貳)
自之爲因甲三乘冪因寅 甲巾丁巾己巾 (肆)
冪一十六段卯冪正二十二位負 甲巾戊巾己巾 (肆)
一十八位寄左氏位房位相 乙巾丙巾丁巾 (壹)
乘正一十八位負一十八位與寄左相 乙巾丙巾戊巾 (參)
消遍省甲冪各以四約之 乙巾丁巾戊巾 (參)
得式正一十位負一十二位擇其相 乙巾戊巾己巾 (陸)
乘相親者各分赤黑符同名 丙巾丁巾戊巾 (四)
相幷異名相減得寄消也 丙巾丁巾己巾 (貳)

戊黑赤　乙黑赤　冪之赤　冪之赤　冪之赤　冪之赤
冪符符　冪符符　去符　去符　去符　去符
己陸陸　己伍伍　甲肆　乙參　丙貳　丙壹
冪餘內　冪餘內　冪各　冪各　冪各　冪各
去減　去減　己幷　戊幷　己幷　丁幷

右黑符共五位爲寄數赤符共六位爲消數但依各斜長短其形屈伸則或有偏正傾倒或有稜反入于界斜內者是皆雖形勢不同而諸斜所號亦異恃準此形借舊名而相乘別其傍書之同異而各得變形相乘加減之名也

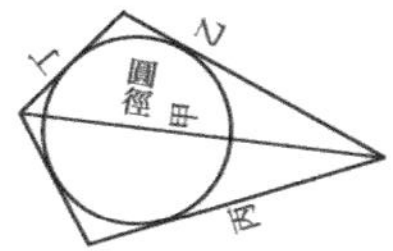

假如有四斜甲二尺一寸乙二尺丙一尺七寸丁一尺三寸問圓徑

答曰圓徑一尺四寸

法曰丙一尺七寸與丁一尺三寸相并共得三尺內減乙二尺餘一尺爲戊自是依前得四斜形積二百一十寸置積四之得八百四十寸爲實列并乙丙丁戊得圍六尺爲法除之得圓徑

二十九

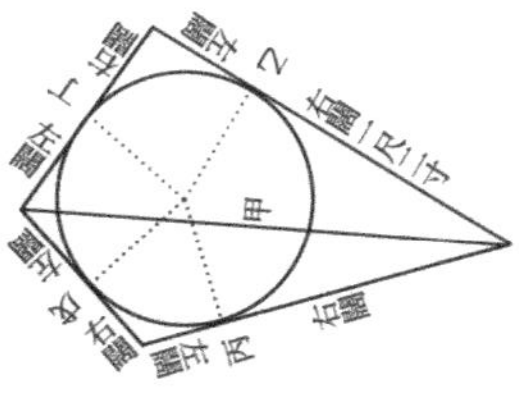

解曰自圓中心至每斜所交而取直視各左右闊乙丙兩交右闊各等乙左丁右兩交闊各等丁戊兩交左闊各等戊右丙左兩交闊各等若丙交左闊與丁交左闊相并者與戊全同故丙丁和內減乙則得戊乃六斜八斜十斜等皆如此外一斜不言而於術中自得故各代內斜一件而言之自是用甲乙丙丁戊依四斜法求積以圍除四段積得圓徑也

假如有五斜甲二尺一寸乙二尺丙一尺八寸丁一尺七寸戊一尺四寸己一尺三寸庚一尺問辛

答曰辛二尺八寸二分八釐一毫四三四强

法曰立天元一爲辛自之加入乙冪與
丁冪以丙冪相乘辛冪內減丁冪餘以
乙冪相乘二位相幷共得內減丙冪內
減戊冪餘以丙冪相乘辛冪內減丁冪
餘以戊冪相乘二位相幷數餘寄子位

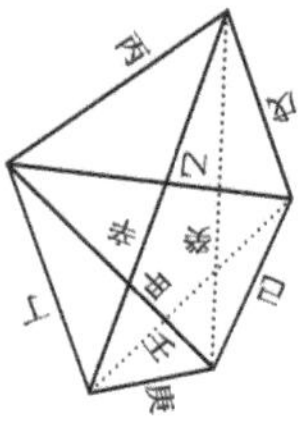

列幷乙冪與己冪及庚冪以甲冪相乘乙冪內
減己冪餘以庚冪相乘二位相幷內減甲冪內減
丁冪餘以甲冪相乘乙冪內減己冪餘以丁冪相
乘二位相幷數餘寄丑位　列幷乙冪與辛冪以
乙冪辛冪相乘列幷丁冪與戊冪以丁冪戊冪相
乘乙冪內減丁冪餘以丙冪戊冪相乘丙冪內減
戊冪餘以丁冪辛冪相乘四位相幷共得內減列

三十

幷丙冪與丁冪以乙冪辛冪相乘列幷丁冪與辛
冪以乙冪戊冪相乘二位相幷數餘寄寅位　列
幷乙冪與庚冪以乙冪庚冪相乘列幷丁冪與己
冪以丁冪己冪相乘甲冪內減乙冪餘以丁冪庚
冪相乘己冪內減庚冪餘以甲冪乙冪相乘四位
相幷共得內減列幷甲冪與乙冪以丁冪己冪相
乘列幷乙冪與丁冪以己冪庚冪相乘二位相幷
數餘寄卯位　子位甲冪相乘內減丑位丙冪相
乘數餘寄辰位　寅位甲冪相乘內減卯位丙冪
相乘數餘寄巳位　丑位寅位相乘得內減子位
卯位相乘數餘以辰位相乘得數寄左　列巳位
自乘之與寄左相消得開方式七乘方飜法開之

得辛　求壬者立天元一爲壬自之加甲冪又以甲冪壬冪相乘列幷乙冪與庚冪以乙冪庚冪相乘列幷丁冪與己冪以丁冪己冪相乘甲冪內減己冪餘以丁冪庚冪相乘己冪內減庚冪餘以甲冪乙冪相乘五位相幷共得數寄左　列幷甲冪與乙冪以丁冪己冪相乘列幷甲冪與己冪以丁冪壬冪相乘列幷丁冪與己冪以乙冪庚冪相乘列幷己冪與庚冪以甲冪壬冪相乘甲冪內減丁冪餘以乙冪壬冪相乘乙冪內減己冪餘以庚冪壬冪相乘六位相幷與寄左相消得開方式三乘方翻法開之得壬　求癸者立天元一爲癸自之加乙冪又以乙冪癸冪相乘列幷甲冪與戊冪以

三十一

甲冪戊冪相乘列幷丙冪與己冪以丙冪己冪相乘癸冪內減己冪餘以甲冪丙冪相乘四位相幷共得數寄左　列幷甲冪與己冪以乙冪癸冪相乘列幷丙冪與己冪以甲冪戊冪相乘列幷戊冪與癸冪以丙冪己冪相乘列幷己冪與癸冪以乙冪丙冪相乘甲冪內減己冪餘以戊冪癸冪相乘戊冪內減己冪餘以甲冪乙冪相乘癸冪內減丙冪餘以乙冪戊冪相乘七位相幷與寄左相消得開方式三乘方翻法開之得癸各式之畫略

解曰此形內有當求辛癸壬三斜故若問壬者以甲乙丁己庚乃不用丙戊兩斜者得壬問癸者以甲乙丙戊己不用丁庚兩斜者得癸各依四斜法直起眞術而

求之問辛者直求之則難輒
施術故以之卽擬眞數別立
虛術于壬而爲界以乙丙丁
戊辛及壬[或立虛術于癸爲界以甲丙丁以辛]
[及癸得前式又甲爲界者甲乙丙戊己及癸得後式甲亦]
[同]假書舊名[乃以辛假甲以戊假乙以丙依]
[舊以壬假丁以丁假戊以乙假己而各相乘之若以形長減短則雖術中之加減反覆到得式]
[各無差也然於眞術中有以多減少之數者於理不稱故寄者反屬消消者反屬寄而求之也]
[後皆做此]依四斜法立天元一爲壬[卽假丁]○｜自之
加[假]丙[乃丙]冪以丙冪丁冪相乘○○[丙三]○[丙巾]列
并[假]甲[乃辛]冪與[假]己[乃乙][辛乙三巾]列并[假]乙[乃戊]冪與[假]
冪以甲冪己冪相乘[辛乙巾三]戊[乃丁]冪以乙冪戊冪

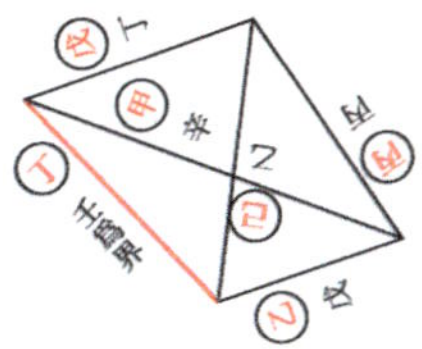

三十二

相[戊丁巾三]甲冪內減丁冪餘以丙冪戊冪相乘[辛丁丙巾巾巾]○
乘[戊丁三巾][丁丙巾巾]丁冪內減戊冪餘以甲冪乙冪相乘[辛戊丁巾巾巾]
○[辛戊巾巾]五位相并[式略之]寄左
列并甲冪與乙冪○○[戊丙巾巾]列
以丙冪丁冪相乘[辛丙巾巾]并
甲冪與丁冪以丙冪己冪相
乘[辛丙乙巾巾巾]○[丙乙巾巾]列并丙冪與丁冪
以乙冪戊冪相乘[戊丁丙巾巾巾]○[戊丁巾巾]列
并丁冪與戊冪以甲冪己冪
相乘[辛丁乙巾巾巾]○[辛乙巾巾]乙冪內減丁冪
餘以戊冪己冪相乘[戊丁乙巾巾巾]○[丁乙巾巾]
甲冪內減丙冪餘[辛戊乙巾巾巾]六位相

｜辛乙巾三	｜辛乙三巾	｜戊丙乙巾巾巾	｜辛丁丙巾巾巾	｜戊丁巾三	｜戊丁三巾	｜辛丙乙巾巾巾	｜戊丁乙巾巾巾	｜辛丁乙巾巾巾	｜辛戊乙巾巾巾
㊀	㊀	㊂	㊃	㊁	㊁	㊄	㊅	㊄	㊅
｜丁乙巾巾	｜丙三	｜辛戊巾巾	｜丙乙巾巾	｜戊丁巾巾	｜辛乙巾巾	｜丁丙巾巾	｜戊丙巾巾	｜辛丙巾巾	
㊁	㊂	㊃	㊀	㊃	㊁	㊀	㊂	㊀	
｜丙巾									

以乙冪己冪相乘〔戊丙乙巾巾巾〕幷之式略

與寄左相消得前式

戊丁丙巾巾巾	辛戊丁巾巾巾
三	四

又乙爲界以甲乙丁己庚及壬假書舊名乃以甲乙各依舊以丁假丙以己假丁以庚假戊以壬假己而各相乘之依四斜法如前求之列幷

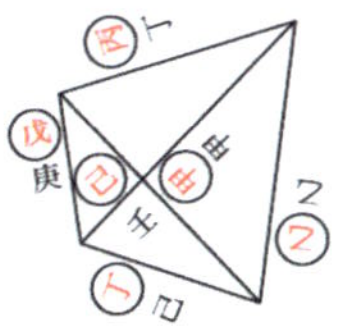

假甲乃甲冪與假己乃壬冪以甲冪己冪相乘。。〔甲三〕。〔甲巾〕列幷假乙乃乙冪與假戊乃庚〔庚乙巾三〕列幷假丙乃丁冪與假丁乃己冪以乙冪戊冪相乘〔庚乙三巾〕丁冪以丙冪丁冪相〔己丁巾三〕甲冪內減丁冪餘〔庚丁甲巾巾巾〕丁冪內減戊冪餘〔己乙甲巾巾巾〕乘〔己丁三巾〕以丙冪戊冪相乘〔庚己丁巾巾巾〕以甲冪乙冪相乘〔庚乙甲巾巾巾〕五位相幷之式略寄左

列幷甲冪與乙冪以丙冪丁冪〔己丁甲巾巾巾〕列幷甲冪與丁相乘〔己丁乙巾巾巾〕冪以丙冪己冪相。。〔丁甲巾巾〕列幷丙冪與丁乘〔己丁巾巾〕冪以乙冪戊冪相〔庚丁乙巾巾巾〕列幷丁冪與戊冪。乘〔庚己乙巾巾巾〕以甲冪己冪相乘。〔己甲巾巾〕甲冪內減丙冪餘。〔庚甲巾巾〕以乙冪己冪相乘。〔乙甲巾巾〕乙冪內減丁冪餘。〔丁乙巾巾〕以戊冪己冪相乘。〔庚乙巾巾〕六位相幷之式略與寄〔庚己巾巾〕左相消得後式

三十三

己乙甲巾巾巾	庚丁甲巾巾巾	庚乙巾三	庚乙三巾	己丁巾三	己丁三巾	庚乙甲巾巾巾	己丁甲巾巾巾	己丁乙巾巾巾	庚丁乙巾巾巾	庚己乙巾巾巾	庚己丁巾巾巾
肆	參	壹	壹	貳	貳	肆	伍	伍	參	陸	陸
甲三	丁乙巾巾	庚己巾巾	乙甲巾巾	己甲巾巾	丁甲巾巾	庚甲巾巾	庚乙巾巾	己丁巾巾			
參	肆	貳	壹	壹	參	壹	貳	肆			
甲巾											

兩式各縮空級而如符括之乃單位隅故級如者各舊先前
式方級赤圈一去丙幕赤圈二去乙幕共二負位黑
圈三去丙幕黑圈四去戊幕共二正位各乘所去之
某名負位二幷內減正位二餘方負眞術號子　後式
方級赤圈壹去甲幕赤圈貳去庚幕共二負位黑圈
參去甲幕黑圈肆去丁幕共二正位各乘所去之某
名負位二幷內減正位二餘方負眞術號丑　前式實
級黑圈一去乙幕辛幕黑圈二去丁幕戊幕黑
圈三去丙幕戊幕黑圈四去丁幕辛幕共四正位赤
圈五去乙幕辛幕赤圈六去乙幕戊幕共二負位各
乘所去之某名正位四幷內減負位二餘實正眞術名
寅　後式實級黑圈壹去乙幕庚幕黑圈貳去

三十四

丁幕己幕黑圈參去丁幕庚幕黑圈肆去甲幕
乙幕共四正位赤圈伍去丁幕己幕赤圈陸去己幕
庚幕共二負位各乘所去之某名正位四幷內減負位二
餘實正眞術號卯
括前式寅子丙巾前式遍乘甲幕後式遍乘［甲寅巾　甲子巾］
後式卯丑甲巾丙幕相減之得換一式［丙卯巾　丙丑巾］
前式遍乘丑加一式後式［寅丑　甲寅巾］一式方級括之
遍乘子減一式得換二式［卯子　丙卯巾］正眞術號辰
實級括之負眞術號巳　二式方級負與一式
實級負全同故卽巳　實級正於術中不號之
直乘一式方級辰爲眞術寄左　一式實級巳
與二式方級巳相乘相消也

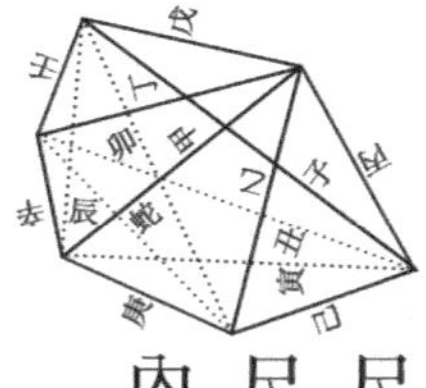

假如有六斜甲二尺一寸乙二尺丙一尺八寸丁一尺七寸戊一尺六寸己一尺四寸庚一尺三寸辛一尺壬七寸問內係諸斜

答曰得內係斜

法曰求子者立天元一爲子自之加入戊冪以丁冪相乘列幷壬冪與子冪內減戊冪餘以戊冪相乘二位相幷共得內減子冪內減丙冪餘以壬冪相乘丁冪內減戊冪餘以丙冪相乘二位相幷數餘寄角位　列幷乙冪與丁冪以己冪相乘列幷丙冪與丁冪以乙冪相乘二位相幷共得內減丙冪丁冪相乘乙三乘冪二位相幷數餘寄亢位

三十五

列幷乙冪與辛冪以甲冪相乘列幷甲冪與丁冪內減辛冪餘以庚冪相乘二位相幷共得內減甲冪內減丁冪餘以甲冪相乘丁冪內減辛冪餘以乙冪相乘二位相幷數餘寄氐位　列幷丙冪與辛冪以甲冪丙冪相乘列幷甲冪與丁冪內減辛冪餘以丙冪庚冪相乘乙冪內減己冪餘以甲冪丁冪相乘三位相幷共得內減列幷甲冪與己冪以甲冪丙冪相乘丁冪內減辛冪餘以乙冪丙冪相乘二位相幷數餘寄房位　列幷丙冪與壬冪以丙冪壬冪相乘列幷丁冪與子冪以丁冪子冪相乘丙冪內減丁冪餘以戊冪子冪相乘三位相幷共得內減列幷戊冪與子冪以丙冪壬冪相乘

列扞壬冪與子冪以丙冪丁冪相乘子冪内減戊冪餘以丁冪壬冪相乘三位相扞數餘寄心位　列扞乙冪與辛冪以乙冪辛冪相乘列扞丁冪與庚冪以丁冪庚冪相乘庚冪内減辛冪餘以甲冪乙冪相乘三位相扞共得内減列扞乙冪與丁冪以庚冪辛冪相乘列扞庚冪與辛冪以乙冪丁冪相乘庚冪内減辛冪餘以甲冪丁冪相乘三位相扞數餘寄尾位　列扞甲冪與乙冪及辛冪以丙冪丁冪庚冪相乘列扞丁冪與庚冪以乙冪丙冪辛冪相乘列扞丁冪與己冪内減乙冪餘以甲冪丁冪己冪相乘列扞己冪與辛冪内減庚冪餘以甲冪乙冪丙冪相乘四位相扞共得内減列扞乙

三十六

冪與辛冪以乙冪丙冪辛冪相乘列扞丁冪與庚冪以丙冪丁冪庚冪相乘列扞己冪與辛冪以甲冪丙冪丁冪相乘三位相扞數餘寄箕位　列扞乙冪與丙冪以尾位相乘列扞乙冪與丁冪以己冪氐位相乘乙冪内減丁冪餘以丙冪氐位相乘三位相扞共得内減列扞乙冪與丁冪以乙冪氐位相乘己冪尾位相乘二位相扞數餘寄斗位　列扞丁冪與己冪内減丙冪餘以丁冪己冪氐位相乘丙冪内減丁冪餘以乙冪己冪氐位相乘丙冪内減丁冪餘以丙冪尾位相乘三位相扞共得内減列扞丙冪與丁冪以己冪尾位相乘丙冪内減丁冪餘以乙冪尾位相乘二位相扞數餘寄牛

位　列丼乙冪與丙冪以氐位相乘倍甲冪以亢
位相乘二位相丼共得內減己冪氐位相乘數餘
寄女位　列丼甲冪亢位冪相乘與己冪斗位相
乘以甲冪相乘列丼倍甲冪箕位相乘與氐位房
位相乘以乙冪相乘二位相丼共得內減列丼乙
冪與丙冪以甲冪斗位相乘得數餘寄虚位　列
丼乙冪與丙冪以甲冪牛位相乘倍甲冪以亢位
箕位相乘二位相丼共得內減甲冪己冪牛位相
乘房位斗位相乘二位相丼數餘寄危位　列丼
甲冪乙冪角位相乘與戊冪女位相乘數寄室位
　列丼甲冪乙冪心位相乘與角位女位相乘數
寄壁位　列丼甲冪乙冪心位女位相乘與角位

虚位相乘數共得內減戊冪危位相乘數餘寄奎
位　列丼房位牛位相乘與箕位冪以戊冪相乘
得內減心位虚位相乘數餘寄婁位　列丼房位
牛位相乘與箕位冪以角位相乘得內減心位危
位相乘數餘寄胃位　甲三乘冪乙三乘冪角位
冪心位冪女位冪相乘段一甲三乘冪乙三乘冪心
位冪壁位冪相乘段一甲冪乙冪戊冪角位心位女
位婁位相乘段二甲冪乙冪戊冪心位壁位婁位相
乘段二甲冪乙冪戊冪心位冪女位奎位相乘段二甲
冪乙冪角位冪室位胃位相乘段一戊三乘冪奎位
胃位相乘段一戊三乘冪婁位冪相乘段一八位相丼
共得數寄左　甲三乘冪乙三乘冪角位心位冪

女位壁位相乘段二甲冪乙冪戊冪角位壁位胃位
相乘段二甲冪乙冪角位心位室位婁位相乘段二甲
冪乙冪心位冪室位奎位相乘段一四位相幷與寄
左相消得開方式一十五乘方翻法開之得子
求丑者立天元一爲丑自之加入乙冪與丁冪以
丙冪相乘丑冪內減丁冪餘以乙冪相乘二位相
幷共得內減丙冪內減己冪餘以丙冪相乘丑冪
內減丁冪餘以己冪相乘二位相幷數餘寄角位
　列幷乙冪與庚冪及辛冪以甲冪相乘乙冪內
減庚冪餘以辛冪相乘二位相幷共得內減甲冪
內減丁冪餘以甲冪相乘乙冪內減庚冪餘以丁
冪相乘二位相幷數餘寄亢位　列幷乙冪與丑

三十八

冪以乙冪丑冪相乘列幷丁冪與己冪以丁冪己
冪相乘乙冪內減丁冪餘以丙冪己冪相乘丙冪
內減己冪餘以丁冪丑冪相乘四位相幷共得內
減列幷丙冪與丁冪以乙冪丑冪相乘列幷丁冪
與丑冪以乙冪己冪相乘二位相幷數餘寄氐位
　列幷乙冪與辛冪以乙冪辛冪相乘列幷丁冪
與庚冪以丁冪庚冪相乘甲冪內減乙冪餘以丁
冪辛冪相乘庚冪內減辛冪餘以甲冪乙冪相乘
四位相幷共得內減列幷甲冪與乙冪以丁冪庚
冪相乘列幷乙冪與丁冪以庚冪辛冪相乘二位
相幷數餘寄房位　甲冪角位相乘得內減丙冪
亢位相乘數餘寄心位　甲冪氐位相乘得內減

丙幂房位相乘數餘寄尾位　亢位氏位相乘得
內減角位房位相乘數餘以心位相乘得數寄左
　列尾位自乘之與寄左相消得開方式七乘方
翻法開之得丑　求寅者立天元一爲寅自之加
入乙幂以乙幂寅幂相乘列并甲幂與己幂以甲
幂己幂相乘列并丙幂與庚幂以丙幂庚幂相乘
寅幂內減庚幂餘以甲幂丙幂相乘四位相并共
得數寄左　列并甲幂與庚幂以乙幂寅幂相乘
列并丙幂與庚幂以甲幂己幂相乘列并己幂與
寅幂以丙幂庚幂相乘列并庚幂與寅幂以乙幂
丙幂相乘甲幂內減庚幂餘以己幂寅幂相乘此一位本屬消數雖爲寄數而有反減之理故求之也
己幂內減庚幂餘以甲

三十九

幂乙幂相乘寅幂內減丙幂餘以乙幂己幂相乘
七位相并與寄左相消得開方式二乘方翻法開
之得寅　求卯者立天元一爲卯自之加入甲幂
與戊幂以乙幂相乘卯幂內減戊幂餘以甲幂相
乘二位相并共得內減乙幂內減庚幂餘以乙幂
相乘卯幂內減戊幂餘以庚幂相乘二位相并數
餘寄角位　列并甲幂與辛幂及壬幂以丁幂相
乘甲幂內減辛幂餘以壬幂相乘二位相并共得
內減甲幂內減辛幂餘以戊幂相乘丁幂內減戊
幂餘以丁幂相乘二位相并數餘寄亢位　列并
甲幂與卯幂以甲幂卯幂相乘列并戊幂與庚幂
以戊幂庚幂相乘甲幂內減戊幂餘以乙幂庚幂

相乘乙冪內減庚冪餘以戊冪卯冪相乘四位相
并共得內減列并乙冪與戊冪以甲冪卯冪相乘
列并戊冪與卯冪以甲冪庚冪相乘二位相并數
餘寄氐位　列并甲冪與壬冪以甲冪壬冪相乘
列并戊冪與辛冪以戊冪辛冪相乘辛冪內減壬
冪餘以甲冪丁冪相乘三位相并共得內減列并
甲冪與丁冪以戊冪辛冪相乘列并甲冪與戊冪
以辛冪壬冪相乘甲冪內減丁冪餘以戊冪壬冪
相乘（是一位舊雖爲加數依相反屬減數也）三位相并數餘寄房位
　丁冪角位相乘得內減乙冪亢位相乘數餘寄
心位　丁冪氐位相乘得內減乙冪房位相乘數
餘寄尾位　亢位氐位相乘內減角位房位相乘

四十

數餘以心位相乘得數寄左　列尾位自乘之與
寄左相消得開方式七乘方翻法開之得卯　求
辰者立天元一爲辰自之加入丁冪以丁冪辰冪
相乘列并甲冪與壬冪以甲冪壬冪相乘列并戊
冪與辛冪以戊冪辛冪相乘丁冪內減辛冪餘以
戊冪壬冪相乘辛冪內減壬冪餘以甲冪丁冪相
乘五位相并共得數寄左　列并甲冪與丁冪以
戊冪辛冪相乘列并丁冪與辛冪以戊冪辰冪相
乘列并戊冪與辛冪以甲冪壬冪相乘列并辛冪
與壬冪以丁冪辰冪相乘甲冪內減辛冪餘以壬
冪辰冪相乘丁冪內減戊冪餘以甲冪辰冪相乘
六位相并與寄左相消得開方式三乘方翻法開

之得辰　求蛇者立天元一爲蛇自之加入甲冪
以甲冪蛇冪相乘列幷乙冪與辛冪以乙冪辛冪
相乘列幷丁冪與庚冪以丁冪庚冪相乘甲冪內
減庚冪餘以丁冪辛冪相乘庚冪內減辛冪餘以
甲冪乙冪相乘五位相幷共得數寄左　列幷甲
冪與乙冪以丁冪庚冪相乘列幷甲冪與庚冪以
丁冪蛇冪相乘列幷丁冪與庚冪以乙冪辛冪相
乘列幷庚冪與辛冪以甲冪蛇冪相乘甲冪內減
丁冪餘以乙冪蛇冪相乘乙冪內減庚冪餘以辛
冪蛇冪相乘六位相幷與寄左相消得開方式三
乘方翻法開之得蛇
解曰此形內有當求子丑寅卯辰蛇六斜故若問丑者

四十一

以丁爲界戊壬不用用甲乙丙丁己庚辛乃甲乙丙丁各依舊
名己假戊庚假己辛假庚丑假辛依四斜法問寅者以甲爲界
丁戊辛壬不用用甲乙丙己庚乃甲假戊乙假己丙依舊己假乙庚假丁寅假
甲依五斜法問卯者以乙爲界丙己不用用甲乙丁
戊庚辛壬乃甲假乙乙假丙丁假甲戊假丁庚假戊辛假己壬假庚卯假辛依
四斜法問辰者以甲爲界乙丙己庚不用用甲丁戊辛
壬乃甲假乙丁假甲戊假丙辛假丁壬假戊辰假己依四斜法問蛇者
以乙丁爲兩界丙戊己壬不用用甲乙丁庚辛乃甲乙各依舊
丁假丙庚假丁辛假戊蛇假己依四斜法各直起眞術得所問
之斜問子者難輒起術故卽擬眞數別立虛術
于丑或立虛術于寅者以甲爲界用甲乙丙己庚依四斜法得前式又以寅爲界用甲丙
丁戊辛壬子依五斜法得後式或立于卯者以之爲界用乙丙戊己子依四斜法得前式又以

乙爲界用甲乙丁戊庚辛壬依五斜法得後式或立于辰者以甲爲界用甲丁戊辛壬依四斜法得前式又以辰爲界用甲乙丙戊己庚子依五斜法得後式若立于蛇者雖得前式蛇本末各離于子而不能起虛術故不得後式也以之爲界

用丙丁戊壬子及丑假書舊名乃丙戊各依舊丁假己壬假丁子假甲丑假乙依四斜

法立天元一爲丑卽假乙。|自

之加假戊同眞冪以乙冪戊冪相

乘。。|戊三。|戊巾列幷假甲子眞冪

與假己丁眞冪以|子丁巾三列幷假丙同眞冪與假丁|壬丙巾三甲

甲冪己冪相乘|子丁三巾壬眞冪以丙冪丁冪相乘|壬丙三巾冪

內減丁冪餘以|子戊丙巾巾巾丁冪內減戊冪餘。。|子壬巾巾五

丙冪戊冪相乘|壬戊丙巾巾巾以甲冪乙冪相乘|子戊巾巾位

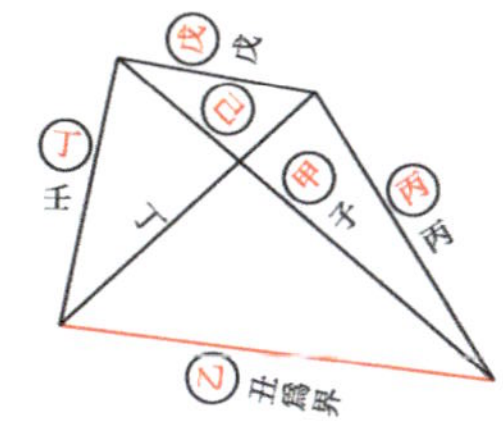

四十二

相幷之式略寄左　列幷甲冪

與乙冪以丙冪丁冪相乘|子壬丙巾巾巾冪

。|壬丙巾巾列幷甲冪與丁冪以丙

冪己冪|子丁丙巾巾巾列幷丙冪與丁冪

相乘|壬丁丙巾巾巾以乙冪戊冪相乘

。。|戊丙巾巾列幷丁冪與戊冪|子戊丁巾巾巾

|壬戊巾巾以甲冪己冪相乘|子壬丁巾巾巾

甲冪內減丙冪餘。。|子丁巾巾乙

以乙冪己冪相乘|丁丙巾巾冪

內減丁冪餘以戊冪己冪相

乘|壬戊丁巾巾巾。|戊丁巾巾六位相幷之式略與

寄左相消得前式

|壬丙巾三	|壬丙三巾	|子戊丙巾巾巾	|子丁巾三	|壬戊丁巾巾巾	|子丁三巾	|壬丁丙巾巾巾	|子丁丙巾巾巾	|壬戊丙巾巾巾	|子壬丙巾巾巾	|子戊丁巾巾巾	|子壬丁巾巾巾
㈠	㈠	㈢	㈡	㈢	㈡	㈡	㈡	㈠	㈠	㈢	㈢
|丁丙巾巾	|戊三	|子壬巾巾	|戊丙巾巾	|壬丙巾巾	|戊丁巾巾	|子丁巾巾	|壬戊巾巾	|子戊巾巾			
㈡	㈡	㈠	㈡	㈠	㈠	㈠	㈡	㈡			
|戊巾											

又以丁爲界用甲乙丙丁己庚辛及丑假書舊
名乃甲乙丙丁各依舊己假戊庚假己辛假庚丑假辛依
五斜法如前求之列并假乙同眞
冪與假丁同眞冪及假辛丙乙巾巾。丙巾
丑眞冪以假丙同眞冪相乘丁丙巾巾
辛冪內減丁冪餘以乙冪相乘
丁乙巾巾。乙巾二位相并得內減丙冪
內減假戊己眞冪丙三辛冪內減丁冪餘以戊冪相
餘以丙冪相乘己丙巾巾乘己丁巾巾。己巾二位相并數餘
丙乙巾巾。乙巾寄子位　列并乙冪與乙甲巾巾乙冪內減己
丁丙巾巾丙巾假己庚眞冪及假庚辛眞冪辛甲巾巾冪餘以庚辛乙巾巾
己丙巾巾己巾以假甲同眞冪相乘　庚甲巾巾冪相乘辛庚巾巾

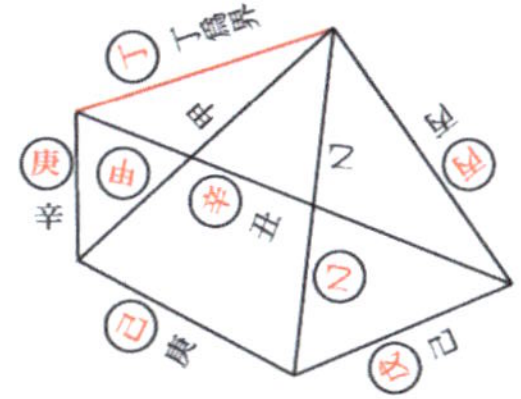

四十三

己丁巾巾二位相并共得內減甲冪內減丁冪餘
丙三以甲冪甲三乙冪內減己冪丁乙巾巾二位相并
丁乙巾巾相乘丁甲巾巾餘以丁冪相乘庚丁巾巾數餘冪乃甲乙
冪相乘甲冪丁冪相乘甲冪庚冪相乘甲冪辛
冪相乘乙冪辛冪相乘丁冪乘庚冪相乘六位各
正甲三自乘乙三冪丁冪相乘括之正即眞得
庚冪辛冪相乘三位各負術號氏氏
寄丑位　列并乙冪與辛冪以乙冪辛冪相乘
。。乙三。乙巾列并丁冪與己丁三乙冪內減丁冪餘
戊冪以丁冪戊冪相乘　己丁三以丙冪戊冪相乘
己丙乙巾巾丙冪內減戊冪餘。。丁丙巾巾四位相并內減列
丁丙己巾巾以丁冪辛冪相乘　己丁巾巾并丙冪與丁冪以
乙冪辛。。丙乙巾巾列并丁冪與辛冪以乙冪戊冪
冪相乘丁乙巾巾相乘己丁乙巾巾。己乙巾巾二位相并數餘

（己丙乙巾巾巾）○（乙三）○（乙巾）寄寅位　列幷乙冪與（辛乙巾三）列幷丁
（己丁巾三）（丁丙巾巾）庚冪以乙冪庚冪相乘（辛乙三巾）冪與己
（己丁三巾）（丙乙巾巾）冪以丁冪（庚丁巾三）甲冪內減乙冪餘（辛丁甲巾巾巾）
（己丁丙巾巾巾）（丁乙巾巾）己冪相乘（庚丁三巾）以丁冪庚冪相乘（辛丁乙巾巾巾）
（己丁乙巾巾巾）（己乙巾巾）己冪內減庚冪餘（庚乙甲巾巾巾）四位相幷共
（己丁巾巾）以甲冪乙冪相乘（辛乙甲巾巾巾）得內減列幷
甲冪與乙冪以（庚丁甲巾巾巾）列幷乙冪與丁冪（辛庚乙巾巾巾）二位相
丁冪己冪相乘（庚丁乙巾巾巾）以己冪庚冪相乘（辛庚丁巾巾巾）幷數餘
乃甲冪乙冪庚冪相乘甲冪丁冪辛冪相乘乙
三乘冪辛冪相乘乙冪辛三乘冪相乘丁三乘
冪庚冪相乘丁冪庚三乘冪相乘六位各正甲
冪乙冪辛冪相乘甲冪丁冪庚冪相乘乙冪丁
冪庚冪相乘乙冪丁冪辛冪相乘乙冪庚冪
辛冪相乘丁冪庚冪辛冪相乘六位各負　括
正即眞
之術號尾得（尾）寄卯位　甲冪子位相乘得內

四十四

減丙冪丑位相乘數餘實級甲三乘冪丙冪相
甲冪丁冪己冪相乘乙乘甲冪丙冪己冪相乘
庚冪辛冪相乘五位各冪丙冪丁冪相乘丙冪
甲冪丙三乘冪相乘甲正甲冪乙冪丁冪相乘
丙冪辛冪相乘乙冪丙冪丙冪庚冪相乘甲冪
庚冪相乘六位負　方冪辛冪相乘丙冪丁冪
相乘甲冪丙冪相乘二級空　廉級甲冪乙冪
位　位正甲冪己冪相乘一
負實級括（房）○（乙甲巾巾）寄辰位　甲冪寅位相乘得
負即眞
之（丙甲巾巾）內減丙冪卯位相乘數餘實
術號房　級
廉級得（己甲巾巾）甲冪乙冪丙冪己冪相乘甲
如舊　冪乙冪丙冪辛冪相乘甲冪
丙冪丁冪庚冪相乘甲冪丁三乘冪己冪相乘
甲冪丁冪己三乘冪相乘乙冪丙冪丁冪庚冪
相乘乙冪丙冪丁冪辛冪相乘乙冪丙冪庚冪
辛冪相乘丙冪丁冪庚冪辛冪相乘九位各正
甲冪乙冪丙冪庚冪相乘甲冪乙冪丁冪己冪
相乘甲冪丙冪丁冪己冪相乘甲冪丙冪丁冪
辛冪相乘乙三乘冪丙冪辛冪相乘乙冪丙冪
辛三乘冪相乘丙冪丁三乘冪庚冪相乘丙冪
丁冪庚三乘冪相乘八位各負　方級空　上
廉級甲冪乙三乘冪相乘甲冪丙冪丁冪相乘

二位各正甲幂乙幂丙幂相乘甲幂乙幂丁幂相乘甲幂乙幂己幂相乘甲幂丁幂己幂相乘
四位各負下廉級空隅級甲幂乙幂相乘一位正括之上廉級負去甲幂眞術號
亢却書甲幂實級正號箕得|箕○|亢甲巾○|乙甲巾巾寄巳位丑位寅
位相乘得內減子位卯位相乘數餘實級乙幂丙幂己幂
氐相乘乙幂丁幂尾相乘丙三乘幂尾相乘丁三乘幂己幂氐相乘丁幂巳三乘幂氐相乘五
位各正乙幂丙幂尾相乘乙幂丁幂己幂氐相乘丙幂丁幂己幂氐相乘丙幂丁幂尾相乘丙
幂己幂尾相乘丁幂己幂尾相乘六位各負方級空上廉級乙三乘幂氐相乘丙幂丁幂
氐相乘己幂尾相乘三位各正乙幂丙幂氐相乘乙幂丁幂氐相乘乙幂己幂氐相乘乙幂尾
相乘丙幂尾相乘丁幂己幂氐相乘六位各負下廉級空隅級乙幂氐相乘一位正括
之上廉級負卽眞術號斗實級正號牛得|牛○|斗○|氐乙巾亦以辰位
相乘之式略得數寄左
列巳位自乘之之式略與
寄左相消得得後式

四十五

兩式各縮空級前式爲平方後式爲三乘方而括之乃隅級者各單
位故如舊先前式方級黑圈一去丁幂合之黑圈二
去戊幂合之又赤圈一去壬幂相減異名者皆相減也後
徼此赤圈二去丙幂相減却乘所去之某名而後
各如次序幷之以赤符減黑符餘負眞術號角
實級赤圈一去丙幂壬幂合之赤圈二去丁

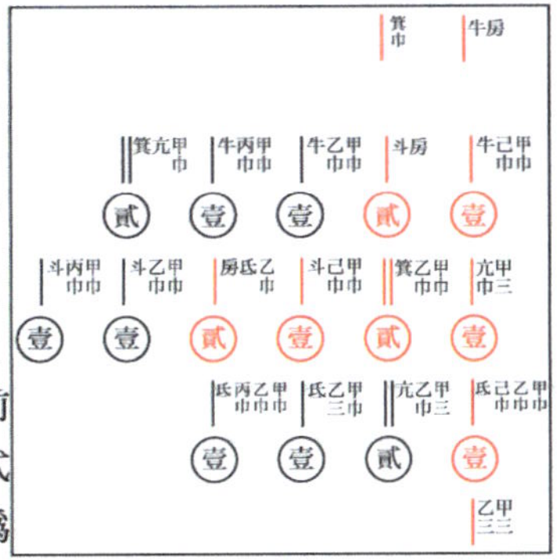

冪子冪合之赤圈三去戊冪子冪相減又黑圈
一去丙冪合之黑圈二去丙冪丁冪合之
黑圈三去丁冪壬冪相減却乘所去之某名而
後各如次序幷之以黑符減赤符餘正眞術號
心後式下廉級先遍去甲冪乙冪而後黑圈壹去氏合
之却乘氏幷黑圈貳而內減赤圈壹餘負眞術
號女却以遍去之甲冪乙冪書之上廉級赤圈壹去甲冪
合之赤圈貳去乙冪合之又黑圈壹去甲冪合
之却乘所去之某名而後赤符幷內減黑符餘
正眞術號虛方級黑圈壹去甲冪合之却乘
甲冪幷黑圈貳而內減赤圈貳壹餘負眞術號危
實級正依舊作兩位而布之

四十六

此式級數不均若疊之則雖開方之乘數加舊諸級悉增三次之乘數故不疊之唯借空二級于前式之下而準三乘方也

前式 心 角 戊巾 ○ ○

括
後式 牛房 危 虛 女乙甲巾巾 乙甲三三

箕巾

以後式隅級遍乘前式

去空一級而得換一式

去空一級而減

一式得換二式

虛上乃廉後級式遍

乘前式加二

式得換三式

箕戊巾巾	牛房戊巾	虛心
女心乙甲巾巾	虛角	危戊巾
	女角乙甲巾巾	心乙甲三三
		角乙甲三三

	女心乙甲巾巾
女角乙甲巾巾	心乙甲三三
女戊乙甲巾巾巾	角乙甲三三
	戊乙甲巾三三

心乙甲三三
角乙甲三三
戊乙甲巾三三
○

以後式下廉級遍乘前式

以戊冪隅乃級前式遍乘

後式以減二式又以

以角方前級式遍乘後式加

三式以危方乃級後式遍乘

前式減三式得換四式

一式遍縱省甲三乘冪與乙三乘冪

牛房角	箕角巾
箕戊巾巾	虛心
	女心乙甲巾巾
	心乙甲三三

又二式三式四式下級各橫省甲三

危心	牛房戊巾		

乘冪與乙三乘冪而後括之
一式皆單位故各循舊　二式隅級與一式廉
級全同　廉級去甲冪與乙冪而括之負眞術
號室　方級去甲冪與乙冪而括之正眞術號
壁各却以甲冪與乙冪書之　實級負單位故
如舊　三式隅級與一式方級全
同　廉級與二式方級全同　方
級括之負眞術號奎　實級括之
負眞術號婁　四式隅級與一式
實級全同　廉級與二式實級全
同　方級與三式實級全同　實
級括之正眞術號胃　如此相對

一	二	三	四
(房) 心	(斗) 女心乙甲巾巾	(危) 婁	(婁) 胃
(氐) 角	(箕) 壁乙甲巾巾	(虛) 奎	(危) 婁
(亢) 戊巾	(尾) 室乙甲巾巾	(箕) 壁乙甲巾巾	(斗) 女心乙甲巾巾
(角) ○	(亢) 戊巾	(氐) 角	(房) 心

之級其號相同故假書舊諸級名而依三乘方變乘
法相乘之乃當一式下級者爲空故相乘數無之矣

角尾虛婁相乘一段加　空
亢冪危冪相乘一段加　‖婁戊巾三　寄　(八)
亢房斗虛相乘二段加　‖奎女心戊乙甲巾巾巾巾　寄　(五)
氐房尾危相乘二段加　‖婁室心角乙甲巾巾　消　(參)
角尾危冪相乘一段減　空

角箕斗危相乘二段加　空
亢氐箕婁相乘二段加　‖胃壁角戊乙甲巾巾巾　消　(貳)
氐冪斗冪相乘一段加　|女心角乙甲巾巾巾三三　寄　(一)
房冪箕冪相乘一段加　|壁心乙甲巾巾三三　寄　(二)
角箕冪婁相乘一段減　空

角斗冪虛 相乘段一減 空

亢氏斗危 相乘段二減 ‖婁女心角戊乙甲 巾巾巾 寄 ㊂

氏冪尾婁 相乘段一減 |胃室角乙甲 巾巾巾 寄 ㊅

房冪尾婁 相乘段一減 |奎室心乙甲 巾巾巾 消 ㊃(肆)

亢冪虛婁 相乘段一減 |胃奎戊 三 寄 ㊆

亢房箕危 相乘段二減 ‖婁壁心戊乙甲 巾巾巾 寄 ㊃

氏房箕斗 相乘段二減 ‖壁女心角乙甲 巾 三三 消 ㊀(壹)

右各如次序赤圈八位相幷爲眞術寄左數黑圈四位相幷爲眞術相消數也七斜已上皆準此例而宜求之矣

大成算經卷之十終

四十八

Notes on Collation of Complete Book of Mathematics Vol. 10: Geometry

The following are the call numbers of manuscripts made use of in these notes:

NK: T20 / 70 in the Main Library of University of Tokyo. It used to be in the Nan-Ki Library [南葵文庫] of Kii Tokugawa family [紀伊徳川家].

DT: Date Library [伊達文庫] 1682 KD090-セ 5 / The Later Bestowals of Seki's School [關算後伝] no. 45 in Miyagi Prefectural Library [宮城県図書館].

1v. l. 8 : 反 is written 及 in DT.
1v. l. 9 : 下 is missing in DT.
1v. l. 9 : byscript 方巾 is missing in DT.
1v. l. 10 : 如前以一半積湊下 is missing in DT.
1v. lower figure : 下 is missing in NK.
2v. l. 1 : 于 is 千 in DT.
3r. l. 9 : 有 before 直 is missing in both NK and DT.
3v. l. 4 : 横 is written 縦 in both NK and DT.
3v. l. 4 : 七寸 is written 一尺二寸 in both NK and DT.
3v. l. 5 : 縦 is written 横 in both NK and DT.
5r. 6r. and 7r upper figures : 股 is missing in both NK and DT.
5r. l. 11 : 加 is written 如 in DT.
8r. left figure and 8v. both figures: 弦 is missing in both NK and DT.
8v. l. 12 : 曰 is inserted between 者 and 置 in DT.
9r. l. 4 : 問 is written 開 in DT.
9v. l. 4 : 四百三 is written 四百二 in DT.
10r. l. 8 : 三寸 is written 三十 in DT.
10v. l. 4 : 箇 is of normal size in DT.
11r. l. 6 : 十五 is written 千五 in DT.
11v. upper figure : 勾, 方, 股 are written in DT.
13r. left figure : 勾 is missing in DT.
13v. l. 3 : 遞 is replaced by a space in DT.
15r. l. 12 : 八寸 is written 八十 in NK.
16r : 差一寸 is missing in DT.
16r. 差三寸, l. 2 : 勾三寸 is written 勾五寸 in DT.
16r. 差四寸, l. 2 : 勾二寸 is written 三寸 in DT.
16v. 差五寸, l. 5 : 三分 is written 二分 in DT.
16v. 差六寸, l. 2 : 之二 is written 之 in DT.
16v. 差六寸, l. 10 : 寸 is written 七 in DT.
16v. 差七寸, l. 13 : 寸之九 is written 之九 in DT.
17r. l. 10 : 一十四 is written 一寸四 in DT.
17v. figure : 後 is missing at the right end in DT.
17v. ll. 8, 9 & 12 : 右 looks like 石 in NK.
17v. l. 9 : 斉直問 is replaced by 3 spaces in DT.
18r. figure : A pine tree is missing in DT.
18v. ll. 1 & 7 : 間 is written 問 in DT.

19r. figure : 退行三丈 is written 退行二丈.
20v. figure : 引去五尺 and 出水一尺 are missing in DT.
20v. l. 11 : 已 is written 己 in NK, and 巳 inDT.
21r. l. 4 : 至 is written 互 in DT.
21v. l. 3 : Two 于 are written 千 in DT.
21v. l. 6 : 末 is written 未 in DT.
21v. l. 9 : 偶 is written 隅 in NK.
21v. l. 12 : 起其術 is missing in NK.
22r. l. 1 : 自 is written 留 in DT.
22r. l. 1 & 22v. ll. 1 & 4 : 巳 is written 己 in both NK and DT.
22r. l. 5 : 乃分 is inserted before 作四斜 in NK and 2 spaces remain in DT.
23r. l. 9 : 短 is written 矩 in NK.
23r. l. 9 : 于 is written 干 in both NK and DT.
24r. l. 9 : 千九百 is written 十九百 in DT.
24r. l. 10 : 十三寸 is written 寸三寸 in DT.
24v. l. 2 : 伸 is written 仲 in DT.
25r. l. 12 : 容 is written 客 in DT.
25r. figure : 大斜 is missing in NK.
26r. figure : 右闊 is written 右交闊 in both NK and DT.
26r. l. 11 : 同 is written 日 in DT.
26r. l. 12 : 内 is overlapping in DT.
26v. figure : 小斜 is written 斜 in DT.
27r. l. 2 : 己 is written 巳 in DT.
27r. l. 7 : 丁 is written 下 in DT.
27r. l. 9 : The counting rod 0 of the second medium class is missing in DT.
27v. ll. 1–2 : The counting rods representing 118629 is missing in DT.
27v. ll. 1–4 : The counting rods representing $60800X^2$ is missing in DT.
27v. l. 9 : 末 is written 未 in DT.
27v. l. 11 : 己 is written 巳 in DT.
28r. l. 11 : 以減 is erroneously written 内減 in both NK and DT.
28v. l. 3 : 丙冪己 is missing in DT.
28v. l. 7 : 一十八位 is erroneously written 一十七位 in both NK and DT.
29r. l. 11 : 幷 is written 第 in DT.
30r. figure : 壬 is missing in NK.
30r. ll. 6 & 8 : 己 is written 巳 in DT.
30r. l. 6 : There is a space after 己冪 in DT.
30v. l. 3 : 以 is written 與 in DT.
30v. ll. 5 & 7 : 己 is written 巳 in DT.
30v. ll. 10 & 11: 巳 is written 己 in NK.
31r. ll. 3×2, 4, 6 & 8 : 己 is written 巳 in DT.
31v. ll. 4, 5 & 11: 己 is written 巳 in DT.
32r. l. 7 : 丁 is written 下 in DT.
32r. l. 8 : 反 is written 及 in DT.
32r. l. 12 : 甲 and 己 in 甲冪己冪 are actually the pseudonym 假甲 of 辛 and 假己 of 乙. Similarly, 乙冪戊冪 means 戊冪丁冪, etc.

32v. l. 1 : 丁 is written 下 in DT.
32v. l. 1 : The counting rod 0 is missing in DT.
32v. ll. 6, 9 & 11 : 己 is written 巳 in DT.
33r. ll. 1, 3 & 4 : 己 is written 巳 in DT.
33r. l. 1 : byscript 乙巾 is written 巳巾 in DT.
34r. l. 2 : 位 is written 住 in DT.
34r. l. 4 : 某 is written 其 in DT.
34v. ll. 9, 10, 11 & 12 : 巳 is written 己 in NK.
34v. ll. 10, 11 & 12 : 巳 is written 己 in DT.
35r. figure : 辰 is missing in DT.
35r. 1. 2 : 己 is written 巳 in DT.
35r. l. 7 : 減 is written 戊 in DT.
35v. ll. 6 & 7 : 己 is written 巳 in DT.
36r. l. 2 : 三 is written 二 in DT.
36v. l. 6 : 列幷乙冪與丁冪 is written 乙冪内減與丁冪餘 in NK.
38r. l. 8 : 己 is written 巳 in DT.
39v. l. 1: 己 is written 巳 in DT.
41v. l. 2 : 名己 is written 名乙 in both NK and DT.
41v. ll. 2 & 6 : 四 is written 五 in both NK and DT.
41v. l. 4 : 五 is written 四 in in both NK and DT.
41v. l. 8 : 己 is written 巳 in DT.

42r. l. 7 : 乙 in 乙冪 is actually the pseudonym 假乙 of 丑. Similarly, 甲冪己冪 on line 10 means 子冪丁冪, etc.

42r. l. 9 : The byscript 壬巾 is written 子巾 in DT.
42r. l. 10 : 己 is written 巳 in DT.
42v. ll. 7, 9 & 10 : 己 is written 巳 in DT.
42v. Former Equation, l. 8 : byscript 子巾 is written 壬巾.
42v. l. 9 : 丙巾 is written 戊巾 in DT.
44r. l. 5 : byscript 甲巾 is written 丁巾 in DT.
44r. l. 12 : The counting rod 1 besides byscript 尾 is missing in DT.
44v. l. 7 : 如 is written 加 in DT.
44v. ll. 8 & 10×2 : 己 is written 巳 in DT.
44v. l. 11: The uppermost 冪 is written 并 in DT.
45r. ll. 3 & 11: 巳 is written 己 in NK.
45r. l. 9 : 己 is written 巳 in DT.

45v. ll. 1 & 2 : The color of the counting rod 1 in the first medium class is turned into black in DT.

45v. ll. 5 & 6 : byscript 斗 in the third class is turned into a counting rod −1 in DT.

46r. l. 8 : 貳 is of normal size in DT.
46v. l. 1 : 加 is written 如 in DT.
46v. l. 3 : 箕巾 is missing in DT.
48r. l. 10 : 巳 is written 己 in both NK and DT.

Seki's Theory of Elimination as Compared with the Others'

Takefumi Goto and Hikosaburo Komatsu*

Abstract This is an enlarged version of the authors' papers published in Journal of Northwest University (Natural Science Edition), 33 Nos.3 and 4 (2003). Added is Section 3, where proofs are given of the main properties of determinants and resultants as originally introduced by Seki Takakazu (1642?–1708), Étienne Bézout (1739–83), James J. Sylvester (1814–97) and Arthur Cayley (1821–95).

1 Determinants, Resultants and Discriminants in Japan in the Seventeenth Century and in Europe in the Eighteenth and Nineteenth Centuries

Abstract It is by now well known that Japanese mathematicians introduced determinants in the seventeenth century but it is not necessarily understood well why and how they made use of determinants. We follow their calculations and show that what they did is essentially the same as the elimination method of auxiliary variables from systems of algebraic equations with more than one unknowns as developed in Europe in the eighteenth and nineteenth centuries.[1]

Takefumi Goto
Edogawa-ku Mizue Middle School, 4-16, Edogawa, Edogawa-ku, Tokyo 134-0013, Japan
e-mail: white-wing@msd.biglobe.ne.jp

Hikosaburo Komatsu
Graduate School of Mathematical Sciences, University of Tokyo, 3-8-1, Komaba, Meguro-ku, Tokyo 153-8914, Japan e-mail: komatsu@ms.u-tokyo.ac.jp

* The authors thank the editors of the Journal of Northwest University, first for publishing the original papers, the Japanese version of which had been rejected by the History of Science Society of Japan in 2002, and secondly for permitting their reproduction in these Proceedings. The second author is supported in part by JSPS Grant-in-Aid for Scientific Research (C) no. 20540107.

[1] This section was published in Journal of Northwest University (Natural Science Edition), 33 No.3, pp. 363–367 (2003). Footnotes are added on this occasion.

1.1 Chinese mathematics as the background

Japan imported mathematics from China twice in her history. In the ancient time it was the practical mathematics represented by the textbook Nine Chapters of Mathematics [九章算術 Jiǔ-Zhāng Suàn-Shù] [1] for schools of government officials since the Han [漢] dynasty.

The second importation took place around the year 1600 and Japanese learned algebraic equations from books Yang Hui's Methods of Mathematics [楊輝算法 Yáng Huī Suàn-Fǎ] [2] and Introduction to Mathematics [算学啓蒙 Suán-Xué Qǐ-Méng] [3] by Zhū Shì-Jié [朱世傑]. In these books an algebraic equation

$$f(x) = a_0 + a_1 x + \cdots + a_n x^n = 0 \tag{1}$$

is denoted by a column

$$\begin{pmatrix} a_0 \\ a_1 \\ \vdots \\ a_n \end{pmatrix} \tag{2}$$

of calculating rods representing the numerical coefficients a_i. Since this arrangement comes from the Chinese way of writing, we will here denote the same by a row

$$\begin{pmatrix} a_0 & a_1 & \cdots & a_n \end{pmatrix} \tag{3}$$

of coefficients. The first place for the constant is called res [実 shí], the second for the linear term modulus [方 fāng], the third for the quadratic term medium [廉 lián], $\cdots$, and the last for the term of the highest degree ultima [隅 yú].

In this terminology a polynomial $f(x)$ could not be distinguished from the equation $f(x) = 0$. It seems that this caused sometimes carelessness of Japanese mathematicians in the signs of polynomials.

Given a number ξ, they could calculate the coefficients a_i' of the shifted equation

$$f(\xi + x) = a_0' + a_1' x + \cdots + a_n' x^n = 0 \tag{4}$$

very quickly by manipulating rods. Thus they were able to compute an approximate solution ξ digit by digit as accurate as they wished by making the res a_0' smaller and smaller. We note that they knew that

$$a_1' = f'(\xi) = a_1 + 2a_2\xi + \cdots + na_n\xi^{n-1}. \tag{5}$$

The seventeenth century was the time when the mass education started in Japan. There was, therefore, a large demand for textbooks of mathematics and accordingly more advanced books were also published. Some of them contained a list of unsolved problems. Successors published their solutions and left their own problems. Soon the problems became very complicated. A problem Seki solved needed an equation of degree 1458 !

The books [2] and [3] deal with only algebraic equations of one unknown with numerical coefficients. No symbols were used except for verbal analysis of problems. The expression (2) didn't mention the indeterminate x at all. Seki Takakazu [関孝和] (1642?–1708) was the first in Japan to consider systems of algebraic equations with more than one unknowns. He invented expressions of polynomials of any number of letters and admitted them as coefficients a_i in (2). Later Seki, Tanaka Yoshizane [田中由真] (1651–1719) and other Japanese mathematicians developed a systematic theory of elimination with use of determinants. Izeki Tomotoki [井関知辰] published the first book [6] on elimination in 1690.

1.2 Determinants and resultants

Seki classified mathematical problems into the following three classes: the explicit problems [見題 kendai] which can be solved with arithmetic, the implicit problems [隠題 indai] which needs an algebraic equation of one unknown, and the concealed problems [伏題 fukudai] which needs algebraic equations with more than one unknowns, and wrote for each class a book of solution. Determinants and resultants appeared for the first time in Methods of Solving Concealed Problems [解伏題之法 Kaihukudai no Hō][4] of these trilogy dated 1683 and were discussed in detail in Volume 17 of Complete Book of Mathematics [大成算経 Taisei Sankei] [8] written in 1683 through 1711 by Seki and his pupils Takebe Kataakira [建部賢明] (1661–1716) and Takebe Katahiro [建部賢弘] (1664–1739). Unfortunately neither of these books was published but copies were inherited in Seki's school of mathematicians.

Without giving a precise terminology Seki [4] defines the *determinant* of n equations of degree $n-1$

$$\begin{pmatrix} c_{10} & c_{11} & \cdots & c_{1,n-1} \\ c_{20} & c_{21} & \cdots & c_{2,n-1} \\ & & \vdots & \\ c_{n0} & c_{n1} & \cdots & c_{n,n-1} \end{pmatrix}, \text{ that is, } \begin{cases} c_{10}+c_{11}x+\cdots+c_{1,n-1}x^{n-1}=0, \\ c_{20}+c_{21}x+\cdots+c_{2,n-1}x^{n-1}=0, \\ \qquad \cdots\cdots \\ c_{n0}+c_{n1}x+\cdots+c_{n,n-1}x^{n-1}=0, \end{cases} \tag{6}$$

to be the constant term of the single equation, or the equation itself, obtained by multiplying the first equation by the cofactor of c_{10}, the second equation by the cofactor of $c_{20}, \cdots$, the last equation by the cofactor of c_{n0}, and then adding all equations (see Corrections at the end of §13 in Mikami [24])[2]. The constant term of the equation is equal to the determinant and the other terms vanish. Therefore, his determinant is the same as in Europe. It is clear from this definition that the determinant vanishes whenever system (6) has a common root.

Seki's book [4] makes an error in the expansion of determinants of order ≥ 5 as pointed out by Mikami [23, pp.13–24], Horiuchi [28, pp.192–194] and many others since the year 1715. We remark, however, that the error had already been

[2] See §§2.1, 2.2, 3.1 and 3.2 for the definition of determinants and their main properties.

corrected in Izeki [6] and Complete Book [8]. We will further discuss on this issue in a forthcoming paper[3] with a new proposal of corrections.

To eliminate a variable x, we usually start with two equations:

$$f(x) = a_0 + a_1 x + \cdots + a_n x^n = 0, \tag{7}$$

$$g(x) = b_0 + b_1 x + \cdots + b_m x^m = 0, \tag{8}$$

where we may assume that $n \geq m$, $a_n \neq 0$ and $b_m \neq 0$.

Seki [4, 8] transforms these equations into n equations (6) of degree less than n and then applies the determinant.

As exhibited in Mikami [23, p.10] and Horiuchi [28, p.190] the first transformed equation [換式 kansiki]

$$h_1(x) = d_{10} + d_{11} x + \cdots + d_{1,n-1} x^{n-1} = 0 \tag{9_1}$$

is obtained by eliminating the top terms of $f(x)$ and $x^{n-m} g(x)$, that is,

$$h_1(x) = b_m f(x) - a_n x^{n-m} g(x). \tag{10_1}$$

Then, the i-th transformed equation

$$h_i(x) = d_{i,0} + d_{i,1} x + \cdots + d_{i,n-1} x^{n-1} = 0 \tag{9_i}$$

for $1 < i \leq m$ is defined by

$$h_i(x) = x h_{i-1}(x) + b_{m-i+1} f(x) - a_{n-i+1} x^{n-m} g(x), \tag{10_i}$$

in which the coefficient of x^n vanishes. The remaining $n-m$ transformed equations are the shifts[4]

$$h_i(x) = x^{i-m-1} g(x) = 0, \quad m+1 \leq i \leq n. \tag{11}$$

The resulting determinant

$$\mathscr{R}(f,g) = \begin{vmatrix} d_{10} & d_{11} & & \cdots\cdots & & & d_{1,n-1} \\ d_{20} & d_{21} & & \cdots\cdots & & & d_{2,n-1} \\ & \cdots & & \cdots\cdots & & & \\ d_{m,0} & d_{m,1} & & \cdots\cdots & & & d_{m,n-1} \\ b_0 & b_1 & \cdots & b_m & 0 & \cdots & 0 \\ 0 & b_0 & b_1 & \cdots & b_m & 0 & 0 \\ 0 & & \ddots & & & \ddots & 0 \\ 0 & \cdots & 0 & b_0 & b_1 & \cdots & b_m \end{vmatrix} \tag{12}$$

---

[3] That is, §2 of this paper.

[4] To be exact, this is the definition of transformed equations by Bézout [13] and, in case $n > m$, is different from that of Seki [4]. See §3.4 of this paper.

is easily shown[5] to be equal to

$$\begin{vmatrix} a_0 & a_1 & \cdots & \cdots & \cdots & a_n & 0 & \cdots & 0 \\ 0 & a_0 & a_1 & \cdots & \cdots & \cdots & a_n & 0 & 0 \\ & & \ddots & & & & & \ddots & \\ 0 & \cdots & 0 & a_0 & a_1 & \cdots & \cdots & \cdots & a_n \\ b_0 & b_1 & \cdots & \cdots & b_m & 0 & \cdots & & 0 \\ 0 & b_0 & b_1 & \cdots & \cdots & b_m & 0 & \cdots & 0 \\ & & \ddots & & & \ddots & & & \\ & & & & & & & & 0 \\ 0 & \cdots & \cdots & 0 & b_0 & b_1 & \cdots & \cdots & b_m \end{vmatrix}, \tag{13}$$

which is formally the same as the standard definition of the *resultant* in Europe, but is actually multiplied by $(-1)^{nm}$ because the terms of a polynomial are written in the descending order in Europe.

Let ξ_i and η_j be the roots of (7) and (8), respectively. Then we have

$$\begin{aligned} \mathscr{R}(f,g) &= a_n{}^m b_m{}^n \prod_{i=1}^{n} \prod_{j=1}^{m} (\eta_j - \xi_i) \\ &= b_m{}^n \prod_{j=1}^{m} f(\eta_j) = (-1)^{mn} a_n{}^m \prod_{i=1}^{n} g(\xi_i) \end{aligned} \tag{14}$$

as shown by Cauchy [16] and many others. Therefore, the eliminated equation

$$\mathscr{R}(f,g) = 0 \tag{15}$$

gives a necessary and sufficient condition in order that the system of two equations (7) and (8) has a solution in an algebraically closed field containing all coefficients of the equations in the system.[6]

Formula (13) of resultants is attributed to Sylvester [15] published in 1840 but an essentially same formula of elimination had already been given by Euler [12] in 1764[7]. Seki's formula (12) was also discovered in the same year by É. Bézout (1739–1783) in [13, pp. 319–323]. This should have been the quickest way of eliminating an unknown from two general algebraic equations. Earlier in 1748 Euler [10] employed a more primitive method, due to Newton, and also the resultant defined by (14), to estimate the number of the intersection points of two algebraic curves on a plane, which is known today as Bézout's theorem.

[5] See §3.3.

[6] under the generic condition that $a_n b_m \neq 0$.

[7] See the editor's note to Knobloch's article in these Proceedings on p. 238 f.

In Japan, Izeki [6] contains a list of entries of formula (12) for $n = m \leq 6$ and the complete expansion of determinants of order ≤ 5. Therefore, at the end of the 17th century, Japanese mathematicians were able to compute the resultant of at least quintic equations.

On the other hand, Cayley [20] calculated by a different method of Hirsch [14] the resultant of two general algebraic equations of degrees $m \leq n \leq 4$. In case $n = m = 4$, it has 219 terms.

1.3 Discriminants

Seki introduced the *discriminant* of an algebraic equation (7) in his book Methods of Equation Modifications [開方翻変之法 Kaihō Honpen no Hō] [5]. It is also discussed in Volume 3 of Complete Book of Mathematics [8]. Seki's terminology exactly vanishing condition of the modulus class [適尽方級法 tekijin hōkyū hō] means the "vanishing condition of modulus (5) exactly at a root" ξ. Since this is equivalent to saying that the original equation $f(x) = 0$ and the equation $f'(x) = 0$ for modulus have a common root, we have for this condition

$$\mathscr{R}(f, f') = 0. \tag{16}$$

It follows from (14) that if

$$f(x) = a_n \prod_{i=1}^{n} (x - \xi_i), \quad \text{then} \quad f'(x) = a_n \sum_{i=1}^{n} \prod_{j \neq i} (x - \xi_j),$$

so that we have

$$\mathscr{R}(f, f') = (-1)^{\frac{n(n-1)}{2}} a_n^{2n-1} \prod_{i<j} (\xi_i - \xi_j)^2. \tag{17}$$

This is divisible by a_n as seen from (13). In Europe the *discriminant* of $f(x)$ is

$$D(f) = (-1)^{\frac{n(n-1)}{2}} a_n^{-1} \mathscr{R}(f, f') = a_n^{2n-2} \prod_{i<j} (\xi_i - \xi_j)^2 \tag{18}$$

but Seki and his pupils paid little attention to the sign and called

$$\mathscr{D}(f) = a_n^{-1} \mathscr{R}(f, f') = n^{2-n} \mathscr{R}(nf - xf', f')[= 0] \tag{19}$$

the exactly vanishing condition of the modulus class.

Seki computed in [5] the discriminants $\mathscr{D}(f)$ for equations $f(x) = 0$ up to degree 4, and in Volume 3 of [8] there is an almost correct list of 59 terms of the discriminant $\mathscr{D}(f)$ of a quintic equation.

The subjects of Methods of Equation Modifications [5] are first to count the number of positive or real roots, and secondly when there are no real roots, to modify a coefficient so that the modified equation have real roots. The discriminant was employed to know the extreme values of the coefficient of an equation with real roots. Yet, Seki does not seem to have noticed the fact that the sign of the discriminant decides the number of real roots of a real quadratic equation, which everybody knows today.

2 A Correction of Seki's Error in the Expansion of Determinants

Abstract Seki Takakazu (1642?–1708) is a Japanese mathematician who introduced determinants and resultants as a means of eliminating an auxiliary unknown from systems of algebraic equations with more than one unknowns. Unfortunately his theory has not received due respects because of his error in the expansion of determinants of order 5. Since none of the so-called corrections are persuasive, the authors propose a new one.[8]

2.1 Seki's determinants

As shown in our previous paper[9] Seki [4] denotes by an $n \times n$ matrix

$$\begin{pmatrix} c_{10} & c_{11} & \cdots & c_{1,n-1} \\ c_{20} & c_{21} & \cdots & c_{2,n-1} \\ & \cdots\cdots & & \\ & \cdots\cdots & & \\ c_{n0} & c_{n1} & \cdots & c_{n,n-1} \end{pmatrix} \tag{1}$$

the system of n equations of degree less than n:

$$\begin{cases} c_{10} + c_{11}x + \cdots + c_{1,n-1}x^{n-1} = 0, \\ c_{20} + c_{21}x + \cdots + c_{2,n-1}x^{n-1} = 0, \\ \cdots\cdots \\ c_{n0} + c_{n1}x + \cdots + c_{n,n-1}x^{n-1} = 0, \end{cases} \tag{2}$$

where coefficients c_{ij} are polynomials in unknowns $y, z, \cdots$ other than x.

His theory of elimination in Methods of Solving Concealed Problems [4] starts with the observation that the determinant[10]

[8] This section was published in Journal of Northwest University (Natural Science Edition), 33 No.4, pp. 376–380 (2003). Footnotes are added on this occasion.

[9] That is, §1 of this paper.

[10] See §3.1.

$$\begin{vmatrix} c_{10} & c_{11} & \cdots & c_{1,n-1} \\ c_{20} & c_{21} & \cdots & c_{2,n-1} \\ & \vdots & & \\ c_{n0} & c_{n1} & \cdots & c_{n,n-1} \end{vmatrix} = \begin{vmatrix} c_{10}+c_{11}x+\cdots+c_{1,n-1}x^{n-1} & c_{11} & \cdots & c_{1,n-1} \\ c_{20}+c_{21}x+\cdots+c_{2,n-1}x^{n-1} & c_{21} & \cdots & c_{2,n-1} \\ \vdots & & \vdots & \\ c_{n0}+c_{n1}x+\cdots+c_{n,n-1}x^{n-1} & c_{n1} & \cdots & c_{n,n-1} \end{vmatrix} \tag{3}$$

vanishes whenever system (2) has a common root. He shows this by induction on n.

In case $n=2$, we have

$$\begin{pmatrix} B & A \\ D & C \end{pmatrix} \quad \text{or} \quad \begin{cases} B+Ax=0, \\ D+Cx=0. \end{cases} \tag{4}$$

He multiplies the first equation by C, the second by A and subtracts the products to get the equation

$$\begin{vmatrix} B & A \\ D & C \end{vmatrix} = BC-DA=0. \tag{5}$$

This procedure is known already in Nine Chapters of Mathematics [1] and is called a cross multiplication [維乘 wéi chéng].

If $n=3$, the determinant of the system

$$\begin{pmatrix} C & B & A \\ F & E & D \\ I & H & G \end{pmatrix} \quad \text{or} \quad \begin{cases} C+Bx+Ax^2=0, \\ F+Ex+Dx^2=0, \\ I+Hx+Gx^2=0, \end{cases} \tag{6}$$

is *defined* to be the *equation* obtained as the sum of the first equation multiplied by the cofactor $\begin{vmatrix} E & D \\ H & G \end{vmatrix}$ of C, the second by $-\begin{vmatrix} B & A \\ H & G \end{vmatrix}$ of F, and the third by $\begin{vmatrix} B & A \\ E & D \end{vmatrix}$ of I.

Seki [4] did not introduce a symbol for a determinant nor for a cofactor. What he actually showed is the fact that with a suitable choice of the signs $\pm$ of six equations

$$\begin{aligned} (C+Bx+Ax^2)EG&=0, & \pm(C+Bx+Ax^2)HD&=0, \\ \pm(F+Ex+Dx^2)BG&=0, & \pm(F+Ex+Dx^2)HA&=0, \\ \pm(I+Hx+Gx^2)BD&=0, & \pm(I+Hx+Gx^2)EA&=0, \end{aligned} \tag{7}$$

the total sum has only the constant term which is equal to the determinant of today.

2.2 Seki's error and our proposal of correction

He continues a similar construction up to order $n=4$. Then, to evade the increased complexity according to the order, he states another method, called formula exchanges [交式 kōshiki] and oblique multiplications [斜乘 shajō], which generalize Sarrus' expansion in the case of order 3. However, if we calculate by the latter method as he wrote, we always have 0 as its value in case $n=5$, which is obviously an error. Many corrections have been proposed, by Matsunaga Yoshisuke [松永良弼] in 1715, and by Sugano Mototake [菅野元健] and by Ishikuro Nobuyoshi [石黒信由] both in 1798. The modern historians follow these old interpretations (cf. Hayashi [21], Mikami [23] et al.). Since they seem to be far from Seki's original intention, we will give here another correction. To do so, we first reproduce Seki's calculation of the determinant when $n = 4$:

Consider

$$\begin{pmatrix} D & C & B & A \\ H & G & F & E \\ L & K & J & I \\ P & O & N & M \end{pmatrix} \quad \text{or} \quad \begin{cases} D+\ Cx+\ Bx^2+\ Ax^3 &= 0, \\ H+\ Gx+\ Fx^2+\ Ex^3 &= 0, \\ L+\ Kx+\ Jx^2+\ Ix^3 &= 0, \\ P+\ Ox+\ Nx^2+\ Mx^3 &= 0, \end{cases} \tag{8}$$

where the coefficients are replaced from Seki's original letters, which represent 28 lunar mansions, into Roman capitals in the same order. Then, he writes 24 equations, successively, starting with $(DGJM$ creative $\mid 0 \mid CGJM \mid BGJM \mid AGJM)$ which is obtained by multiplying the first equation by the term GJM of the cofactor of D, where creative [生 sei] means a term to be added, while annihilative [剋 koku] to be subtracted. The next 0 means that the following coefficients of x, x^2 and x^3 cancel out with two terms of the same factors and with the opposite signs. The original list has a numbering of the canceling terms, which is omitted here. Then, we arrange all equations in the following way with the omission of the linear terms on, where the order of the equations is a little different from that in the text of Collected Works [4] but is restored to what we believe is in his original calculation[11]:

$$\begin{array}{llc} DGJM \text{ crea} & DOJE \text{ anni} & \\ HKNA \text{ anni} & HCNI \text{ crea} & [1 \quad 2 \quad 3 \quad 4] \\ LOBE \text{ crea} & LGBM \text{ anni} & \\ PCFI \text{ anni} & PKFA \text{ crea} & \\ & & \\ DOFI \text{ crea} & DGNI \text{ anni} & \\ HCJM \text{ anni} & HKBM \text{ crea} & [1 \quad 3 \quad 4 \quad 2] \\ LGNA \text{ crea} & LOFA \text{ anni} & \\ PKBE \text{ anni} & PCJE \text{ crea} & \\ & & \\ DKNE \text{ crea} & DKFM \text{ anni} & \\ HOBI \text{ anni} & HOJA \text{ crea} & [1 \quad 4 \quad 2 \quad 3]. \\ LCFM \text{ crea} & LCNE \text{ anni} & \\ PGJA \text{ anni} & PGBI \text{ crea} & \end{array} \tag{9}$$

[11] See these Proceedings "Methods of Solving Concealed Problems" Sheets 14–15 (pp. 484–485) for the original text and the Amended Sheets 1–2 (pp. 490–491) to our restoration.

In almost all texts of [4] including the one in Collected Works the arrangement in each block is the same as above but the places of blocks are interchanged between the upper right and the middle left and between the middle right and the lower left. In spite of this difference we believe that ours is the original arrangement because we obtain a Sarrus like figure when we illustrate the products in the upper blocks on the matrix (8), where the left block are composed of the principal diagonal and its parallels and the left of the anti-principal diagonal and its parallels. Similarly, the middle and the lower blocks represent the same kind of products for the matrices

$$\begin{pmatrix} D & B & A & C \\ H & F & E & G \\ L & J & I & K \\ P & N & M & O \end{pmatrix} \quad \text{and} \quad \begin{pmatrix} D & A & C & B \\ H & E & G & F \\ L & I & K & J \\ P & M & O & N \end{pmatrix}, \tag{10}$$

which are composed of the 1st, 3rd, 4th and 2nd columns, and the 1st, 4th, 2nd and 3rd columns of matrix (8), respectively. In order to remark this fact we added the sequences on the right of (9), which are not given in the original text. The current text says that it was revised for the second time in 1683. In the unrevised original the six blocks might have been set linearly. Then, the disorder could have happened on a revision.

After Table (9) the text says: "In each of the above, we make alternating multiplications of enumerated equations [逐式交乗 chikushikikôjô] to obtain the creative and the annihilative terms[12]. When the number of terms to be multiplied becomes large, it is not easy to see, and therefore, we employ formula exchanges [交式 kōsiki] and oblique multiplications [斜乗 shajō], instead [右各逐式交乗而得生剋也雖然相乗之数位繁多而不易見故以交式斜乗代之]."

The following section Formula Exchanges starts with the paragraph "The transformed three equations give rise to the transformed four equations. The transformed four equations give rise to the transformed five equations, and so on. The transformed two equations and the transformed three equations need no formula exchanges. The ordinary and the opposite [parities of the principal diagonal and its parallels] are both obtained recursively by adding 1. That is, if the number of equations is odd, all are ordinary; if it is even, the ordinary and the opposite [parities] alternate [從換三式起換四式從換四式起換五式逐如此 換二式換三式者不及交式也順逆共遞添一得次乃式数奇者皆順偶者順逆相交也]."

Then, the following table is given:

[12] This sentence should have been translated as "In the above, we have successively multiplied each equation alternatingly to obtain the creative and the annihilative terms. For this purpose our alteration on pp. 492–493 in these Proceedings would be more appropriate than the original table on Sheets 14–15.

Transformed 3 eqs.			Transformed 4 eqs.				Transformed 5 equations					
ord	ord	ord	ord	opp	ord	opp	[ord]	[ord]	[ord]	[ord]	[ord]	
1	2	3	1	2	3	4	1	2	3	4	5	
			1	3	4	2	1	3	2	5	4	
			1	4	2	3	1	4	5	2	3	
							1	5	4	3	2	
							1	2	4	5	3	
							1	4	2	3	5	(11)
							1	5	3	2	4	
							1	3	5	4	2	
							1	2	5	3	4	
							1	5	2	4	3	
							1	3	4	2	5	
							1	4	3	5	2	

This is the most enigmatic part of [4]. Our understanding is that "the transformed n equations" means the determinant of order n and the following part defines the signs of cofactors. The traditional interpretation due to Matsunaga (1715) regards this as a description of the rule for calculating the sequences in table (11) and even corrects the list. The editors of Collected Works [4] adopt this interpretation. But it is absurd. The sequences in table (11) represent the even permutations with 1 fixed. The middle sentence may mean the rule. In that case, we obtain (a generic part of) the next list by adding 1 to each number in the permutations corresponding to the creative products in the development of the previous determinant.

The following section Oblique Multiplications says "Applying each formula exchange [to the original matrix], we make oblique multiplications from left and from right, and then, obtain the creative and the annihilative terms. If a product hits an empty number, then delete it. If the number of transformed equations is odd, then the left oblique products [parallel to the principal diagonal in our notation] are creative, and the right oblique products annihilative, and if the number is even, then in both left and right oblique products the creative and the annihilative alternate [交式各布之從左右斜乘而得生剋也 若當空級者除之 換式数奇者以左斜乘為生以右斜乘為剋偶者左斜乘右斜乘共生剋相交也]".

Then, Seki illustrates the oblique products of the transformed two to five equations with an indication of the creative and the annihilative terms. See [4, pp. 156–157], [23, pp.12–13] or [28, pp.192–193] If, in particular, $n = 5$, i.e., for the general equations

$$\begin{pmatrix} E & D & C & B & A \\ J & I & H & G & F \\ O & N & M & L & K \\ T & S & R & Q & P \\ Y & X & W & V & U \end{pmatrix} \quad \text{or} \quad \begin{cases} E + Dx + Cx^2 + Bx^3 + Ax^4 = 0, \\ J + Ix + Hx^2 + Gx^3 + Fx^4 = 0, \\ O + Nx + Mx^2 + Lx^3 + Kx^4 = 0, \\ T + Sx + Rx^2 + Qx^3 + Px^4 = 0, \\ Y + Xx + Wx^2 + Vx^3 + Ux^4 = 0, \end{cases} \tag{12}$$

he explains that the counterparts of the upper blocks of (9) are

$$\begin{matrix} EIMQU \text{ crea} & EXRLF \text{ anni} \\ JNRVA \text{ crea} & JDWQK \text{ anni} \\ OSWBF \text{ crea} & OICVP \text{ anni} \\ TXCGK \text{ crea} & TNHBU \text{ anni} \\ YDHLP \text{ crea} & YSMGA \text{ anni} \end{matrix} \qquad 1\ \ 2\ \ 3\ \ 4\ \ 5, \tag{13}$$

but this is not true, because all products in the right block are also creative. Checking only the cases $n = 2,3$, and 4, Seki thought that there were always a creative and an annihilative oblique products starting at each place of the leftmost column in the uppermost blocks. This is no more the case when $n = 5$ or 6 or more generally if $n \equiv 1$ or 2 mod 4 as pointed out by Sugano and Ishikuro in 1798. Descartes' warning of precipitation applies to Seki.

In order to correct this error, it is enough to replace each product in the right block by a product from the annihilative terms with the first factor from E,J,O,T,Y fixed. To save the labor of computation, we can leave the first three factors as in the left block and choose the following[13] :

$$\begin{matrix} EIMQU \text{ crea} & EIMVP \text{ anni} \\ JNRVA \text{ crea} & JNRBU \text{ anni} \\ OSWBF \text{ crea} & OSWGA \text{ anni} \\ TXCGK \text{ crea} & TXCLF \text{ anni} \\ YDHLP \text{ crea} & YDHQK \text{ anni} \end{matrix} \qquad 1\ \ 2\ \ 3\ \ 4\ \ 5. \tag{14}$$

Other choices are also possible. Then, we make the same computations for all matrices obtained by the exchanges of columns from the original (12) according to the permutations in list (11). By adding all, we get the determinant of order 5.

In our interpretation formula exchanges meant the exchanges of columns, or the coefficients of the same degrees. We are led to this interpretation by Seki's way of computations. Moreover, there are two other reasons. In the preceding section in [4] Seki allows to factor out a common factor in the coefficients of the same degree. This is easily explained by our construction. Secondly, he had, at least, to convince himself of the fact that the coefficients of $x, x^2, \cdots, x^{n-1}$ cancel out when he added the equations to get the determinant. To establish a theorem, a formal proof was not demanded at that time, but Japanese mathematicians had to show sufficient evidence, e.g. general examples, to convince people. The vanishing of the coefficients of the linear term on is equivalent to the vanishing of the determinant with two identical columns, which is recursively proved by showing that the determinant changes its sign if we transpose the first and the second columns. Seki did not prove it but he could easily check it whenever he calculated determinants.[14]

---

13 See these Proceedings "Methods of Solving Concealed Problems" Amended Sheet 2 on p. 491 .

14 We give a proof in §3.1.

The Chinese character 式 [shiki] Seki used for formula in formula exchanges usually meant equation. Actually there were Japanese mathematicians who understood Seki's formula exchanges as we do, but for most of them it meant equation exchanges or permutation of rows. Fortunately or unfortunately, determinants change the sign by a transposition of two rows as well as of two columns. Therefore, they were able to compute correct values of determinants by the expansion with respect to the first row, without understanding Seki's idea. The first volume of Izeki's book [6] is a manual for computing the resultants by the latter method. There are no explanations why we are able to eliminate an indeterminate in this way[15]. Volume 17 of Complete Books [8] is essentially the same.

3 Mathematical Notes

Abstract Determinants and resultants were introduced by many mathematicians in many ways independently under quite different cultural backgrounds. Therefore, it is not necessarily easy to understand what determinants meant in a historical setting in question. To ease these difficulties, the authors develop here rudiments of the theory of determinants based only on the definition of Seki [4] (1683) and Cayley [17] (1841) who were the inventors of the present notation of determinants and both intended to apply them to solve problems in geometry.

In the first section of this paper the authors claimed that Seki's resultants were the same as Bézout's. But this is not true when the degrees of two equations are different. A comparison of these two is given in the last subsection.

3.1 Determinants

When Seki [4] and Bézout [13] introduced determinants they did so as a means to represent resultants. They didn't even name determinants and were interested only in the equation that a determinant was equal to 0. For example, the first statement Lemme 1 of Bézout [13] says that the system of linear equations

$$\begin{cases} c_{10}x_0 + c_{11}x_1 + \cdots + c_{1,n-1}x_{n-1} = 0, \\ c_{20}x_0 + c_{21}x_1 + \cdots + c_{2,n-1}x_{n-1} = 0, \\ \qquad \cdots\cdots \\ c_{n0}x_0 + c_{n1}x_1 + \cdots + c_{n,n-1}x_{n-1} = 0, \end{cases} \tag{1}$$

[15] Actually Izeki [6] has a proof that the determinant = 0 is a consequence of the three quadratic equations under the condition that the constant term of the first equation is not 0, which is not a generic condition. We do not know how to generalize this proof to the higher order case. In doing so, we should take it into account that no documents have ever been found in Wasan which claim that the determinants keep the same values under the transposition of raws and columns.

has a nontrivial solution $(x_0, x_1, \cdots, x_{n-1}) \neq 0$ if and only if the determinant

$$\begin{vmatrix} c_{10} & c_{11} & \cdots & c_{1,n-1} \\ c_{20} & c_{21} & \cdots & c_{2,n-1} \\ & & \cdots & \\ & & \cdots & \\ c_{n0} & c_{n1} & \cdots & c_{n,n-1} \end{vmatrix} = 0. \tag{2}$$

Seki (see Mikami [24, Corrections at the end of §13]), Bézout and many others, including Cayley [17] as late as 1841, defined determinants inductively by

$$|d| = d, \tag{3}$$

and

$$\begin{vmatrix} d_1 & c_{11} & \cdots & c_{1,n-1} \\ d_2 & c_{21} & \cdots & c_{2,n-1} \\ & & \cdots & \\ & & \cdots & \\ d_n & c_{n1} & \cdots & c_{n,n-1} \end{vmatrix} = d_1 \begin{vmatrix} c_{21} & \cdots & c_{2,n-1} \\ c_{31} & \cdots & c_{3,n-1} \\ & \cdots & \\ c_{n1} & \cdots & c_{n,n-1} \end{vmatrix} - d_2 \begin{vmatrix} c_{11} & \cdots & c_{1,n-1} \\ c_{31} & \cdots & c_{3,n-1} \\ & \cdots & \\ c_{n1} & \cdots & c_{n,n-1} \end{vmatrix}$$

$$+ \cdots + (-1)^{n-1} d_n \begin{vmatrix} c_{11} & \cdots & c_{1,n-1} \\ c_{21} & \cdots & c_{2,n-1} \\ & \cdots & \\ c_{n-1,1} & \cdots & c_{n-1,n-1} \end{vmatrix}, \quad n > 1. \tag{4}$$

First of all, it is then evident that the determinant is linear in each column.

Secondly, the determinant changes its sign under a transposition of two different columns. In fact, in the case of the transposition of the first and the second columns, we get a proof by applying the same development as above to the $(n-1)$ minors of the right hand side of (4). Then we obtain the Laplace (or Vandermonde) development (1772):

$$\begin{vmatrix} c_{10} & c_{11} & \cdots & c_{1,n-1} \\ c_{20} & c_{21} & \cdots & c_{2,n-1} \\ & & \cdots & \\ & & \cdots & \\ c_{n0} & c_{n1} & \cdots & c_{n,n-1} \end{vmatrix} = \sum_{1 \leq p < q \leq n} (-1)^{p+q-1} \begin{vmatrix} c_{p0} & c_{p1} \\ c_{q0} & c_{q1} \end{vmatrix} \begin{vmatrix} c_{12} & c_{13} & \cdots & c_{1,n-1} \\ & & \cdots & \\ \hat{c}_{p2} & \hat{c}_{p3} & \cdots & \hat{c}_{p,n-1} \\ & & \cdots & \\ \hat{c}_{q2} & \hat{c}_{q3} & \cdots & \hat{c}_{q,n-1} \\ & & \cdots & \\ c_{n2} & c_{n3} & \cdots & c_{n,n-1} \end{vmatrix}, \tag{5}$$

where the p-th and q-th rows of the right hand side are deleted. In Wasan this was shown by Kurushima Yoshihiro [久留島義太] (? –1757) in his bequeathed manuscript [9]. The factor

$$\begin{vmatrix} c_{p0} & c_{p1} \\ c_{q0} & c_{q1} \end{vmatrix} = c_{p0}c_{q1} - c_{q0}c_{p1}$$

clearly changes its sign by the transposition.

Since any two neighboring columns are the first and the second columns during the process of the recursive definition (4) of determinants, it follows that the determinant changes its sign by the the transposition of neighboring columns. And any transposition of two different columns is realized by an odd number of transpositions of neighboring columns.

Thirdly, it is immediately shown by our definition that

$$\begin{vmatrix} 1 & 0 & \cdots & 0 \\ 0 & 1 & \cdots & 0 \\ & & \ddots & \\ 0 & \cdots & 0 & 1 \end{vmatrix} = 1; \tag{6}$$

and these three properties characterize the determinant.

Namely, if a polynomial $D(\mathbf{a}_1, \cdots, \mathbf{a}_n)$ in the entries a_{ij} of n columns $\mathbf{a}_j = (a_{ij})$ of n variables is (i) linear in each column $\mathbf{a}_j$, (ii) skew symmetric in any transposition of different columns $\mathbf{a}_j$ and $\mathbf{a}_k$, and (iii) satisfying (6), then we have

$$D(\mathbf{a}_1, \cdots, \mathbf{a}_n) = \sum_{\sigma \in \mathbf{S}} \operatorname{sign} \sigma \, a_{1\sigma 1} a_{2\sigma 2} \cdots a_{n\sigma n} = \sum_{\sigma \in \mathbf{S}} \operatorname{sign} \sigma \, a_{\sigma^{-1} 1 1} \cdots a_{\sigma^{-1} n n}, \tag{7}$$

where $\mathbf{S}$ denotes the group of all permutations of n letters and sign σ is $+$ if σ is an even permutation and $-$ if σ is an odd permutation. That is the modern definition of the determinant. For a proof see any textbook of linear algebra, e.g. Takagi [27].

Since sign σ = sign σ^{-1}, the latter part of (7) says that the determinant remains the same under the transposition of columns and rows.

If equation (1) has a non-trivial solution (x_j), then the column $\mathbf{a}_j$ with a non-zero x_j is a linear combination of the other columns and hence the determinant vanishes. For the proof of its converse the reader is referred to any textbook of linear algebra.

If, in particular, a system of n algebraic equations of degree less than n has a common solution ξ, then the system (1) has the non-trivial solution $x_j = \xi^j$ and hence the determinant (2) should vanish.

We also note that if A and C are square matrices, and 0 is a rectangular matrix composed only of 0, then we have

$$\begin{vmatrix} A & B \\ 0 & C \end{vmatrix} = |A||C| \tag{8}$$

as is easily proved from the above definition.

3.2 Cayley's multiplication theorem and his proof of the quadrilateral method

The product

$$\begin{pmatrix} c_{11} & c_{12} & \cdots & c_{1,n} \\ c_{21} & c_{22} & \cdots & c_{2,n} \\ & & \cdots & \\ & & \cdots & \\ c_{n1} & c_{n21} & \cdots & c_{n,n} \end{pmatrix} = \begin{pmatrix} a_{11} & a_{12} & \cdots & a_{1,n} \\ a_{21} & a_{22} & \cdots & a_{2,n} \\ & & \cdots & \\ & & \cdots & \\ a_{n1} & a_{n2} & \cdots & a_{n,n} \end{pmatrix} \times \begin{pmatrix} b_{11} & b_{12} & \cdots & b_{1,n} \\ b_{21} & b_{22} & \cdots & b_{2,n} \\ & & \cdots & \\ & & \cdots & \\ b_{n1} & b_{n2} & \cdots & b_{n,n} \end{pmatrix} \tag{9}$$

of two $n \times n$ matrices is defined by $c_{ik} = \sum_{j=1}^{n} a_{ij} b_{jk}$.

Cayley [17] was one of the first to state the *multiplication theorem*

$$\begin{vmatrix} c_{11} & c_{12} & \cdots & c_{1,n} \\ c_{21} & c_{22} & \cdots & c_{2,n} \\ & & \cdots & \\ & & \cdots & \\ c_{n1} & c_{n21} & \cdots & c_{n,n} \end{vmatrix} = \begin{vmatrix} a_{11} & a_{12} & \cdots & a_{1,n} \\ a_{21} & a_{22} & \cdots & a_{2,n} \\ & & \cdots & \\ & & \cdots & \\ a_{n1} & a_{n2} & \cdots & a_{n,n} \end{vmatrix} \times \begin{vmatrix} b_{11} & b_{12} & \cdots & b_{1,n} \\ b_{21} & b_{22} & \cdots & b_{2,n} \\ & & \cdots & \\ & & \cdots & \\ b_{n1} & b_{n2} & \cdots & b_{n,n} \end{vmatrix}, \tag{10}$$

though his definition of the product of matrices is with the second factor transposed.

This is proved from the facts that the left hand side of (10) satisfies properties (i) and (ii) of determinants as a function of the columns $\mathbf{b}_j$ and that its value at the identity matrix is the determinant of (a_{ij}).

He applied it to get the relation that exists between the distances of five points in space, etc. Here we limit ourselves to the proof of the algebraic relation of distances between four points on a plane, which was used by Seki Takakazu in his solutions of Sawaguchi's problems Nos. 12 and 14 under the name of quadrilateral method. Let $P_1(x_1, y_1, z_1), P_2(x_2, y_2, z_2), P_3(x_3, y_3, z_3)$ and $P_4(x_4, y_4, z_4)$ be four points in space with coordinates (x_i, y_i, z_i). By the multiplication theorem we have

$$\begin{vmatrix} x_1^2+y_1^2+z_1^2, & -2x_1, & -2y_1, & -2z_1, & 1 \\ x_2^2+y_2^2+z_2^2, & -2x_2, & -2y_2, & -2z_2, & 1 \\ x_3^2+y_3^2+z_3^2, & -2x_3, & -2y_3, & -2z_3, & 1 \\ x_4^2+y_4^2+z_4^2, & -2x_4, & -2y_4, & -2z_4, & 1 \\ 1, & 0, & 0, & 0, & 0 \end{vmatrix}$$

$$\times \begin{vmatrix} 1, & 1, & 1, & 1, & 0 \\ x_1, & x_2, & x_3, & x_4, & 0 \\ y_1, & y_2, & y_3, & y_4, & 0 \\ z_1, & z_2, & z_3, & z_4, & 0 \\ x_1^2+y_1^2+z_1^2, & x_2^2+y_2^2+z_2^2, & x_3^2+y_3^2+z_3^2, & x_4^2+y_4^2+z_4^2, & 1 \end{vmatrix}$$

$$= \begin{vmatrix} 0, & \overline{P_1P_2}^2, & \overline{P_1P_3}^2, & \overline{P_1P_4}^2, & 1 \\ \overline{P_2P_1}^2, & 0, & \overline{P_2P_3}^2, & \overline{P_2P_4}^2, & 1 \\ \overline{P_3P_1}^2, & \overline{P_3P_2}^2, & 0 & \overline{P_3P_4}^2, & 1 \\ \overline{P_4P_1}^2, & \overline{P_4P_2}^2, & \overline{P_4P_3}^2, & 0, & 1 \\ 1, & 1, & 1, & 1, & 0 \end{vmatrix}. \tag{11}$$

If the four points P_j are on the same plane, we can choose a coordinate system such that the plane is $\{z=0\}$, so that the right hand side has to vanish. Writing

$$\begin{aligned} a &= \overline{P_1P_2} = \overline{P_2P_1}, \quad b = \overline{P_1P_3} = \overline{P_3P_1}, \quad c = \overline{P_1P_4} = \overline{P_4P_1}, \\ p &= \overline{P_3P_4} = \overline{P_4P_3}, \quad q = \overline{P_2P_4} = \overline{P_4P_2}, \quad r = \overline{P_2P_3} = \overline{P_3P_2}, \end{aligned}$$

we have twice

$$\begin{aligned} &a^2p^2(-a^2+b^2+c^2-p^2+q^2+r^2) \\ &+b^2q^2(a^2-b^2+c^2+p^2-q^2+r^2) \\ &+c^2r^2(a^2+b^2-c^2+p^2+q^2-r^2) \\ &-a^2b^2r^2-b^2c^2p^2-c^2a^2q^2-p^2q^2r^2=0. \end{aligned} \tag{12}$$

This may look tricky but is the shortest proof of the quadrilateral method (12). He gave proofs of many other geometric theorems in the same way. Similarly Takagi [27] gives a proof of the fact that the left-hand side of (12) represents 144 times the square of volume of the tetrahedron with a, b, c, p, q and r as the lengths of the edges, which is attributed to Euler (1758) and Arima Yoriyuki [有馬頼徸] (1766).

3.3 Bezout's and Sylvester's Resultants

Let

$$\begin{cases} f(x) = a_0 + a_1x + \cdots + a_nx^n = 0, \\ g(x) = b_0 + b_1x + \cdots + b_mx^m = 0, \end{cases} \tag{13}$$

be a system of two algebraic equations with $n \geq m$, $a_n \neq 0$ and $b_m \neq 0$. Then its common root is also a common root of

$$\begin{cases} a_0 + a_1x + \cdots \quad \cdots \quad \cdots \quad + a_nx^n = 0, \\ \quad a_0x + a_1x^2 + \cdots \quad \cdots \quad \cdots \quad + a_nx^{n+1} = 0, \\ \quad\quad \ddots \quad\quad\quad\quad \ddots \\ \quad\quad a_0x^{m-1} + a_1x^m + \cdots \quad \cdots \quad + a_nx^{n+m-1} = 0, \\ b_0 + b_1x + \cdots \quad \cdots \quad + b_mx^m = 0, \\ \quad b_0x + b_1x^2 + \cdots \quad \cdots + b_mx^{m+1} = 0, \\ \quad\quad \ddots \quad\quad\quad\quad \ddots \\ \quad\quad b_0x^{n-1} + b_1x^n + \cdots \quad \cdots + b_mx^{n+m-1} = 0. \end{cases} \tag{14}$$

The determinant of order $n+m$

$$\mathscr{R}_{\text{Sylvester}}(f,g)=\begin{vmatrix} a_0 & a_1 & \cdots\cdots\cdots & a_n & 0 & \cdots & 0 \\ 0 & a_0 & a_1 & \cdots\cdots\cdots & a_n & 0 & 0 \\ & & \ddots & & & \ddots & \\ 0 & \cdots & 0 & a_0 & a_1 & \cdots\cdots\cdots & a_n \\ b_0 & b_1 & \cdots\cdots & b_m & 0 & \cdots & 0 \\ 0 & b_0 & b_1 & \cdots\cdots & b_m & 0 \cdots & 0 \\ & & \ddots & & \ddots & & \\ & & & & & & 0 \\ 0 & \cdots\cdots & 0 & b_0 & b_1 & \cdots\cdots & b_m \end{vmatrix} \tag{15}$$

of the coefficients of the latter equations is the resultant Sylvester [15] introduced in 1840 up to the sign $(-1)^{mn}$.

As we claimed in Section 1 this coincides with Bézout's resultant

$$\mathscr{R}_{\text{Bézout}}(f,g)=\begin{vmatrix} d_{10} & d_{11} & & \cdots\cdots & & d_{1,n-1} \\ d_{20} & d_{21} & & \cdots\cdots & & d_{2,n-1} \\ & \cdots & & \cdots\cdots & & \\ d_{m,0} & d_{m,1} & & \cdots\cdots & & d_{m,n-1} \\ b_0 & b_1 & \cdots & b_m & 0 \cdots & 0 \\ 0 & b_0 & b_1 \cdots & b_m & 0 & 0 \\ 0 & & \ddots & & \ddots & 0 \\ 0 & \cdots & 0 & b_0 & b_1 \cdots & b_m \end{vmatrix}, \tag{16}$$

where d_{ij} are the coefficients of the transformed equations

$$h_i(x)=d_{i,0}+d_{i,1}x+\cdots+d_{i,n-1}x^{n-1}=0, \quad 1\le i\le m, \tag{17_i}$$

defined by

$$h_1(x)=b_mf(x)-a_nx^{n-m}g(x) \tag{18_1}$$

for $i=1$ and

$$h_i(x)=xh_{i-1}(x)+b_{m-i+1}f(x)-a_{n-i+1}x^{n-m}g(x), \tag{18_i}$$

for $1<i\le m$.

In order to prove the coincidence we multiply the following determinant on the left of (15):

$$
\begin{vmatrix}
b_m & 0 & \cdots & \cdots & 0 & \cdots & 0 & -a_n & 0 & \cdots & 0 \\
b_{m-1} & b_m & 0 & \cdots & 0 & \cdots & 0 & -a_{n-1} & -a_n & 0 & 0 \\
\vdots & \ddots & \ddots & & & & & \vdots & \ddots & \ddots & \vdots \\
b_1 & \cdots & b_{m-1} & b_m & 0 & \cdots & 0 & -a_{n-m+1} & \cdots & -a_{n-1} & -a_n \\
0 & 0 & \cdots & \cdots & 1 & 0 & \cdots & 0 & \cdots & & 0 \\
0 & 0 & 0 & \cdots & \cdots & 1 & 0 & 0 & & \cdots & 0 \\
 & & & & & & 1 & & & & \\
 & & 0 & & & & & & \ddots & & \\
 & & & & & & & & & 1 & 0 \\
0 & \cdots & \cdots & 0 & 0 & & 0 & \cdots & \cdots & 0 & 1
\end{vmatrix}, \tag{19}
$$

whose value is equal to $b_m{}^m \neq 0$. If we carry out the multiplication as matrices, then we obtain as the the first m rows the coefficients of transformed equations $h_i(x)$, $1 \leq i \leq m$ and as the following n rows those of $x^{i-m-1}g(x) = 0$, $m+1 \leq i \leq m+n$.

We have the commutator of

$$
\begin{pmatrix}
b_m & 0 & \cdots & 0 \\
b_{m-1} & b_m & 0 & 0 \\
\vdots & \ddots & \ddots & \vdots \\
b_1 & \cdots & b_{m-1} & b_m
\end{pmatrix}
\quad \text{and} \quad
\begin{pmatrix}
a_n & 0 & \cdots & 0 \\
a_{n-1} & a_n & 0 & 0 \\
\vdots & \ddots & \ddots & \vdots \\
a_{n-m+1} & \cdots & a_{n-1} & a_n
\end{pmatrix} \tag{20}
$$

as the $m \times m$-minor of the product at the upper right corner.

Since these two matrices are polynomials in the nilpotent matrix N with 1 under the diagonal, they commute each other, so that the commutator is 0. That is the reason why the transformed equations are all of degree less than n,

The right hand side of (16) is therefore the principal minor of order n, on the right of which all entries are zero in the product determinant. The remaining principal minor is the determinant of the left member of (20). Hence we have the required identity

$$
\mathscr{R}_{\text{Sylvester}}(f,g) = \mathscr{R}_{\text{Bézout}}(f,g) \tag{21}
$$

by canceling the non-zero factor $b_m{}^m$.

3.4 Seki's and Bézout's resultants

In Section 1 we wrote that Seki's resultants and Bézout's were the same but, when the degree m of $g(x)$ is less than the degree n of $f(x)$, this is no more true. In that case, Seki set

$$
b_{m+1} = b_{m+2} = \cdots = b_n = 0 \tag{22}
$$

and defined his transformed equations $h_i(x)$, $1 \le i \le n$, as if $g(x)$ were of degree n. Thus, his resultant is

$$\mathscr{R}_{\text{Seki}}(f,g) = \begin{vmatrix} d_{10} & d_{11} & \cdots\cdots & d_{1,n-1} \\ d_{20} & d_{21} & \cdots\cdots & d_{2,n-1} \\ & \cdots & \cdots\cdots & \\ d_{n,0} & d_{n,1} & \cdots\cdots & d_{n,n-1} \end{vmatrix}, \tag{23}$$

where d_{ij} are the coefficients of Seki's transformed equations

$$h_i(x) = d_{i,0} + d_{i,1}x + \cdots + d_{i,n-1}x^{n-1} = 0, \quad 1 \le i \le n, \tag{24_i}$$

defined by

$$h_1(x) = b_n f(x) - a_n g(x) \tag{25_1}$$

for $i = 1$ and

$$h_i(x) = x h_{i-1}(x) + b_{n-i+1} f(x) - a_{n-i+1} g(x), \tag{25_i}$$

for $1 < i \le n$.

If $m = n$, Seki's resultant is obviously equal to Bézout's. If $m < n$, we have

$$\mathscr{R}_{\text{Seki}}(f,g) = \pm {a_n}^{n-m} \mathscr{R}_{\text{Bézout}}(f,g). \tag{26}$$

For the proof we need only the identity

$$\begin{vmatrix} 1 & 0 & \cdots & \cdots & 0 & & 0 & 0 \\ 0 & 1 & 0 & \cdots & \cdots & \cdots & 0 & 0 \\ & & \ddots & & & & & \\ 0 & \cdots & 0 & 1 & 0 & \cdots & \cdots & 0 \\ b_n & 0 & \cdots & 0 & -a_n & 0 & \cdots & 0 \\ b_{n-1} & b_n & 0 & \cdots & -a_{n-1} & -a_n & 0 & 0 \\ \vdots & \ddots & \ddots & 0 & \vdots & & \ddots & 0 \\ b_1 & \cdots & b_{n-1} & b_n & -a_1 & \cdots & -a_{n-1} & -a_n \end{vmatrix} \times \begin{vmatrix} a_0 & a_1 & \cdots & a_{n-1} & a_n & 0 & \cdots & 0 \\ 0 & a_0 & a_1 & \cdots & \cdots & a_n & 0 & \\ & & \ddots & \ddots & & & \ddots & 0 \\ 0 & \cdots & 0 & a_0 & a_1 & \cdots & \cdots & a_n \\ b_0 & b_1 & \cdots & b_{n-1} & b_n & 0 & \cdots & 0 \\ 0 & b_0 & b_1 & \cdots & b_{n-1} & b_n & 0 & \cdots \\ & & \ddots & & & & \ddots & \\ 0 & \cdots & \cdots & b_0 & b_1 & \cdots & \cdots & b_n \end{vmatrix}$$

$$= \begin{vmatrix} a_0 & a_1 & \cdots & a_{n-1} & a_n & 0 & \cdots & 0 \\ 0 & a_0 & a_1 & \cdots & \cdots & a_n & 0 & \\ & & \ddots & \ddots & & & \ddots & 0 \\ 0 & \cdots & 0 & a_0 & a_1 & \cdots & \cdots & a_n \\ & & & & 0 & 0 & 0 & 0 \\ & & & & 0 & \cdots & \cdots & 0 \\ & \mathscr{R}_{\text{Seki}}(f,g) & & & 0 & & 0 & \\ & & & & 0 & 0 & \cdots & 0 \end{vmatrix} = (-a_n)^n \mathscr{R}_{\text{Seki}}(f,g). \tag{27}$$

Hence we have

$$\mathscr{R}_{\text{Seki}}(f,g) = \begin{vmatrix} a_0 & a_1 & \cdots & a_{n-1} & a_n & 0 & \cdots & 0 \\ 0 & a_0 & a_1 & \cdots & \cdots & a_n & 0 & \\ & & \ddots & \ddots & & & \ddots & 0 \\ 0 & \cdots & 0 & a_0 & a_1 & \cdots & \cdots & a_n \\ b_0 & b_1 & \cdots & b_{n-1} & b_n & 0 & \cdots & 0 \\ 0 & b_0 & b_1 & \cdots & b_{n-1} & b_n & 0 & \cdots \\ & & \ddots & & & & \ddots & \\ 0 & \cdots & \cdots & b_0 & b_1 & \cdots & \cdots & b_n \end{vmatrix}. \quad (28)$$

by canceling the non-zero factor $(-a_n)^n$ from both sides.

If $m = n$, then the right hand side is by definition equal to $\mathscr{R}_{\text{Sylvester}}(f,g) = \mathscr{R}_{\text{Bézout}}(f,g)$. If $m < n$, then because of (22), we can factor out a_n one by one by expanding the determinant on the right hand side of (28) with respect to the last column $(n-m)$ times and obtain (26).

In spite of these extra factors, Seki's resultant has the advantage that the entries d_{ij} of (23) are symmetric with respect to the anti-principal diagonal, i.e. $d_{10} = d_{n.n-1}, d_{11} = d_{n-1,n-1}, d_{n-1,0} = d_{n,1}$, etc. Complete Book of Mathematics [8] Vol. 17 has a list of expansions of determinants of order ≤ 5 with such symmetry under the title queer multiplication procedure [変乗法 henjōhō].

References

1. Lǐ Jì-Mǐn [李継閔]: *Nine Chapters of Mathematics, Introduction and Commentaries* [《九章算術》導読与譯注 "Jiǔ Zhāng Suàn-Shù" Dǎo-Dú yǔ Yì-Zhù], Shanxi Science Technology Publ. [陝西科学技術出版社], Xi'an (1998).
2. Yáng Huī [楊輝]: *Yang Hui's Methods of Mathematics* [楊輝算法 Yáng Huī Suàn-Fǎ] (1275); transcribed by Seki Takakazu in 1661, a later copy of which by Ishikuro Nobuyoshi [石黒信由] is preserved in Shinminato Museum, Toyama.
3. Zhū Shì-Jié [朱世傑]: *Introduction to Mathematics* [算学啓蒙 Suàn-Xué Qǐ-Méng] (1299); published with commentaries by Takebe Katahiro, Ibaraki Hōdō [茨木方道], Kyoto (1690).
4. Seki Takakazu [関孝和]: *Methods of Solving Concealed Problems* [解伏題之法 Kaihukudai no Hō] (1683); A. Hirayama, K. Shimodaira and H. Hirose [平山諦, 下平和夫, 広瀬秀雄] ed.: *Takakazu Seki's Collected Works edited with Explanations* [關孝和全集], Osaka Kyōiku Tosho [大阪教育図書], Osaka, pp.142–158 (1974) ; These proceedings, pp. 470–493 .
5. Seki Takakazu [関孝和]: *Methods of Equation Modifications* [開方翻変之法 Kaihō Honpen no Hō] (1685) ; *Takakazu Seki's Collected Works edited with Explanations*, pp. 159–170.
6. Iseki Tomotoki [井関知辰]: *An Exhibition of Mathematical Methods* [算法発揮 Sanpō Hakki], Nagano Hikoemon, Osaka (1690).
7. Tanaka Yoshizane [田中由真]: *Resolution of Entanglements in Mathematics* [算学紛解 Sangaku Funkai] (ca 1690); Manuscript 618/204 in the Nakanoshima Library, Osaka.
8. Seki Takakazu, Takebe Kataakira and Takebe Katahiro [関孝和, 建部賢明, 建部賢弘]: *Complete Book of Mathematics* [大成算経 Taisei Sankei], 20 vols. (1683–1711); Manuscripts 物-423∼425 preserved in the Modern Science Museum, Tokyo University of Science.
9. Kurushima Yoshihiro [久留島義太]: *Bequeathed Manuscripts, vol. Heaven* [久氏遺稿天之巻], preserved in Japan Academy; Xu Zelin ed. [徐澤林 譯注]: *Selected Texts from Wasan* [和算選粹], Science Press, Beijing [科学出版社, 北京], pp. 370–387 (2008).

10. L. Euler: Démonstration sur le nombre des points où deux lignes des ordres quelquonques peuvent se couper, *Opera Omnia*, Ser. I. 26, Orell Füssli Turici, Lausanne, pp. 46–59 (1953).
11. Gabriel Cramer: Appendices I, II de l'évanouissement des inconnues, *Introduction à l'analyse des lignes courbes algébriques*, Frères Cramer & Cl. Philibert, Genève, pp. 656–676 (1750).
12. L. Euler: Nouvelle méthode d'éliminer les quantités inconnues des équations, *Opera Omnia*, Ser. I, 6, Teubner, Berlin, pp. 197-211 (1921).
13. E. Bézout: Recherches sur le degré des équations résultantes de l'évanouissement des inconnues, et sur les moyens qu'il convient d'employer pour trouver ces équations, *Mémoires Acad. Royale Sci. Paris*, pp. 288–338 (1764).
14. Meier Hirsch: *Sammlung von Beispielen, Formeln und Aufgaben aus der Buchstabenrechnung und Algebra*, Duncker und Humblo, Berlin (1804).
15. J. J. Sylvester: A method of determining by mere inspection the derivatives from two equations of any degree, *Phil. Magazine*, vol. 16, pp. 132–135 (1840).
16. A. Cauchy: Mémoire sur l'élimination d'une variable entre deux équations algébriques, *Exercices d'analyse et de physique mathématique*, vol. 1, pp. 91–104 (1840).
17. A. Cayley: On a theorem in the geometry of position, *Cambridge Math. J.*, vol. 2, pp. 267–271 (1841).
18. A. Cayley: The theory of determinants, *Trans. Cambridge Philos. Soc.*, vol. 8, pp. 1–16 (1843).
19. J. J. Sylvester: On a theory of the syzygetic relations of two rational integral functions, comprising an application to the theory of Sturm's functions, and that of the greatest algebraical common measure, *Phil. Trans. Royal Soc. London,* vol. 143, pp. 407–548 (1853).
20. A. Cayley: Memoir on the resultant of a system of two equations, *Phil. Trans. Royal Soc. London*, vol. 147, pp. 703–715 (1857).
21. Tsuruichi Hayashi [林鶴一]: The "Fukudai" [伏題] and determinants in Japanese mathematics, *Proc. Tokyo Mathematico-Physical Soc.*, Ser. 2, vol. 5, pp. 254–271 (1910).
22. Yoshio Mikami [三上義夫]: *The Development of Mathematics in China and Japan*, Teubner, Leipzig (1913).
23. Yoshio Mikami: On the Japanese theory of determinants, *Isis*, vol. 2, pp. 9–36 (1914).
24. Yoshio Mikami: Relations and comparisons of Seki Takakazu's works with methods of mathematics by mathematicians in Kyoto and Osaka and those in China [関孝和の業績と京阪の算家並びに支那の算法との關係及び比較 Seki Takakazu no gyōseki to Keihan no sanka narabini Sina no sanpō tono kankei oyobi hikaku], Part 2, *Tōyō Gakuhō* [東洋学報], vol. 20, pp. 543–566 (1932).
25. Heizaemon Kato [加藤平左エ門]: On the expansion of determinants in the Wasan [和算ノ行列式展開ニ就テノ検討], *Tohoku Math. J.*, vol. 45, pp. 338–353 (1939).
26. Heizaemon Kato [加藤平左エ門]: *Study of Wasan, Determinants and Theory of Circles* [和算ノ研究 行列式及圓理], Kaiseikan [開成館], Tokyo (1944).
27. Teiji Takagi [高木貞治]: *A Course of Algebra* [代数学講義], Kyōritsu Shuppan [共立出版], Tokyo (1948).
28. Annick Horiuchi: *Les mathématiques japonaises à l'époque d'Edo (1600–1868): une étude des travaux de Seki Takakazu (? –1708) et de Takebe Katahiro (1664–1739)*, J. Vrin, Paris (1994).

Index

abacus, 11, 99–103, 120, 123, 174, 181, 246, 250
accelerated approximation method [増約術 *zōyaku jutsu*], 26, 315
Adam Shall von Bell Johann [湯若望], 363
Agicki [阿直岐], 93
Aida, Yasuaki [会 (會) 田安明], 102, 349, 354, 355, 365
Ajima, Naonobu [安島直圓], 439
algebraic, → lopsided
algebraic number, → lopsided
Alternating Multiplication [交乗], 264
alternating multiplications of enumerated equations [逐式交乗 *chikushikikōjō*], 562
alternative algorithm [又術 *you shu*], 90, 91
alternative quotient formula [交商式 *kōshō shiki*], 323
Analects [論語], 93
analysis [分析 *fenxi*], 75, 81
Andō, Yoshiharu [安藤吉治], 257, 264, 265
Andō, Yūeki [安藤有益], 249
angles [角 *kaku*], 324
annihilative [剋 *koku*], 561–563
Annual Records of Tokugawa Shōgunate [年録 *Nenroku*], 4, 7, 13, 18
Answers to the Clarification of Mathematics [勿憚改答術 *Futsudankai Tōjutsu*], 324, 325
Answers to the Elucidation of Mathematics [闕疑抄答術 *Ketsugishō Tōjutsu*], 324, 325
Aoki, Konyō [青木昆陽], 364
Applied Measurements, → Complicated Configuration of Figures
Ara, Shijū [荒至重], 371
Arai, Hakuseki [新井白石], 12, 355
Araki, Murahide [荒木村英], 15–17
Arc Rates [弧率 *Koritsu*], 334, 340, 341
Archimedes, 49, 51, 53, 56–62
area of a circle, 75
Arima, Yoriyuki [有馬頼徸], 569
Arithmetic by Western Numeral and Notation [洋算用法 *Yōsan yōhō*], 373
arithmetic for right-angled triangles [勾股算術 *gou-gu suan-shu*], 111, 115
Arithmetics, 68
Art of Computer Programming, 308
art of cutting bamboo [剪 (翦) 管術 *jianguan-shu*], 208, 210, 211, 214, 320, 324, 325, 388
art to leave remainder one [剰 (剩) 一術 *jōichi jutsu*], 210, 320, 321
Ashikaga, Yoshimitsu [足利義満], 174
Astronomical Instruments and Star Maps [儀象考成 *Yi Xiang Kao Cheng*], 301

Babylonian, 31–46
Ban of books in Kan'ei [寛永の禁書令 *Kan'ei no kinsho rei*], 359
barycenter, 196, 200, 202–205
bequeathed problems [遺題 *Idai*], 14, 16, 101, 121, 184, 185, 249, 250, 252, 283
Bernoulli numbers, 27, 188, 291
Bernoulli, J., 188, 291
Bézout, É., 233, 237–240, 268, 556, 557, 565, 566, 570–572
Bézout's theorem, 557
Bhāskara, 135
Binet, J. P. M., 233, 305
Biographies of Mathematicians and Astronomers [疇人傳], 50
Biographies of Mathematicians and Astronomers, Series iii [疇人傳三編], 60

Biography of the Takebe [建部氏伝記 *Takebe-shi denki*], 15, 17, 18, 322, 326, 327, 387
Blomhoff, J. C., 365
Boethius, 96
Book of Chongzhen Calendar [崇禎暦書 *Chongzhen Li Shu*], 363, 373
Book of Division, → Book on Division
Book of Mathematics in Nine Chapters, → Mathematical Treatise in Nine Chapters
Book of Music and Calendar [律暦志 *Lulizhi*], 315
Book of Western Watersupply [泰西水法 *Taixi Shuifa*], 360
Book of Yi [易経 *ekikyō*], 96, 97
Book on Computations with Counting Rods [算数書 *Suan Shu Shu*], 158
Book on Division [割算書 *Warizansho*], 100, 120, 175
Book on Things Old and New, → Jinkōki
Book on Things Small and Large, → Jinkōki
Brown, W. G., 305
Bruins, E. M., 37
Budo [夫道], 93
Burin Inkenroku [武林隠見録], 161
byscript method [傍書法 *bōsho hō*], 16, 101, 129–131, 254, 265, 313, 314, 441
byscript notation, → byscript method

Calculation for using trigonometric function table [八線表算法解義 *Hassenhyō Sanpō Kaigi*], 363
calculation of sharing oil [油はかり分け算 *abura hakari wake san*], 183
Cao Mien The Thu [高綿世次], 151
Cardano, G., 233
Catalan, E. C., 303, 305–308
Cauchy, A. L., 230, 233, 239, 557
Cayley, A., 230, 257, 558, 566, 568
celestial element [天元 *tengen*], 102, 246, 279, 457
celestial element method [天元術 *tengen jutsu*], 26, 27, 97, 99, 101–103, 107, 108, 111, 119–123, 126, 129, 130, 246, 249, 251, 255, 313, 337, 457
center, 196, 200
Chakun-sul, → root-extraction method
Challenge problems, → bequeathed problems
Chaotic Interpolation Method [混沌招差術 *konton shōsajutu*], 26
Chemla, K., 78, 79
Chenciner, A., 379
Cheng Dawei [程大位], 14, 120, 152, 157, 158, 175, 208, 217, 246, 278
Chi Minh Lap Thanh Toan Phap [指明立成算法], 150–153, 157, 159, 162, 164, 165, 167
Chiljongsan [七政算], 95
Chinese abacus, → counting boards
Choi Hanki [崔漢綺], 97
Choi Sukjung [崔錫鼎], 96, 97, 100
Cholsul, → Zhuì Shù
Chongzhi legacy method [冲之遺法], 50
Chonon-sul, → celestial element method
Chonwon, → celestial element
Chwijae [中人], 98
Cicero, 61
circle dividing method [割圓術 *Geyuan Shu*], 22, 25
circle theory [円 (圓) 理 *enri*], 317, 318, 321, 324–326, 328, 388
Circle Theory and the Circular Arc Length [円理弧背術 *Enri Kohai Jutsu*], 338
Circle-town measurement from remoteness [遥度圓城 *yao-du yuan-cheng*], 115
Circular Arc and its multi-sections [弧背截約集 *Kohai Setsuyaku Shū*], 333
circular field method [圓田術 *yuantian shu*], 65, 66, 78
circular principle [円理 *enri*], 129
Classic of Mo [墨経 *Mojing*], 69
Clavius, C., 358
Clearly Explained Methods of Mathematics [算法明解 *Sanpō Meikai*], 264
Clever Man's Booklet on Mathematics [勘者御伽雙紙 *Kanja Otogi Zōsi*], 210
coarse rice [糲米 *limi*], 67
Collected Ancient Arithmetic Classics [緝古算経 *Ji Gu Suan-Jin*], 108
Collected Diagrams and Explanations of Wonderful Machines from the Far West [遠西奇器圖説], 56, 57
Collection of document concerning Kai [甲斐志料集成 *Kai-shiryō-shūsei*], 12
Colloquial Commentaries of Introduction to Mathematics, → Complete Colloquial Commentaries on Introduction to Mathematics
Colloquial Commentaries of Operations in the Mathematical Methods for Exploring Subtle Points, → Commentaries in Japanese on Procedures in Mathematical Methods without Secrets
Colloquial Commentary on the Operations in the Mathematical Methods for Exploring Subtle Points, → Commentaries in Japanese on Procedures in Mathematical Methods without Secrets

Combined Learning Mathematical Indicator-General Part [同文算指通編], 57
Commentaries in Japanese of Procedures in Mathematical Methods without Secrets [発 (發) 微算法演段諺解 *Hatsubi Sampō Endan Genkai*], 15, 16, 122, 129, 131, 254, 264, 284, 313, 315
Commentaries in the Vernacular on the Hatsubi Sanpō, → Commentaries in Japanese on Procedures in Mathematical Methods without Secrets
Commentary [評林 *Hyōrin*], 355
Commentary book of ellipse [側円規矩例弁解 *Sokuen kikurei benkai*], 365
Commentary on the Nine Chapters on the Mathematical Art [九章算術注 *Jiuzhang Suanshu Zhu*], 329
Comments to the Diagram of Right-angled Triangle with its Inscribed Circle and Square [勾股方圓圖注 *Gou-gu fang-yuan tu-zhu*], 112
compasses for drawing ellipse [側円規 *Sokuen ki*], 365
Compendium of Calendrical Science and Astronomy [曆象考成 *Li xiang Kao cheng*], 364, 373
Compendium of Mathematics, → Essential Mathematics
Complete book of ellipse problems [算法側円集 *Sanpo sokuen shū*], 365
Complete Book of Mathematical Methods, → Complete Collection of Mathematics
Complete Book of Mathematics [大成算経 *Taisei Sankei*], 5, 17, 130, 210, 212–214, 224, 227, 256, 264, 266, 272, 312, 313, 317, 321–323, 326, 336, 387, 438, 439, 495, 555, 556, 558, 565, 573
Complete Collection of Mathematics [算法大成 *Sampō Taisei*], 17, 162, 163, 322
Complete Colloquial Commentaries on Introduction to Mathematics [算学啓蒙諺解大成 *Sangaku Keimō Genkai Taisei*], 14, 122, 124, 126, 129, 130, 313
Complete Explanation of Measurements [測量全義], 50, 51
Complete Library in Four Branches of Literature [四庫全書 *Si Ku Quan Shu*], 155
Complete Treatise on Calendar and Computation [暦算全書 *Li Suan Quan Shu*], 314, 361–364, 373
complicated, → fractional
Complicated Configuration of Figures [形巧 *Keikō*], 324, 388
Computational Algorithms of Nine Categories, → Nine Chapters of Mathematics
concealed problem [伏題 *fukudai*], 324, 325, 388, 438, 555
Concise Collection of Mathematical Methods, → Essential Mathematics
Concise Collection of Mathematical Science, → Essential Mathemtaics
Configurations for the Extraction of Roots [開方算式 *Kaihō Sanshiki*], 278
Conics [円錐曲線], 51
counting boards [算盤 *suàn pán*], 123, 152, 164, 166–168, 246, 392
counting rods [算木 *suàn mù*], 99, 101–103, 123–127, 152, 153, 162–168, 246, 285, 294, 441, 497
covered problem, → submerged problem
Cramer, G., 236
creative [生 *sei*], 561–563
Critical Studies of Problems [題術辯議之法 *Daijutsu Bengi no Hō*], 324–326, 388
cross multiplication [維乗 *wéi chéng*], 560

Da Cruz, J., 161
Dai Thanh Toan Phap [大成算法], 163
Dāmodara, 134
Dan Yangi [段楊爾], 94
Dang Dinh Ang, 381
Davis, C., 379
De Fermat, P., 320
De Ursis, S., 359
Death Register [過去帳 *Kako-chō*], 7, 8, 10
deductive reasoning, 69, 70, 75
definitions and deductive reasoning, 69
degraded, → residual
Dehn, M., 84
Descartes, R., 27–29, 106, 251, 269, 272, 303, 564
determinant, 230, 232–234, 236–238, 240, 264–266, 268, 272, 469, 498, 555, 556, 558–561, 563–571, 573
diagram of self-multiplying of the four elements [四元自乘演段之圖 *si-yuan zi-cheng yan-duan zhi-tu*], 112
dilemma reasoning, 75
diminishing and stretching method [消長法 *shōchō hō*], 264
Diophantine' Arithmetics, 68
Diophantus, 277
discriminant, 387, 558, 559
disjunctive reasoning, 74

distance of the barycenter of the arc-figure from the chord, 199
dividend [実], 158, 178
Dividing Method by Compasses [規矩分等集 *Kiku buntō shū*], 360
divisor [法], 158, 178
Dong Yuan [洞淵], 109
Double Differences [重差 *Chongcha*], 89
Draft for Deformations of Sphere Segments [毬闕変形草解 *Kyūketsu Henkei Sōkai*], 219, 221, 223, 227
Drasin, D., 376
draw an analogy [触類而長 *chulei erzhang*], 70
Drawing Method of Ellipse [側円周規法 *Sokuen shūki hō*], 365
Du Shiran [杜石然], 131
Du Zhigeng [杜知耕], 50
dynamic, → provisional

earth [地 *di*], 24
Elemental Physics [格物質学], 53, 58, 59
Elementary Introduction to Mathematical Methods [算法発蒙集 *Sanpō Hatsumō-shū*], 121
Elements [幾何原本], 51, 58, 360, 361, 373
Elements of Analytical Geometry and of the Differential and Integral Calculus [代微積拾級], 51, 371
Elements of Geometry, → Elements
Elements of Plane Trigonometry [平三角挙要 *Ping Sanjiao Juyao*], 362
Elements of Right-angled Triangles [勾股法義 *Cougu Fayi*], 360, 361
Elements of Spherical Trigonometry [弧三角挙要 *Hu Sanjiao Juyao*], 362
Elements of Surveying [測量法義 *Celiang Fayi*], 360, 361
Elimination by Mutual Multiplications [互乗相消], 26
elimination by Out-In Mutual Multiplication [内外互乗相消 *Zhongwai Hucheng Xiangxiao*], 26
ellipse [側円 *sokuen*], 198, 365, 366
Elucidation of Mathematics with Author's Comment [増補算法闕疑抄 *Zōho Sanpō Ketsugishō*], 318
Encyclopedia of Calendrical Mathematics [暦算全書 *Lisuan Quanshu*], 293
equation rule for finding differences [方程招差法 *hōtei shōsa hō*], 321, 322, 326
Essential Mathematics [括要算法 *Katsuyō Sampō*], 5, 16, 26, 129, 187, 195, 208, 210, 213, 214, 217, 227, 249, 323–325, 332, 336
Euclid, 210
Euclid's Elements, 68
Euler, L., 237–240, 305, 557, 569
exactly vanishing condition of modulus class [適尽方級法 *tekijin hōkyū hō*], 320, 323, 558
excess and deficiency method [盈不足術 *Ying Buzu Shu*] , 22
excess radius [余径 *Yüjing*], 76
Excess-Deficiency [盈不足 *Yingbuzu*] Chapter, 66–68
excess-deficiency procedure [盈不足術 *yingbuzu shu*], 66, 67, 70, 87
excessive [背 *hai*], 323, 390, 391
Exhibition of Mathematical Methods [算法発揮 *Sampō hakki*], 266
explicit problem [見題 *kendai*], 324, 325, 388, 438, 555
Extinct Family Trees [断家譜 *Dankafu*], 6, 8, 11, 13, 18
extracting roots [開方 *kaihō*], 323
Extraction of Roots [開方 *Kaihō*], 388
Extraction of Roots from Equations [開方算式 *Kaihō Sanshiki*], 323, 325
extreme [極 *kyoku*], 323, 390, 391

Fair Labor [均輸 *Junshu*] Chapter, 66–68
fair labor procedure [均輸術 *junshu zhu*], 66, 87
Family Tree of the Ogiwara [荻原家系譜 *Ogiwara Kakei-fu*], 11, 18
Family Tree of the Takebe [建部家系譜 *Takebe Kakei-fu*], 7, 18, 20
Family Tree of the Toda [戸田家系譜 *Toda Kakei-fu*], 12, 18
Family Tree of the Uchiyama [内山家系譜 *Uchiyama Kakei-fu*], 6–8, 11, 18, 19
figure [形 *kei*], 224, 323, 389, 391
finger ring, 197–204
Five Sums and Five Differences [五和五較 *Wu-He Wu-Jiao*], 107, 112, 116
Five Techniques [五技 *Gogi*], 329, 387
fixed circle number [定周 *teishū*], 15
flat-headed stack [芻童 *chutong*], 66, 220, 221
Flow and Ebb [満 (滿) 干 *mankan*], 224, 225, 323, 388–391, 393
formula exchanges [交式 *kōshiki*], 560, 562–565
fractional [繁 *han*], 225, 323, 391–393
From Trifling Dust to Eternal Time, → Jinkōki
frustum cut by an oblique plane [圓臺斜截 *endai shasetsu*], 201

Fryer, J. [傅蘭雅], 51, 52
Fujita, Teishi [藤田貞資], 11, 363
Fujiwara, Matsusaburō [藤原松三郎], 7, 224, 265, 344, 388
Fukuzawa, Yukichi [福沢諭吉], 354
Fukyū Tetsujutsu, → Fukyū's Technique of Linkage
Fukyū's Technique of Linkage [不休綴術], 130, 293
full quotient formula [全商式 *zenshō shiki*], 323
Fundamental Mathematics [算法根源記 *Sanpō Kongenki*], 250
Fundamental Method to find resultants of two equations of different degrees [双式異乗之陰陽率并求根源術], 266
Fundamental Principles of Calculus [微積溯源], 51
Fundamentals for Astronomy [天学初函 *Tianxue Chuhan*], 359, 360

Gauss, C. F., 84, 230, 497
general method of extraction [開出總法 *kaishutsu sōhō*], 323
Genkō [玄光], 176
Geometrical and Algebraic Interpretation to the Treatise Yi-Gu [益古演段 *Yi-Gu Yan-Duan*], 108
geometrical constants [形率 *keiritsu*], 324
geometry [形法 *Keihō*], 324
Go-Duk [固德], 94
Goodman, G. K., 292, 295
Gotō, Takefumi [後藤武史], 469
great art of general remainders [大衍總数術 *dayan zongshushu*], 208, 320
great art on the remainder one, → great art to seek remainder one
great art to seek remainder one [大衍求一術 *dayan quyi shu/daien kyūichi jutsu*], 210, 214, 320
great art to seek the total from remainders, → great art of general remainders
Great Canon of the Yongle Era [永樂大典 *Yong Le Da Dian*], 158
Great Compendium of Mathematical Methods, → Complete Collection of Mathematics
Great Learning [大學 *Da Xue*], 155, 158
Gregory, J., 303
Gros, F., 151, 153, 157
Grothendieck, A., 379, 380
Gu Yingxiang [顧應祥], 116
Guide book of new natural science and astronomy [暦象新書 *Rekishō shinsho*], 367
Guide towards Understanding of Computational Methods [指明算法 *Chi Minh Toan Phap*], 149, 150, 156, 159, 162, 164, 167
Guo Shoujing [郭守敬], 322
Guo Shuchun [郭書春], 78, 79
Gwon-Rok [勸勒], 94

Ha Gukju [何国柱], 99, 101
Habuto, Seizaemon [羽太清左衛門], 12
Hado [河圖], → Systematic Treatise of Arithmetic
Haji, Dōun [土師道雲], 14, 121
Hanabusa, Heizaemon [花房平左衛門], 20
Hangul, Korean phonetics, 103
Hashiguchi, Kōnosuke [橋口侯之介], 353
Hashimoto, Masakazu [橋本正数], 121, 249
Hayashi, Kichizaemon [林吉左衛門], 291
Hayashi, Takao [林隆夫], 133
Hayashi, Tsuruichi [林鶴一], 561
Hazama, Shigetomi [間重富], 366
Hazama, Shigetoshi [間重新], 366, 367
He Chengtian [何承天], 320
He Shaogeng [何紹庚], 308
heaven [天 *tian*], 24
heavenly element, → celestial element
heavenly element method algebra, → celestial element method
heavenly element method, → celestial element method
Henkel, 68
higher interpolation [招差術 *zhaocha shu*], 22
Hilbert, D., 84
Hindenberg, K. F., 232
Hirata, Atsutane [平田篤胤], 373
Hirayama, Akira [平山諦], 254
Hirsch, M. W., 558
Hisada, Gentetsu [久田玄哲], 14, 121, 249
History of Nihon [日本書紀 *Nihon Shoki*], 93
History of the Sui Dynasty [隋書 *Suishu*], 315
History of the Sui Dynasty, Record of Books [隋書経籍志 *Sui Shu, Jingji Zhi*], 89
History of three Kingdoms [三国史記], 94
Hoang Tuy, 377, 378, 381
Hoe, J., 251, 497
Hōjō, Gengoemon [北条源五衛門], 313
Hon-ami, Kō'etsu [本阿弥光悦], 175
Honda, Toshiaki [本多 (田) 利明], 9, 369
Hong Daeyong [洪大容], 97
Hong Jungha [洪正夏], 98, 99
Horiuchi, A., 122, 129, 276, 280, 281, 293, 555, 556
Horner, W. G., 457
Hoshina, Masayuki [保科正之], 249

Hoshino, Sanenobu [星野実 (實) 宣], 14, 121, 215, 249
Hosoi, Kōtaku [細井広澤], 360, 361
Hsu Yu-Fang [許瑜芳], 155
Hua Hengfeng [華衡芳], 51
Hua Lo-Keng, 378
human being [人 *ren*], 24
Ḫunzû, 38
Hwang Yoonsuk [黄胤錫], 97
hypotenuse [弦 *xian*], 24, 110
hypothetical reasoning, 72

Ichino, Shigetaka [市野茂喬], 369–371
Ikeda, Masaoki [池田正意], 335
Iksan [翼算], 100
Illustration of A Circle Town [圓城圖式 *Yuan-Cheng Tu-Shi*], 107–109, 115
Illustrative Hydraulics [水学圖説], 52, 58
Imamura, Tomoaki [今村知商], 175, 280, 281, 318, 334
implicit problem [隠題 *indai*], 324, 325, 388, 438, 555
Inalterable Treatise, → Jinkōki
Inaqibit-Anu, 38, 39
incremental method of accelerated approximation, → accelerated approximation method
infinite inferences, 75
infinite series, 26, 27, 99, 101, 280, 293, 303–308, 317, 318, 331, 332, 337
initial quotient [初商 *shou shang*], 23
Initiatory Textbook of Natural Sciences [格致啓蒙], 54, 58
integer, → whole
interpolation, 22, 26, 27, 37, 280, 281, 285, 292, 317, 321, 336
Introduction to Algebra and Calculus [代微積拾級 *Dai Wei Ji Shi Ji*], 358, 371, 372
Introduction to Mathematical Methods [算法入門 *Sampō Nyūmon*], 313
Introduction to Mathematics [算学啓蒙 *Suànxué Qǐméng*], 5, 14, 95, 97, 99, 100, 103, 119–126, 129, 246, 249, 251, 276, 278, 313, 323, 325, 457, 554
"Introduction to Mathematics" and Comments, → New Commentary on Introduction to Mathematics
inverse proportional distribution [非衰分 *Feicuifeng*], 68
irrational, → residual
Ishikuro, Nobuyoshi [石黒信由], 14, 365, 561, 564
Isomura, Yoshinori [礒村吉徳], 14, 281, 318, 354
Isu-Sinpyon [理数新編], 97
Itakura, Shigenori [板倉重矩], 360
Izeki, Tomotoki [井関知辰], 266, 555, 556, 558, 565

Jacobi, 321
Jade and Stone Hide Each Other [玉石互隱 *Yushi Huyin*], 68
Jade Mirror of Four Elements [四元玉鑑 *Si yuán Yù jiàn*], 24, 107, 111, 112, 115, 120, 246, 251, 497
Jadelike Examples of Four Unknowns, → Jade Mirror of Four Elements
Jami, C., 299, 308
Japanese traditional mathematics [和算 *Wasan*], 101, 119, 120, 122–124, 126, 129, 131, 197, 286, 337, 354, 355
Jartoux, P., 293, 303, 304
Jia Xian [賈憲], 108
Jin Shizhen [金始振], 121
Jinkōki [塵劫記], 5, 14, 100, 120, 122, 176–182, 184–186, 209, 249, 278, 323, 325
Jisan [地算], 95
Jōchi, Shigeru [城地茂], 7
Johann Terrenz [鄧玉函], 56
Jōkyō calendar [貞享暦 *Jōkyō reki*], 246
Joseph's problem [継子立 *mamako date*, 計子 *keishi*], 182, 324, 388
Ju Segul, → Zhu Shijie
Jugai's Record [竪亥録 *Jugairoku*], 280, 281, 318, 334
Jugairoku, → Jugai's Record
Juhae-suyong [籌解需用], 97
Juhak-Iphyukan [籌学入格案], 98
Jungjong [中宗], 121

Kaempfer, E., 290
Kalinowski, M., 166
Kanda, Takahira [神田孝平], 371
Kangxi [康熙], 300
Kansei Family Trees [寛政重修諸家譜 *Kansei-chōshū-shokafu*], 6–8, 11, 13, 18
Kantorovich, L. V., 378
Katsu, Kaishū [勝海舟], 373, 374
Katz, V. J., 32
Kawahara, Hideki [川原秀城], 79
ken-ichizan [見一算], 178
Kepler's apples, 204
Kepler's lemons, 204
Kepler, J., 204, 319, 364
Key of Mathematics [数学鑰], 50
Kimura, Kinji [木村欣司], 254, 288, 289, 498
King She [赦王], 51

Knobloch, E., 557
Knuth, D. E., 308
Kobayashi, Tatsuhiko [小林龍彦], 11, 13
Kobayashi, Yoshinobu [小林義信], 291
Köngker, I., 364
Kōfu Diary [甲府日記 *Kōfu-Nikki*], 4, 6, 10, 11, 18
Kōfu Palace Diary [甲府御館記 *Kōfu-Oyakata-ki*], 4, 12, 18
Kōfu Samurai Directory [甲府分限帳 *Kōfu bugenchō*], 6, 12, 18–20
Kōfu samurai Directory of Genroku 8 (1695) [甲府様御人衆中分限帳 *Kōfu-sama-Goninjūchū Bugen-chō*], 8, 9, 12, 18
Kōfu samurai Directory of Kōmon [甲府黄門次郎様臣下録 *Kōfu-kōmon-jirō-sama-shinka-roku*], 18
Kōfu samurai Directory of Saishō [甲府宰相綱重卿之御事並御分限帳 *Kōfu-Saishō Tsunashige-kyō-no-on-koto narabini go-Bugen-chō*], 18
Komatsu, Hikosaburō [小松彦三郎], 17, 106, 285, 287–289, 388, 389, 469, 496, 497
Korean double entry system of Book Keeping, 103
Kuilchip [九一集], 99
Kujang [九章 *Kushō*], 94
Kujang-Sansul, → Nine Chapters of Mathematics
Kurushima, Yoshihiro [久留島義太], 566
Kusuba, Takanori [楠葉隆德], 133
Kusuryak [九数略], 96
Kyemong-Sanbop, → Introduction to Mathematics
Kyungguk-Daejon [経国大典], 98

L'Hospital, M., 231, 235
Land Surveying by Trigonometry [三角惑問 *Sankaku wakumon*], 373
Laplace, P. S., 233, 236, 237
Larcombe, P. J., 305, 306, 308
Laurent Schwartz, 379
law of right triangles [勾股弦の法 *kōkogen no hō*], 197
Le Dung Trang, 381
Le Van Thiem, 375–379, 381
Leader of the Imperial Observatory [欽天監監正], 301
Learning by analogy [挙一反三 *juyi fansan*], 70
Learning by consequence [告往知来 *gaowang zhilai*], 70
Lee Sanghyuk [李尚爀], 99
left place [寄左], 257
Leibniz, G. W., 27, 230–238, 240–242, 291
Length of Arcs [弧率], 388
Li Di [李迪], 107, 299, 308
Li Rui [李鋭], 50
Li Shanlan [李善蘭], 59, 60, 371
Li Wenlin [李文林], 28, 106
Li Yan [李儼], 79, 89, 90, 131, 299, 303, 308
Li Ye [李冶], 106–112, 115–116
Li Zhizao [李之藻], 57, 359
Library of Kai [甲斐叢書 *Kai-sōsho*], 12
license of concealed problems [伏題免許 *fukudai menkyo*], 438
license of explicit problems [見題免許 *kendai menkyo*], 438
license of implicit problems [隠題免許 *indai menkyo*], 438
Lin Hung-Chun [林虹君], 154
Liu Hui [劉徽], 22, 24, 25, 65, 69–79, 81, 83, 84, 87, 89–91, 159, 194, 277, 278, 282, 315, 329, 497
Liu Hui Principle, 78
Liu Yi [劉益], 108
longer leg [股 *gu*], 24, 114, 496
Loomis, E., 371
lopsided [畸 *ki*], 225, 323, 391–393
loss and gain [損益 *Sunyi*], 67
lü [率], 69
Lucid Methods of Mathematics [明元算法 *Meigen Sampō*], 257
Łukasiewicz, J., 251
Lulofs, J., 367–369
Luo Jianjin [羅見今], 308
Luo Shilin [羅士琳], 121
Luong The Vinh [梁世榮], 162–163

Maclaurin, C., 235
Mādhava, 134–147
Maeda, Toshiie [前田利家], 174
magic circle [円攢 *ensan*], 224, 388
magic square [方陣 *hōjin*], 97, 224, 388
Majima, Hideyuki [真島秀行], 7
Malgrange, B., 379
Mao, Tokiharu [万尾時春], 360
Martineau, A., 379
Martzloff, H.-C., 131
Master Araki's Comments [荒木彦四郎村英先生茶談 *Araki Hikoshirō Murahide Sensei Sadan*], 15
matching cover [合蓋 *hegai*], 194
matching square cover [牟合方蓋 *Mohe Fanggai*], 71, 72
Mathematical Chopping-Board [算俎 *Sanso*], 5, 14, 249, 281, 288, 315, 318, 332

Mathematical Cooking Boards, → Mathematical Chopping-Board
Mathematical Cuneiform Texts, 32
mathematical definitions, 69
mathematical induction principle, 75
Mathematical Method for Clarifying Subtle Points, → Mathematical Methods without Secrets
Mathematical Methods for Clarifying Slight Signs [研幾算法 *Kenki Sampō*], 15, 129–131, 313, 315, 326, 335
Mathematical Methods for Exploring Subtle Points, → Mathematical Methods without Secrets
Mathematical Methods of Exploring Subtle Points, → Mathematical Methods without Secrets
Mathematical Methods Old and New [古今算法記 *Kokon Sampōki*], 16, 121, 249, 252, 283, 286, 287, 289
Mathematical Methods without Doubts, → Unquestionable Method in Mathematics
Mathematical Methods without Secrets [発(發) 微算法 *Hatsubi sampō*], 5, 16, 121, 129, 187, 252, 264, 283, 289, 312–314, 324, 325, 497
Mathematical Textbook on Multiplication and Division [数学乗除往来 *Sūgaku Jōjo Ōrai*], 313, 335
mathematical theory, 69
Mathematical Treatise in Nine Chapters [数書九章 *Shùshū Jiǔzhāng*], 23, 208–210, 214, 215, 246
Mathematical Treatise of Technique of Linkage, → Mathematical Treatise on Technique of Linkage
Mathematical Treatise on the Technique of Linkage [綴術算経 *Tetsujutsu Sankei*], 130, 314, 317, 319, 320, 324–326, 328, 331, 497
Mathematics of Gnomon in Zhou [周髀算経(經) *Zhou Bi Suan-jing*], 111, 112, 496
mathematics of symbols [象数学 *shō sūgaku*], 323
Mathematics of Zhou Gnomons, → Mathematics of Gnomon in Zhou
Mathematics with Humble Determination, → Mathematical Methods without Secrets
Matsuda, Masanori [松田正則], 313
Matsudaira, Sadanobu [松平定信], 373
Matsudaira, Tsunashige [松平綱重], 11
Matsudaira, Tsunatoyo [松平綱豊], 13
Matsumoto, Nobuhiro [松本信廣], 150
Matsunaga, Sadatoki [松永貞辰], 438, 439
Matsunaga, Yoshisuke [松永良弼], 363, 561, 563
Matsushita, Kenrin [松下見林], 360
Matteo Ricci [利瑪竇], 57, 58, 358, 360
Measurements [求積 *Kyūseki*], 193, 197, 205, 219, 220, 222, 226, 324, 325, 388, 496
measurements [度量 *duliang*], 84
Measuring of height of the sun and the moon [日月高測 *Nichigetsu kōsoku*], 363
Mei Kucheng [梅瑴成], 293
Mei Wending [梅文鼎], 116, 293, 314, 361, 363
Menshov, D., 377
Method for Restoring Defective Problems [病題明致之法 *Byōdai meichi no hō*], 324, 325
method of approximate fractions [零約術 *reiyaku jutsu*], 318, 321, 324
method of calling the roll by General Hanxin [韓信點兵 *hang xin dian bing*], 208
method of magic circles [円攢之法 *ensan no hō*], 324, 325
method of magic squares [方陣之法 *hōjin no hō*], 324, 325
method of one-time accelerated approximation [一遍増約術 *ippen zōyaku jutsu*], 315, 317
method of regular polygons [角術 *Kaku jutsu*], 324, 325
method of repeated fractions [重約術 *Jūyaku jutsu*], 321
method of residues [累約術 *Ruiyaku-jutsu*], 321, 328
Method of seeking the volume of a sphere [求立円積術], 195
method of side writing, → byscript method
method of solving Josephus problem [算脱之法 *sandatsu no hō*], 324, 325
method of successive accelerated approximation [累遍増約術 *ruihen zōyaku jutsu*], 315, 317, 318, 324, 328
method of successive divisions [累約術 *Ruiyaku jutsu*], 343
method of the check of sign [験符之法 *kenpu no hō*], 324, 325
Methods of Equation Modifications [開方翻変之法 *Kaihō honpen no hō*], 323, 325, 558
Methods of Mathematics, Old and New, → Mathematical Methods Old and New
Methods of Solving Concealed Problems [解伏題之法 *Kaifukudai no hō*], 17, 129, 130, 251, 264, 265, 267, 324, 438–440, 555, 559, 561, 564

Methods of Solving Explicit Problems [解見題之法 *Kaikendai no hō*], 130, 222, 324, 325, 438–440, 496
Methods of Solving Implicit Problems [解隠(隱) 題之法 *Kaiindai no hō*], 129, 130, 324, 438–440
Methods of Symbols [象法 *Shōhō*], 323, 388
Mikami, Yoshio [三上義夫], 63, 79, 97, 102, 149–154, 159–164, 166–168, 289, 291, 293, 303, 308, 555, 556, 561, 566
Millet and Rice [粟米 *Sumi*] Chapter, 66
Mimata, Hachizaemon Hisanaga [三俣八左衛門久長], 15, 16
Ming Antu [明安圖], 300–309
Mitaki, Shirōemon Gunchi [三滝四郎右衛門郡智], 15, 16
mixed rule of finding differences [混沌招差法 *konton shōsa hō*], 321
Miyachi, Shingorō [宮地新五郎], 15
Miyagi, Kiyotsura [宮城清行], 249, 257, 264, 265
Mo Ruo [莫若], 111
moat-end [塹堵 *Qiandu*], 72, 73, 78, 79, 81, 83
Modification Techniques [変技], 264
Momokawa, Jihē [百川治兵衛], 175
Monopolized Mathematical Method [一極算法 *Ikkyoku Sampō*], 257
Morimoto, Mitsuo [森本光生], 130, 281
Motoori, Norinaga [本居宣長], 373
Mo-tse [墨子], 58
Mōri, Shigeyoshi [毛利重能], 100, 101, 120, 174–175
Muihae [無異解], 100
Muramatsu, Shigekiyo [村松茂清], 14, 249, 253, 281, 282, 315, 318, 332
Muroi, Kazuo [室井和男], 33
mutual cancellation [相消 *sōshō*], 457
mutual multiplication [互乗 *gojō*], 323, 388

Nakamura, Hajime [中村元], 102
Nakane, Genjun [中根彦循], 210
Nakane, Genkei [中根元圭], 258, 264, 265, 314, 321, 343–345, 348, 361, 363
Nam Byungchul [南秉哲], 99
Nam Byunggil [南秉吉], 99
Nam Su Tap Bien [南史輯編], 163
Needham, J., 299, 308
Neugebauer, O., 32, 34, 35, 38, 39, 46
New chapters for plane trigonometry [勾股新編 *Kōko shinpen*], 368
New Commentary on Introduction to Mathematics [新編算学啓蒙註解 *Shimpen Sangaku Keimō Chūkai*], 14, 121
New Japanese version of Complete Treatise on Calendar and Computation [新写訳本暦算全書], 314, 361, 363
New Version of Complete Treatise on Calendar and Computation, → New Japanese version of Complete Treatise on Calendar and Computation
Newton's method, 285, 457
Newton, I., 26, 27, 238, 303, 557
Nguyen Thi Duong [阮氏楊], 151
Nguyen Canh Toan, 377
Nguyen Dinh Ngoc, 381
Nguyen Hoan [阮俒], 163
Nguyen Thuc Hao, 377
Nīlakhaṇṭha, 134
Nindintu-Anu, 38
Nine Chapters, → Nine Chapters of Mathematics
Nine Chapters of Mathematics [九章算術 *Jiŭ Zhāng Suàn Shù*], 22–24, 64–73, 75, 78, 81, 87, 89–91, 99, 108, 109, 111, 112, 119, 124, 125, 129, 156, 158, 173, 194, 221, 245, 246, 277, 279, 282, 497, 554, 560
Nine Chapters of the Mathematical Arts, → Nine Chapters of Mathematics
Nine Chapters on Arithmetic Art, → Nine Chapters of Mathematics
Nine Chapters on the Mathematical Art, → Nine Chapters of Mathematics
Nine Chapters on the Mathematical Procedures, → Nine Chapters of Mathematics
Noi Cac Thu Muc [内閣書目], 150
non-algebraic number, → residual
Notes of Various Calculations [諸勘分物 *Shokanbumono*], 175
Notes on Arithmetic [算用記 *Sanyōki*], 174
Notes on Division, → Book on Division
null quotient formula [無商式 *mushō shiki*], 323
Numbers [数 (數) *sū*], 224, 323, 388, 389, 391–393

oblique multiplications [斜乗 *shajō*], 560, 562, 563
Oda, Nobunaga [織田信長], 101
Oda, Tadao [小田忠雄], 438
Ogawa, Tsukane [小川束], 15, 131, 312
Ogiwara, Magoshirō [荻原孫四郎], 11
Ogura, Kinnosuke [小倉金之助], 99
Ōhashi, Yukio [大橋由紀夫], 149, 150
Ojosan [五曹算], 95
Okumura, Sakuzaemon [奥村作左衛門], 12

On Defective Problems [病題議 *Byōdaigi*], 324
Ono, Eijū [小野栄重], 365
open problems, → bequeathed problems
operation of sphere volume [開立圓 *kai liyuan*], 71
operations of celestial element [天元演段 *tengen endan*], 328
ordinary [全 *zen*], 323, 390, 391
Ore, O., 32, 35
Origin of Dutch trigonometry [法蘭三角形起元 *Hōran sankakukei kigen*], 368
Origin of ellipse [楕円起源 *Daen kigen*], 366
Origin of Mathematical Methods [算法根源記 *Sanpō Kongen-ki*], 121
Origin of Universe [万象一原], 52
Origin of Western Learning [西学原始考], 59
Origins of Western Science, 61
Ōtaka, Yoshimasa [大高由昌], 16, 17
Ōtsuki, Seijun [大槻清準], 368
outer arc ring [外弧環 *gaikokan*], 221, 222
Outline of Western Learning [西学考略], 56, 59
Ozaki Fumiaki [尾崎文秋], 388

Pappos-Guldin formula, 196, 199, 203
Pappus-Guldinus theorem, 219, 220, 224
Pascal's onglet, 201–203
Pascal, B., 188, 201
pentagonal method [五斜術], 272
Pham Huu Chung [范有鍾], 163, 165, 166
Pham, F., 381
Phan Dinh Dieu, 378
Phan Huy Khuong [潘輝框], 150, 153, 154, 156, 157, 159, 164, 168
Phenomenon and Figure [象形 *shōkei*], 224, 225
pile sum [垜 (朶) 積 *daseki*], 323, 325, 388
Pile Sum and Rule of Finding Differences [朶積招差 *Daseki Shōsa*], 323
Platter of Mathematics, → Mathematical Chopping-Board
polished rice [粺米 *baimi*], 67
Polya, G., 106
Polynomial Fitting [畳乗 *jōjō*], 323, 388
positive-negative root-extraction method [正負開方術 *Zhengfu Kaifang Shu*], 23
power [冪 *Mi*], 69
power procedures [冪乗演式 *bekijō enshiki*], 257, 265, 267, 268
Powered Roots [冪商], 264
Preliminary Introduction to Mechanics [重学淺説], 56, 59
Principle of Machines, 53
Problems on Remainder [大衍類 *dayan lei/daien rui*], 214
procedure [術 *shu*], 22, 64, 328, 329
procedure in operations of celestial element [天元演段術 *tengen endan jutsu*], 314
procedure of celestial element, → celestial element method
Procedure of the Seventh Powers [七乗冪演式 *Shichijōbeki enshiki*], 258
Proclus, 32
proportional distribution [衰分 *Cuifen*], 67
provisional [動 *dō*], 224, 391–393
Pythagoras, 96
Pythagorean theorem, 111, 197, 250, 255, 318, 325, 496, 497
Pythagorean triples, 32, 34, 37, 46, 290

Qian Baocong [銭宝琮], 79, 89, 90, 299, 308
Qin Jiushao [秦九韶], 23, 24, 107, 108, 115, 208, 214–215, 246, 320
quadrilateral method [四斜法 *shishahō*], 253, 255, 257, 272, 568, 569
questions associated with procedures as examples, 65, 68
Quick Methods for Accurate Values of Circle Segments [割圜密率捷法, *Ge Yuan Mi Lu Jie Fa*], 301, 306, 307
quotient [商], 178

Rakusho [洛書], 97
rational, → fractional
rational number, → fractional
Ready-Made Computational Methods of Nine Categories [九章立成算法 *Cuu Chuong Lap Thanh Toan Phap*], 153, 157, 162–164, 166
Ready-made Methods of Addition of Nine Categories [九章立成併法 *Cuu Chuong Lap Thanh Tinh Phap*], 164, 165, 167
Ready-made methods of Computations [立成筭法], 157
reasoning by contradiction, 75
Record of Jugai, → Jugai's Record
Record to guidebook in computational procedures [算法點竄指南録 *Sanpo tenzan shinanroku*], 369
rectangular array [方程 *Fangcheng*], 22, 67
rectangular array method [方程術 *Fangcheng Shu*], 22, 26, 27
Rectangular Field [方田 *Fangtian*] Chapter, 65, 66
Register of Jugai, → Jugai's Record
Regular Polygons [角法 *Kakuhō*], 388
Relational Reasoning, 70

repeated rule of finding differences [累裁招差法 *ruisai shōsa hō*], 321
replacement of numbers [替数 *taisū*], 323
residual, → lopsided
residual [零 *rei*], 225, 323, 392, 393
Resolution of Entanglements in Mathematics [算学紛解 *Sangaku Funkai*], 264
respective reduction [逐約 *chikuyaku*], 212
Restoration of Defective Problems, 325, 326
Restoring Defective Prolblems [病題擬 *byōdaigi*], 388
resultant, 230, 232, 234, 237–241, 264–268, 555, 557–559, 565, 570–573
Résumé of Works on Mathematics, → Essential Mathematics
Reyneau, C., 232
Rho, J., 363
Richardson, L., 317
Right Triangle, → Right-angled Triangles
right-angled triangle containing a circle [勾股容圓 *gou-gu rong-yuan*], 109
right-angled triangle containing a square [勾股容方 *gou-gu rong-fang*], 109
right-angled triangle theorem [勾股之法 *gou-gu zhi fa*], 111
Right-angled Triangles [勾 (句) 股 *Gougu*], 34, 37, 38, 45, 66, 68, 89, 109, 184, 205, 250, 496
Robson, E., 32, 33, 37, 46
Romberg, W., 317
root extraction [開方 *kaihō*], 457
root-extracting [開方 *kaifang*], 66
root-extracting method, → root-extraction method
root-extraction method [開方術 *kāifāng shù*], 22, 24, 97, 246, 251
Ruan Yuan [阮元], 50
rule of augmentation [増術 *zōjutsu*], 281
Rybnikov, K. A., 27

Sachs, A., 32, 34, 46
Saga [謝家], 94
Saji, Kazuhira [佐治一平], 313
Sakabe, Kōhan [坂部廣胖], 369
Samgae [三開], 94
Sanbup-Tongjong, → Systematic Treatise of Arithmetic
Sangmyon-Sanbop [詳明算法], 95, 97, 99
Sanhak Bonwon [算学本原], 97
Sanhak Ipumun [算学入門], 97
Sanhak-Kemon, → Introduction to Mathematics
Śankara, 134–136, 138–147
Sansom, G., 290, 293
Sansul Gwangyun [算学正義], 100
Sasaki, Chikara [佐々木力], 28
Satō, Ken'ichi [佐藤健一], 131
Satō, Ken'ichi [佐藤賢一], 7, 15–17
Satō, Masaoki [佐藤正興], 121, 250
Satō, Masayasu [佐藤政養], 373, 374
Sawaguchi, Kazuyuki [沢 (澤) 口一之], 16, 121, 249–252, 254, 255, 257, 264, 283, 286–289, 294, 497
Sea Island Mathematical Canon [海島算経 (經) *Haidao Suanjing*], 89, 90, 497
Sea Islands Mathematics, → Sea Island Mathematical Canon
Sea-Mirror of Circle Measurement [測圓海鏡 *Ce-Yuan Hai-Jing*], 107, 108, 110–112
Seajong [世宗], 95–97, 120
Seajong Chronicles [世宗実録], 95
Searching Arc Length [探弧術 *tan-ko jutsu*], 326
Searching Circle Number [探円数 *tan-ensu*], 326
Secret treatise on trigonometric calculation [三角提要秘算 *Sankaku teiyō hisan*], 368
Seeking the Total [求総数 *kyūsōsū*], 213, 214
Seki, Gorōemon [関五郎右衛門], 11
Seki, Gorōzaemon [関五郎左衛門], 11
Seki, Heizō [関平蔵], 4, 8
Seki, Jūrōemon [関十郎右衛門], 4, 10, 11, 19
Seki, Shinshichi(rō) [関新七 (郎)], 4, 8
Seki, Shinshichi(rō) Hisayuki [関新七 (郎) 久之], 13
Seki, Shinsuke [関新助], → Seki, Takakazu
Seki, Shinsuke Fujiwara Takakazu [関新助藤原孝和], → Seki, Takakazu
Seki, Shinsuke Takakazu [関新助孝和], → Seki, Takakazu
Seki, Takakazu [関 (關) 孝和], 3, 4, 8–17, 19, 20, 26, 27, 101, 102, 121, 122, 129–131, 161, 185, 187, 188, 193–198, 201–205, 208, 210, 215, 217, 219–224, 226–227, 239, 249, 251–257, 264–267, 272, 276, 278, 281, 283–294, 312–315, 317–328, 332–333, 335–338, 355, 387, 388, 438–440, 495–497, 555–573
Seki, Yaemon [関弥右衛門], 10, 11
Seongjong [成宗], 121
Septenary [七部書 *Shichibu Sho*], 219, 495, 496
Sequel of Calendars and Astronomical Phenomena, → Sequel of Compendium of Calendrical Science and Astronomy
Sequel of Compendium of Calendrical Science and Astronomy [暦象考成後

(后) 編 *Li xiang Kao cheng hou bian*], 301, 364, 365, 373
Sequel of desultory essay by Konyō [続昆陽漫録補 *Zoku konyō manroku ho*], 365
Sequel to Wahon for Beginners [続和本入門], 353
Seven Documents, → Septenary
Several methods on mathematics, 3 vols. [算法諸約術, 上中下], 349
Shang Gao [商高], 111
Shibukawa, Kagesuke [渋川景佑], 365
Shibukawa, Harumi [澁川春海], 246, 249
Shiduki, Tadao [志筑忠雄], 367–369
Shimada, Sadatsugu [島田貞継], 249
Shimodaira, Kazuo [下平和夫], 102
shorter leg [勾 *Gou*], 24, 114
Si-Duk [施徳], 94
side writing method, → byscript method
sign procedure (rule) [正負術 *zhengfu shu*], 67, 124
Six Books of Selected Western Knowledge [西学輯存六種], 53, 59
six obliques procedure [六斜術], 253
Small Width [少広 *Shaoguang*] Chapter, 66
Smith, D. E., 289, 291, 293
Solution of Whole Problems [全題解 *Zendaikai*], 264, 324, 388
Solutions of Well-posed Problems, → Solution of Whole Problems
Source of Tuning and Calendars [律暦淵源, *Lu Li Yuan Yuan*], 301
sphere segment, 196, 197, 200
square town [邑方 *yi-fang*], 109
Standard Procedure for two equations [双式定格術], 265
static, → stationary
stationary [静 *sei*], 224, 391–393
Stories of Mathematicians by Fujita Teishi (1734–1807) [藤定資算家談 *Tō-Teishi-Sanka-Dan*], 6
Studies in Indian Mathematics [インド数学研究 *Indo Sūgaku Kenkyū*], 133, 145
submerged problem [潜題 *sendai*], 324, 325, 388
subsidiary cone [旁錐 *bōsui*], 194
subtraction by one hundred and five [百五減算 *hyakugo genzan*], 210
succeeding quotient [續商 *Xu Shang*], 23
succession of bequeathed problems [遺題継承 *idai keishō*], 249
Sugano, Mototake [菅野元健], 561, 564
Sugimoto, Toshio [杉本敏夫], 197, 221
Sugiyama, Sadaharu [杉山貞治], 121
Suminokura, Ryōi [角倉了以], 175, 180
Summarized Fundamentals of Computational Methods, → Systematic Treatise of Arithmetic
Summary of Mathematical Methods, → Systematic Treatise of Arithmetic
summation of higher order arithmetic series [垜積術 *Duoji Shu*], 22, 27
sums of powers, 27
Sung Confucianism [宋儒 *sōju*], 96, 97
Sunzi's Arithmetical Canon, → Sunzi's Mathematical Canon
Sunzi's Mathematical Canon [孫子算経 *Sunzi Suanjing*], 183, 189, 207–210, 215, 216
'suppose' procedure [今有術 *jinyou shu*], 66, 67, 74, 87
Supsan Jinbol [習算津筏], 97
Syllogism, 70
Sylvester, J. J., 237–240, 557, 570
symbol [象 *shō*], 323, 389, 391
Symbols and Figures [象形 *shōkei*], 323, 388–391, 393
synthesis [綜合 *Zonghe*], 75
Systematic Treatise of Arithmetic [算法統宗 *Suànfǎ Tǒngzōng*], 5, 14, 97, 100, 103, 119, 120, 152, 153, 156–158, 168, 175, 176, 184, 208–210, 215, 217, 246, 278, 323, 325
Systematic Treatise on Arithmetic, → Systematic Treatise of Arithmetic
Systematic Treatise on Computational Methods [統宗算法 *Thong Tong Toan Phap*], 157

Ta Huu Thuong [謝有常], 157
Ta Quang Buu, 376, 377, 379
Takagi, Teiji [高木貞治], 267, 567, 569
Takebe, Kataakira [建部賢明], 5, 15, 17, 212, 213, 224, 264, 272, 311–313, 322, 323, 326, 327, 336, 387, 438, 495, 497, 555
Takebe, Katahiro [建部賢弘], 5, 7, 14–17, 20, 26, 121, 122, 127–131, 212, 213, 224, 227, 254, 256, 264, 272, 276, 278–282, 284, 285, 290, 291, 293–295, 311–315, 317, 318, 320–323, 326–329, 331–341, 343–345, 348, 361, 363, 387, 392, 438, 439, 495, 497, 555
Takebe, Katamitsu [建部賢充], 311
Takebe, Katao [建部賢雄], 311, 312
Takebe, Naomasa [建部直昌], 311
Takebe, Naotsune [建部直恒], 311, 312
Takebe, Shōkō [建部昌興], 311
Tanaka, Yoshizane [田中由真], 249, 264–268, 555
technique of linkage, → Zhuì Shù

Tegus [特古斯], 308
Ten Classics of Mathematics [算経十書 *Suan Jing Shi Shu*], 106, 209
ten quotients [十商 *jūshō*], 323
Textbook of Mechanics [力学課編], 55, 60
Textbook of Modern Physics [近世物理学教科書], 56
theory of circles [圓理 *enri*], 27, 249, 250
Thousand-Character Text [千字文], 93
Three Essentials [三要 *San'yō*], 224, 225, 323, 325, 326, 329, 388, 389, 393, 496
three formulas [三式 *sanshiki*], 130, 323, 333, 334, 337, 341
Three methods for land surveying [量地三略 *Ryōchi sanryaku*], 371
Toan Phap Chi Minh [筭法指明], 155, 159
Toda, Kahei [戸田加 (嘉) 兵衛], 11
Tokugawa, Iemitsu [徳川家光], 311, 312
Tokugawa, Ienobu [徳川家宣], 312, 313
Tokugawa, Ietsugu [徳川家継], 312, 313
Tokugawa, Ietsuna [徳川家綱], 11
Tokugawa, Ieyasu [徳川家康], 311, 359
Tokugawa Samurai Directory [御家人分限帳 *Gokenin Bugen-chō*], 7, 13, 18
Tokugawa, Tsunashige [徳川綱重], 312
Tokugawa, Tsunatoyo [徳川綱豊], 312, 313
Tokugawa, Tsunayoshi [徳川綱吉], 11, 13, 312
Tokugawa, Yoshimune [徳川吉宗], 312–314, 361
Tokugawa, Yoshinao [徳川義直], 359
tomb tunnel [羨除 *yanchu*], 72
Toyotomi, Hideyoshi [豊臣秀吉], 100, 175
track of the barycenter of the arc-figure, 199, 200
Tran Dai Nghia, 376
Tran, N., 151, 153, 157
transformed equation [換式], 266, 469, 556, 563, 570–572
Treatise of Ancient and Modern Mathematics, → Mathematical Methods Old and New
Trilogy [三部抄 *Sambushō*], 130, 325, 388, 438, 439, 495, 496
truth from the fact [實事求是], 102
Tulling, N., 366
turtle-forelimb [鼈臑 *Bienao*], 73, 74, 78, 79, 83, 84
Tycho Brahe [第谷], 301, 364

Uchida, Gokan [内田五観], 358, 371
Uchiyama, Nagasada [内山永貞], 7
Uchiyama, Shichibei Nagaakira [内山七兵衛永明], 6, 8
Uchiyama, Shichibei Nagasada [内山七兵衛永貞], 19
Uchiyama, Shichinosuke [内山七之助], 6, 7
Uchiyama, Shichinosuke Nagasada [内山七之助永貞], 6
Uchiyama, Shingorō (or Shōken) Nagayuki [内山新五郎 (松軒) 永行], 4, 7, 8
Ueno, Kenji [上野健爾], 14
umbrella-shaped solid, 204, 205
Unified Foundations of Mathematics, → Systematic Treatise of Arithmetic
Unilateral Procedure for two equations [双式一貫之術 *Sōshiki Ikkan no jutu*], 265
unknown total [物不知其総数 *wu bu zhi qi zong shu*], 208
Unlocking of Roots [開方釋鎖 *kaifang shisuo*], 276
Unquestionable Methods in Mathematics [算法闕疑抄 *Sampō Ketsugishō*], 5, 14, 281, 354

Valiron, G., 376
Van Ceulen, L, 61
variable quotient formula [変商式 *henshō shiki*], 323
Various Techniques [変技 *Hengi*], 323, 388
Verbiest, F., 364
Vitruvius, 57
volume of a moat-end, 73
volume of a sphere, 25, 51, 61, 71, 192, 195, 199, 249, 318, 319
volume of a yang-ma, 73
Volume of Spherical Segments [球闕率 *Kyūketsuritsu*], 388
Vreserik, J., 365

Waerden, B. L., 267
Wagner, D. B., 79
Wang Boson [王保孫], 94
Wang Doryang [王道良], 94
Wang In [王仁], 93
Wang Tao [王韜], 53, 56, 59
Wang Zheng [王徵], 56
wasan, → Japanese traditional mathematics
Weaving Methods in Mathematics, → Mathematical Treatise on the Technique of Linkage
Weierstrass, K., 230
Western Mirror of European Learning [欧羅巴西鏡録], 57
whole [全 *zen*], 225, 323, 391–393
Work Discussing [商功 *Shanggong*] Chapter, 66
Works and Days Calendar [授時暦 *Shoushili*], 320, 321

Works on Right Triangles [股勾弦鈔 *Kōkogen Shō*], 215
Wu Wen-Tsun [呉文俊], 29, 30, 105, 106, 272
Wylie, A. [偉烈亞力], 56, 371

Xavier, F., 120
Xia Luanxiang [夏鸞翔], 52
Xu Guangqi [徐光啓], 50, 363
Xu Shen [許慎], 152
Xu Youren [徐有壬], 59
Xu Zelin [徐澤林], 388, 497
Xuān Míng Lì [宣明暦 *Senmyō reki*], 246

Yabuuchi, Kiyoshi [薮内清], 14
Yamagata, Bantō [山片蟠桃], 373
Yamaji, Nushizumi [山路主住], 438, 439
Yamaji, Yukitaka [山路諧孝], 365
Yanagawa, Shunsan [柳川春三], 373
Yang Hui [楊輝], 129, 208, 210, 211, 216, 246, 320, 325
Yang Hui's Arts on Arithmetic, → Yang Hui's Methods of Mathematics
Yang Hui's Methods of Mathematics [楊輝算法], 5, 14, 95, 97, 103, 119, 129, 208, 211, 215, 216, 246, 325, 554
yang-ma [陽馬 *Yangma*], 72, 78, 79, 81, 83
Yanghee-Sanbop, → Yang Hui's Methods of Mathematics
Yáng Huī Suàn Fǎ, → Yang Hui's Methods of Mathematics
Yano, Michio [矢野道雄], 133
Yin-Yang [陰陽 *in'yō*], 97, 102
Yokotsuka, Hiroyuki [横塚啓之], 333
Yōrō Legal Codes [養老律令 *Yōrō Ritsuryō*], 94, 173
Yōrō Ritsurei, → Yōrō Legal Codes
Yoshida, Mitsuyoshi [吉田光由], 14, 100, 120, 175, 176, 182, 184, 209, 249, 271, 278
Yoshida, Shūan [吉田周庵], 175
Yu Soosuk [劉寿錫], 99
Yukjang [六章 *Rokushō*], 94

Zhang Xingcheng [張行成], 157
Zhao Shuang [趙爽], 112
Zhou Gong [周公], 111
Zhu Shijie [朱世傑], 14, 24, 26, 100, 107, 111–116, 120, 122, 124–127, 246, 251, 276–278, 282, 313, 457, 497, 554
Zhuì Shù [綴術 *tetsu jutsu*], 94, 99, 130, 174, 245
Zu Chongzhi [祖冲之], 25, 50, 245, 249, 282, 315, 320, 328

MIX
Papier aus verantwortungsvollen Quellen
Paper from responsible sources
FSC® C105338

If you have any concerns about our products,
you can contact us on
ProductSafety@springernature.com

In case Publisher is established outside the EU,
the EU authorized representative is:
Springer Nature Customer Service Center GmbH
Europaplatz 3, 69115 Heidelberg, Germany

Printed by Libri Plureos GmbH
in Hamburg, Germany